Chemical Engineering: Processes, Synthesis and Analysis

Chemical Engineering: Processes, Synthesis and Analysis

Edited by Lewis Allison

Clanrye International
www.clanryeinternational.com

Clanrye International,
750 Third Avenue, 9th Floor,
New York, NY 10017, USA

ISBN: 978-1-63240-791-7

Cataloging-in-Publication Data

Chemical engineering : processes, synthesis and analysis / edited by Lewis Allison.
p. cm.
Includes bibliographical references and index.
ISBN 978-1-63240-791-7
1. Chemical engineering. 2. Chemical processes. 3. Chemistry, Technical. I. Allison, Lewis.
TP155 .C44 2019
660--dc23

For information on all Clanrye International publications
visit our website at www.clanryeinternational.com

CLANRYE INTERNATIONAL

Printed in China.

Contents

Preface

This book presents the technological advancements in the processing, synthesis and analysis of chemical compounds. It details the different approaches, evaluations and methodologies. Chemical engineering is concerned with the manufacture of commercially viable products from basic raw materials or chemicals by applying the principles of chemistry, physics, biochemistry and microbiology. Applications of chemical engineering are prevalent in the areas of polymers, pharmaceuticals, alternative energy sources and semiconductors among others. This book elucidates the concepts and innovative models of chemical engineering in reference to the processes, synthesis and analysis. The topics included herein are of utmost significance and bound to provide incredible insights to readers. This book attempts to assist those with a goal of delving into the field of chemical engineering.

This book is a result of research of several months to collate the most relevant data in the field.

When I was approached with the idea of this book and the proposal to edit it, I was overwhelmed. It gave me an opportunity to reach out to all those who share a common interest with me in this field. I had 3 main parameters for editing this text:

1. Accuracy – The data and information provided in this book should be up-to-date and valuable to the readers.
2. Structure – The data must be presented in a structured format for easy understanding and better grasping of the readers.
3. Universal Approach – This book not only targets students but also experts and innovators in the field, thus my aim was to present topics which are of use to all.

Thus, it took me a couple of months to finish the editing of this book.

I would like to make a special mention of my publisher who considered me worthy of this opportunity and also supported me throughout the editing process. I would also like to thank the editing team at the back-end who extended their help whenever required.

Editor

Adsorptive Recovery of Gold (III) from Acidic Chloride Media Using Carbonized Microalgal Residue

Khunathai K[1,2], Inoue K[1], Ohto K[1*], Kawakita H[1], Kurata M[2], Atsumi K[2], Fukuda H[2] and Alam S[3]

[1]*Department of Applied Chemistry, Faculty of Science and Engineering, Saga University, Honjo-machi 1, Saga 840-8502, Japan*
[2]*Research Laboratories, DENSO CORPORATION , Minamiyama 500-1, Komenoki, Nisshin, Aichi 470-0111, Japan*
[3]*Department of Chemical and Biological Engineering, College of Engineering, University of Saskatchewan, 57 Campus Drive, Saskatoon, SK, S7N 5A9, Canada*

Abstract

Adsorption gel was prepared by carbonizing a residual waste of microalgae generated in the biofuel conversion process. The prepared gel exhibited high affinity and selectivity for Au (III) over other precious and base metal ions in hydrochloric acid. From the adsorption isotherms, the maximum loading capacity of Au (III) on the carbonized microalgae was evaluated as 1.20 mmol/g (236 mg/g), which is much higher than that on the original microalgal residue. Observations by microphotographs, scanning electron microscope, and X-ray diffraction confirmed the formation of metallic Au, suggesting that a redox reaction had taken place between the adsorbent and Au (III) ions during adsorption. Adsorption kinetics was interpreted in terms of pseudo-first order rate expression to evaluate the activation energy of this system as 33.7 kJ/mol.

Keywords: Adsorption; Gold (III); Microalgal residue; Carbonization; Acidic chloride media

Introduction

In recent years, recycling and recovery of valuable metals from spent electronic and electric appliances has become an emerging important research subject from both of environmental and economical viewpoints [1]. This is due to the facts that recycling of these metals reduces the cost of metallurgical treatments associated with their mining activity, and sharply decreases the destructive impacts on the environment. Adsorption techniques using activated carbons have played important roles in their recovery [2-6]. In recent years, biomass wastes have found their uses for the separation and recovery of toxic and precious metal ions from industrial solutions [7,8].

On the other hand, also in recent years, new energy or green energy has been strongly required as alternative to fossil fuels like petroleum and coal as well as atomic energy. One of such new energy is biodiesel oil produced from microalgae, on which much interests have been attracted all over the world [9]. However, the present production cost of such biodiesel oil is much higher than commercially available diesel oil produced from petroleum. One of the candidate methods to lower the total production cost of biodiesel oil is the effective use of the residue of microalgae after extracting biodiesel oil.

From such backgrounds, the authors attempted to develop a new technology for recovering gold from various industrial wastes using adsorption gel prepared from the above-mentioned microalgal residue only by treating in boiling sulfuric acid in the previous work [10]. Although the preparation of this gel is very simple, it suffers from a problem; that is, it leaves waste sulfuric acid containing polysaccharides and other organics after the preparation, which needs some costs for the post-treatment.

Activated carbons or the carbonized materials are carbonaceous material having well-developed porous structure consisting of hydrophobic graphene layer surface and hydrophilic surface functional groups. Main applications of activated carbon to the recovery of precious metals involve the adsorptive recovery of gold and silver from cyanide media in a so-called carbon-in-pulp (CIP) process in the gold industry [11]. Extensive research works have been conducted in this area [2-6]. Activated carbon is a good adsorbent as it has an extended surface area, porous structure, high adsorption capacity and a certain degree of surface reactivity. These properties are derived by both the nature of the raw materials and the processes employed for the production of the activated carbon.

In the present work, the authors attempted to prepare the new adsorption gel alternative to the above-mentioned gel at a low cost by means of the carbonization method. The microalgal residue mainly contains proteins and carbohydrates as its major components. The carbonization would convert such components to disorganized graphite-liked material. It was reported by Chand et al. [12] that carbonization of biomass waste at moderate temperature without prior chemical activation is attractive since the high specific surface area is retained and use of chemicals can be avoided. Also, due to its high surface area and porous structure, high adsorption capacity is also expected for the recovery of precious metals. The objective of the present work is to fundamentally investigate the adsorption behaviors of the prepared carbon material for the recovery of gold from acidic chloride media because, in hydrometallurgical processes for precious metals, feed materials are leached in acidic chloride media like aqua regia or hydrochloric acid containing chlorine gas.

Experimental

Materials and analytical methods

Analytical grade $HAuCl_4{\cdot}4H_2O$, $H_2PtCl_6{\cdot}6H_2O$, $PdCl_2$, $FeCl_3{\cdot}6H_2O$ (Wako Pure Chemical Industries, Ltd., Osaka, Japan), $ZnCl_2$ (Sigma Aldrich, St Louis, MO, USA), $CuCl_2$, and $NiCl_2{\cdot}6H_2O$ (Katayama

***Corresponding author:** Keisuke Ohto, Department of Applied Chemistry, Faculty of Science and Engineering, Saga University, Honjo-machi 1, Saga 840-8502, Japan, E-mail: ohtok@cc.saga-u.ac.jp

Chemicals, Osaka, Japan) were employed to prepare test solutions of Au (III), Pt (IV), Pd (II), Fe (III), Zn (II), Cu (II), and Ni (II), respectively. All other chemicals used for the preparation of the adsorbent and for the adsorption tests were of analytical grade and were used without further purification.

To measure the specific surface area and pore size distribution of the prepared carbonized gel, 0.2 g of sample was pretreated to remove water content using a model BELPREP-vacII powder/particle characterization analyzer (MicrotracBEL Inc., Osaka, Japan) at 150°C for 42 h and N_2 adsorption-desorption isotherms at 77 K was measured using a model BELSORP-miniII-SP powder/particle characterization analyzer (MicrotracBEL Inc., Osaka, Japan). The data were analyzed using BET-Plot and t-Plot to evaluate the specific surface area. The classical pore size model developed by Barret-Joyner-Halande (BJH) was used for the pore size distribution calculation.

The infrared spectra of the microalgal residue were recorded using a FT-IR spectrophotometer, JASCO FT/IR-410 (JASCO Corp., Tokyo, Japan). The C, H, and N contents were determined by elemental analysis. The surface morphology of the adsorbent was recorded using a digital microscope KEYENCE model VHX-1000 (KEYENCE Corp., Osaka, Japan) and a scanning electron microscope (SEM), JEOL model JCM-5100 (JEOL Ltd., Tokyo, Japan) at an acceleration energy of 20 kV. The X-ray diffraction (XRD) analysis was carried out by using an X-ray diffractometer, Rigaku model L-094 (Rigaku Corp., Tokyo, Japan). The metal ion concentration in the solution was measured by using a Shimadzu model ICPS-8100 inductively coupled plasma atomic emission spectrometer (ICP-AES) (Shimadzu Corp., Kyoto, Japan).

The cyclic voltammogram was recorded to qualitatively investigate the electrochemical property of tested samples. The working electrode (WE) was prepared by grinding 2.0 mg of adsorbent with 1.0 mg of teflonized acetylene black (TAB-2) with works as a conductive binder. The equipment was set up by combination of WE together with a Ag/AgCl electrode as the reference electrode (RE) and a Pt plate as the counter electrode (CE) and it was carried out in 1.0 M (M = mol/L) $NaNO_3$ solution (electrolyte solution). The measurement was conducted with a two-electrode configuration using a model HSV-100 Automatic Polarization System (Hokuto Denko Corp., Tokyo, Japan) at a scan speed of 50 mV/s.

Preparation of the carbonized adsorption gel

Dried sample of the residue of *Pseudochoricystis ellipsoidea*, the microalga, generated in the biofuel extraction process was kindly provided by DENSO CORPORATION, Aichi, Japan, to be employed as the raw material of the adsorption gel. This sample was carbonized by the same method proposed by Chand et al. [12,13] as follows. Five g of dried sample was put in an alumina crucible, which was then put in a larger crucible and covered with graphite powder and capped with an alumina cover in order to avoid oxidation. This crucible was placed in a muffle furnace (Nitto Kagaku Co, Ltd., Aichi, Japan) for carbonization at heating rate of 5°C/min. Temperature was kept constant at 800°C for 1 h for carbonization. Then, the temperature was slowly lowered to the ambient temperature. The product was washed with distilled water and followed by hot water until the pH of filtrate reached neutral pH and was dried in a convection oven at 70°C for 2 days. The obtained carbon material was ground and sieved to uniform particle size before adsorption test. Thus, prepared carbon material was abbreviated as CMR (carbonized microalgal residue), hereafter. It was compared with a couple of other carbon materials, a commercially available activated carbon purchased from Wako Pure Chemical Industries, Ltd., Japan, which is termed as Wako activated carbon, hereafter, and a carbonized orange waste prepared by the same method from a dried sample of orange juice residue kindly donated from JA Beverage Saga Co. Ltd., Japan.

Batch-wise adsorption tests

All adsorption tests were carried out batch-wise to observe the adsorption behavior of various metal ions on the tested adsorbents from individual metal solutions at varying HCl concentrations. Mixtures of 20 mg of adsorbent and 15 mL of 0.2 mM metal solutions were prepared in sample vials, which were then shaken in a thermostated incubator (THOMAS AT24R) at 30°C for 24 h to attain equilibrium. Then, the samples were immediately filtered and the metal ion concentrations in the filtrate were measured. The percentage adsorption was calculated according to the following equation.

$$\text{Adsorption (\%)}=[(C_i-C_e)/C_i] \times 100 \quad 1$$

where C_i and C_e are the initial and equilibrium concentrations of the tested metal ion (mM), respectively. The amount of metal adsorbed on the adsorbent (q) was calculated according to the following equation.

$$q=[(C_i-C_e)/M] \times V \quad 2$$

where V is the volume of the test solution (L), and M is the dry weight of the adsorbent (g).

Adsorption isotherm tests were carried out by shaking 10 mg of adsorbent together with 10 mL of test solutions containing varying concentrations of Au (III) in 0.1 M HCl in a thermostated incubator at 30°C for 100 h to attain equilibrium.

Adsorption rate of Au (III) on the carbonized microalgae was measured at varying time intervals at 30, 40 and 50°C from 0.1 M HCl containing 0.6 mM Au (III).

Results and Discussion

Characterization of the carbonized microalgal residue

The infrared (IR) spectra of the original microalgal residue, the feed material, and CMR as well as WAKO activated carbon are shown in Figure 1 for comparison. In this figure, the spectrum of the original microalgal residue indicates the presence of O–H stretching at 3313 cm^{-1} and C–H stretching at 2910 and 2842 cm^{-1}. The sharp absorption at 1623 cm^{-1} indicates the presence of C=O groups. These spectra well described the presence of various functional groups in microalgal residue. On the other hand, the spectra of CMR indicate that most of OH groups and other functional groups existed in the microalgal residue were disappeared by heating during the carbonization process, whilst the presence of hydroxyl group was not observed in WAKO activated carbon.

From the elemental analysis, it was found that microalgal residue, the feed material, contains 6.33% H, 43.6% C, 4.04% N, and 46.1% O, while CMR contains 0.56% H, 49.8% C, 4.50% N, and 45.2% O as its major elemental compositions. The product yield of the carbonized microalgae was approximately 30% of the initial weight of the raw material.

From the measurement using scanning electron microscope (SEM) of CMR shown in Figure 2, it was observed that the morphological surface of CMR has a high degree of porosity which was considered to be well developed by the carbonization process.

From the measurement of the adsorption-desorption isotherms of nitrogen on CMR and WAKO activated carbon shown in Figure 3,

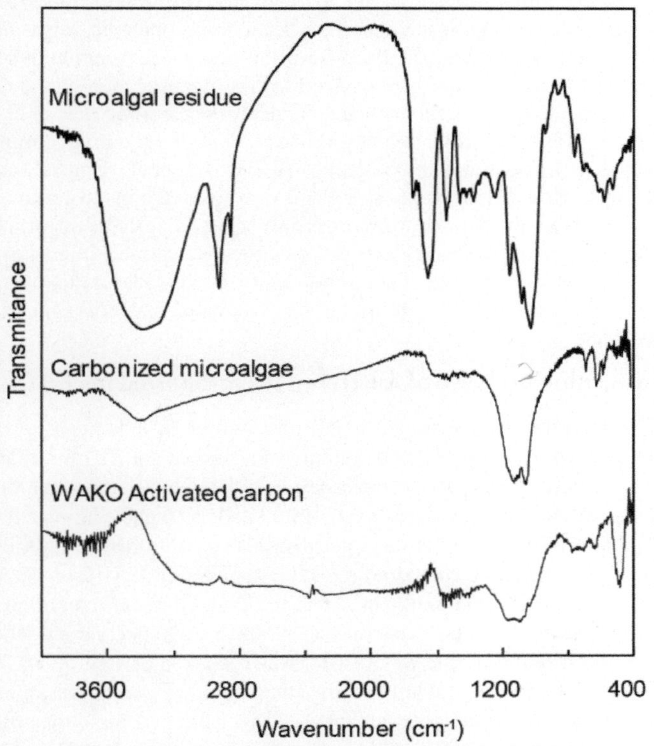

Figure 1: FT-IR spectra of the microalgal residue and CMR as well as WAKO activated carbon.

Figure 2: SEM images of CMR.

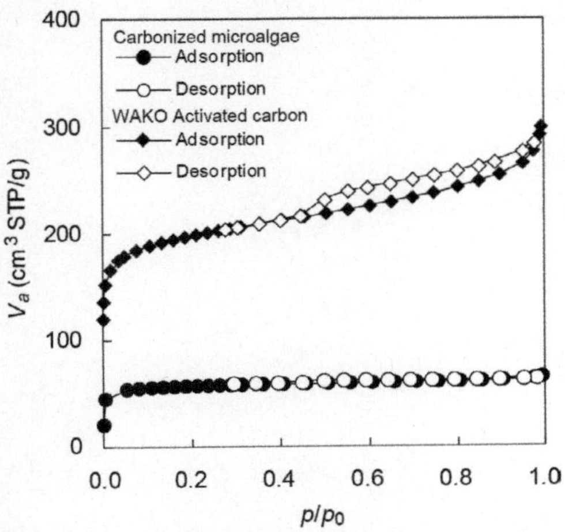

Figure 3: Nitrogen adsorption-desorption isotherms of CMR and WAKO activated carbon; V_a=Volume of N_2 adsorbed; p/p_0=Partial pressure of N_2.

it was found that the adsorption isotherms for both samples suggest the characteristic of microporous materials, while small hysteresis loop observed in the adsorption-desorption isotherms indicates the existence of mesopores [14,15]. The surface area of CMR was measured as approximately 204 to 272 m²/g, whilst that of WAKO activated carbon is 704 to 900 m²/g. Pore volumes of CMR and WAKO activated carbon were measured as 0.02 and 0.20 cm³/g, respectively. Figure 4 shows the pore size distribution of CMR and WAKO activated carbon. It is obvious that the majority of pores of CMR are microporous having average pore size of 1.29 nm which is very close to that of WAKO activated carbon (1.21 nm).

Electrochemical property was investigated by recording cyclic voltammograms curves. Figure 5 shows the cyclic voltammograms curves of CMR and WAKO activated carbon. Both curves represent rectangular-like shapes without any redox current peaks on both positive and negative sweeps in the potential range, indicating good capacitive behavior similar to the supercapacitor [14,15]. In addition, the voltammetric charges on the positive and negative sweeps were approximately equal, which demonstrates that the charge-discharge behavior was highly reversible [16-18].

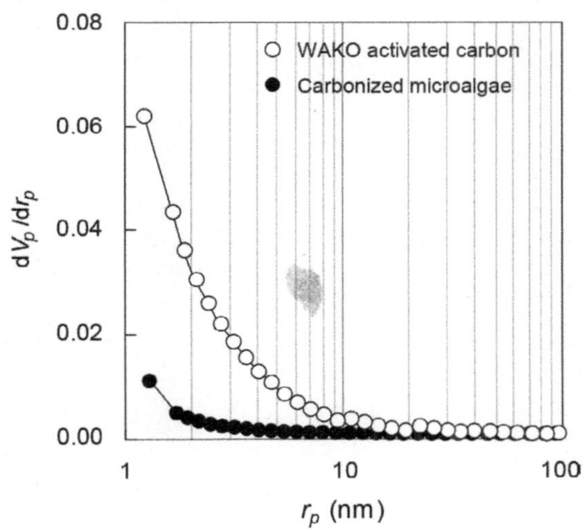

Figure 4: Pore size distribution diagrams of CMR and WAKO activated carbon. V_p: pore volume of adsorbent, r_p: pore radius of adsorbent.

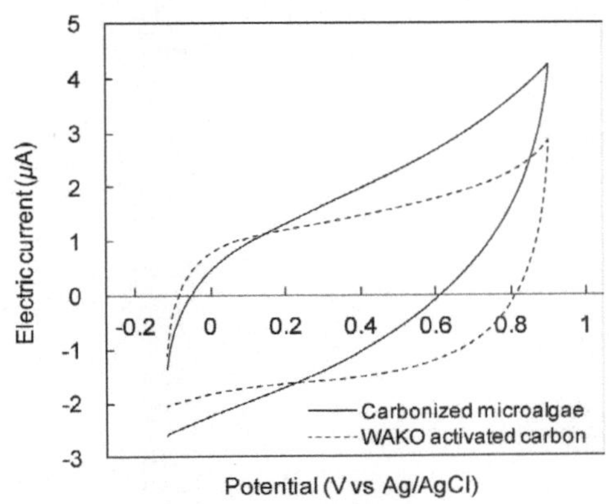

Figure 5: Cyclic voltammograms of CMR and WAKO activated carbon. Conditions: RE=Ag/AgCl, CE=Pt plate, electrolyte solution=1.0 M $NaNO_3$, scan speed=50 mV/s.

Effect of hydrochloric acid concentration on the adsorption of metal ions

In acidic chloride medium, Au (III), Pd (II), and Pt (IV) give rise to stable anionic chloride complexes such as $AuCl_4^-$, $PdCl_4^{2-}$, and $PtCl_6^{2-}$, respectively, while majority of base metal ions like Ni (II), Zn (II), and Cu (II) exist as cationic or neutral species and the concentration of anionic chloro-complexes is significant only at higher chloride concentration in the case of Fe (III). Figures 6a-6d show % adsorption of some precious and base metal ions on microalgal residue as it is, CMR, carbonized orange waste, and WAKO activated carbon, respectively, at varying HCl concentration.

From the comparison of Figures 6a and 6b, it is apparent that adsorption of Au (III) was greatly enhanced by the carbonization. Further, CMR exhibits high selectivity for Au (III) whereas adsorption of Pd (II) and Pt (IV) was much less than that of Au (III) over the whole HCl concentration range. Under the same experimental conditions, a group of base metal ions such as Ni (II), Fe (III), Zn (II), and Cu (II) were inert to be adsorbed on CMR, indicating high affinity and selectivity of this adsorbent towards precious metal ions.

From the comparison with the carbonized orange waste, it is seen that the similar adsorption behavior was observed for both carbonized materials prepared by the same method. On the other hand, although WAKO activated carbon exhibits high affinity for Au (III) over a wide range of HCl concentration, it exhibited higher degree of adsorption also for Pd (II) and Pt (IV) than CMR and carbonized orange waste, suggesting that the selectivity for Au (III) over Pd (II) and Pt (IV) is higher for CMR and carbonized orange waste than WAKO activated carbon. On the other hand, adsorption of base metals such as Fe(III) is insignificant over the whole concentration range of HCl on all tested adsorbents except for WAKO activated carbon.

It is well known that the carbon surfaces of activated carbon are composed of graphene layer surfaces and functional groups on the edges. The functional groups on the activated carbon are mostly of oxygen functional groups, i.e., carboxylic, lactonic, phenolic, carbonyl, and ether groups. Although, the adsorption of organic compounds on activated carbon is mostly controlled by physical adsorption due to its porous characteristics, the majority of adsorption of ionic species like metal ions is considered to be governed by surface functional groups such as hydroxyl and carboxyl groups [19]. Chand et al. reported that the high selectivity of barley straw and rice husk carbons to precious metal ions is attributable to its positively charged surface, on which negatively charged species are adsorbed by electrostatic interaction [12]. In the cases of base metal ions, due to unfavorable cationic and neutral species, their adsorption on the carbonized materials is nearly negligible.

Adsorption isotherm of Au(III) on the carbonized microalgae

Adsorption isotherm of Au (III) on CMR at 30°C is depicted in Figure 7, which shows that the amount of adsorbed Au (III) increases with increasing Au (III) concentration and tends to approach the plateau region at high concentration of Au (III), suggesting that the Langmuir type adsorption takes place in this adsorption system. From this plateau region, the maximum loading capacity of Au (III) on CMR was evaluated as 1.20 mmol/g or 236 mg/g. Under the same conditions, the maximum loading capacities of WAKO activated carbon and carbonized orange waste for Au (III) was evaluated as 1.67 and 0.70 mmol/g, respectively. Table 1 lists the comparison of loading capacities of various carbonized materials for Au (III). Although the maximum loading capacity on CMR is lower than WAKO activated carbon, it is higher than other carbonized biomass wastes such as the carbonized orange waste and rice husk carbon.

Observation of the formation of metallic gold particles

In the previous work on the adsorptive recovery of Au (III) using waste microalgal gel prepared by the treatment in boiling sulfuric acid, we found the formation of metallic gold by means of the observation

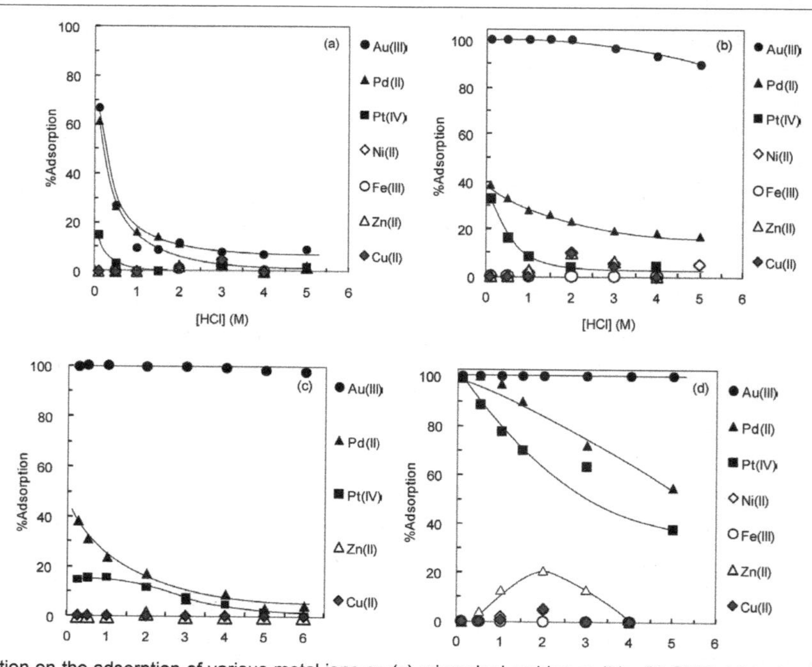

Figure 6: Effect of HCl concentration on the adsorption of various metal ions on (a) microalgal residue as it is, (b) CMR, (c) carbonized orange waste, and (d) WAKO activated carbon. Conditions: metal concentration=0.2 mM, dry weight of adsorbent=20 mg, volume of test solution=15 mL, shaking time=24 h, temperature=30°C.

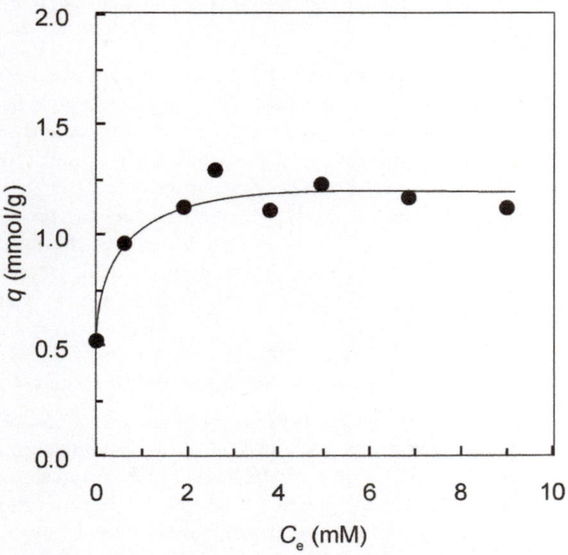

Figure 7: Adsorption isotherm of Au (III) on the carbonized microalgae. Conditions: HCl concentration=0.1 M, weight of adsorbent=10 mg, volume of test solution=10 mL, shaking time=100 h, temperature=30°C.

Adsorbent	Loading capacity (mmol/g)	Condition	References
Microalgal residue as it is	0.34	0.1 M HCl	This work
CMR	1.2	0.1 M HCl	This work
Orange waste carbon	0.7	0.1 M HCl	This work
WAKO activated carbon	1.67	0.1 M HCl	This work
Rice husk carbon	0.76	0.1 M HCl	[12]
Barley straw carbon	1.47	0.1 M HCl	[12]
Heated immobilized *C. vulgaris*	0.14	pH 3.0	[20]
Heated rice husk	0.26	pH 3.0	[20]
Activated carbon (Sigma C-5261)	0.37	pH 3.0	[20]

Table 1: Comparison of maximum loading capacities of various carbonized materials for Au (III).

by optical microscope and SEM as well as XRD [10]. Consequently, the formation of metallic gold was confirmed also in this system by means of SEM and XRD. The optical microphotograph shown in Figure 8a and a SEM image shown in Figure 8b clearly reveal the formation of metallic gold deposited on the surface of CMR. It is apparent from these figures that Au(III) was reduced to metallic Au (0) during the adsorption also under the present experimental conditions according to the following equation.

$$AuCl_4^- + 3e^- = Au\ (0) + 4Cl^- \quad E^0 = 1.00\ \text{V vs SHE} \qquad 3$$

Based on these results, it can be concluded that CMR is an excellent electrical conductor, it functions as the electrode of redox reaction in the adsorption of Au (III) from solution. In the same way as electrochemical cementation reduction, there are two respective half-cell areas on the carbon surface. In the cathodic region, Au (III) and Au (I) consume electrons and are reduced to its metallic state (Au (0)).

Figure 9 shows the XRD pattern of CMR after the adsorption of Au (III), which indicates the presence of four identical peaks at 38, 44, 64, and 77° 2θ, which are assigned to Au (111), Au (200), Au (220), and Au (311) of fcc (face-centered cubic) gold crystal phase, respectively.

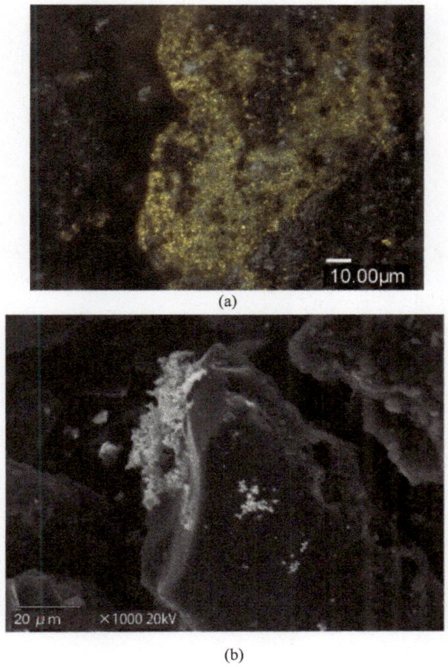

Figure 8: (a) Image of the optical microphotograph of CMR with deposited metallic Au on its surface taken at 2000X magnification. (b) SEM image of CMR with deposited metallic Au.

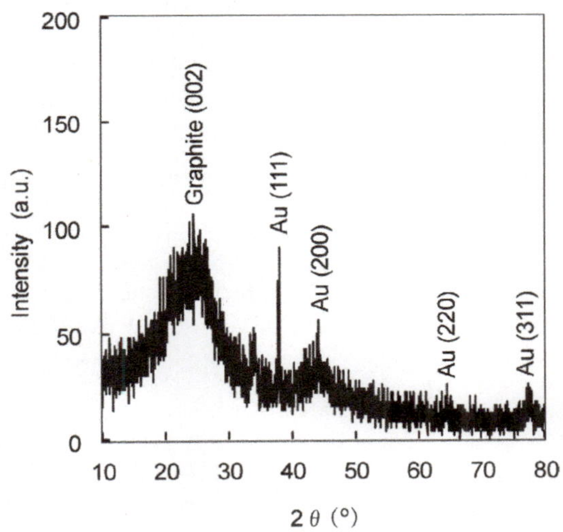

Figure 9: XRD pattern of CMR after Au (III) adsorption.

The standard reduction potentials for $AuCl_4^-$ species, the predominant species of Au (III) in this study, is 1.00 V which is relatively higher than other precious metal complexes like $PdCl_4^{2-}$ (0.62 V) and $PtCl_6^{2-}$ (0.68 V). Unlike the case of Au adsorption from alkaline cyanide solution by the activated carbon [20,21], Au adsorption from acidic halide solution involves the electrochemical reduction. This is attributable to the relative instability of $AuCl_4^-$ complexes and the ability of the activated carbon to function as a reducing agent. Figure 10 shows a comparative cyclic voltammograms of CMR before and after saturated adsorption of Au (III). It is obvious that the positive and negative sweep of CMR becomes notably small after Au (III) adsorption, which

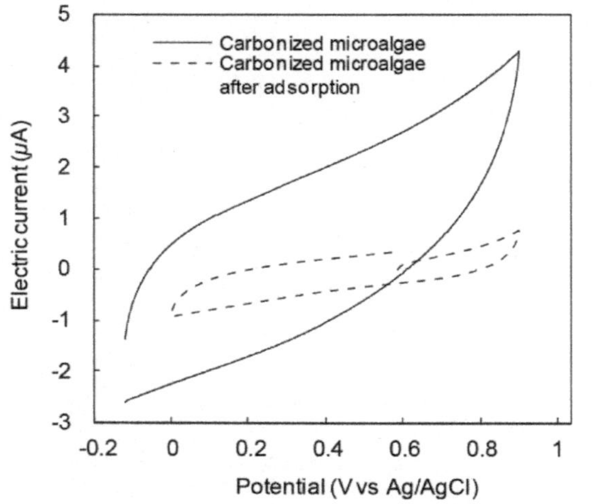

Figure 10: Cyclic voltammograms of CMR before and after Au(III) adsorption. Conditions: RE=Ag/AgCl, CE=Pt plate, electrolyte solution=1.0 M $NaNO_3$, Scan speed=50 mV/s.

indicates that CMR have lost their electron donation ability due to the slow redox reactivity.

Jia and Demopoulos suggested that hydroquinone and phenolic groups is mainly responsible for the redox characteristic of carbon [19]. Due to the reduction of Au (III) to Au (0), hydroquinone simultaneously undergoes the oxidation and form the quinones. In addition, graphene layers exhibit the characteristic of Lewis bases and behave as electron donors due to the enrichment of π-electrons on the basal plane surfaces. Therefore, it is also considered that the reduction of Au(III) to Au (0) on CMR might partly occur on the graphene layer surfaces.

Adsorption kinetics of Au(iii) on the carbonized microalgae

Since it was observed to take long time to reach equilibrium in the adsorption of Au (III) on the adsorption gels prepared from various biomass wastes including the microalgal residue by treating in boiling sulfuric acid, the adsorption kinetics of Au (III) on the carbonized microalgae was investigated in detail.

Figure 11a shows the time variation of the amount of Au (III) adsorbed on CMR at varying temperatures. This observed kinetic data were interpreted according to the pseudo-first order rate equation described by Equation 4 as depicted in Figure 11b.

$$\ln (C_t/C_i)=-k_1 t \qquad 4$$

where C_t and C_i are the concentration of Au (III) in the aqueous phase at time (t) and its initial concentration, respectively, while k_1 is the pseudo-first order rate constant of the adsorption reaction.

As expected from Equation 4, the plots lie on straight lines corresponding each temperature in this figure. From the slopes of each straight lines, the pseudo-first order reaction rate constants were evaluated at each temperature and plotted according to the Arrhenius equation described by Equation 5 as shown in Figure 11c to evaluate the activation energy (E_a).

$$\ln k=\ln A-(E_a/RT) \qquad 5$$

where A is the frequency factor and R is the ideal gas constant (8.314 J/(mol·K)).

As expected from Equation 5, the plots in Figure 11c lie on a straight line, from the slope of which the activation energy in this adsorption reaction was evaluated as 33.7 kJ/mol. This value is in the range of interaction controlled for the chemical adsorption (30-120 kJ/mol) according to Saeed et al. [22]. Similar to other heterogeneous reactions between solids and fluids, it is inferred that the process of gold chloride adsorption on activated carbon includes the diffusion of $AuCl_4^-$ through the liquid film surrounding the carbon particles followed by the reduction of Au (III) to Au (I) and finally to metallic gold on the carbon surface [19].

Conclusion

It can be concluded that CMR prepared in the present work was effective for the selective recovery of Au(III) from acidic chloride media. The adsorption of Au (III) is nearly independent of HCl concentration, which is an advantage for the practical applications. The reduction of Au (III) to metallic Au (0) was found to be involved in the adsorption of Au (III) on the surface of CMR. The results of the present work suggest a promising application of CMR prepared by a simple treatment for the selective and effective recovery of Au (III).

Acknowledgement

The present work was financially supported by the Ministry of Agriculture, Forestry, and Fisheries, Japan, for the research project on Development of Innovative Total Technologies for Carbon Separation, Biomass Production, and Biomass Utilization. The authors are indebted to Prof. Takanori Watari and Dr. Hirokichi Nakamura of Department of Applied Chemistry of Saga University for supporting the equipment for the preparation of CMR and assist in electrochemical measurement, respectively.

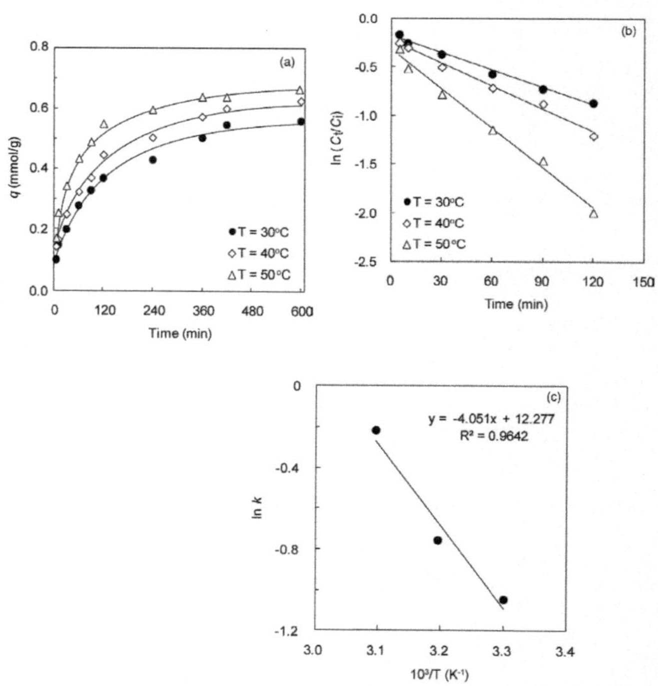

Figure 11: Kinetic aspect of the adsorption of Au (III) on CMR. Conditions: weight of adsorbent=10 mg, volume of test solution=10 mL, initial Au (III) concentration=0.6 mM, HCl concentration=0.1 M. (a) Time variation of amount of Au (III) adsorbed on CMR. (b) Plots of the kinetic data according pseudo-first order rate expression expressed by Equation 4. (c) Arrhenius plot of Au (III) adsorption on CMR.

References

1. Park JY, Fray DJ (2009) Recovery of high purity precious metals from printed circuit boards. J Hazard Mater. 164: 1152-1158.
2. Juarez CM, Oliveira JF (1993) Activated carbon adsorption of gold from thiourea solutions. Miner Eng. 6: 575-583.
3. Ibrado AS, Fuerstenau DW (1995) Infrared and X-ray photoelectron spectroscopy studies on the adsorption of gold cyanide on activated carbon. Miner Eng. 8: 441-458.
4. Jia YF, Steele CJ, Hayward IP, Thomas KM (1998) Mechanism of adsorption of gold and silver species on activated carbons. Carbon 36: 1299-1308.
5. Yalcin M, Arol AI (2002) Gold cyanide adsorption characteristics of activated carbon of non-coconut shel origin. Hydrometallurgy. 63: 201-206.
6. Soleimani M, Kaghazchi T (2008) Adsorption of gold ions from industrial wastewater using activated carbon derived from hard shell of apricot stones-An agricultural waste. Biores Technol. 99: 5374-5383.
7. Volesky B, Holan ZR (1995) Biosorption of heavy metals. Biotechnol Prog. 11: 235-250.
8. Mack C, Wilhelmi B, Duncan JR, Burgess JE (2007) Bio-sorption of precious metals. Biotechnol Adv. 25: 264-271.
9. Dubinsky Z, Aaronson S, Berner T (1980) Some economic considerations in the mass culture of microalgae: Algae Biomass. Elsevier/North-Holland Biomedical Press, Amsterdam, Netherlands, 819-832.
10. Khunathai K, Xiong Y, Biswas BK, Adhikari BB, Kawakita H, et al. (2012) Selective recovery of gold by simultaneous adsorption-reduction using microalgal residues generated from biofuel conversion processes. J Chem. Technol Biotechnol. 87: 393-401.
11. www.miningeducation.com/2012/10/extracting-gold-by-carbon-in-pulp-with.html
12. Chand R, Watari T, Inoue K, Kawakita H, Lutel HN, et al. (2009) Selective adsorption of precious metals from hydrochloric solutions using porous carbon prepared from barley straw and rice husk. Miner Eng. 22: 1277-1282.
13. Chand R, Watari T, Inoue K, Torikai T, Yada M (2009) Evaluation of wheat straw and barley straw carbon for Cr(VI) adsorption. Sep Purif Technol 65: 331-336.
14. Bansal RC, Goyal M (2005) Activated Carbon Adsorption. CRC Press, Boca Raton, FL, USA, pp: 7-36.
15. Groen JC, Peffer LAA, Perez-Ramirez J (2003) Pore size determination in modified micro- and meso-porous materials. Pitfalls and limitations in gas adsorption data analysis. Micropor Mesopor Mat. 60: 1-17.
16. Nakamura M, Nakanishi M, Yamamoto K (1996) Influence of physical properties of activated carbons on characteristics of electric double-layer capacitors. J Power Sources. 60: 225-231.
17. Pushparaj VL, Shaijumon MM, Kumar A, Murugesan A, Ci L, et al. (2007) Flexible energy storage devices based on nanocomposite paper. PNAS 104: 13574-13577.
18. Taer E, Deraman M, Talib IA, Awitdrus A, Hashmi SA, et al. (2011) Preparation of a highly porous binderless activated carbon monolith from rubber wood sawdust by a multi-step activation process for application in supercapacitors. Int J Electrochem Sci. 6: 3301-3315.
19. Jia Y, Demopoulos GP (2003) Adsorption of silver onto activated carbon from acidic media: nitrate and sulfate media. Ind Eng Chem Res. 42: 72-79.
20. Nakbanpote W, Thiravetyan P, Kalambaheti C (2002) Comparison of gold adsorption by Chlorella vulgaris, rice and activated carbon. Miner Eng 15: 549-552.
21. Sibrell PL, Miller JD (1992) Significance of graphitic structural features in gold adsorption by carbon. Miner Metall Proc 9: 189-195.
22. Saeed MM, Rusheed A, Ahmed N (1996) Kinetics and thermodynamic aspect of the adsorption of Fe(III) on HTTA-immobilized polyurethane foam. J Radioanal Nucl Chem. 211: 293-303.

Effects of Controlled Atmosphere Storage and Temperature on Quality Attributes of Mango

Hailu Z*

Addis Ababa Institute of Technology, School of Chemical and Bio Engineering, Food Engineering, Addis Ababa University, Addis Ababa, Ethiopia

Abstract

This study was conducted to analyze the effects of Controlled Atmosphere Storage (CAS) and low temperatures on quality attributes of cv. Keitt mango. Mango is one of the choicest fruits in the world and popular due to its delicious taste, pleasant aroma and nutritional value. To date, only 3% of the total world production is being exported due to the perishable nature of the product, lack of proper storage and other postharvest related diseases. Post-harvest loss of mango in Ethiopia is estimated to be between 25-40% and would occur during handling, transportation and storage. Storage loss could arise owing to atmospheric conditions and temperature. CAS is agricultural technologies which monitor and adjust constantly the level of CO_2 and O_2 within storage to reduce the rate of physiological and biochemical changes, ethylene sensitivity and incidence of decay development of mangoes and also inhibit pathogen growth. Thus, this study was under taken to determine the optimum storage conditions of Ethiopian mango in order to reduce considerable postharvest mango losses. In this work, randomized block design experimental was employed for optimization of the influencing factors mainly CAS and temperatures. Accordingly, experiments were carried out at constant storage temperature of 7°C with different atmospheric conditions of A (10% CO_2+6% O_2+84% N_2), B (8% CO_2+6% O_2+86% N_2), C (5% CO_2+5% O_2+90% N_2) and Control (0.03% CO_2+21% O_2+79% N_2). Besides, fruits also stored at different constant temperatures of 7°C, 10°C, 13°C and control (21-24°C) respectively. The result obtained showed that effect of CAS on quality attributes (weight loss and firmness) of cv. Keitt mango significantly improved the storage life as compared to that stored at ambient atmosphere with list significant difference values being at 0.05=3.12 and 1.708 respectively. In addition, the effect of temperature on Total soluble solid, Titrable acidity, pH, weight loss, firmness and skin color was found to be significant in prolonging the storage life of the fruit with List significant difference values of 0.05 (0.317, 6.58, 3.46, 4.148, 6.50 and 2.41). The study revealed that the fruit stored at 7°C under 10% CO_2+6% O_2 storage condition, as confirmed by sensory analysis, was given better preference to extend cv. Keitt mango up to 6 weeks. On the other hand, the fruits stored at 13°C with no CAS conditions showed better storage time of up to 21 days without chilling injury effect. Finally, it can be conclude that application of CAS technology can significantly enhance the quality and storage life thereby minimizing post-harvest losses of mango fruit.

Keywords: CV Keitt; Controlled atmosphere storage (CAS); Temperature; List significance difference

Introduction

Mango (*Mangifera indica* L.) is native to south Asia and known as king of the fruits due to its excellent flavor, delicious taste and high nutritive values. There are over thousands varieties around the world, the most common are Keitt, Kent and Tommy Atkins and they are also grow in Ethiopia. Mango is produced mainly in-west and east of Oromia, SNNPR, Benishangul and Amhara [1], which are specific areas of Ethiopia and they have a potential to produce wide variety of mango fruit. According to Ethiopian central statistical agency report in 2011/12 report about 12-14% of total fruit production is mango. But, it is exported less than 2% [2]. This is due to perishable nature of the product, lack of proper handling and lack of proper storage [3]. Besides, the post-harvest loss of mango is 26.3% [4]. Thus, Maintenance of fruit quality for a specific period of time before its consumption is important factor in the post-harvest life of the fruit. The physiological changes may occur in harvested fruit due to unfavorable atmospheric conditions especially fluctuation in temperature and humidity. Fresh mango fruits have a short storage life of 4 to 12 days at room temperature and can also suffer low temperature injury (chilling injury) during refrigerated storage [5]. Therefore, it is necessary to develop improved methods of mango storage in fresh state that can enhance its shelf life without any detrimental effect on the quality. Various methods of extending the storage life of mangoes have been tested but little research has been conducted on controlled atmosphere (CA) techniques. Controlled atmosphere Storage (CAS) has been shown to be beneficial technique in reducing the rate of physiological and biochemical changes, ethylene sensitivity and incidence of decay development of perishable fruit products [6] and as a result of it senescence is delayed. The response of mangoes to CA condition reported in the literature has been shown to vary. Nakamura et al. reported that CA storage having 5-10% CO_2 is effective to suppress the respiration rate of ripe mango. Lalel et al. found that CA comprised of 2% O_2 and 2% CO_2 is better for maintaining the aromatic compounds of ripen fruit. Similarly, Mitra and Baldwin stated that low temperature, hypobaric and CA storage can keep the mango fruit for about two to four weeks however, they suggested further research in this area. Lalel et al. [7] reported that CA treatments reduced respiration rate, de-greening of the skin and fruit softening of mangoes but yellowness of the skin, total soluble solids (TSS), total sugars content and taste of the ripe fruit were maintained. In another

***Corresponding author:** Hailu Z, Addis Ababa Institute of Technology, School of Chemical and Bio Engineering, Food engineering, Addis Ababa University, Addis Ababa, Ethiopia, E-mail: zinisheh@gmail.com research Palding et al. [8] recommended that shelf life of keitt mangoes can be increased up to 35 days without any significant quality effect by storing them at 13°C in 5% CO_2 and 5% O_2. Different varieties of mangoes showed different responses to CA storage. Therefore, the study focused with effects of controlled atmosphere storage and temperature on quality attributes of Ethiopian mango to reduce the post-harvest loss and maintain desired quality. The experiment was carried out to evaluate the effects of CAS and temperature, determine sensory evaluation and optimum storage condition of cv. Keitt mango.

Materials and Methods

The mango cv. Keitt sample used in this experiment was collected in plastic bags from Addis Ababa Et-fruit. The fruits were sorted according to their size, weight and color. Then the sorted fruits were washed by tape water, and dry through blowing dried air stored in CA storage, and ambient storage condition. To determine the effects of temperature and CAS, Completely Randomized Design (CRD) was used with below treatments: Storing fruits at constant temperature of 7°C under CAS conditions of A-(6% O_2+10% CO_2), B-(6% O_2+8% CO_2), C-(5% O_2+5% CO_2) and D-Control (21% O_2+0.03% CO_2). Besides, the effect of temperatures without considering CAS effect was also studied fruits stored at different storage temperature of 7°C, 10°C, 13°C and Control (21-24°C) respectively. About 112 uniform cv. Keitt mango fruits samples were used for studying out of which 42 samples were stored at CAS, with gas tight plastic bag in order to control the atmosphere by using closed system, 14 samples were stored in ambient atmosphere, while the remaining 56 samples were investigated for temperature effects. Each Samples was randomly taken within five day intervals to evaluate the effect of Controlled Atmosphere Storage (CAS). CAS chambers were calibrated to establish the specified gas composition by a gas blending flow system. The gas blending system generated CA conditions using external supplies of gases from, pressurized gas cylinders fitted with double-stage regulators and outlet controlling devices. These outlets were connected to the inlet flexible pipes that were inserted into the gas tight plastic containers in which the mangoes were stored. Fruits were then stored at specified temperature with different controlled atmosphere which were evaluated every 5 days for total soluble solids (TSS), titratable acidity (TA), pH, weight loss, firmness, and skin color. To investigate effects of different temperatures, sample was taken within 7 days interval using I Button. The One-Wire Viewer is a Java™-based software package to explore Maxim's 1-Wire and I Button devices with a personal computer. The 1-Wire and I Button devices communicate over a single data line plus ground reference, using the 1-Wire protocol. The skin color was measured by using skin indicator chart 1-deep green, 2-light yellow-green, 3-yellow-light green, 4-yellow-orange, 5-golden orange. Flash firmness was measured (kg/mm^2) using FACCHINI-48011-Model (Italy) penetrometer with 8 mm probe diameter. Total soluble solid (TSS) was measured by using Model-RF18 (India) handheld sugar refractometer. Titrable Acidity of the fruit was measured by means of an acid base titration method using a juice sample (10 ml juice+10 ml distilled water) and 0.1N Sodium hydroxide (NaOH) with phenolphthalein color indicator and pH was measured using digital pH meter. Weight loss; also measured by using digital weight balance. For the sensory evaluation tests, ten panelists were selected from the university and the tests involved individuals in isolated tasting conditions. Five of panelists were asked to give their individual ratings on all quality attributes of stored fruits including color, aroma, taste, flavor, firmness and overall acceptability of the stored fruits with different conditions while the other five panelists were asked to evaluate quality of the effects of temperatures with a 9-point structured hedonic scale to conduct the preference test: 1- dislike extremely and 9- like extremely. The scores marked by panelists were collected and an average was calculated for each parameter. One way Analysis of variance (ANOVA) was performed on the data collected using Microsoft excel and Origin Pro 8 soft wares. Comparison between treatment means at 5% level of significance was calculated.

Results and Discussion

The effects of controlled atmosphere storage (CAS) and low temperatures analysis for quality attributes of cv. Keitt mango and the optimum conditions for long storage were studied. In addition, sensory analysis for the final product using nine hedonic scale also carried out. The results of samples analyzed in this work are presented in tabular and graphical form.

The effect of CAS on quality attributes of cv. Keitt mango

Skin color: as the skin indicator chart showed that(1-deep green, 2-light yellow-green, 3-yellow-light green, 4- yellow orange, 5- golden orange), Skin color of fruits stored at ambient atmosphere changed their color from deep green to yellow orange after 15 days and to golden-orange after 30 days of storage (Table 1), whereas the skin color of fruits stored at controlled changed their color after 25 days from deep green to yellow orange and to golden orange colors after 40 days of storage (Table 2). The loss of green color was the most obvious change in mango which was probably due to the physiochemical change by degradation of the chlorophyll structure and increased in carotenoids pigments during storage. The principal agents responsible for this degradation might be oxidative system, pH change, and enzymes like chlorophyllase [9]. The experimental result observed, showed in significant of CAS on skin color (new LSD at 0.05=1.36) between fruits stored in CAS and ambient atmosphere as presented in Figure 1 and the effect of CAS and ambient atmosphere had similar mean result (5-golden orange) after 40 days storage.

Flash firmness: Mangoes exposed to CAS used in this work remained greener and more firm than those stored in ambient atmosphere (Figure 2). The current investigations confirmed the view that CAS conditions delay fruit ripening and softening [10]. Ripening and senescence rates in many climacteric fruits like mangoes, can be affected by control of the availability of O_2 and CO_2 to the fruit during respiration and that these two compounds can have a significant inhibitory effect on ability of ethylene to initiate ripening [11]. Thus, flash firmness of fruits kept in air significantly decreased very rapidly from 14.1 kg/mm^2 to 0.099 kg/mm^2 after 35 days of storage (Table 1) and new LSD at 0.05=1.708. The flesh firmness of fruits which were kept under (CAS) with treatment of (10% CO_2+6% O_2) had high firmness than the other treatments after 6 weeks storage with mean firmness result being 5.541 kg/mm^2.

Days	Firmness (kg/mm^2)	TSS(°Brix)	TAA (%)	pH	Weight loss (%)	Skin Color
0	14.1	4.1	1.665	2.40	0	1
5	12.1	8.9	1.152	2.83	6.289	2
10	9.2	14	0.7541	3.24	5.370	3
15	5.4	16	0.3328	3.54	3.272	4
20	1.2	17	0.128	4.56	4.384	4
25	0.25	18	0.0828	4.96	3.144	5
30	0.1	18.5	0.0256	5.27	12.182	5
35	0.099	14.8	0.0234	5.48	12.234	5

NB: 1-deep green, 2-light yellow-green, 3-yellow-light green, 4- yellow orange, 5-golden-orange

Table 1: Mean results of mango quality attributes stored in ambient atmosphere for 0-35 days.

Days	Firmness (kg/mm²)	TSS (°Brix)	TAA (%)	pH	Weight loss (%)	Skin color
0	14.13	4.1	1.66	2.40	0	1
5	12.16	8.8	1.151	2.45	0.0869	1
10	9.43	13.13	0.754	2.46	0.603	2.6
15	6.33	14.36	0.373	2.79	0.533	3.3
20	4.8	15.03	0.293	3.43	0.457	3.6
25	2.6	15.46	0.198	3.42	0.623	3.6
30	1	16.36	0.127	5.01	1.452	4
35	0.67	17.23	0.091	5.08	1.457	4.6
40	0.48	17.86	0.056	5.15	1.475	5
45	0.29	18.33	0.0391	5.16	1.491	5

NB: 1-deep green, 2-light yellow-green, 3-yellow-light green, 4- yellow orange, 5-golden-orange

Table 2: Mean results of mango quality attributes stored in CAS for 0- 45 days.

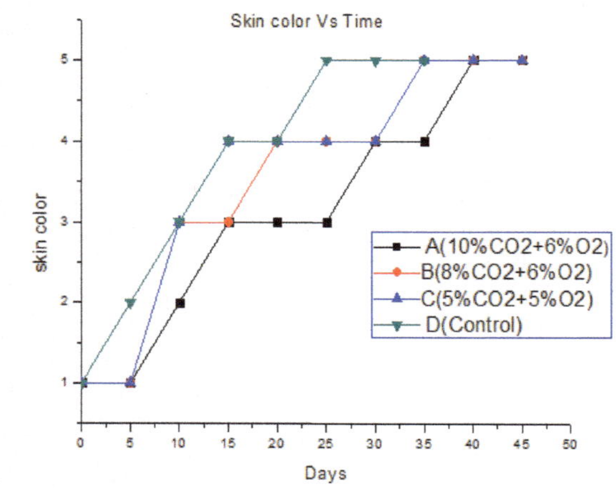

Figure 1: Skin color of mangoes at 7°C after storage under controlled atmosphere.

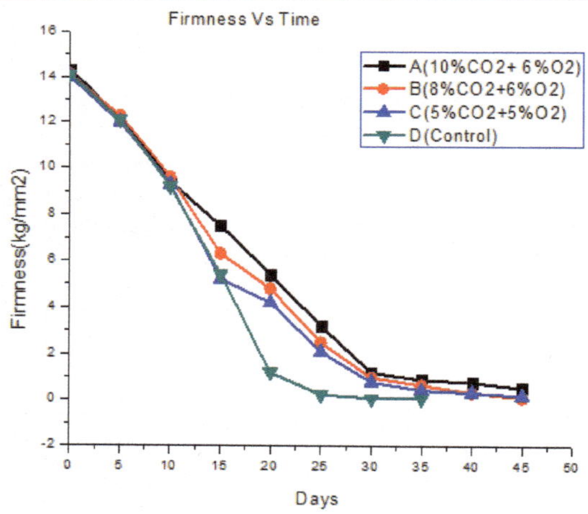

Figure 2: Flesh firmness of mangoes at 7°C after storage under controlled atmosphere.

Total Soluble Solid (TSS): it was observed that, TSS of fruits stored in ambient atmosphere were increased from 4.1 to 18.86° Brix after 30 days and decreased to 14.8° Brix after 35 days storage (Table 1). The increase in TSS might be due to alteration in cell wall structure and break down of complex carbohydrates into simple sugars. As showed the graph below Figure 3 the total soluble solid of fruits stored at ambient atmosphere sharply decreased in between 30 to 35 storage time, this is due to metabolic process as the temperature become high microbial and respiration activities become increased, conversion of sugar into alcohol(ethanol)become exist. The acidity of the fruit also becomes high. This increase and decrease in TSS are directly correlated with hydrolytic changes in starch and conversion of starch to sugar being important index of ripening process in mango fruits and other climacteric fruits and further hydrolysis decreased the TSS during storage [12]. Besides, fruits stored at controlled atmosphere storage also increased the TSS content from 4.1 to 18.25° Brix after 45 days of storage (Table 2). As the result showed, it would stay in CAS further days without significant quality effects. From the CAS treatments; treatment- A (10% CO_2+6% O_2) had less TSS content with average (13.5° Brix) than of the other treatments as shown in Figure 2. Hence, this treatment has a potential to extend the storage life of fruits beyond 45 days of storage time. No significant change was observed in TSS between the fruit samples kept at ambient atmosphere and CAS (new LSD at 0.05=4.406) (Figure 3).

Titratable Acidity (TAA): The change in Titratable acidity of the mango (cv. Keitt) recorded during storage at ambient temperature of (21-24°C) is presented in Table 1. The results revealed that percent Titratable acidity of treated cv. Keitt mango ranged from 1.665% to 0.0234% with an average of 0.52% during storage and the pH also increased from 2.4 to 5.16 after 35 day of storage. the results observed, indicated that percent of Titratable acidity showed decreasing trend during 35 days of storage period might be due to the degradation of citric acid which could be attributed to ripening or reduction in acidity may be due to their conversion into sugars and their further utilization in metabolic process of the fruit. These results coincided with those of Doreyappa-Gowda and Huddar [13] who reported similar pattern in different varieties of mango fruit stored at 18-34°C under gone a series of physico-chemical changes during ripening and the major changes were considerably increased in pH from 2.85 to 4.38 and decreased in acidity from 2.71 to 0.04% during ripening. Thus, no significant change in TAA and pH; between fruits kept at ambient atmosphere and CAS were observed (new LSD at 0.05=0.536 and 1.118) respectively (Figure 4).

Weight loss (%): The reduction in weight is attributed to the physiological loss in weight (PLW) due to respiration, transpiration of water through peel tissue and other biological changes taking place in the fruit. The effect of CAS was highly significant on weight loss of stored fruit (new LSD at 0.05=3.12). The mean weight loss percentage of fruits stored with treatment A (10% CO_2+6% O_2) was less (0.75%), whereas fruits stored at ambient atmosphere was high (5.87%) after 4-6 weeks of storage (Figure 5).

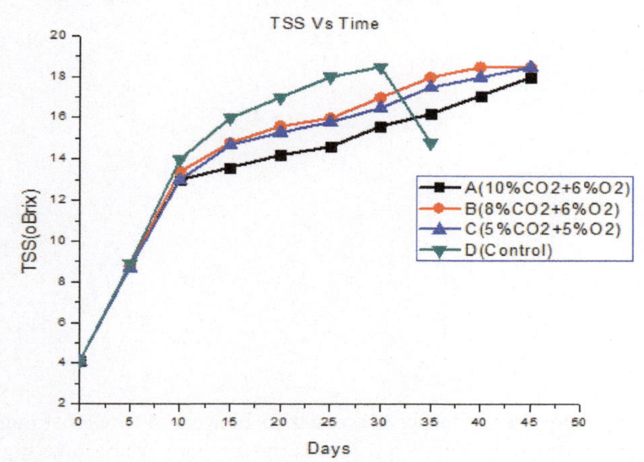

Figure 3: Total soluble solid content of mangoes at 7°C after storage under controlled atmosphere.

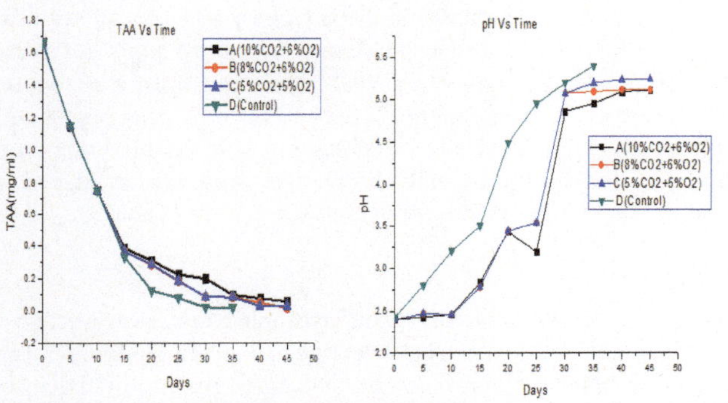

Figure 4: Titrable acidity and pH of mangoes at 7°C after storage under controlled atmosphere.

Effect of temperatures on quality attributes of stored cv. Keitt mango

Total Soluble Solid (TSS): Significant increase in sucrose content of mango has been observed during ripening and this has been attributed to an increase in total soluble solids during ripening. This is due to transformation of starch into soluble sugars as the carbohydrates in the fruit are broken down under the action of phosphorylase enzyme during ripening into simple sugars [14]. Storage temperature significantly (new LSD at 0.05=0.317) affected the TSS of fruits. As the mean storage temperature decreased TSS values obtained were 14.07° Brix, 12.87° Brix, and 9.87° Brix for the fruit stored 21-24°C, 13°C and 7°C respectively after 3 weeks storage time (Figure 6).

Titratable Acidity (TAA): Titrable Acidity of the stored fruit were significantly decreased as the temperature of the storage increased (new LSD at 0.05=6.58) after three weeks of storage (Table 3) with the mean values of 1.2865 mg/10 mg juice at temperatures of 7°C and 0.937 mg/10 mg juice at ambient temperature (21-24°C) after 3 weeks storage period. The decline in acidity could be due to susceptibility of citric acid to oxidative destruction as impacted by the ripening environment, and also it is a consequence of starch hydrolysis leading to an increase in total sugars and a reduction in acidity [15].

Increase in pH during ripening of mango fruits has been reported by other authors and was similar to what was observed in the present study. According to the authors, there is an inverse relationship between Titratable acidity and pH. The increase in pH (decline in acidity) could be due to utilization of acids as respiration substrates [16]. The pH of the fruits also significantly affected by storage temperatures (new LSD at 0.05=3.46); as the storage temperature increased from 7°C to ambient temperature (21-24°C). The pH of the fruits also increased from 3.18 to 4.01 after three weeks storage as shown in Table 3.

Skin color: The average skin color of fruit samples stored at 7°C, 10°C, 13°C and 21-24°C were light yellow green light yellow green, yellow-green and yellow orange respectively (Table 3). According to the result, the effect of temperatures on skin color of the stored fruits was not significance changes fruits which was exposed to 7, 10, and 13°C, whereas fruits stored in between 7°C and ambient temperature (21-24°C) was significant (new LSD at 0.05=2.41).

Flash firmness: Effect of storage temperatures on firmness of cv. Keitt showed significant difference between fruits stored at temperature of 7°C and ambient (21-24°C) temperature (Table 3) with new LSD at 0.05=6.50).

Weight loss (%): Maximum average weight loss percentage loss observed for fruits stored at ambient temperature (4.44%) after three weeks storage (new LSD at 0.05=4.148). As temperature of storage increased the average weight loss also increased (Table 3). It was observed that Mangoes stored at low temperature leads to certain physiological disorder observed such as Chilling Injuries (CI). The primary cause of CI is thought to be the damage of cell membrane that initiates a cascade of secondary reaction. CI is a time and temperature problem. Mango fruits are subjected to CI when stored below 10°C. The symptoms include grayish scald- like discoloration of the skin, skin pitting, uneven ripening, and reduction in the level of carotenoids, aroma and flavor during ripening. Most of mangoes cultivars show injury below 10°C (50°F), especially if fruit have just reached maturity and Tolerance to chilling increases during ripening [6]. Chilling injury (CI) has been reported to occurs in mango fruit at temperatures below about 10-13°C, although some cultivars (Dasheri, Langara) were reported to be safely stored at 7-8°C for up to 25 days. Storage at 10 to 13°C (50 to 55°F) with 85 to 90% relative humidity should give a shelf-life of 14 to 28 days for mature green fruit, depending upon variety. The result of the experiment showed that fruits stored at temperature of 7°C after three weeks storage time, were exposed to CI starting from 7 days storage, for the fruit stored at 7°C with symptoms of unripening, skin pitting, discoloration, poor aroma and flavor during ripening and is in agreement with Jobin Decor result (Figures 7 and 8).

Effect of CAS and temperatures on Sensory Quality Attributes (SQA)

Color changes in mango fruit are due to the disappearance of chlorophyll and appearance of other pigments [17]. The observed results of skin color of fruits stored with treatment (10% CO_2+6% O_2) showed maximum preference with it mean and standard deviation result was 8.4 ± 0.89 (Table 4). Fruit flavor depends upon taste (balance between sweetness and sourness or acidity, and low to no astringency) and aroma (concentrations of odor-active volatile compounds). Taste development is due to a general increase in sweetness, which is the result of increased gluconeogenesis, hydrolysis of polysaccharides, especially starch, decreased acidity, and accumulation of sugars and organic acids resulting in an excellent sugar/acid blend Sourness or acidity is determined by the concentrations of the predominant organic acids, which are citric acid, malic acid, and tartaric acid. Some amino acids, such as aspartic and glutamic acid may also contribute to sourness.

Temperatures	Firm(kg/mm²)	TAA (%)	TSS(°Brix)	Wet loss (%)	pH	Skin color
7°C	11.5	1.2865	9.875	2.6	3.18	1
10°C	10.125	1.202	11.125	2.9	3.3975	2
13°C	8.2	0.98975	12.875	4.25	3.81	3
Control	6.255	0.9374	14.075	4.44	4.015	4

NB: 1-deep green, 2-light yellow-green, 3-yellow-light green, 4-yellow orange, 5-golden orange

Table 3: Mean effect of temperatures on mango quality attributes after 3 weeks storage.

SQA	A	B	C	D
Color	8.4 ± 0.89	7.8 ± 0.83	7.6 ± 1.14	7.8 ±1.09
Flavor	7.4 ± 1.14	7.4 ± 0.89	6.8 ± 1.30	7 ± 2.00
Taste	8 ± 0.70	8.6 ± 0.54	6 ± 1.58	5.8 ± 2.38
Firmness	7.4 ±1.14	7.4 ± 0.54	7.4 ± 0.54	8 ± 0.00
Over all acceptance	8 ± 0.70	7.8 ± 0.44	7.2 ± 0.83	6.8 ± 1.78

NB: A (10%CO_2+6%O_2), B (8%CO_2+6%O_2), C (5%CO_2+5%O_2) and D (Control)

Table 4: Effect of CAS on Sensory Quality Attributes of stored fruits (Means ± SD).

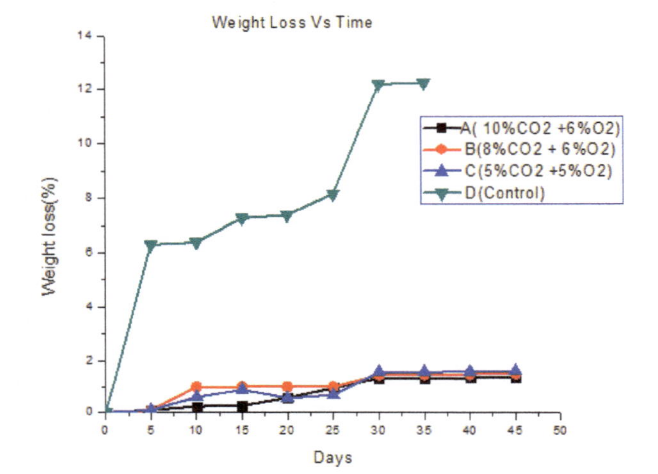

Figure 5: Weight loss of mangoes at 7°C after storage under controlled atmosphere.

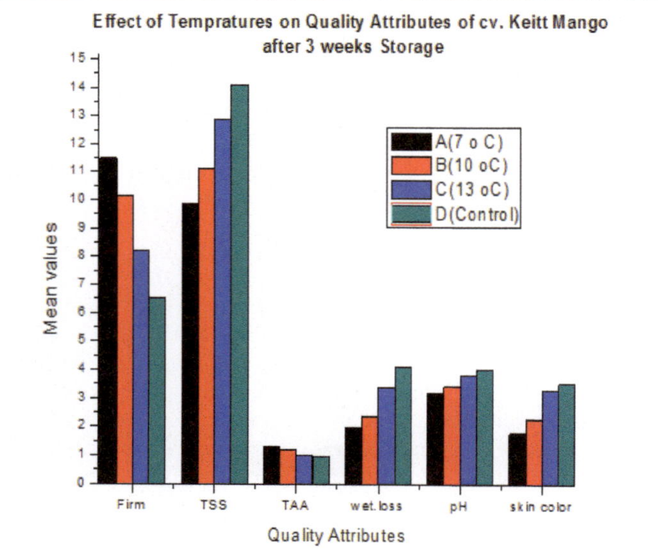

Figure 6: Effects of temperatures on cv. Keitt mango quality attributes after 3 weeks storage.

Several factors affect sugar and acid contents in mango, including cultivar [18]. The flavor of Mangoes stored under treatment conditions of A (10% CO_2+6% O_2) and B (8% CO_2+6% O_2) with average values of 7.4 was preferred by panelists. But, the overall acceptance was high for fruits stored with treatment A (10% CO_2+6% O_2) mean result of 8.0 (Table 5). Similarly, the effects of temperatures on quality of stored fruits was evaluated by sensory panelists, the color of the fruits stored at temperature of 13 and ambient (21-24°C) had high preferences with their mean values got 7.6 (Table 5). Fruit samples stored at temperature of 13°C got maximum overall acceptance, while fruits stored at temperature of 7°C got minimum overall acceptance (Table 5).

Conclusions

Mangoes are highly nutritious and favorable fruits. However, post-harvest losses and quality deterioration due to exposure to temperature and improper handling management challenged the productivity and wider usage of the fruit. In Ethiopia, the post-harvest loss (transporting, handling and storage) of mango is estimated as to be about 23.6% from this total loss about 7-10% is storage loss. The major storage loss factors are atmospheric conditions and temperature. Thus, study was carried out by using CAS technology which is constantly monitor and adjust the CO_2 and O_2 levels within gas tight stores or containers to reduce storage loss and extending the storage life of mangoes for better duration without loss of its quality. The Controlled Atmosphere Storage (CAS) positively affected the quality parameters and significantly improved the shelf life of mangoes. It extends storage life of cv. Keitt mangoes for up to 6 weeks without significant quality loss with treatment at 7°C under 10% CO_2+6% O_2+84% N_2. This condition is better to store the mango for long time without significance loss relative the other conditions and quality of the fruit after storage also confirmed by sensory analysis, thus the above condition is best. On the other hand, the effect of storage temperature on quality attributes of mango was significant. The cv. Keitt mangoes stored at 13°C had better quality after 3 weeks storage time as confirmed by sensory analysis than of fruits stored with other treatments. In general, extending storage duration more than 6 weeks is extending marketing period and maintained fruit.

Acknowledgements

Above of all, I thank the almighty God who gave me strength and encouragement and led me through all the rough and difficult times to complete this study. I wish to express my deepest appreciation to my advisors, Mr. Teshome Worku for his endless help in correcting, commenting, and encouraging accomplishing this work. I also like to appreciate their patience in following up this work from the very beginning to the end. I am very grateful to Mr. Mekidim Asefa, for his assistance, invaluable comments and excellent supervision throughout my research work. I also extend my appreciation to Addis Ababa Institute of Technology, School of

Figure 7: Photograph of Keitt variety mango fruit before storage.

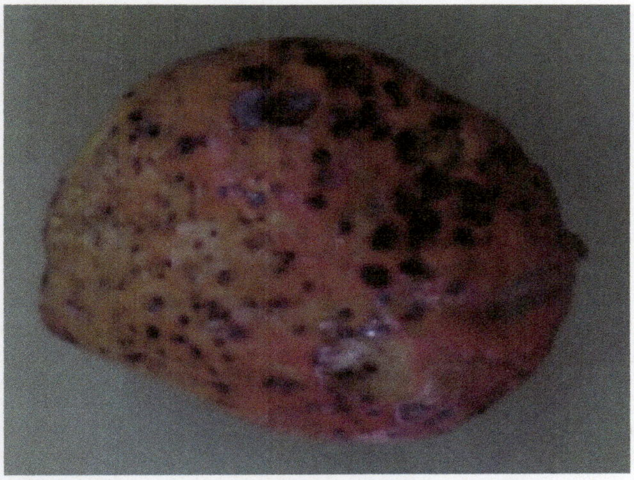

Figure 8: Photograph of Keitt type mango fruits stored at temperatures of 7°C.

SQA	7°C	10°C	13°C	Control (21-24°C)
Color	4 ± 1.58	6.4 ± 1.51	7.6 ± 1.14	7.6 ± 1.14
Flavor	3.2 ± 1.48	5.8 ± 0.83	8 ± 0.70	7.2 ± 0.83
Taste	3.4 ± 0.89	5.2 ± 0.83	7.6 ± 1.14	5.8 ± 0.83
Firmness	3.8 ± 3.8	6 ± 1.22	7.2 ± 0.83	5.6 ± 1.14
Over all acceptance	3.6 ± 3.6	6.4 ± 0.54	8 ± 0.70	6.4 ± 0.54

Table 5: Effect of Temperatures on Sensory Quality attributes (Means ± SD).

Chemical and Bio Engineering Laboratory staff members for their technical assistance during experimental work. Finally, I would like to express my sincere gratitude to my families and friends who have been providing their support and encouragement. And also all others who directly and indirectly contributed to this manuscript are highly acknowledged for their voluntary participation.

References

1. Desta H (2005) Export potential of Ethiopia processed fruit and vegetables, export promotion department of English, Jaffee PS and J.J. Okello.2006.
2. Joosten F (2007) Development Strategy for Export Oriented Horticulture in Ethiopia. Wageningen, Netherlands.
3. Leon DM, De la Cruz J, Parkin KL, Garcia HS (1997) Effect of controlled atmospheres containing low O_2 and high CO2 onchilling susceptibility of Manila mangoes. Acta Horticulturae. 455: 635-642.
4. Tadesse F (1991) Postharvest losses of fruits and vegetables. Journal of Horticulture 270: 261-27.
5. Lakshminarayana S, Subhadra NV, Subramanyam H (1970) Some aspect of developmental physiology of mango fruit. Journal of Horticultural Science 45: 133-142.
6. Medlicott AP, Sigrist JM, Sy O (1990) Ripening of mangoes following low temperature storage. J. Amer. Soc. Hort. Sci. 115:430-434.
7. Lalel HJ, Singh Z (2004) Biosynthesis of aroma volatile compounds and fatty acids in 'Kensington Pride' mangoes after storage in a controlled atmosphere at different oxygen and carbon dioxide concentrations. J Hort Sci Biotechnol 79: 343-353.
8. Spalding DH, Reeder WF (1977) Low pressure (hypobaric storage) of mangoes. Journal of the American Society for Horticultural Science 102: 367-369.
9. Wills RH, Lee TH, Graham D, McGlasson WB, Hall EG (2007) Postharvest; An introduction to the Physiology and Handling of Fruit, Vegetables and Ornamentals Pp: 83-99.
10. Kader AA (1986) Biochemical and physiological basis for effects of controlled and modified atmospheres on fruits and vegetables. FoodTechnol 40: 99-104.
11. Ben-Yehoshua S (1985) Individual seal-packaging of fruits and vegetables in plastic film new postharvest technique. Journal of Horticultural Science 20: 32-37.
12. Kittur FS, Saroja N, Tharanathan R (2001) Polysaccharide-based composite coating formulations for shelf-life extension of fresh banana and mango. Eur Food Res Tec 213: 306-311.
13. Doreyappy-Gowda IND, Huddar AG (2001) Studies on ripening changes in mango (Mangiferaindica L.) fruits. J Food Sci Tec 38: 135-137.
14. Mitra SK, Baldwin E Z (1997) Mango. In: Postharvest Physiology and Storage of Tropical and Subtropical Fruits (Mitra S K, ed), CAB International, West Bengal, India
15. Aina JO (1990) Physico-chemical changes in African Mango (Irvingia gabogensis) durin normal storage ripening. J Food Chem 36: 205-212.
16. Dadzie BK, Orchard JE (1997) Routine Post Harvest Screening of Banana/ Plantain Hybrids: Criteria and Methods. International Plant Genetic Resources Institute, Wageningen, the Netherlands pp: 63.
17. Lakshminarayana S, Shetty MS, Krishnaprasad CA (1975) Accelerated ripening of Alphonso mangoes by the application of ethrel. Tropical Science 17: 95-101.
18. Kapse BM, Rane DA, Khedkar DM (1989) Correlation between biochemical parameters and organoleptic evaluation in mango varieties. ActaHort. 231:756-762.

3

Chemical, Structural and Elemental Characterization of Biosorbents Using FE-SEM, SEM-EDX, XRD/XRPD and ATR-FTIR Techniques

Sohan Shrestha*

Nepal Academy of Science and Technology, Lalitpur, Bagmati Zone, Nepal

Abstract

The chemical, structural, and elemental composition of any adsorbents plays undoubtedly crucial role and has paramount significance in adsorption process. The present study of chemical, structural, and elemental characterization of powdered *Pinus densiflora* pine cones, lignite and coconut shell-based activated carbon fiber (ACF), and powdered oyster shell prior employed for anionic surfactant sodium dodecyl sulfate (SDS) adsorption was undertaken using field emission scanning electron microscope (FE-SEM), scanning electron microscope coupled with energy dispersive x-ray (SEM-EDX), X-ray diffraction/X-ray powder diffraction (XRD/XRPD), and multiple internal reflectance (MIR) or attenuated transform reflectance – fourier transform infrared spectroscopy (ATR-FTIR) techniques. The FE-SEM imaging observed at three different magnifications depicted the change in surface texture of aforementioned biosorbents before and after SDS adsorption. SEM-EDX analysis indicated alteration in the concentration and distribution of different compositional elements. The XRD analysis suggested the amorphous character in lignite and coconut shell-based ACF and powdered *Pinus densiflora* cones, whilst indicating the sample of powdered oyster shell to be crystalline in nature and entirely composed of $CaCO_3$ with the very trivial amount of impurities. Further, infra-red (IR) spectroscopy demonstrated variation in IR spectra of the samples prior to and after surfactant adsorption and also further suggested occurrence of complexing or (complexation) process between ionizable functional groups and DS^- or Na^+ ions of SDS. In addition, iodine number of the lignite and coconut shell-based ACF, powdered *Pinus densiflora* cones, and powdered oyster shell were found to be 285.86 mg/g, 25.01 mg/g, and 28.48 mg/g, respectively.

Keywords: Biosorbents; FE-SEM; SEM-EDX; FTIR; XRD

Introduction

Given the increased emphasis on the use of lignite and coconut shell-based activated carbon fiber (ACF), powdered oyster shell, and acidified powdered *Pinus densiflora* (Japanese red pine) pine cones as the prominent biosorbents for anionic surfactant sodium dodecyl sulfate (SDS) adsorption, there have been surprisingly few studies undertaken to characterize the chemical, structural, and elemental composition of aforementioned samples. The main aim of this study, therefore was to investigate the chemistry, structure, and mineralogy of lignite and coconut shell-based activated carbon fiber (ACF), powdered oyster shell, and powdered *Pinus densiflora* pine cones samples using multiple characterization techniques including x-ray diffraction (XRD)/x-ray powder diffraction (XRPD) analysis, field emission scanning electron microscopy (FE-SEM), multiple internal reflectance (MIR) or attenuated transform reflectance – fourier transform infrared spectroscopy (ATR-FTIR), scanning electron microscopy coupled with energy dispersive x-ray (SEM-EDX). In addition, the samples were treated to determine the iodine number or iodine index for the characterization of the three samples. Results will provide an evaluation of the key differences in structure, chemical, and elemental composition of aforementioned samples (adsorbents) prior to and after anionic surfactant adsorption.

Materials and Methods

Materials

The lignite and coconut shell-based activated carbon fiber (ACF) used in the present study was procured from Korea ACF Co. Ltd. Similarly, the naturally occurring cones of *Pinus densiflora* were collected from the premises of Kumoh National Institute of Technology (KIT), Gumi. Meanwhile, the oyster shells were fetched from the beach of Pohang, South Korea.

Biomodification of the samples

The lignite and coconut shell-based activated carbon fiber was used in the experimentation without further processing as procured, while the oyster shells were pre-treated before use by washing thoroughly with the distilled water and dried at the temperature of 110 ± 2°C for 24 h in an oven. Further, the raw biosorbent was crushed and ground into the powder form using grinder mixture. The biosorbent was again washed thoroughly with the distilled water and dried naturally. The resulted shell particles were stored in sterile, sealed plastic bottle and used in the experimentation. On the other hand, the cones was also pre-treated prior to its use by washing thoroughly with the distilled water and then dipped into the 2.0% concentrated HCl solution for 12 h. The acidified pine cones were then dried at the temperature of 110 ± 2°C for 24 h in an oven. Further, the cones were crushed and ground into the powder form. The powdered pine cones were kept in sealed plastic bottle and employed for all the adsorption experiments.

Methods

Iodine number (or iodine index) determination for characterization of adsorbents: In determining the iodine number of lignite and coconut shell-based acivated carbon fiber, 0.1 g of adsorbent was placed with 25 mL of iodine solution in a 250 mL conical flask and was shaken for 1 min. After that the solution was filtered and 10 mL of

***Corresponding author:** Sohan Shrestha, Nepal Academy of Science and Technology, Lalitpur, Bagmati Zone, Nepal, E-mail: scientist.nepalensis@gmail.com the filtrate was taken in a conical flask. The solution was titrated with 0.01 N sodium thiosulfate solution until it become clear. The iodine number of the activated carbon was determined by using Eq. (1) which represents the number of milligrams of iodine adsorbed by one gram of activated carbon [1]:

$$Iodinenumber = \frac{V \times (T_i - T_f) \times C_i \times M_i}{T_i \times g} \quad (1)$$

where, V is the volume of iodine solution (25 mL); T_i is the volume of sodium thiosulfate solution used for the titration of 10 mL iodine solution; T_f is the volume of sodium thiosulfate solution used for the titration of 10 mL filtrate; g represents the weight of adsorbent (0.1 g); M_i is the molar weight of iodine (126.9044 g/mol); and C_i is the concentration of iodine solution (0.01 N).

Similarly, 1.0 g of adsorbent was placed with 25 mL of iodine solution in a 250 mL conical flask and was shaken for 1 min for determining the iodine number in the present study, whilst the other procedures are same as aforementioned in case of ACF.

Infra-red (IR) spectroscopy (ATR-FTIR): In order to detect the functionality present in lignite and coconut shell-based ACF, powdered oyster shell, and powdered *Pinus densiflora* pine cones prior to and after anionic surfactant, i.e., sodium dodecyl sulfate (SDS) adsorption, FTIR spectroscopy studies using attenuated total reflectance (ATR) or multiple internal reflectance (MIR) apparatus were carried out by VERTEX 80v, Fourier Transform Infrared Spectrometer, Bruker Optics Korea Co., Ltd (Germany). Adsorption in the infrared (IR) region takes place due to the rotational and vibrational movement of the molecular groups and chemical bond of a molecule [2].

Scanning electron microscopy (FE-SEM): The surface morphology of lignite and coconut shell-based ACF, powdered oyster shell, and powdered *Pinus densiflora* pine cones before and after SDS adsorption were observed using Field Emission Scanning Electron Microscope (FE-SEM). The samples for FE-SEM observation were first cut into small pieces in accordance with the requirement, before placed on the sample – stubs. Prior to scanning, the samples were attached to the sample holder and coated with gold using sputter – cutting (SEM Coating System Machine). Samples were scanned using JEOL JSM – 6701F FE-SEM operating at 15 kV.

Scanning electron microscope and energy dispersive X-ray microanalysis (SEM-EDX): All the samples were observed from SEM, with elemental compositions being analyzed by EDX. After air-dried, the samples were passed through a 0.65 micrometer of filter paper. Then, samples were mounted on stubs with conductive carbon tape and coated with carbon. All samples were analyzed for their micro-structure and elemental compositions under JEOL JSM – 6701F FE-SEM operating at 15 kV.

X-ray diffraction (XRD): The X-ray powder diffraction spectra (XRD data) were taken by employing using 10 mm × 10 mm beam size, graphite-monochromated Cu/Kα radiation (λ=1.5406 Å) at 40 kV and 100 mA. The 2θ ranges were from 10° to 90° in the step of 2.0°. The detection of the powder X-ray diffraction pattern was performed by using SWXD Diffractometer, Rigaku Corporation (Japan) operating at 18 kW, which collects the two-dimensional diffraction patterns of the sample. Data were processed using the D/MAX – 2000/PC version 3.0.0.0 ~ software.

Results and Discussion

Determination of iodine number

Iodine number is one of the most essential parameters widely used to characterize the adsorbents [3]. It was found that iodine number of the lignite and coconut shell-based activated carbon fiber used in the present study as an adsorbent was found to be 285.86 mg/g. It was found that iodine numbers of powdered *Pinus densiflora* cones and powdered oyster shell were 25.01 mg/g and 28.48 mg/g, respectively. Thus, lignite and coconut shell-based ACF could be considered far better adsorbent in comparison to powdered *Pinus densiflora* cones and powdered oyster shell on the basis of iodine index as iodine is the most common standard adsorbate.

ATR-FTIR studies

IR spectra of powdered *Pinus densiflora* cones before and after SDS adsorption are shown in Figure 1. Similarly, the main frequencies of vibrations and their corresponding functional groups that interacted with the SDS during the adsorption process are presented in Table 1.

It is apparent that there were some changes in the absorption peak frequencies from infrared (IR) spectra prior to and after SDS adsorption. It was observed that absorption peak frequencies, i.e., 1631.04 cm^{-1}, 1567.09 cm^{-1}, 1770.14 cm^{-1}, and 3482.55 cm^{-1} were devoid in SDS loaded IR spectrum, corresponding to the carboxamide (C=O stretching vibrations), carboxylate, acid (aryl) halide, and aromatic 1° (primary) amine (N-H stretching) functional groups, respectively. Meanwhile, absorption band frequencies 2881.82 cm^{-1} and 2946.66 cm^{-1} corresponding to the C-H stretching in alkanes were not observed in SDS free spectrum (Figure 2).

On the other hand, the SDS loaded spectrum demonstrated minor changes in absorption peak frequencies corresponding to various other functional groups with the exception of major shift in the positions of O-H stretching in alcohol, C=O stretching in aldehydes or ketones, C-N stretching vibration in 2° (secondary) amine, and 1,4-disubstitution (para) in C-H bonding, i.e., from 3842.55 to 3484.73 cm^{-1}, 1706.99 to 1707.49 cm^{-1}, 1139.83 to 1140.06 cm^{-1}, and 848.38 to 843.31 cm^{-1}, respectively (Table 1).

It is well familiar that the electrostatic interactions are not strong enough to shift the absorption band of the functional groups present on the adsorbent surface [4]. Consequently, these shifts suggested complexing or (complexation) process took place between ionizable functional groups and DS^{-}or Na^{+} ions of SDS. In conclusion, the results obtained were suffice enough to draw a bottom line conforming the occurrence of chemical complexation between DS^{-}or Na^{+} ions of SDS and various functional groups present on the surface of powdered cones.

The ATR-FTIR spectroscopy studies were undertaken to elaborate the adsorption mechanism governing the SDS adsorption on lignite and coconut shell-based activated carbon fiber. It was observed that absorption band frequency 2881.72 cm^{-1} corresponding to C-H stretching of methyne and 737.34 cm^{-1} corresponding to C-Cl stretching of aliphatic chloro-compounds were absent in SDS loaded spectrum as depicted in Ref. [5] (Figure 2).

On the other hand, the SDS loaded IR spectrum demonstrated minor changes in absorption peak frequencies corresponding to various other functional groups with the exception of noticeable major changes in the positions of O-H stretching (alcohol), C-N stretching (2° amine), N-H stretching (aromatic 1° amines), C=O stretching (carboxylic acid

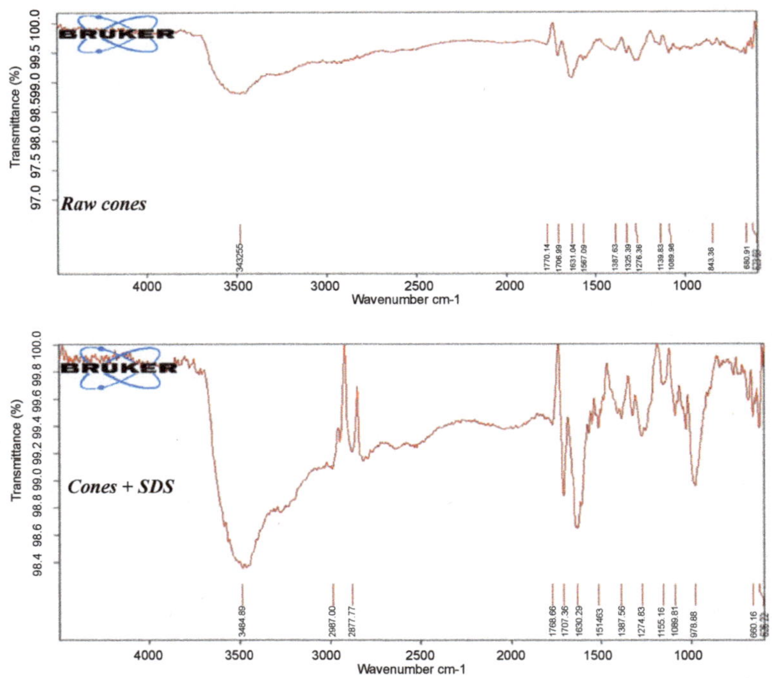

Figure 1: ATR-FTIR spectra absorption band/peak frequencies and corresponding possible functional groups of powdered pine cone before and after SDS adsorption.

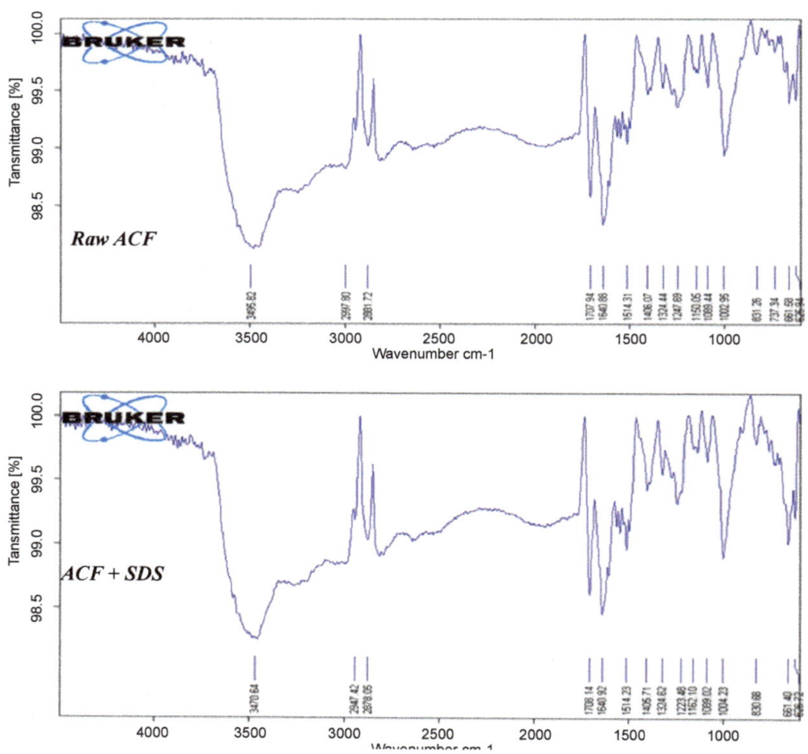

Figure 2: ATR-FTIR spectra absorption band/peak frequencies and corresponding possible functional groups of ACF before and after SDS adsorption.

Wave numbers (cm^{-1})		Corresponding functional groups
Before SDS adsorption	After SDS adsorption	
3482.55	3484.73	Alcohol, hydroxy group (O-H stretching) (broad band)
-	2881.82, 2946.66	Alkane (C-H stretching)
3482.55, 1631.04, 1567.09	3484.73, 1628.41	1° amines (N-H stretching)
1706.99	1707.49	Aldehydes or ketones (R-COR, C=O stretching)
1706.99, 1770.14	1707.49, 1768.63	Carboxylic acid (R-COOH) (C=O stretching)
1631.04	-	Amide (Carboxamide) (R-$CONR_2$) (R = H or C) (C=O stretching)
1631.04	1628.41	Alkene (C=C stretching)
1089.98, 1139.83, 1276.83	1031.20, 1089.38, 1140.06, 1277.82	Alcohol or ether (C-OH)
848.38	843.31	1,4-disubstitution (para) in C-H bonds (aromatic ring bending)
660.91, 623.59	623.42	Aliphatic organohalogen compounds (C-Br stretching)
1089.98, 1139.83	1089.38, 1140.06	Cyanate (-OCN and C-OCN stretching)
1089.98	1031.20, 1089.38	Organic siloxane or silicone (Si-O-Si)
1567.09	-	Carboxylate (carboxylic acid salt)
1770.14	-	Acid (acyl) halide
3482.55	-	Aromatic 1° amine (N-H stretching)
1139.83	1140.06	Secondary amine (C-N stretch)

Table 1: IR absorption bands and corresponding possible functional groups of powdered *Pinus densiflora* (Japanese red pine) cones prior to and after SDS adsorption.

or ketones), and P-O-C stretching (aliphatic phosphates), i.e., 3495.82 to 3470.64 cm^{-1}, 1150.05 to 1162.1 cm^{-1}, 3495.82 to 3470.64 cm^{-1}, 1707.94 to 1708.14 cm^{-1}, and 1002.95 to 1004.23 cm^{-1}, respectively as shown in Table 2 (Coates). Consequently, these shifts suggested that the complexing process took place between the ionizable functional groups present on the adsorbent surface and DS^{-}or Na^{+} ions [4]. However, minor changes in absorption peak frequencies corresponding to various other functional groups after SDS adsorption corroborated that there was no significant and strong coordination between DS^{-}or Na^{+} ions and functional groups at those sites.

It is apparent that there were some changes in absorption peak frequencies or wave numbers from IR spectra in powdered oyster shell prior to and after SDS adsorption as shown in Figure 3 [5,6]. As shown in Table 3 [5], the SDS loaded spectrum demonstrated minor changes in absorption peak frequencies corresponding to various other functional groups with the exception of major shift or changes in the positions of C-H stretching in methyne, C-N stretching in 2° or 3° amine, C=O stretching in carboxyl (carboxylic acid) or carbonyl (ketones) groups, and C=O stretching vibrations in acyl halide (or haloformyl group), i.e., 2883.08 to 2878.84 cm^{-1}, 1150.78 to 1139.74 cm^{-1}, 1707.24 to 1706.83 cm^{-1}, and 1798.11 to 1797.73 cm^{-1}, respectively [5-8]. It is well known that the electrostatic interactions are not strong enough to shift the absorption band frequencies or wavenumbers of the functional groups present on the adsorbent surface [4]. Consequently, these shifts suggested that there was complexing process taking place between the ionizable functional groups and DS^{-}or Na^{+} ions of SDS. However, minor changes in absorption peak frequencies corresponding to various other functional groups after SDS adsorption corroborated that there was weak coordination between the DS^{-}or Na^{+} ions and functional groups at those sites [4].

Field Emission Scanning Electron Microscope (FE-SEM) studies

The surface textural characterization of powdered pine cones, lignite and coconut shell-based ACF, and powdered oyster shell prior to and after SDS adsorption were carried out using Field Emission Scanning Electron Microscope (FE-SEM) at various magnifications. Further, the surface textures of powdered pine cones (Figure 4), lignite and coconut shell-based ACF (Figure 5), and powdered oyster shell (Figure 6) prior to SDS adsorption were observed to be smooth and regular with less undulations, i.e., less degree of roughness and waviness, while the surface morphology of powdered pine cones (Figure 7), lignite and coconut shell-based ACF (Figure 8), and powdered oyster shell (Figure 9) were observed to be rough with more undulations, i.e., high degree of roughness and waviness, after the biosorbents were subjected to adsorption process at various magnifications ranging from X5000, X10000, and X30000.

Microanalysis of structural and elemental composition of biosorbents using SEM-EDX technique

Scanning electron microscope and energy dispersive X-ray microanalysis (SEM-EDX) determined the micro-structure and elemental composition of *Pinus densiflora* cones, lignite and coconut shell-based activated carbon fiber, and powdered oyster shell as depicted in aforementioned (Figures 10-12).

EDX detected the emission of elements of carbon (C), oxygen (O), and chlorine (Cl). The percentage (wt%) of all elements before SDS adsorption was investigated as following order: C (62.30%) > O (37.07%) > Cl (0.63%) (Figure 10). Similarly, the percentage (wt%) of all elements present in *Pinus densiflora* cone after adsorption was investigated as following order: C (62.24%) > O (37.65%) > Cl (0.10%) (Figure 10).

As shown in Figure 11, EDX detected the presence of elements of carbon (C), oxygen (O), sodium (Na), and phosphorus (P). The percentage (wt%) of all elements present in ACF was investigated as following order (before adsorption) (Figure 11): C (82.37%) > O (15.79%) > P (1.66%) > Na (0.18%). Similarly, the percentage (wt%) of all elements present in ACF was investigated as following order (after adsorption): C (77.12%) > O (21.39%) > P (1.42%) > Na (0.08%) (Figure 11).

As shown in Figure 12, EDX detected the presence of elements of carbon (C), oxygen (O), magnesium (Mg), sulfur (S), calcium (Ca), and tin (Sn). The percentage (wt%) of all elements present in powdered oyster shell was investigated as following order (before adsorption) (Figure 12): O (55.69%) > Ca (33.10%) > C (9.82%) > Sn (0.84%) > S

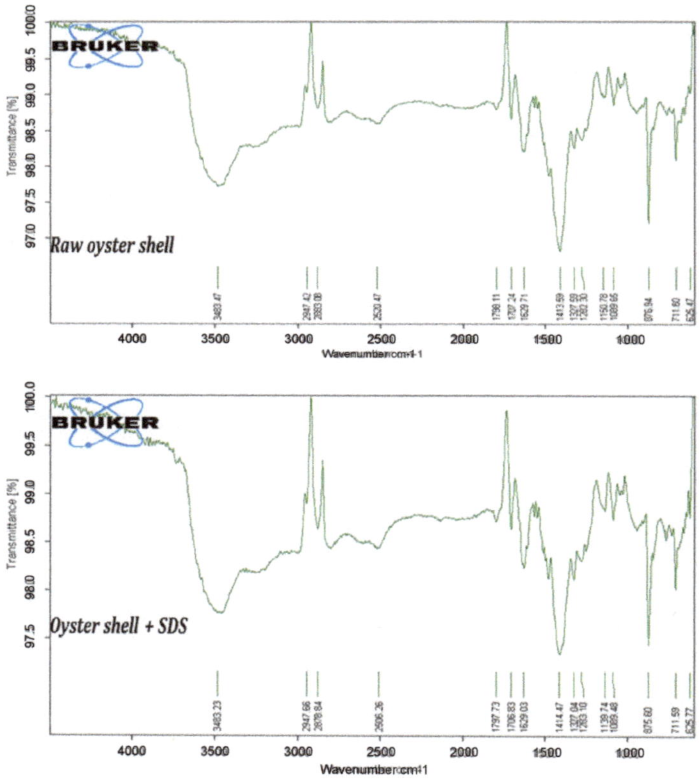

Figure 3: ATR-FTIR spectra absorption band/peak frequencies and corresponding possible functional groups of powdered oyster shell before and after SDS adsorption.

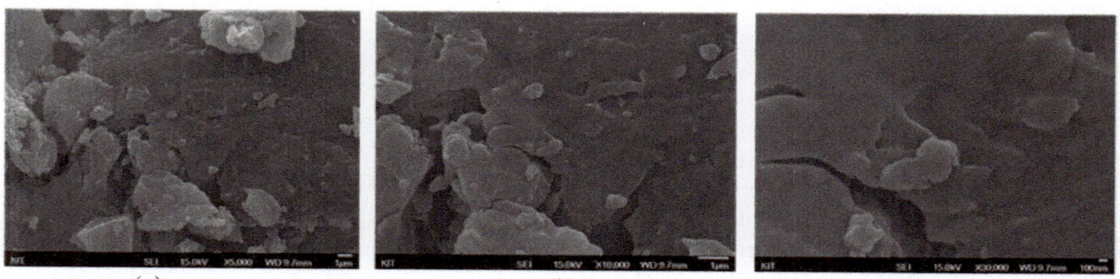

(a) (b) (c)

Figure 4: FE-SEM images of *Pinus densiflora* cone before SDS adsorption at various magnifications, i.e., (a) X5000; (b) X10,000; (c) X30,000.

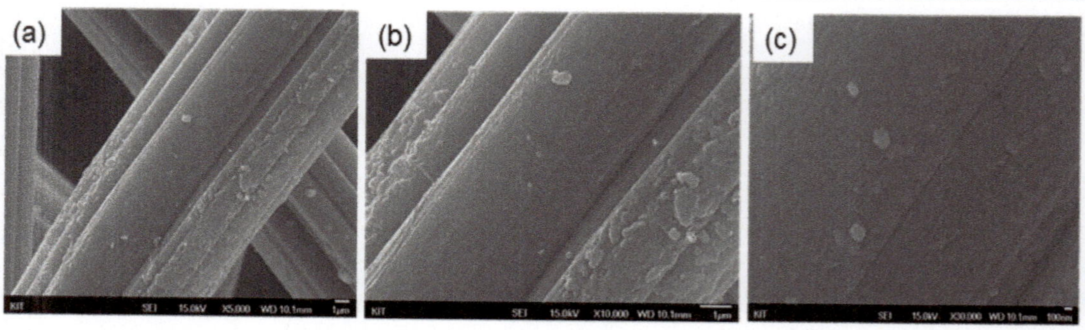

Figure 5: FE-SEM images of lignite and coconut shell-based activated carbon fiber before SDS adsorption at various magnifications, i.e., (a) X5000; (b) X10,000; (c) X30,000.

(a) (b) (c)

Figure 6: FE-SEM images of Oyster shell before SDS adsorption at various magnifications, i.e., (a) X5000; (b) X10,000; (c) X30,000.

(a) (b) (c)

Figure 7: FE-SEM images of *Pinus densiflora* cone after SDS adsorption at various magnifications, i.e., (a) X5000; (b) X10,000; (c) X30,000.

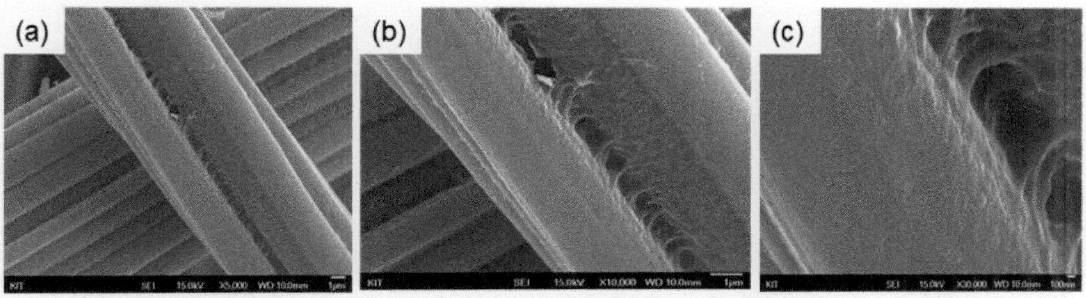

Figure 8: FE-SEM images of lignite and coconut shell-based activated carbon fiber after SDS adsorption at various magnifications, i.e., (a) X5000; (b) X10,000; (c) X30,000.

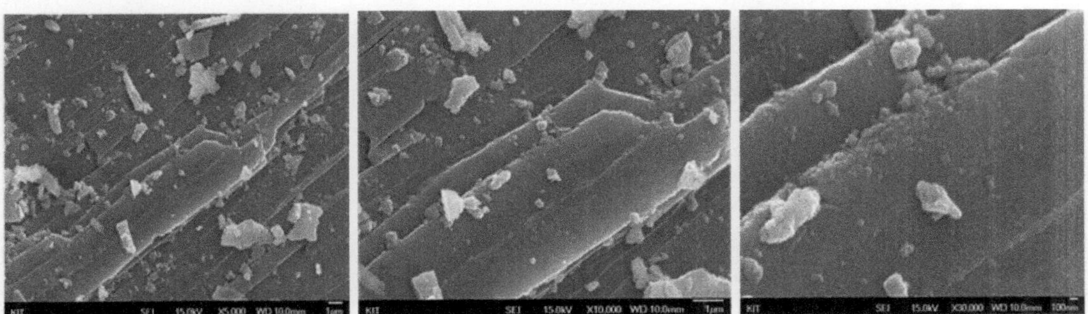

(a) (b) (c)

Figure 9: FE-SEM images of Oyster shell after SDS adsorption at various magnifications, i.e., (a) X5000; (b) X10,000; (c) X30,000.

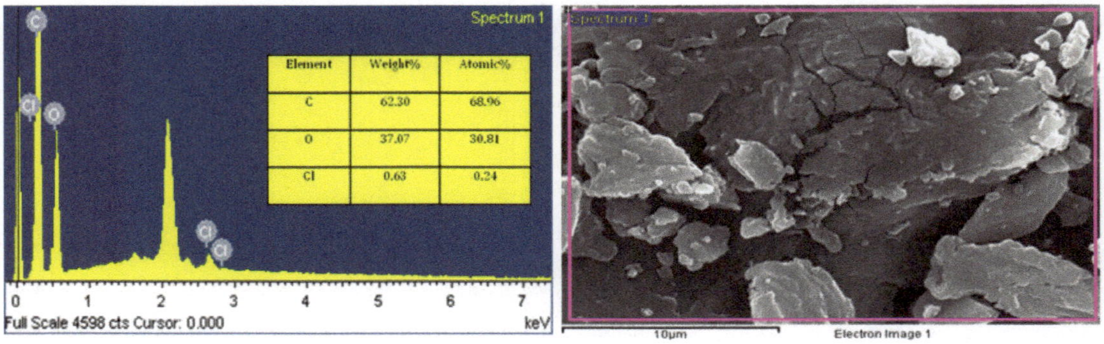

(a)

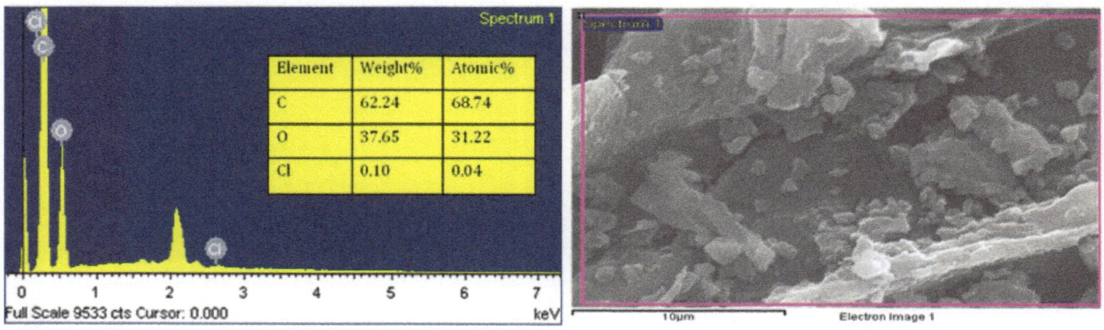

(b)

Figure 10: SEM-EDX image of *Pinus densiflora* cones powder before SDS adsorption (a); and after SDS adsorption (b) at the magnification of X5000.

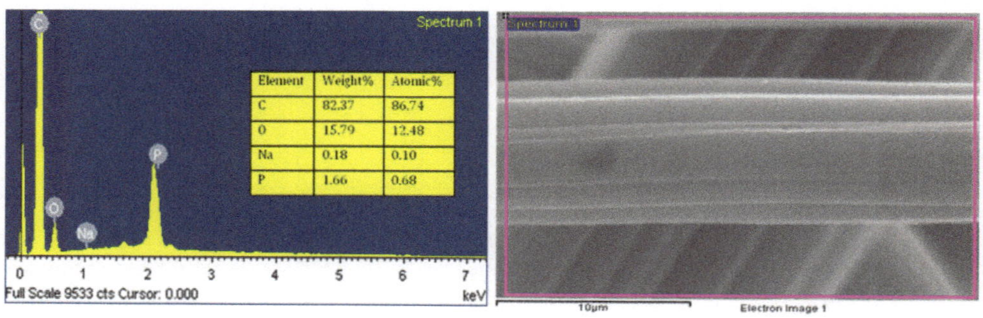

(a)

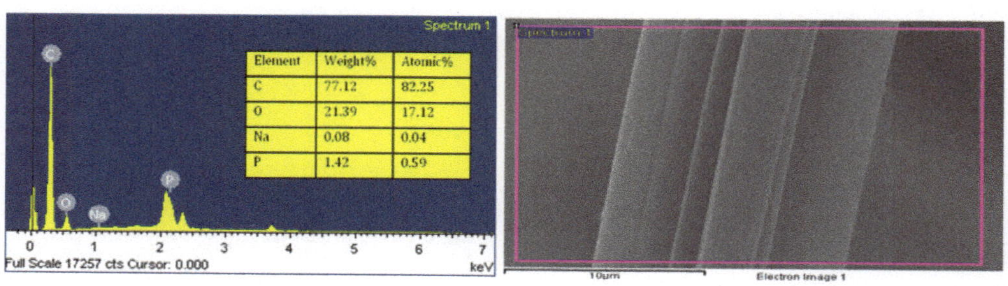

(b)

Figure 11: SEM-EDX image of lignite and coconut shell-based ACF before SDS adsorption (a); and after SDS adsorption (b) at the magnification of X5000.

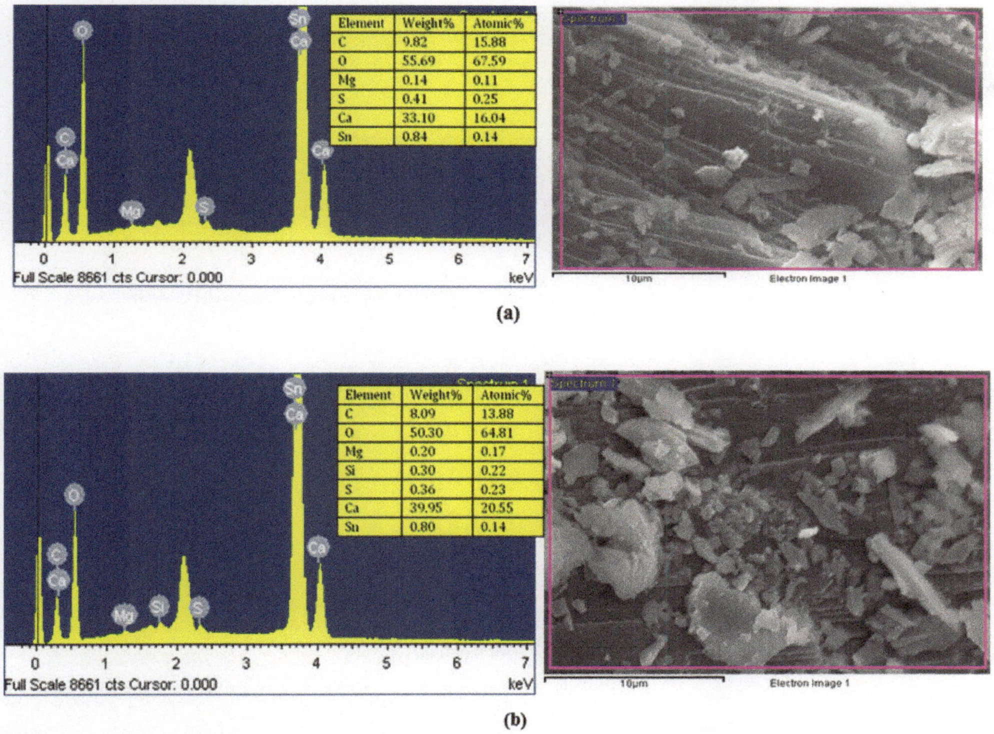

Figure 12: SEM-EDX image of powdered oyster shell before SDS adsorption (a); and after SDS adsorption (b) at the magnification of X5000.

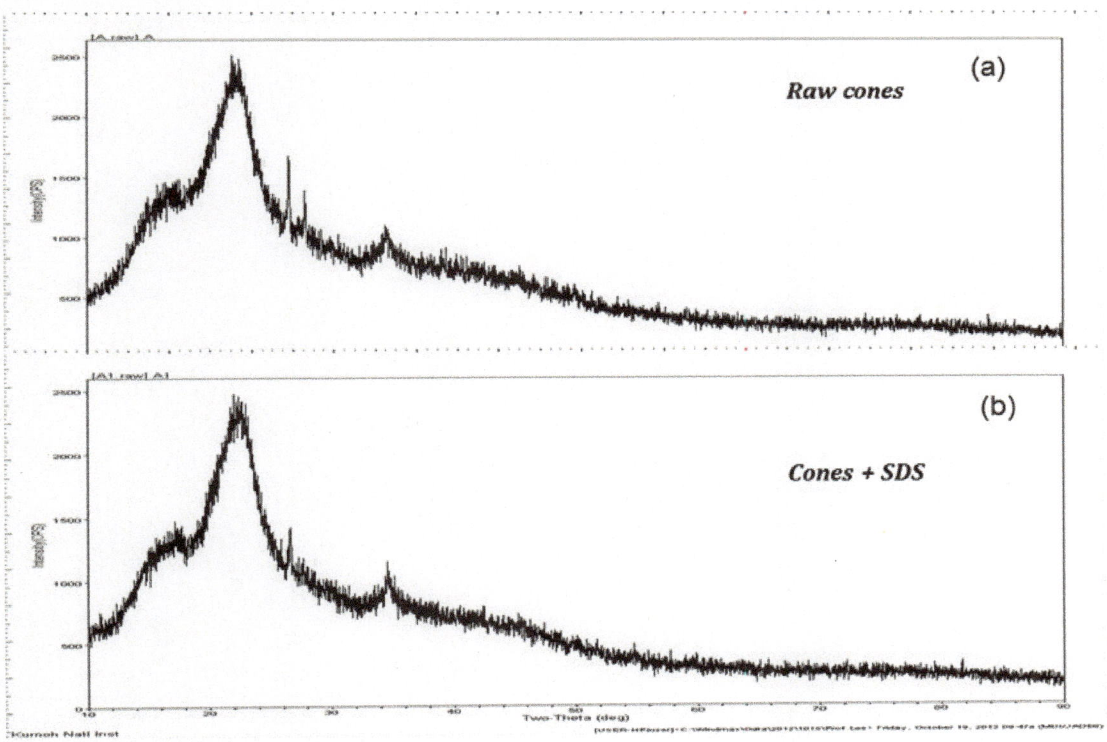

Figure 13: XRD pattern of powdered *Pinus densiflora* cone before SDS adsorption (a) and after SDS adsorption (b).

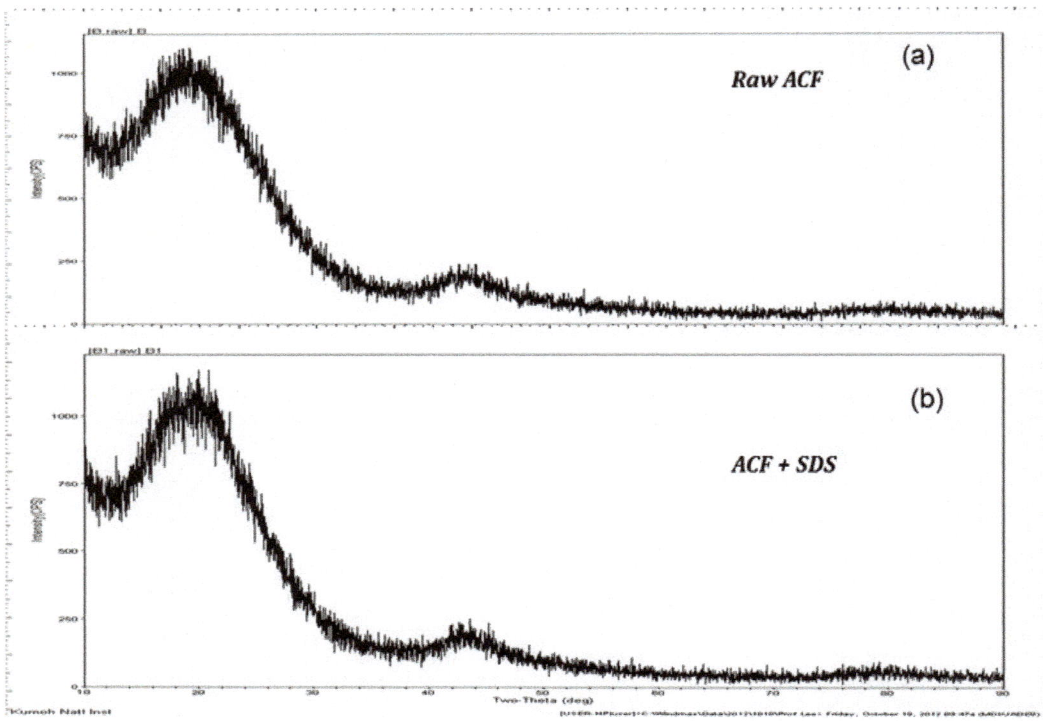

Figure 14: XRD pattern of lignite and coconut shell-based activated carbon fiber before SDS adsorption (a) and after SDS adsorption (b).

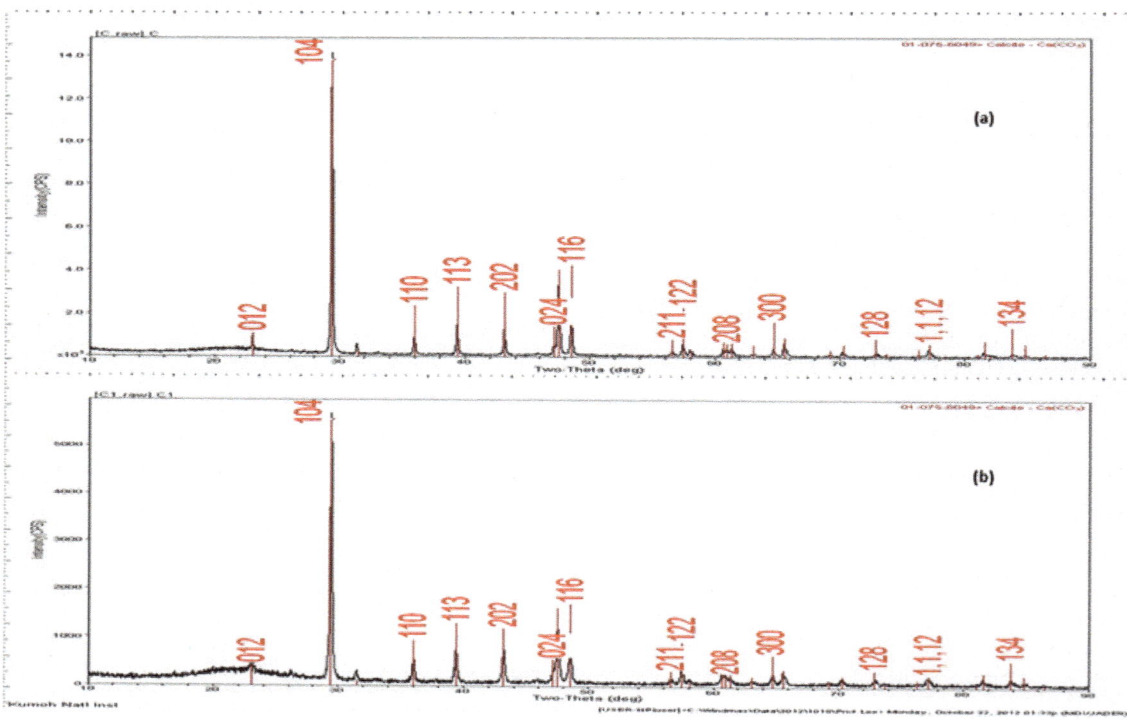

Figure 15: XRPD pattern of powdered oyster shell before SDS adsorption (a) and after SDS adsorption (b).

Wave numbers (cm^{-1})		Corresponding functional groups
Before SDS adsorption	After SDS adsorption	
2881.72	-	Methyne C-H stretching
1640.88	1640.88	Alkenyl C=C stretching
1002.95, 1089.44	1004.23, 1089.02	Aliphatic fluoro compounds (C-F stretching)
625.94, 661.58	626.72, 661.40	Aliphatic bromo compounds (C-Br stretching)
737.34	-	Aliphatic chloro compounds (C-Cl stretching)
1002.95, 1089.44, 1150.05	1004.23, 1089.02, 1162.10	Aromatic C-H in-plane bending
625.94, 661.58, 737.34, 831.26	626.72, 661.40, 830.68	Aromatic C-H out-of-plane bending
3495.82	3470.64	Alcohol, hydroxy group, H-bonded O-H stretching (broad)
1324.44	1324.62	1° or 2°, OH in-plane bending
1406.07, 1324.44	1324.62, 1405.71	Phenol or 3° alcohol, OH bending
625.94, 661.58	626.72, 661.40	Alcohol, OH out-of-plane bending
1089.44	1089.02	Alkyl substituted ether, C-O stretching (C-O-C)
1089.44	1089.02	Cyclic ethers, large rings, C-O stretching
3495.82	3470.64	Aromatic 1° amine (N-H stretching)
1640.88	1640.92	1° amine (N-H bending)
1150.05	1162.10	2° amine (C-N stretching)
1324.44	1324.62	Aromatic 3° amine (C-N stretching)
1324.44, 1406.07	1324.62, 1405.71	Carboxylate (carboxylic acid salt)
1640.88	1640.92	Amide
1707.94	1708.14	Carboxylic acid or ketones
1324.44, 1514.31	1324.62, 1514.23	Aromatic nitro compounds
1002.95	1004.23	Aliphatic phosphates (P-O-C stretching)
1089.44	1089.02	Organic siloxane (silicon-oxy compounds) (Si-O-Si) or (Si-O-C)

Table 2: IR absorption bands and corresponding possible functional groups of lignite and coconut shell-based activated carbon fiber prior to and after SDS adsorption.

(0.41%) > Mg (0.14%). Similarly, the percentage (wt%) of all elements present in powdered oyster shell was investigated as following order (after SDS adsorption): O (50.30%) > Ca (39.95%) > C (8.09%) > Sn (0.80%) > S (0.36%) >Si (0.30%) > Mg (0.20%) (Figure 12).

EDX revealed that, the energy spectra of the x-rays character in powdered pine cones emitted from the elements of C, O, and Cl before and after SDS adsorption indicating no change in elemental composition in the samples after subjected to the surfactant adsorption. Similar condition was observed in case of lignite and coconut shell-based ACF where energy spectra of the x-rays character emitted from the elements of C, O, P, and Na. On the contrary, the energy spectra of the x-rays character in powdered oyster shells emitted from the elements of C, O, Mg, S, Ca, and Sn prior to adsorption, while energy spectra get emitted from the elements of O, Ca, C, Sn, S, Si, and Mg aftermath of surfactant adsorption suggesting the change in elemental composition in the samples after subjected to the surfactant adsorption.

The result from SEM-EDX showed that the concentration and distribution of elements were different prior to and after SDS adsorption in the biosorbents. High percentage of elements (C and O) showed in cluster, while the low percentage of element (Cl) showed wide distribution in powdered cones sample. High percentage of elements (C and O) showed in cluster, while the low percentage of elements (Na and P) showed wide distribution in ACF. Further, SEM-EDX analysis revealed that high percentage of elements (C, O, Ca, and Sn) showed in cluster, while low percentage of the elements (Mg, S) depicted wide distribution in powdered oyster shell before the surfactant adsorption. At the same time, SEM-EDX analysis revealed that high percentage of elements (C, O, Ca, and Sn) showed in cluster, while low percentage of the elements (Mg, S, and Si) depicted wide distribution in powdered oyster shell after the surfactant adsorption [5,6,9].

XRD studies

As seen in Figure 13, partial interference that results the complex pattern as detected by the powder X-ray diffraction suggested amorphous character in the *Pinus densiflora* cone powder. The two broad Bragg or diffraction peaks were observed at around 22° and 34° in both vividly indicated that the adsorption process merely results very little change in surface structure of adsorbent. Similarly, as shown in Figure 14, partial interference that results the complex pattern as detected by the powder X-ray diffraction suggested amorphous character in the lignite and coconut shell-based activated carbon fiber. The two broad Bragg or diffraction peaks were observed at around 20° and 44° in both vividly indicated that the adsorption process merely results very little change in surface structure of adsorbent.

As depicted in the Figure 15, the x-ray powdered diffraction (XRPD) pattern (diffractogram) of all data sets were sharp which indicates that the sample of powdered oyster shell is crystalline material, i.e., crystalline in nature. It was found that oyster shell is entirely composed of $CaCO_3$ and other minerals of trivial amount. The mineral phase of calcium carbonate turns out to be calcite (the mineral $CaCO_3$ of calcite structure is a dominant component of oyster shell).

From the study, it was concluded that the crystal system present in the sample is 'rhombohedral', with Bravais lattice of primitive nature and axial system of a=b=c (axial distance), α=β=γ≠ 90° (axial angles). The lattice parameter of rhombohedral crystal system is 4.988 × 4.988 × 17.061 Å (90° × 90° × 120°).

Conclusion

The lignite and coconut shell-based ACF and two naturally occurring pine cones, including oyster shells were studied by FE-SEM, XRD, ATR-FTIR spectroscopy, SEM-EDX imaging to determine their chemical, elemental, textural, and structural properties. In

Wavenumbers (cm^{-1})		Corresponding functional groups
Before SDS adsorption	After SDS adsorption	
2883.08	2878.84	Methyne C-H stretching
1629.71	1629.03	Alkenyl C=C stretching
1089.65	1089.48	Aliphatic fluoro compounds (C-F stretching)
625.47	625.77	Aliphatic bromo compounds (C-Br stretching)
711.60	711.59	Aliphatic chloro compounds (C-Cl stretching)
1089.65, 1150.78	1089.48, 1139.74	Aromatic C-H in-plane bending
625.47, 711.60, 876.94	625.77, 711.59, 875.60	Aromatic C-H out-of-plane bending
3483.47	3483.23	Alcohol, hydroxy group, H-bonded O-H stretching (broad)
1327.59	1327.04	1° or 2° alcohol, OH in-plane bending
1327.59	1327.04	Phenol or 3° alcohol, OH bending
625.47, 711.60	625.77, 711.59	Alcohol, OH out-of-plane bending
1089.65	1089.48	Alkyl substituted ether, C-O stretching (C-O-C)
3483.47	3483.23	Heterocyclic amine, N-H stretching
3483.47	3483.23	Aromatic 1° amine (N-H stretching)
1629.71	1629.03	1° amine (N-H bending)
1150.78	1139.74	2° or 3° amine (C-N stretching)
1282.30, 1327.59	1283.10, 1327.04	Aromatic 1° or 2° or 3° amine (C-N stretching)
1629.71	1629.03	2° amine, N-H bending
1327.59, 1413.59	1327.04, 1414.47	Carboxylate (carboxylic acid salt)
1707.24	1706.83	Carboxylic acid or ketones
1798.11	1797.73	Acid (acyl) halide (Haloformyl) (RCOX) (X = F, Cl, I, Br)
1089.65, 1150.78	1089.48, 1139.74	Cyanate (-OCN and C-OCN stretching)
1327.59	1327.04	Aromatic nitro compounds
1282.3, 1327.59	1283.10, 1327.04	Organic phosphates (P=O stretching)

Table 3: IR absorption bands and corresponding possible functional groups of powdered oyster shell prior to and after SDS adsorption.

addition, the samples were treated to determine the iodine number. The XRD analysis suggested the amorphous character in lignite and coconut shell-based ACF and powdered *Pinus densiflora* cones, while it indicated that the sample of powdered oyster shell is crystalline in nature and entirely composed of $CaCO_3$ with the very trivial amount of impurities. In addition, XRPD pattern indicated that the adsorption process merely results very little change in surface structure of aforementioned biosorbents.

The SEM-EDX analysis indicated that the concentration and distribution of elements were different prior to and after SDS adsorption in the biosorbents. Similarly, the FE-SEM imaging at three different magnifications depicted the change in surface texture and morphology in all three biosorbent samples prior to and aftermath of surfactant adsorption. The IR spectroscopy demonstrated the variation in IR spectra of the samples prior to and after surfactant adsorption and also further suggested occurrence of complexing or (complexation) process between ionizable functional groups present on the surface of biosorbent samples and DS^{-} or Na^{+} ions of SDS. The chemical, elemental, and micro-structural aspects of the aforementioned samples are likely to play an important role in adsorption process.

Acknowledgements

The present study was financially supported by Korea Research Foundation Grant funded by the Korean Government.

References

1. Chowdhury ZZ, Zain SM, Rashid AK, Ahmed AA, Khalid K, et al. (2012) Application of response surface methodology (RSM) for optimizing production condition for removal of Pb (II) and Cu (II) onto Kenaf Fiber Based Activated Carbon. J Appl Sci Eng Technol 4: 458-465.
2. Özacar M, Sengil İA, Türkmenler H (2008) Equilibrium and kinetic data, and adsorption mechanism for adsorption of lead onto valonia tennin resin. Chem Eng J 143: 32-42.
3. Chowdhury ZZ, Zain SM, Rashid AK, Rafique RF, Khalid K (2012) Breakthrough curve analysis for column dynamics sorption of Mn (II) ions from wastewater by using Mangostana garcinia peel-based granular activated carbon. J Chem.
4. Persson P, Axe K (2005) Adsorption of oxalate and malonate at the water-goethite interface: molecular surface speciation from IR spectroscopy. Geochim Cosmochim Acta 69: 541-552.
5. Coates J (2000) Interpretation of infrared spectra, a practical approach. Encyclopedia of Analytical Chemistry. Meyers RA (ed), John Wiley & Sons Ltd, Chichester, p: 10815-10837.
6. Velasquez P, Leinen D, Pascual J, Ramos-Barrado JR, Cordova R, et al. (2001) XPS, SEM, EDX and EIS study of an electrochemically modifed electrode surface of natural chalcocite (Cu_2S). J Electroanal Chem 510: 20-28.
7. Charalambous FA, Ram R, Pownceby MI, Tardio J, Bhargava SK (2012) Chemical and microstructural characterization studies on natural and heat treated brannerite samples. Miner Eng 39: 275-288.
8. Yoon GL, Kim BT, Kim BO, Han SH (2003) Chemical-mechanical characteristics of crushed oyster-shell. Waste Manag 23: 825-834.
9. Qiu Y, Zhang Q, Tian Y, Zhang J, Cao J, et al. (2011) Composition and structure of luxing coal with different particles sizes. Pet Coal 53: 45-55.

4

Crystallization of Kainite from Processing Solutions of Polymineral Saline Raw Material Marine Type

Kostiv IY[1] and Basystyuk YI[2*]

[1]State Enterprise, Scientific-Research Gallurgi Institute, 5a, Fabrichna Str., 76000 Kalush, Ukraine
[2]Precarpathian National University named after V. Stephanyk 57, Shevchenko Str., 76025 Ivano-Frankivsk, Ukraine

Abstract

This paper presents the results of laboratory researches evaporation processing solutions polymineral salt material that contain sulfates and chlorides of potassium, magnesium and halite. Determined degree of evaporation and temperature of the suspension in which achieved the highest level of separation ions of potassium and sulfate in the solid phase, and the solid phase-the highest content kainite and the lowest-sylvine and carnalite. The lowest content (2.6-5.7%) of carnalite in suspension achieved at temperature of 313-333 K and the degree of evaporation 19.2-26.5%. Cooling of the suspension with the degree of evaporation of 30.1% is not necessary, because it leads to increased degree of separation in the solid phase of carnalite. Regulate the depth of evaporation and cooling can be by concentrate on in the liquid phase of equivalents $MgSO_4$ (0.016-0.032), $MgCl_2$ (0.248-0.289), and the ratio between them (0.0579-0.125).

Keywords: Solution chloride-sulfates potassium-magnesium; Evaporation; Crystallization; Kainite; Carnalite; Sylvine

Introduction

Processing of polymineral potassium-magnesium ore chloride-sulfate type accompanied by reactions conversion of chloride minerals of kainite ($KCl \cdot MgSO_4 \cdot 3H_2O$), sylvine (KCl) in schoenite ($K_2SO_4 \cdot MgSO_4 \cdot 6H_2O$), during which crystallization in the liquid phase increases the concentration of $MgCl_2$. Therefore, for the purpose of regulation part of the solution after separation schoenite derive from the process. With this solution according to potash production thrown out 43.8% of potassium, 44.7%-magnesium and 17.1% of sulfate of their number of processed ore. It is obvious importance of extraction of salt from schoenite solution and returning them to the production of potassium fertilizers. Because these solutions are sent to the phasic evaporation to highlight potassium salt. During evaporation in the industrial conditions is achieved insufficient degree of solid phase extraction of valuable components, including 17.1% potassium, 9.3% magnesium and sulfate and 6.8% of their number in the ore, high dispersion of salts and as a consequence difficulty filtering sediment, and significant loss of liquid phase. Because of insufficient saturation of liquid phases by sulfate salts sulfate potassium-magnesium fertilizer getting low quality with content up to 25% chloride ions. Therefore, the degree of extraction in the solid phase of mineral components excess schoenite solution and their next conversion into schoenite determines the efficiency of processing technology of potash raw materials in general. During the evaporation is crystallized kainite in mixture with NaCl, KCl and other minerals. Kainite and chloride salts differ in their granulometric composition, structure and properties of crystals. The biggest size of crystals kainite is achieved at temperatures 363-368 K [1]. On the size of the crystals kainite significantly affects the depth of evaporation. What makes it more, the more substance for crystal growth is separate from the liquid phase [2]. But thus increases the density of the suspension and evaporation complicated. To decrease density of the suspension part of evaporated solution return to mixing with the initial. This increases the concentration of Mg^{2+} in the liquid phase and after achievement of 8.2% of its value crystallization of kainite stops, and in the liquid phase increases the concentration of SO_4^{2-} [3].

Evaporation of the solution after crystallization of kainite at temperatures 388-403 K makes it possible to separate sulfate ions in the form of finely dispersed kieserite, suspension cooled to a temperature of 303 K in the solid phase crystallized bischofite. Separating of suspension leads to the formation of two layers of sediment: top-finely dispersed sulfate salts and bottom-bischofite. By decanting separate the top layer of sediment and together with liquid phase return to the stage crystallization of kainite. The bottom layer of sediment is filtered on a centrifuge, separating the crystalline bischofite as a commodity product, and the filtrate was returned to the stage crystallization of kainite [4]. Thus from excess solution of schoenite can be almost completely back of sulfate ions in the production of potash fertilizers.

During the cooling of evaporated suspension begins to crystallize carnalite, kainite dissolved in this case [3]. Therefore it was necessary to determine the limit depth of evaporation of solution and temperature of cooling of evaporated suspension for the achievement the greatest degree of extraction in the solid phase kainite and minimum of carnalite because crystallization of carnalite is accompanied by the dissolution of kainite, and the return to production of potassium fertilizers of solid phase which containing $MgCl_2$.

Experimental Section

Research of processes of evaporation solution and crystallization kainite was carried out on a laboratory vacuum evaporating installation and in reactor with paddle mixer. The first included three throat round bottom flask with a capacity of 2500 cm^3, mechanical paddle mixer with electric meter and electric heater, glass refrigerator of vapors, calibrated collection of condensate, vacuum pump and system manually adjust

***Corresponding author:** Basystyuk YI, Precarpathian National University named after V. Stephanyk 57, Shevchenko Str., 76025 Ivano-Frankivsk, Ukraine
E-mail: yaroslava_artus@ukr.net

the depth of vacuum, by which the temperature of evaporation supported with precision of ± 0.5 K. Reactor includes a rubber cover closed double-walled glass vessel with a capacity of 1200 cm^3 elliptical bottom and glass paddle mixer with electric drive. For research use with cooled air compressor UK laboratory 25-1.6 M (SU), consumption of air meter measured the gas and air after the compressor was passed through heat exchanger with cooling water. Research suspension was poured into the reactor or a glass round bottom flask.

The initial solution was prepared in accordance with the composition of industrial hot saturated solution from the stage of evaporation and crystallization of kainite composition by wt.%: 5.50 K^+; 4.39 Mg^{2+}; 2.69 Na^+; 16.74 Cl^-; 7.03 SO_4^{2-}; 63.65 H_2O [5]. In this solution to avert crystallization of during long storage and added 30% water on its mass. Its before each experiment evaporated. Prepared also for mixing with the initial solution kainite solution by evaporation of industrial solution. Its composition (wt.%): 1.46 K^+; 6.59 Mg^{2+}; 0.56 Na^+; 19.73 Cl^-; 2.28 SO_4^{2-}; 69.38 H_2O. For one experiment in the flask to evaporate prepare 1917.5 grams of evaporated initial solution and added 822.4 g kainite solution. The composition of the mixed solution (wt.%): 4.28 K^+; 5.05 Mg^{2+}; 2.05 Na^+; 17.64 Cl^-; 5.60 SO_4^{2-}; 65.38 H_2O, or equivalent indices X=E SO_4^{2-}: (E K^++E Mg^{2+}); X=0.2221; Y=E Mg^{2+}: (E K^++E Mg^{2+}); Y=0.7915. The ratio of X:(Y-X)=E $MgSO_4$:E $MgCl_2$ was 0.390. Evaporation was carried out under vacuum at a temperature of 353 ± 0.5 K for obtaining kainite without impurities langbeinite. The degree of evaporation was determined as ratio of the mass evaporated water to the initial mass of the mixed solution.

From mixed solution was evaporated determined quantities of water to achieve a given degree within the limits 15.5-30.1% transferred evaporated suspension in a thermostatic reactor and through the bubbler turn on the flow air. The samples were periodically taken and filtered under vacuum using heated Schott filter No. 16 fixed on Bunsen funnel. The filtrate trickled in pre-weighed test tube. The resulting liquid and solid phases were analyzed by tetraphenylborate gravimetric analysis to determine K^+ content; volumetric chelatometry-to determine Mg^{2+} and Ca^{2+}; flame photometric analysis-for Na^+; mercurrometry-for Cl^- and gravimetry-for SO_4^{2-}. Water content was determined by the difference. The material balance of the process, from which calculated degree extraction of components in the solid phase, and the contents of basic minerals by chemical composition of solid phases: sylvine, kainite, carnalite and halite whose presence determined by X-ray analysis, calculated on a PC with special software.

Results and Discussion

Figure 1 shows the dependence of concentration of K^+ (a) and SO_4^{2-}(b) in the liquid phase of the temperature during the cooling by air suspension of potassium-magnesium salts, evaporated to varying degrees. As can be seen from Figure 1a, the concentration of K^+ ions in the liquid phase of evaporated suspension at a temperature of 353 K decreases with increasing degree of evaporation. So for the degree of evaporation of 15.5%, it is 4.73% and decreased to 3.21% for the degree of evaporation of 30.1%. During the cooling by air K^+ concentration gradually decreases for all stages of evaporation and achieves the smallest values for temperature 293 K. The lowest concentration of K^+ is equal to 1.42% at 293 K and the degree of evaporation of a solution of 30.1%. For 19.2% degree of of evaporation researched the process of cooling suspension evaporated through the wall of the reactor water. The obtained results are shown in Figure (red curve). As seen from obtained data that K^+ concentration in the liquid phase of suspension during the cooling by air slightly larger than on cooling water. It is explained by the partial drying by air of solution.

From Figure 1a shows that SO_4^{2-} concentration in the liquid phase of evaporated suspension decreases with increasing degree of evaporation. So for his value to 15.5% is equal 6.24% and with an increase to 30.1% decreases to 1.83%. During the cooling suspension by air concentration of SO_4^{2-} decreases, especially for the degree of evaporation 19.2 and 22.8%. At the temperature 293 K its decrease according to 3.37 and 3.04%. For the degree of evaporation 26.5 and 30.1% cooling of suspension has little effect on SO_4^{2-} concentration in the liquid phase of evaporated suspension. So if the temperature of 353 K, it is respectively 2.84 and 1.83%, that the temperature at 293 K-2.36 and 1.74%.

Figure 2 shows the dependence of the ratio of equivalent concentration (λ) $MgSO_4$ and $MgCl_2$ in liquid phase of the suspension potassium-magnesium salts, on the temperature cooling by air to different degree of evaporation. The minimum value of λ in liquid phase indicates to the extraction in the solid phase predominantly kainite. For the suspensions at a temperature of 353 K value λ is gradually decrease with increasing degree of of evaporation from 0.3235 to 0.0659 for the

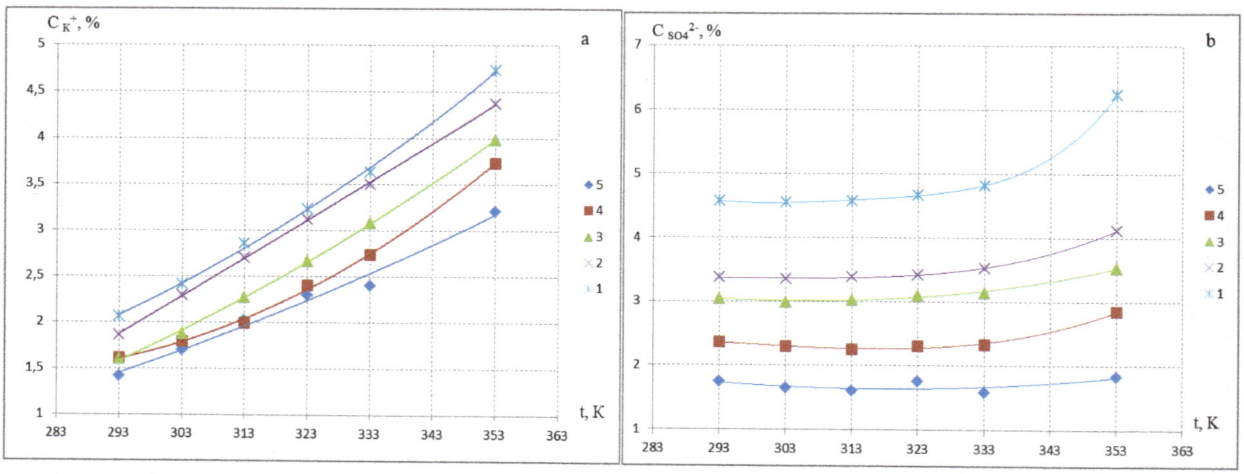

Figure 1: The dependence of concentration (C) K^+ (a) i SO_4^{2-} (b) in the liquid phase vs temperature (T) during the cooling by air suspension of potassium-magnesium salts, evaporated to different degree, %: 1-15.5; 2-19.2; 3-22.8; 4-26.5; 5-30.1.

degrees of evaporation, respectively 15.5 and 30.1%. For the degree of evaporation 30.1% value λ almost unchanged. Therefore, according to the concentration in the liquid phase of ions SO_4^{2-} and value of λ can make a preliminary conclusion that the solutions is expedient evaporate to degree of evaporation 30.1% to achieve a value of λ equal to 0.0659, separate crystallized salts and return them to conversion and crystallization of schoenite. Figure 3 shows data on the degree extraction in solid phase of suspension K^+(a), SO_4^{2-}(b) i Mg^{2+} (c). As seen from the data (a), with increasing the degree of evaporation solution from 15.5% to 30.1% increases the degree of transition in the solid phase of K^+, respectively, from 26.0 to 65.2%. Cooling of suspension for all degree of evaporation contributes to an increase degree of extraction of K^+ ions in solid phase. Most of the value is achieved for the degree of evaporation 30.1% at 293 K and is 91.7%.

The degree of transition in the solid phase ions SO_4^{2-}(b) also increases with the degree of evaporation the solution. For example, for the degree of evaporation 15.5%, it is 22.1%, and for the degree of evaporation 30.1% its 84.5%. Cooling by air of the suspension evaporated to the degree 15.5%, contributes to a significant increase to 42.0% degree of extraction SO_4^{2-} ions in the solid phase to a temperature 333 K. Continued cooling of suspension to the degree of extraction SO_4^{2-} ions in the solid phase affects little. For suspensions obtained with a higher degree of evaporation, cooling is even less effect on the extraction of SO_4^{2-} ions in solid phase. So to the degree of evaporation 30.1% degree of extraction SO_4^{2-} ions in the solid phase is 84.5% at temperature 353 K and 89.8% at temperature 293 K. For the degree of evaporation 22.8 and 26.5% respectively his the value is 61.7 and 71.3% at temperature 353 K and 70.3 and 80.4% at temperature 293 K. During the cooling of suspension by water degree of extraction in the solid phase ions SO_4^{2-} slightly smaller than by air. The degree of conversion ions Mg^{2+} (c) in the solid phase also increases with increasing the degree of evaporation, respectively, from 9.7% to 29.7% for the degree of evaporation 15.5 and 30.1%. Cooling by air of suspension to 293 K has little effect on the degree of separation in the solid phase Mg^{2+} ions for degree of evaporation 15.5, 19.2 and 22.8%. For the suspension evaporated to 30.1%, cooling causes a great increase of the degree extracted ions Mg^{2+} in solid phase. Figure 4 showing the results of calculation content in the solid phase sylvine (a) kainite (c) and carnalite (s). Separation of carnalite causes return of $MgCl_2$ in the crystallization of schoenite and is not desirable. Extraction of sylvine is not desirable too because of its similarity his crystals to the crystals of sodium chloride and complicated of purification kainite from sodium chloride by hydroseparation.

As seen from the obtained results in the solid phase of the evaporated suspension content of sylvine during cooling from 353 to 273 K increases at the degree of evaporation 15.5; 19.2; 22.8%, 24.4% respectively to 19.5 and 18.1%. At the degree of evaporation 26.5% content of sylvine increased from 5.4% at the temperature 353 K to 13.1% at the temperature 313 K. During continued cooling to 303 K and 273 sylvite content of decreases respectively to 7.7 and 5.8% due to its dissolution and crystallization of kainite. Cooling suspension evaporated to 30.1% causes conversion sylvine in kainite is complete at temperature 313 K. The content of kainite in the solid phase (c) of the suspension at a temperature of 353 K increases with the degree of evaporation from 15.5 to 19.2%, respectively, from 34.0 to 64.2%. The next increase the degree of evaporation to 22.8-30.1% decreases content of kainite to 57.3-61.2% at 353 K. Cooling the suspension evaporated to 15.5%, from 353 to 323 K causes increasing the contents of kainite from 34.0 to 53.5%, cooling to 313 K or less leads to decrease of the content kainite to 48.6-46.3%.

In the suspension with degree of evaporation of 22.8 and 26.5% during cooling to 333 K content of kainite increases by 1.5-2.2%, then decreases to 55.1-47.6%. With decreasing temperature of the suspension with the degree of evaporation 30.1% decreases content of kainite from 59.9% to 52.1; 47.2; 41.6; 42.9 and 39.3%, respectively, at a temperature of 353; 333; 323; 313; 303 and 293 K. Therefore, the highest content of kainite 64.2-63.4% the solid phase of the suspension achieved at the degree of evaporation 19.2-26.5% and temperature 333-353 K. The lowest content (2.6-5.7%) of carnalite in suspension is achieved at temperatures 313-333 K and the degree of evaporation 19.2-26.5%. Cooling of the suspension with the degree of evaporation of 30.1% results to a significant increase of carnalite in the solid phase of the suspension (wt.%): 28.6; 35.0; 44.5; 47.1 respectively in temperatures 333; 323; 313 i 293 K.

The degree of evaporation of mixed solution changes in dependence on the quantity returned to the mixing of evaporated solution. Therefore, to obtain of the suspension with a high content in solid phase of kainite and lowest content of sylvine and carnalite the

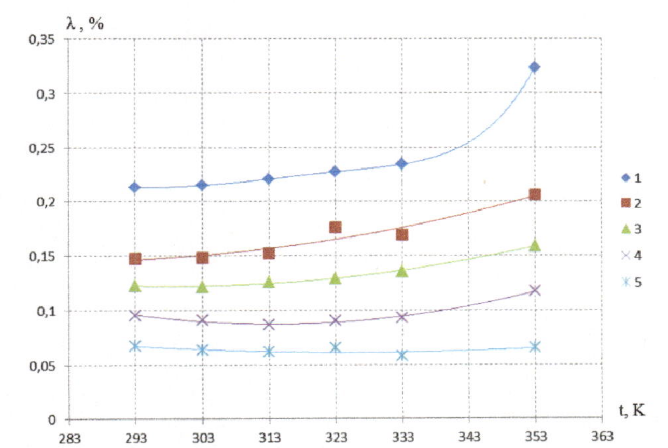

Figure 2: The dependence of ratio the equivalent concentrations of $MgSO_4$ i $MgCl_2$ (λ) in the liquid phase of the suspension potassium-magnesium salts *vs* the temperature (T) by different degree of evaporation: 1-15.5; 2-19.2; 3-22.8; 4-26.5; 5-30.1.

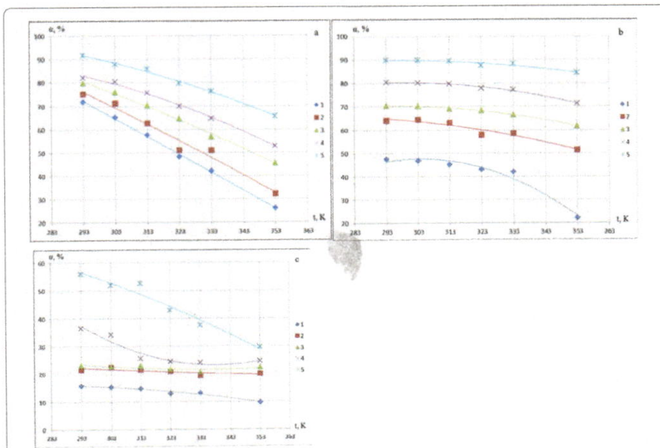

Figure 3: Dependence the degree of transition (α) K^+ (a), SO_4^{2-} (b) i Mg^{2+} (c) in the solid phase vs temperature (T) during the cooling by air of suspension potassium-magnesium salts, evaporated to different concentrations of Mg^{2+} in liquid phase %: 1-24.64; 2-24.48; 3-25.77; 4-26.75; 5-29.34.

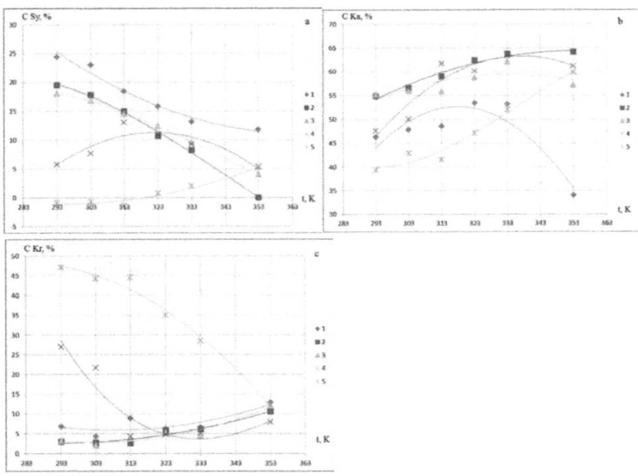

Figure 4: The dependence of content (C) sylvine (a), kainite (b), and carnalite (d) in the solid phase of the suspension vs the temperature (T) during the cooling by air of the suspension potassium-magnesium salts, evaporated to different degree, %: 1-24.64; 2-24.48; 3-25.77; 4-26.75; 5-29.34.

degree of evaporation at a temperature of 353 K should be 22.8-26.5%. Regulate the depth of evaporation and cooling can be by concentration in the liquid phase equivalents $MgSO_4$ (0.016-0.032), $MgCl_2$ (0.248-0.289), and the ratio between them (0.0579-0.125).

Conclusion

The results of research showed that the concentration and degree of extraction of compounds potassium and sulfates in the solid phase increases with increasing degree of evaporation of the solution and decreasing temperature of the suspension. The highest content of kainite and smallest of carnalite in the solid phase is achieved at a temperature of 333-353 K. Without cooling of the suspension can be separated with a high degree of extraction kainite at the degree of evaporation of 30%, equivalent to the concentration of $MgSO_4$ 0.016-0.020, $MgCl_2$ 0.301 i ratio between them 0.035-0.058.

References

1. Kostiv IY (1990) Research process of evaporation solutions in system K^+, Mg^{2+}, Na^+|| Cl^-, SO_4^{2-} - H_2O at temperature 323-383 K. Journal of Applied Chemistry 63: 3-8.
2. Kostiv IY (1999) Untersuchung der Kristallisationskinetik von Kainit und Carnallit aus Lösungen der Verarbeitung von polymineralischen Kalirohsalzen. Freiberger Forschungshefte 853: 70-81.
3. Kostiv IY (1995) Research of technology separate kainite from processing solutions of polymineral potassium ores. Ukraine Chemical Industry 2: 50-56.
4. Basystiuk YI, Kostiv IY (2016) Getting Hydrated Magnesium Chloride from Magnesium Chloride Solutions of Potassium Sulfate Fertilizers Production. J Chem Eng Process Technol 7: 291.
5. Hrebenyuk D, Yaremchuk B (1992) Technological schemes of processing polymineral potassium ores of Precarpathian. Chim Prom Ukrainy 2: 41-45.

Effect of Surface Roughness on the Rate of Mass Transfer at a Vertical Cylinder Under Swirl Flow in Relation to Electrochemical and Catalytic Reactor Design

Mustafa SA, El-Gayar DA* and Sedahmed GH

Chemical Engineering Department, Faculty of Engineering, Alexandria University, Alexandria, Egypt

Abstract

In order to enhance the rate of production of catalytic and electrochemical annular reactors used to conduct diffusion controlled reactions the effect of surface roughness on the rate of mass transfer at the inner cylinder of the annulus was studied under swirl flow. The effect of the following parameters on the rate of mass transfer was studied by the electrochemical technique: solution velocity, degree of roughness, physical properties of the solution, height of the inner cylinder and effect of drag reducing polymers. Roughness was made by cutting longitudinal V grooves in the inner cylinder transverse to swirl flow. The mass transfer data at the rough cylinder were correlated by the dimensionless equation:

$$Sh=0.435\ Sc^{0.328}\ Re^{0.86}\ (L/de)^{-0.53}\ (e/de)^{0.45}$$

Drag reducing polymers were found to decrease the rate of mass transfer by an amount ranging from 5 to 23%. Importance of the present results for the design and operation of high space-time yield electrochemical reactors and biochemical reactors employing immobilized enzymes was pointed out. Also, the importance of the present results in the design and operation of membrane processes employing annular apparatus with a corrugated membrane was highlighted. By virtue of the analogy between heat and mass transfer the present results can be used to design more efficient double tube heat exchangers.

Keywords: Electrochemical reactors; Mass transfer; Catalytic reactors; Heat transfer; Swirl flow; Surface roughness; Diffusion controlled reactions

Introduction

Annular flow reactors are used frequently in industry to conduct liquid-solid diffusion controlled catalytic and electrochemical reactions in view of their advantages which include: (i) Rapid removal of heat generated in case of highly exothermic reactions by passing cold water through the inner side of the inner tube and the cooling jacket surrounding the outer tube. The built in heat transfer facility makes it possible to use the reactor to conduct exothermic reactions involving heat sensitive catalysts or heat sensitive products such as immobilized enzyme catalyzed biochemical reactions. (ii) Uniform current and potential distribution, this property is highly desirable especially in case of electro organic synthesis where a high degree of selectivity is needed. (iii) The reactor occupies a low floor space and can be extended vertically thus saving floor space and capital costs.

Despite these merits the annular flow reactor suffers from the limited surface area which limits the reactor rate of production. To make up for the limited reactor area previous studies tried to increase the rate of mass transfer by a suitable means such as gas sparging [1], vibration and pulsation [2,3], inner tube rotation [4], gas generation [5], surface roughness [6,7], and swirl flow [8-12].

The aim of the present work is to enhance the rate of mass and heat transfer in the annular reactor by the combined swirl flow and surface roughness. Although some work has been done on the mass transfer behavior of smooth annular reactor under swirl flow [8-12] no studies have been reported on the effect of combined surface roughness and swirl flow on the rate of heat or mass transfer in an annular reactors. The effect of surface roughness on the rate of mass transfer has been studied for annular reactors only in axial flow [6,7], surface roughness would not only increase the rate of heat and mass transfer by promoting turbulence but would also increase the effective area of the reactor with a consequent increase in the rate of production. Roughnesses used in the present work are in the form of longitudinal parallel grooves transverse to the direction of swirl flow. The rate of mass transfer was determined by an electrochemical technique which involves measuring the limiting current of the catholic reduction of $K_3Fe(CN)_6$ in a large excess of NaOH as supporting electrolyte [13]. The present mass transfer study is important for the rational design and operation of relatively high space-time yield annular reactor suitable for conducting liquid-solid diffusion controlled reactions such as catalytic reactions which include photo catalytic reactions, immobilized enzyme catalyzed biochemical reactions and electrochemical reactions involved in waste water treatment and electro organic synthesis. The area of membrane technology can benefit from the present work in designing high efficiency annular dialyzers by using swirl flow and corrugated cylindrical membranes. Drag reducing polymers are used in practice to decrease the pumping power requirement of turbulent flow equipments in view of the ability of polymer molecules to damp the small scale high frequency energy dissipating eddies which prevail in the buffer layer of the hydro dynamic boundary

***Corresponding author:** El-Gayar DA, Chemical Engineering Department, Faculty of Engineering, Alexandria University, Alexandria, Egypt
E-mail: dina.elgayar@alexu.edu.eg

layer [14]. Damping of these eddies leads also to decreasing the rate of turbulent flow heat and mass transfer, hence the second aim of the present work is to find the extent to which rates of mass transfer are reduced under swirl flow in view of fact that no previous studies have been reported on the effect of drag reducing polymers on the rate of mass transfer under swirl flow. Polyethylene oxide drag reducing polymer (POLOX WSR-301) a product of Union Carbide was used in the present study.

Experimental Techniques

The Figures 1a and 1b shows the apparatus used in the present study and the swirl inlet nozzle dimensions respectively. The apparatus consisted of Plexiglass cylindrical reactor of length 40 cm and a diameter of 5.0 cm lined with a cylindrical stainless steel sheet which acted as anode and an inner cylinder of nickel plated copper tube of diameter 2.5 cm which acted as cathode. Smooth and rough cylinders were used, peak to valley height of the roughness elements were: 0,1, 2, 3, 4 mm. Roughness was made by cutting V-threads (angle 45°) in the inner cylinder. The annulus was fed with the solution by a tangential feed nozzle of 1.2 cm diameter fitted at the lower end of the reactor to generate swirl flow. A plastic centrifugal pump of 0.27 horse power was used to circulate the solution between a glass storage tank of dimensions 30 × 30 × 30 cm and the reactor. The top of the reactor was fitted with an overflow weir from which the solution return to the storage tank. A bypass with plastic valves was used to regulate the solution velocity in the reactor.

The electrical circuit consisted of 10 volt DC power source with a voltage regulator and a multi-range ammeter connected in series with the reactor. A voltmeter was connected in parallel with the reactor to measure its voltage. Before each run 25 L of the solution were placed in the plastic storage tank, the solution was circulated between the storage tank and the reactor by the centrifugal pump. Solution velocity in the reactor was controlled by the by-pass and was measured volumetrically using a graduated cylinder and a stop watch. The mass transfer behavior of reactor was studied by measuring the limiting current of the catholic reduction of $K_3Fe(CN)_6$ from a solution containing $K_4Fe(CN)_6$ and a large excess of NaOH as supporting electrolyte [13]. Polarization curves from which the limiting current was obtained were plotted by increasing the current stepwise and measuring the corresponding cathode potential against a reference electrode. In view of high anode area compared to the cathode area, the anode was taken as a reference electrode [13].

The following reactions take place during current measurement:

Cathode: $Fe(CN)_6^{-3}$+e$Fe(CN)_6^{-4}$

Anode: $Fe(CN)_6^{-4}$ $Fe(CN)_6^{-3}$+e^-

All solutions used in the present work were prepared using A.R grade chemical and distilled water, temperature was 25 ± 1°C solution composition and their physical properties are shown in Table 1. Solution density and viscosity needed for determined by a density bottle and an Ostwald viscometer [13], while diffusivity of the $Fe(CN)_6^{-3}$ was obtained from the literature [13]. Each experiment was repeated twice under the same conditions, the reproducibility ranged from 3-5%.

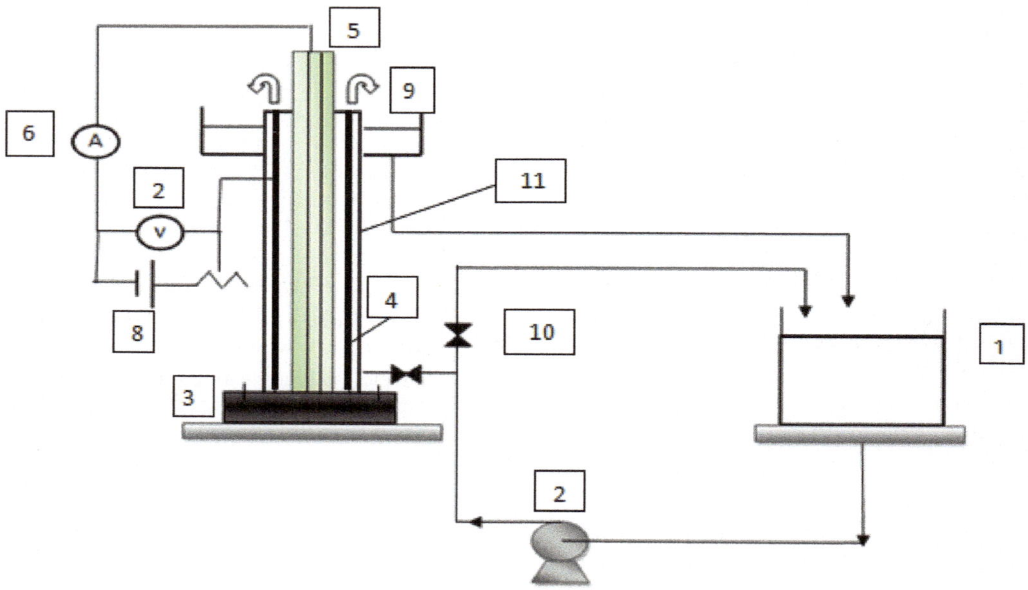

Experimental setup

1-Storage tank
2-Plastic Centrifugal pump
3-Plastic base
4-Stainless steel cylindrical anode
5-Nickel Plated Copper Cylinder cathode
6-Digital ammeter
7-Digital voltmeter
8-DC power supply
9-Overflow weir
10-Plastic valve
11-Plexiglass cylinder lined with cylindrical stainless steel sheet anode

Figure 1a: Apparatus.

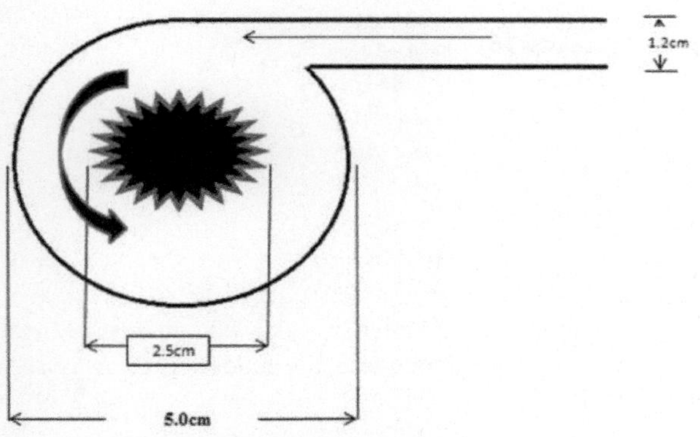

Figure 1b: Swirl inlet nozzle.

Solution Composition	p (g/cm³)	μ × 10² (Poise)	D × 10⁶ (cm²/s)	Sc
0.02M $K_3Fe(CN)_6$+0.2M $K_4Fe(CN)_6$+1N NaOH	1.083	1.228	6.693	1694
0.02M $K_3Fe(CN)_6$+0.2M $K_4Fe(CN)_6$+2N NaOH	1.134	1.326	5.508	2122
0.02M $K_3Fe(CN)_6$+0.2M $K_4Fe(CN)_6$+3N NaOH	1.14	1.599	4.335	3072

Table 1: Physical properties of ferri/ferrocyanide system used at 25°C.

Results and Discussion

Swirl flow mass transfer at smooth cylinder cathode

Polarization curves with a well-defined limiting current plateau were obtained under different conditions as shown in Figure 2, the mass transfer coefficient was calculated from the limiting current using the equation [13]:

$$I_L/ZF = KAC \quad (1)$$

Dimensional analysis which was used to correlate the present data leads to the following functional equation for smooth surface

$$K=f(\rho, \mu, D, V, de, L) \quad (2)$$

Writing the above equation in a dimensionless form gives

$$\frac{Kde}{D} = \alpha\left(\frac{\mu}{\rho D}\right)^{\alpha}\left(\frac{\rho V de}{\mu}\right)^{\beta}(L/de)^{\gamma} \quad (3)$$

i.e.,

$$Sh = a\, Sc^{\alpha} Re^{\beta} (L/de)^{\gamma} \quad (4)$$

The equivalent diameter of the annulus de was calculated from equation

$$de=d_{outer}-d_{inner} \quad (5)$$

The present experimental data will be used to determine a, α, β and γ. The Figure 3 Shows the effect of Re on Sh for different Sc, the data fit the equation

$$Sh=\bar{a}\, Re^{0.96} \quad (6)$$

The Re exponent 0.96 from equation (6) reveals mass transfer by highly turbulent flow mechanism which characterize swirl flow. The Figure 4 shows the effect of the dimensionless active cathode height on Sh, the data fit the equation:

$$Sh=a_1\,(L/de)^{-0.432} \quad (7)$$

The decrease in Sh with cylinder height is attributed to the decay of swirl flow in favor of the less effective axial flow as the solution progresses through the annulus. The Figure 5 shows the effect of Sc on Sh, the data fit the equation:

$$Sh=a_2 Sc^{0.328} \quad (8)$$

The exponent 0.328 agrees with the value 0.33 predicted by the hydrodynamic boundary layer theory [14]. The Figure 6 shows that the mass transfer data for the conditions:

$1694<Sc<3072$, $1130<Re<6328$ and $2.73<L/de<8.2$ fit the eqn:

$$Sh=0.109\, Sc^{0.328}\, Re^{0.96}\,(L/de)^{-0.432} \quad (9)$$

With an average deviation of 10.5%.

The Figure 7 shows a comparison between the present swirl flow data and the axial flow data obtained [15] for mass transfer under developing laminar and turbulent flow in an annulus. The present swirl flow data lie above the axial flow data of Mobarak in the laminar and the turbulent flow region. The Table 2 shows that the mass transfer enhancement ratio Sh_{swirl}/Sh_{axial} ranges from 1.4 to 1.74 depending on the operating conditions.

The higher rates of swirl flow mass transfer compared to axial flow at smooth cylinder shows that for a given set of conditions turbulence generated in case of swirl flow is more intense than that generated in case of axial flow. Figure 8 shows that the present swirl flow data at smooth cylinder agree fairly with the data of legentihomme and Legrand [8] and the data of de Sa et al. [6].

Mass transfer at rough cylinders under swirl flow

Following previous studies in heat and mass transfer at rough surface [16-20] geometrical area rather than true area was used in calculating the mass transfer coefficient at rough surfaces. The Figures 8 and 9 shows that the effect of Re on the mass transfer coefficient at rough cylinders fit the eqn:

$$Sh=a_3\, Re^{0.862} \quad (10)$$

The Figure 10 shows that the effect of rough cylinder dimensionless height (L/de) on the mass transfer coefficient, the data fit the eqn:

$$Sh=a_4\,(L/de)^{-0.53} \quad (11)$$

The Figure 11 shows the effect of the ratio between the dimensionless peak to valley height (e) of the roughness element and the equivalent diameter (de) on the mass transfer coefficient, the data fit the eqn:

$$Sh=a_5\,(e/de)^{0.4} \quad (12)$$

The Figure 12 shows that the swirl flow data at rough cylinders for the conditions:

$1694<Sc<3072$, $1130<Re<6328$, $2.7<L/de<8.2$, $0.04<e/de<0.16$ fit the eqn:

$$Sh=0.435\, Sc^{0.328}\, Re^{0.86}\,(L/de)^{-0.53}\,(e/de)^{0.45} \quad (13)$$

In obtaining the above equation the exponent of Sc was fixed at 0.328 according to equation) of present work. To appreciate the extent to which surface roughness increases the rate of mass transfer compared to smooth cylinders under swirl flow, the enhancement factor Sh_{rough}/Sh_{smooth} was calculated under different conditions. The Table 3 shows that for a given set of conditions surface roughness increases the rate of mass transfer by a factor ranging from 1.02 to 1.94 depending on the operating conditions.

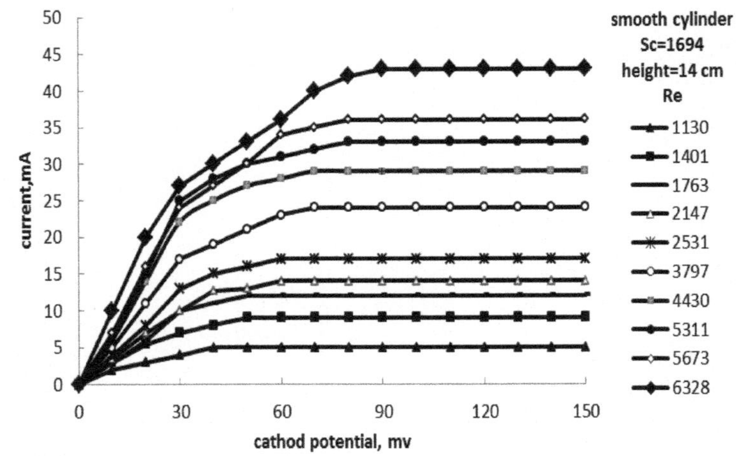

Figure 2: Typical polarization curve for smooth cylinder at different Re.

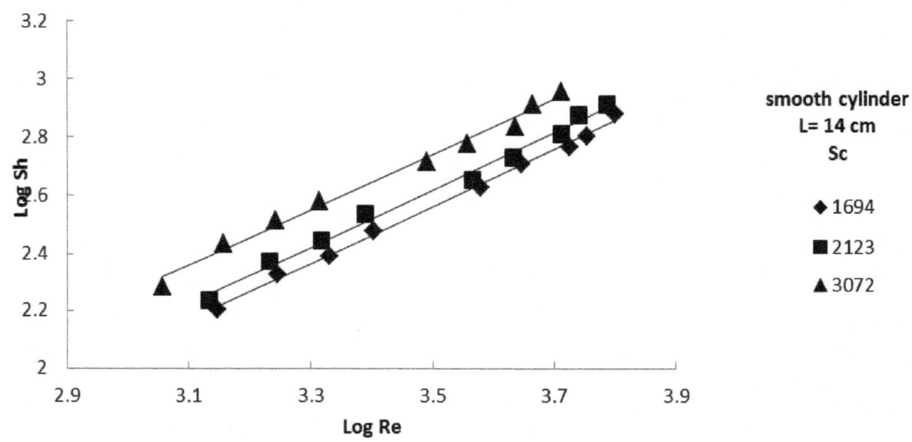

Figure 3: Effect of Re on Sh for a smooth cylinder at different Sc.

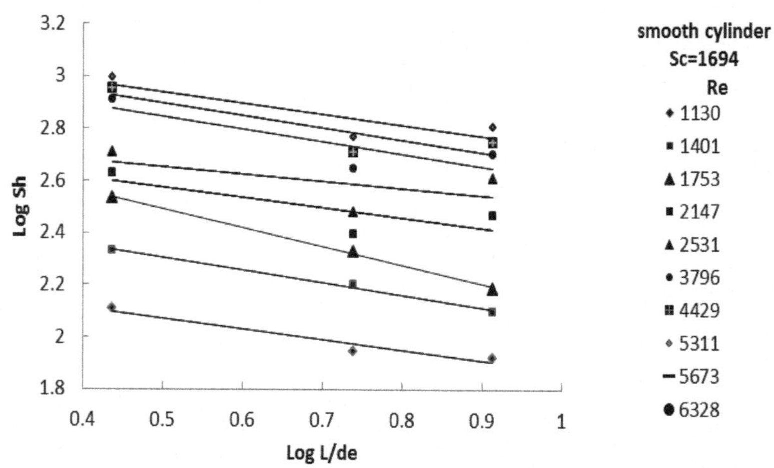

Figure 4: Effect of smooth cylinder height on Sh at different Re.

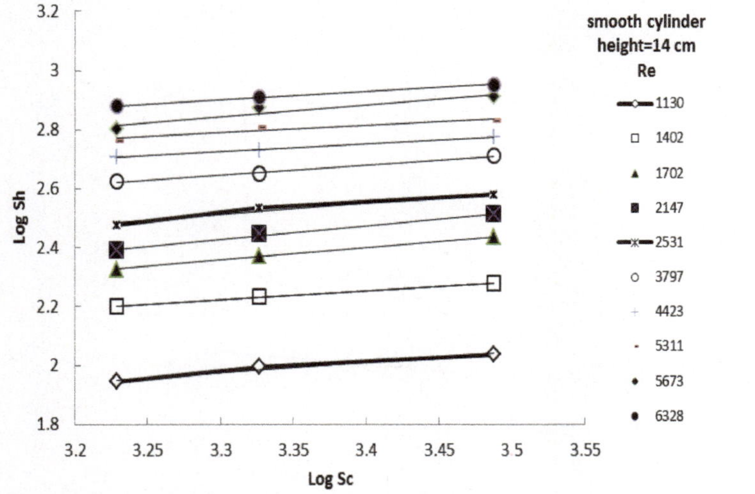

Figure 5: Effect of Sc on Sh at different Re for smooth cylinder.

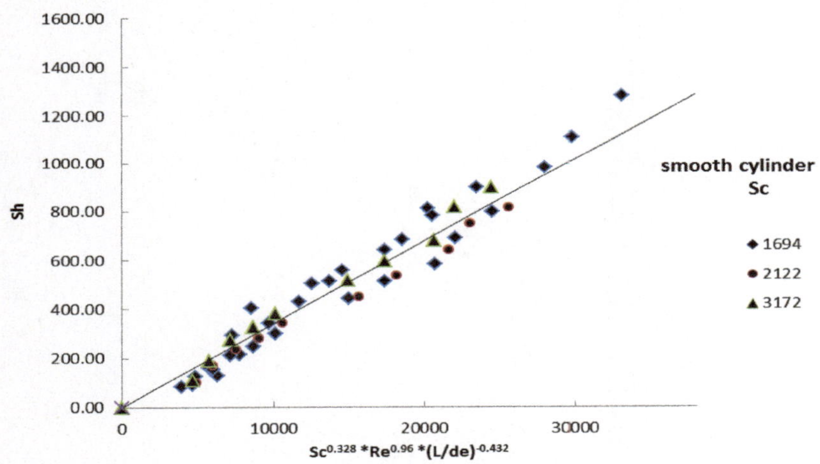

Figure 6: Overall mass transfer correlation at a smooth cylinder under swirl flow.

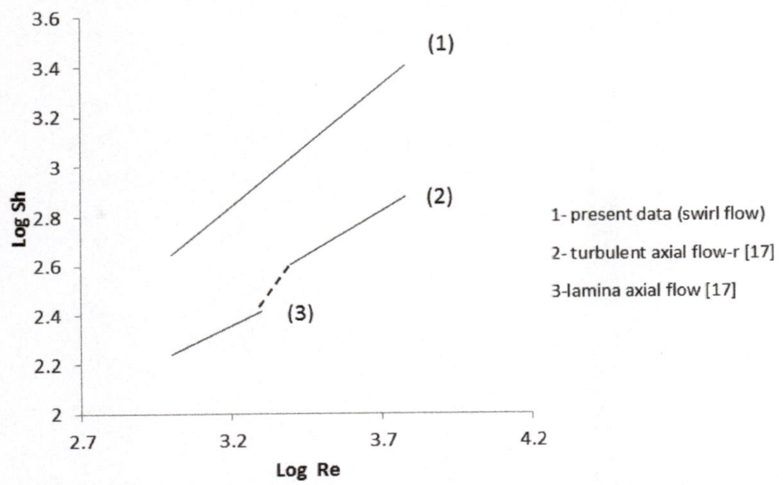

Figure 7: Comparison of the present swirl flow mass transfer coefficient with the value of axial flow in an annulus.

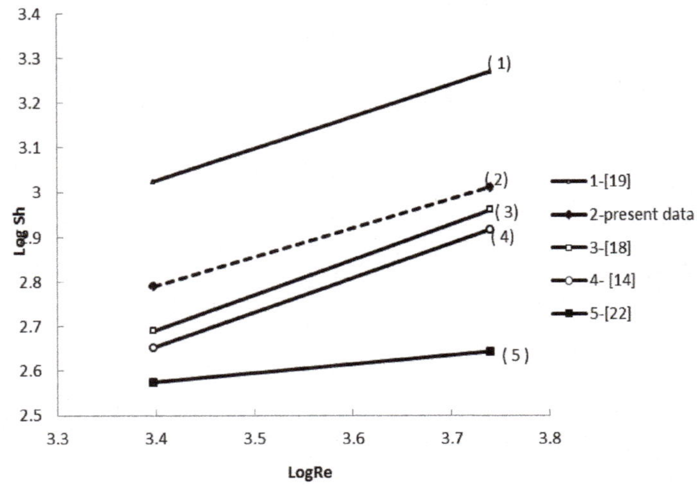

Figure 8: A comparison between the present swirl flow mass transfer data at smooth cylinder with the previous swirl flow studies.

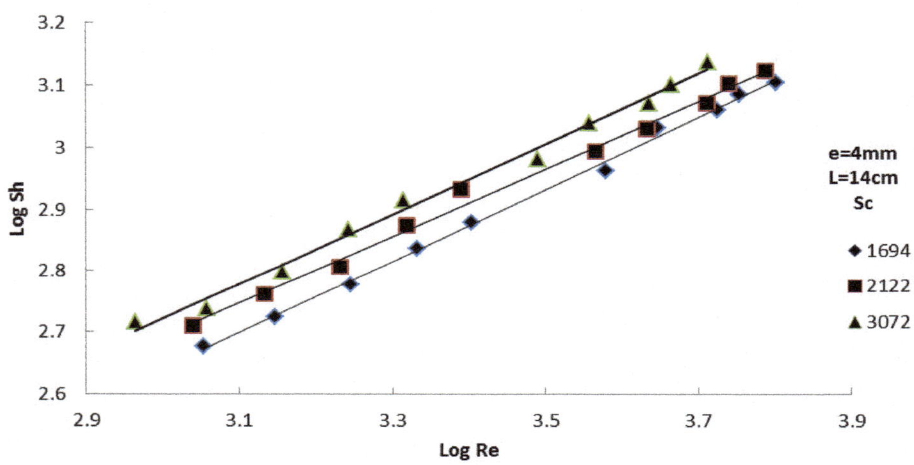

Figure 9: Effect of Re on Sh at rough cylinder (e=4mm) for different Sc.

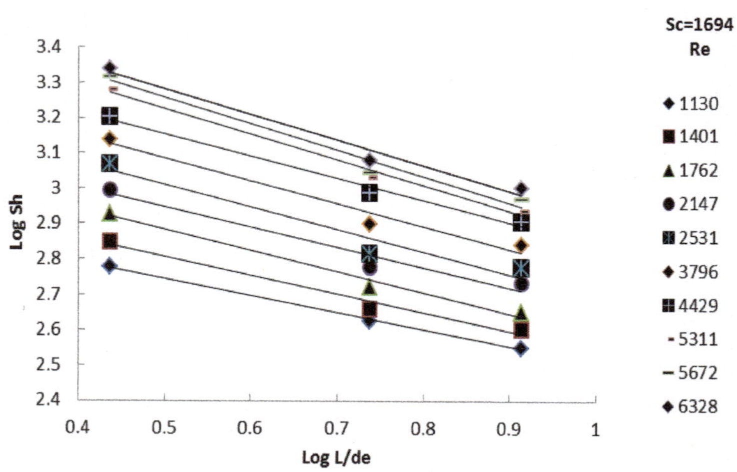

Figure 10: Effect of rough cathode cylinder height L/de on SC at different Re.

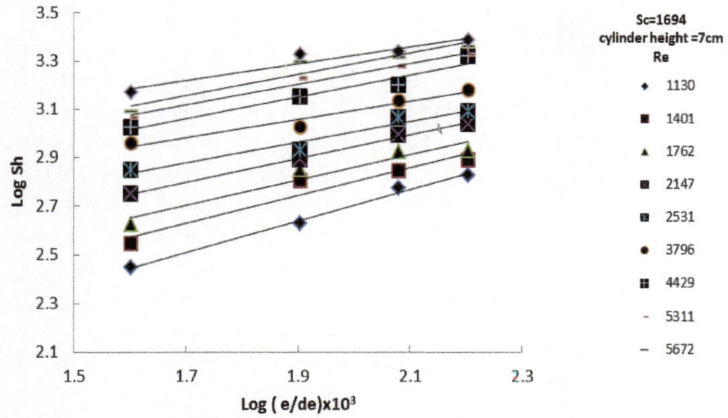

Figure 11: Effect of surface roughness (e/de) on Sh at different Re.

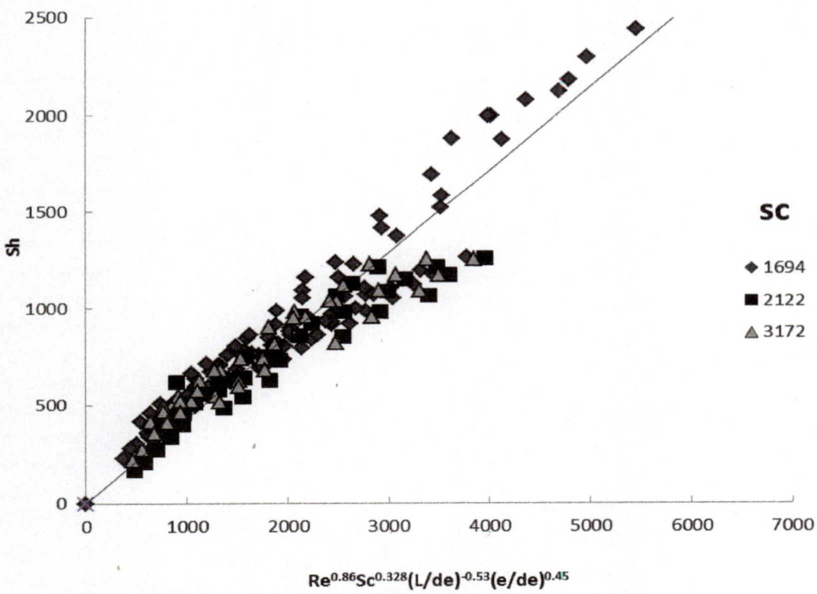

Figure 12: Overall mass transfer correlation at rough cylinders under swirl flow.

Re	L/de=10	Sh_{smooth}/Sh_{axial} L/de=8.2	L/de=5.4	L/de=2.7
1000	1.68	1.66	1.64	1.59
1500	1.61	1.59	1.57	1.53
2000	1.56	1.55	1.52	1.48
2500	1.47	1.46	1.44	1.40
3000	1.52	1.51	1.49	1.45
6000	1.74	1.73	1.70	1.66

Table 2: The enhancement ratio Sh_{smooth}/Sh_{axial} due to swirl flow at a smooth cylinder compared to axial flow [Sc=1500; 2.73<L/de<10].

Re	Sh_{rough}/Sh_{smooth}				
	Smooth	e=1 mm	e=2 mm	e=3 mm	e=4 mm
1000	1	1.02	1.17	1.44	1.68
1500	1	1.06	1.21	1.49	1.73
2000	1	1.08	1.24	1.53	1.77
2500	1	1.10	1.26	1.55	1.80
3000	1	1.12	1.28	1.58	1.83
6000	1	1.18	1.35	1.67	1.94

Table 3: Enhancement ratio Sh_{rough}/Sh_{smooth} at different degrees of roughness (Sc=1500; (L/de)=2.7).

The increase in the rate of mass transfer due to surface roughness may be attributed to:

i) The increased degree of turbulence generated when the tangential flow crosses transversely the vertical roughness elements which protrude outside the laminar sublayer of the hydrodynamic boundary layer, this turbulence penetrate the diffusion layer and enhance the rate of mass transfer [21].

ii) Also surface roughness increases the true surface area over the projected geometrical area with a consequent increase in the rate of mass transfer, the increase in the true area was found to range from 11% to 50% over the projected area. The finding that rough cylinder with peak to valley height of 1 mm behaves almost like smooth cylinder at relatively low Re may be attributed to the fact that a considerable part of these roughness elements (1 mm height) is submerged in the laminar sublayer of the hydrodynamic boundary layer and only a small height (<<1 mm) protrudes outside the laminar sublayer. According to Sclichting [22] for a protrusion in the flowing stream to generate eddies behind it Re should be>100, where R_p is the protrusion Re and is given by $Re_p=\rho ve/\mu$ where e, is the roughness height. The higher the Re the higher the intensity of turbulence and the mass transfer coefficient (Table 3).

Effect of drag reducing polymer on the rate of mass transfer

Drag reducing polymers are used in practice to reduce friction between pipe wall and turbulently moving solution with a consequent decrease in the pumping power required to pump turbulent flow by the ability of polymer molecules to damp the small scale high frequency eddies which prevail in the hydrodynamic boundary layer [14,23]. Despite the beneficial effect of drag reducing polymers on pumping power they may reduce the rate of mass and heat transfer [24,25] taking place at the tube wall in view of damping turbulent eddies which enhance the rate of mass and heat transfer at the tube wall. The aim of this part is to find out the extent to which drag reducing polymers reduces the rate of mass transfer under swirl flow at smooth and rough cylinders.

The Figures 13 and 14 (Tables 4 and 5) show that the presence of polyox drag reducing polymer reduces the rate of mass transfer by a value ranging from 4.9% to 22.86% depending on the operating conditions. The Table 4 shows that the percentage reduction in the rate of mass transfer increases slightly with increasing Re and then decreases with further increase in Re. The initial increase in percentage reduction in k with increasing Re may be attributed to stretching of the polymer molecules under the influence of the high shear stress, stretched polymer molecules are known to be more effective in damping small scale high frequency eddies than coiled polymer molecules [19]. The decline in the percentage reduction in k at high Re may be attributed to mechanical degradation of the polymer molecules under the influence of the high shear stress to ineffective low molecular weight break down products [26].

A comparison between Tables 4 and 5 show that for a given Re rough cylinder produces higher percent reduction in k than smooth cylinder. This is attributed to the ability of rough surfaces to generate a higher proportion of small scale high frequency eddies which are sensitive to the polymer molecules. This result is consistent with the finding of Debrule and Sabersky [23] who studied the effect of drag reducing polymer on the rate of heat transfer in rough tubes, the authors reported higher percent decrease in the rate of heat transfer in rough tubes than in smooth tubes.

Conclusions

The present study has revealed that the combind surface roughness and swirl flow is a powerful tool for enhancing the rate of heat and mass transfer in annular flow reactors. The dimensionless mass transfer correlation obtained in the present work can be used to design and operate annuar electrochemical reactors used to conduct diffusion controlled liquid-solid reactions such as electrochemical removal of heavy metals and organic pollutants from waste water by cathodic deposition and anodic oxidation respectively, and electro organic synthesis. The presence of a builtin heat transfer facility in the annular reactor (outer cooling jacket and the innerside surface of the inner tube) makes it possible to use the annular reactor for conducting exthothermic catalytic diffusion controlled liquid-solid reactions where heat senstive substances are involved such as immoblized enzyme catalyzed biochemical reactions and photochemical reactions. The present results can be used to design and operate more efficient annular membrane dialyzer where a corrugated membrane is used. By virue of the analogy between heat and mass transfer the present results can be used in building more efficient double tube heat exchangers. Drag reducing polymers were found to decrease the rate of mass transfer by a maximum of 22.86%. Accordingly, drag reducing polymers should be used in practice if the benefit of reducing the pumping power outweighs the deleterious effect of reducing the rate of the diffusion controlled process.

List of symbols

A	Geometrical cathode area
$a, \bar{a}, a_1, a_2, a_3, a_4$	Constant
C	Ferricyanide concentration
D	Diffusvity of ferricyanide ion
de	Equivalent diameter of the annulus (do-di)
do	Outer cylinder diameter
di	Inner cylinder diameter
e	Peak to valley height
F	Faraday sconstant
I	Limiting current
K	Mass transfer coefficient
L	Electrode height
V	Solution velocity
Z	Number of electrons invloved in the reaction
Re	Reynolds number(ρVde/μ)
Sc	Schmidt number(μ/ρD)
Sh	Sherwood number (Kde/D)
μ	Solution viscosity
ρ	Solution denscity
α, β and γ	Constants

Re	% Reduction in K at Different Polymer Concentration		
	100 ppm	200 ppm	300 ppm
1130	11.11	18.52	20.30
1401	10.00	13.33	16.67
1762	7.35	11.76	17.65
2147	5.40	11.01	16.20
2531	11.63	16.28	18.60
3796	7.69	12.30	19.40
4429	11.86	15.25	18.50
5311	6.25	14.06	16.20
5672.	5.97	14.93	15.00
6328	8.33	12.50	14.20

Table 4: Effect of drag reducing polymer on the rate of mass transfer at a rough cylinder under swirl flow [Sc=1694, height=14 cm; rough cylinder (e=4 mm)].

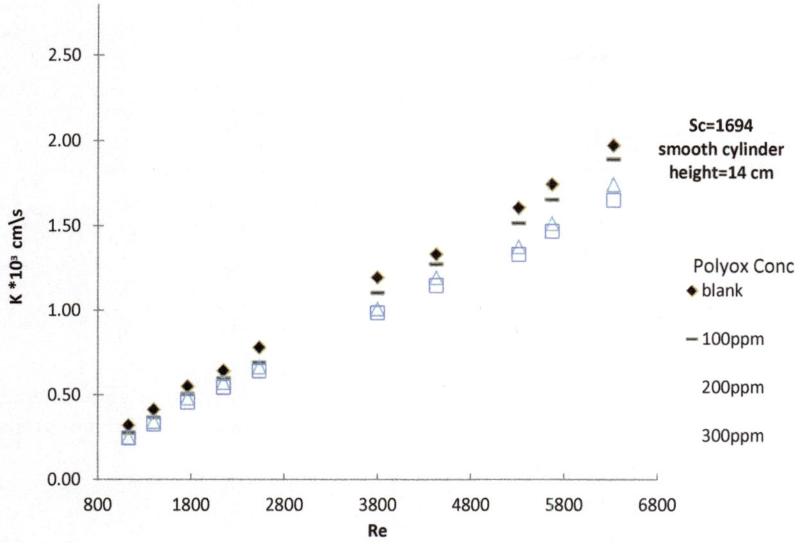

Figure 13: Effect of Re on the mass transfer coefficient at different Polyox concentration (using a smooth cylinder).

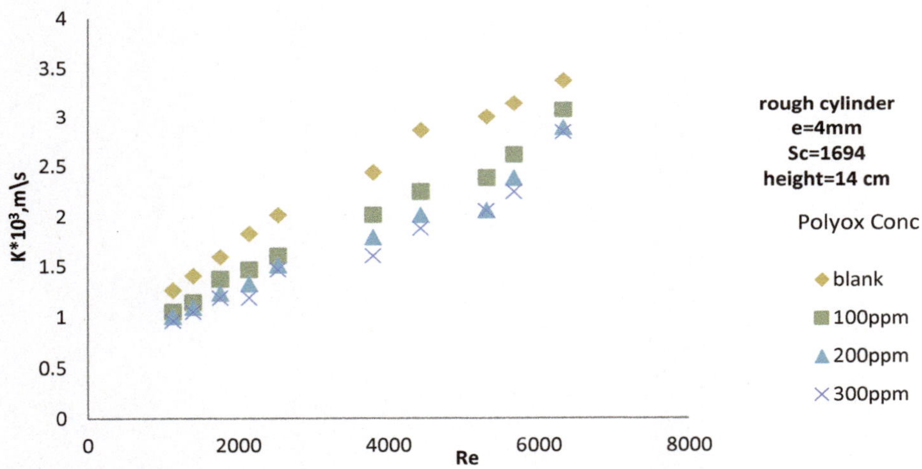

Figure 14: Effect of Re on the mass transfer coefficient at different Polyox concentration using a rough cylinder (e=4mm).

References

1. Sedahmed GH, Hosny AY, Fadally OA, El-Mekkawy IM (2011) Mass transfer at rough gas sparged ducts under conditions with superimposed pulsating flow. Chem Eng Technol 34: 475-481.

2. Coeuret F, Legrand J (1983) Mass and momentum transfer at the rotating surface of a continuous annular electrochemical reactor. Electochem Acta 28: 611-617.

3. El-Sherbiny MF, Zatout AA, Hussien M, Sedahmed GH (1991) Mass transfer at the gas evolving inner electrode of a concentric cylindrical reactor. J Appl Electrochem 21: 537-542.

4. Zaki M, Nirdosh I, Sedahmed GH (1996) Effect of surface roughness induced by woven metallic screens wrapped on the inner surface of an annulus on the rate of turbulent flow mass transfer. Ind Eng Chem Research 35: 4354-4359.

5. Mabarak AA, Farag HA, Sedahmed GH (1997) Mass transfer in smooth and rough annular ducts under developing flow conditions. J Appl Electrochem 27: 201-207.

6. De Sa MS, Shemilt LW, Soegiatro IV (1991) Mass transfer in the entrance region for axial and swirling annular flow. The Canadian J chem Eng 69: 294-299.

7. Shoukry E, Shemilt LW (1985) Mass transfer enhancement in swirling annular pipe flow. Ind Eng chem process Des Dev 24: 53-56.

8. Legentilhomme P, Legrand J (1990) Overall mass transfer in swirling decaying flow in annular electrochemical cells. J Appl Electrochem 20: 216-222.

9. Yapici S, Ozbahar RE (1998) Mass transfer to finite areas at the wall in swirl pipe flow in the transition region. Ind Eng chem Research 37: 643-651.

10. Yapici S, Patricand MA, Wragg AA (1995) Electrochemical study of mass transfer in decaying annular swirl flow. J Appl Electrochem 25: 15-22.

11. Selman JR, Tobias CW (1978) Mass transfer measurements by the limiting current technique. Adv chem Eng 10: 211-318.

12. Hoyt JW (1972) The effect of additives on fluid friction. Trams ASME, J Basic Eng 94: 253-285.

13. Findly A, Kitchner JK (1965) Practical physical chemistry. Longmans, London.

14. Atef NM, Abdel-Aziz MH, Fouad YO, Farag HA, Sedahmed (2015) Mass and heat transfer at an array of horizontal cylinders placed at the bottom of a square agitated vessel. Chem Eng Res and Design 94: 449-455.

15. Sedahmed GH, Nagy soliman M, EL-Kholy NS (1981) Effect of Surface

Roughness on Mass Transfer to Pipe Wall in the mass transfer entry region. The Can J Chem Eng 59: 693-669.

16. Dawson DA, Trass O (1972) Mass transfer at rough surfaces. Int J Heat Mass Transfer 15: 1317-1336.
17. Dipprey DF, Sabersky RH (1963) Heat and momentum transfer in smooth and rough at various Prandtl numbers. Int Heat Mass Transfer 6: 329-353.
18. Abdel Aziz MH, Hassan I, Nirdosh I, Sedahmed GH (2013) Liquid-solid mass transfer behavior of V corrugated surface under two phase flow. Chem Eng Processes 70: 9-16.
19. Zhao W, Trass O (1997) Electrochemical mass transfer measurements in rough surface pipe flow: geometrically similar V shaped grooves. Int J Heat Mass Transfer 40: 2785-2797.
20. Knudsen JG, Katz DL (1958) Fluid dynamics and heat transfer. McGraw Hill, New York, USA.
21. Schlichting H (1979) Boundary layer theory. 7th edn. McGraw Hill, New York, USA.
22. Sellin RH, Hoyt JW, Serivenger O (1982) The effect of drag reducing additives on fluid flow and their industrial applications. J Hydraulic Research 20: 29-67.
23. Debrule PM, Sabersky RH (1974) Heat and fiction coefficients in smooth and rough tubes with dilute polymer solutions. Int J Heat Mass Transfer 17: 529-540.
24. White D, Gordon RJ (1975) The influence of polymer conformation on turbulent drag reduction. AIChE J 21: 1027-1029.
25. White D, Mungal MG (2008) Mechanics and prediction of turbulent drag reduction with polymer additives. Ann Rev Fluid Mech 40: 235-256.
26. Union Carbide Bulletin (1973) Polyox F-44029A.

6

Application Algorithm Development of Pinch Technology in Heat Integration Problem

Ritesh Sojitra*

ITM University, Gwalior, India

Abstract

Pinch technology is developing and its application is widening, reaching new horizons. The original concepts of pinch approach were quite clear and, because of numerous ways, application engineer gets confused among these numerous flexibilities. Therefore, there is a need for a rigorous and robust model which could guide the optimisation engineer on deciding the applicability of the pinch approach and direct sequential step of procedure in predefined workflow, so that the precision of approach is ensured. Exploring the various options of a novice hands-on algorithm development that can be coded and interfaced with GUI and keeping in mind the difficulties faced by designers, an effort was made to formulate a new algorithm for the optimisation activity. As such, the work aims at easing out application hurdles and providing hands-on information to the developer for use during preparation of new application tools. This paper presents a new algorithm, the application which ensures the designer does not violate basic pinch rules. To achieve this, intermittent check gates are provided in the algorithm, which eliminate violation of predefined basic pinch rules, design philosophy, and Engineering Standards and ensure that constraints are adequately considered. On the other side, its sequential instruction to develop the pinch analysis and reiteration promises maximum energy recovery (MER).

Keywords: Pinch analysis; Pinch approach; Robust algorithm; Heat integration; Heat recovery; Integration of heat exchanger network (HEN); Energy optimisation

Introduction

Since its genesis, pinch analysis is continuously evolving and its application is widening, reaching new horizons. The prime reason perhaps is the awareness and willingness amongst industries and environment regulating bodies to reduce carbon footprint in fulfilling their corporate responsibilities in environment conservation. Pinch technology is a powerful tool that can guarantee maximum energy recovery at optimum cost. Hence, its popularity is rising enormously. However, efficacy of the application of pinch technology depends on various parameters. As pinch approach is more of conceptual nature, there are multiple ways to its application. Consequently, it becomes very difficult for the designer to choose from available flexibilities. Often, this leads to the designer ending up with either major flaws in technical integrity of design or capital cost and operating cost. Due to such difficulties in the selection of application of pinch approach, there is a urgent need for conceptual process design team to collaborate with various core engineering domains and develop an integrated and robust process solution for the industries. A small initiative in solving the problem has been done and demonstrated in this paper [1-9].

Objective

Conceptually, the original pinch approach is clear and can definitely guarantee MER. But, its effectiveness depends wholly on the application methodology selected. For pinch approach to be successful and readily endorsed by the industry, there is a need to develop a methodology that assists the designer in selecting the right application on heat integration problem, which would be in-line with basic concept of pinch analysis and guarantees energy savings, if there is any scope of optimisation in the design. Bearing in mind all such requirements of a designer and optimisation engineer, a small initiative in solving the problem has been taken up and demonstrated in this paper [1]. This paper illustrates the algorithm which can guide the coder with sequential steps of procedure. The algorithm encompasses check gates for rules, standards, philosophy and constraints to prevent the design going haywire. The Coded pinch approach can also be interfaced with graphical user interface. This robust tool enables design engineer or optimisation engineer to carry out pinch analysis without very few hassles. The same algorithm can be used for manual application.

Major Procedural Steps of Pinch Analysis

The major procedural steps in formulating an effective pinch approach design are (Figure 1):

Step 1: Formulate process integration problem

Step 2: Assess application feasibility of pinch analysis

Step 3: Construct initial design process flow diagram

Step 4: Identify hot, cold and utility streams in the process

Step 5: Extract thermal data of process and utility streams

Step 6: Construct actual T-H diagram, actual composite curves and actual cascade diagram at $\Delta T_{min}=0$

Step 7: Set optimum energy targets and select initial value ΔT_{min}. Apply pinch rules and other predefined constraints

Step 8: Construct revised design process flow diagram

Step 9: Construct shifted T-H Diagram, feasible cascade diagram, shifted composite curves and grand composite curve

Step 10: Estimate minimum energy cost targets

***Corresponding author:** Ritesh Sojitra, ITM University, Gwalior, India
E-mail: sojiritesh2003@gmail.com

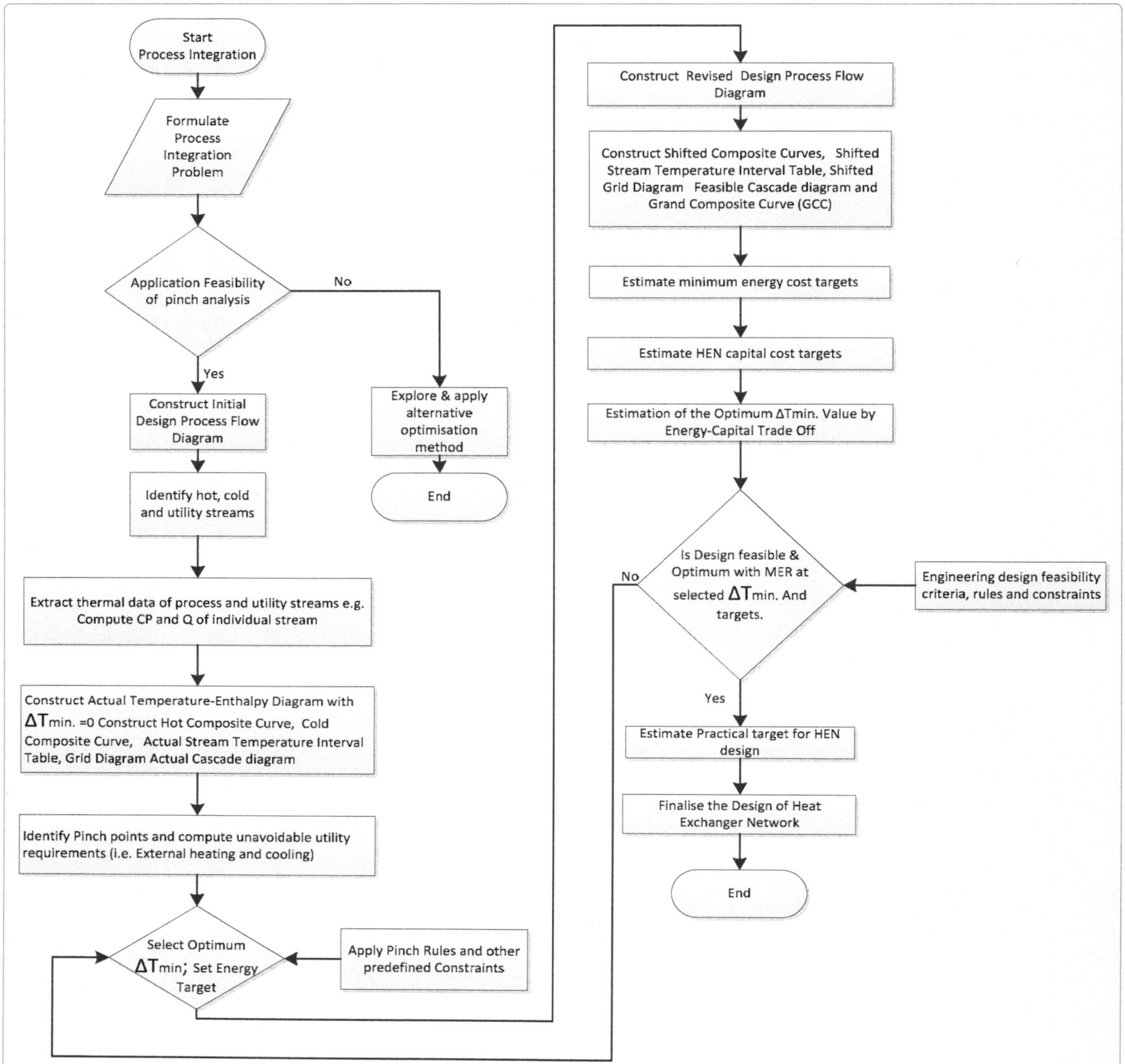

Figure 1: Application algorithm for pinch analysis.

Step 11: Estimate HEN capital cost targets

Step 12: Estimate optimum ΔT_{min} value by energy-capital trade off

Step 13: Assess the optimum design feasibility with regard to engineering design criteria, rules and constraints

Step 14: Estimate practical target for heat exchanger network design

Step 15: Finalise the design of the heat exchanger network

Step 1: Formulate process integration problem

Process integration involves integration of various sections of plants or units. This generally includes understanding concepts, study of feasibility, durability and economic aspects of the application. The process integration problem should be formulated in consideration of the components mentioned above. Pertaining to scope of thesis heat integration problem will be focused.

Step 2: Assess application feasibility of pinch analysis

This step involves assessment of applicability of pinch technique based on available data, nature of the problem and expectation from the application. Additionally, past experience of application may aid in decision making. Pinch analysis is mostly applicable to all process integration problems, however, sometimes other optimisation techniques may prove to be more appropriate than pinch analysis. If it is inferred that pinch analysis is applicable, move to step 3. If not, terminate the analysis and search for alternate method.

Step 3: Construct initial design process flow diagram

The process flow diagram should represent existing plant/design which needs to be optimized [5].

Step 4: Identify cold, hot and utility streams in the process

Identify the cold streams, hot streams and utility streams from initial design process flow diagram.

Hot stream-process stream which is to be cooled.

Cold stream-process stream which is to be heated.

When it is not economically feasible to exchange heat across process streams, external utilities-steam, hot oil, coolant, etc. are employed to mitigate the process requirement of heating or cooling.

Step 5: Extraction of thermal data of process and utility streams

Extract the following thermal data from identified cold streams, hot streams and utilities.

- Stream supply temperature - T_S °C
- Stream target temperature - T_T °C
- Heat capacity flowrate -(CP kW/K)

It is a product of mass flowrate (m kg/s) of the fluid stream and specific heat capacity C_p (kJ/kg K)

$$CP = \mathrm{m} \times C_p \quad (1)$$

As per first law of thermodynamics (in terms of heat flow),

$$Q = \int_{TS}^{TT} CPdT = CP\left(T_T - T_S\right) = \Delta H \quad (2)$$

During heat exchange, no mechanical work is done on or by the system. So, $\Delta W=0$

Step 6: Construct actual T-H Diagram, actual composite curves, actual cascade diagram at ΔT_{min} (minimum temperature difference)=0

Construct actual T-H diagram, actual composite curves, actual cascade diagram at ΔT_{min}=0. This helps visualise underlying opportunities in the application of pinch analysis.

Step 7: Set optimum energy targets and select initial value ΔT_{min} (minimum temperature difference) apply pinch rules and other predefined constraints

The design of heat transfer equipment always follows 2nd law of thermodynamics which restricts any crossover of temperature between cold fluid stream and hot fluid stream i.e., a minimum heat transfer driving force is always allowed to ensure feasible heat transfer design. Exchangers must always have a minimum temperature difference (ΔT_{min}) between hot and cold fluid streams. This ΔT_{min} value represents the threshold limit of heat recovery [10].

The following heat transfer equation should be used to set energy target at selected ΔT_{min}

$$Q = U \times A \times \Delta T_{LM} \quad (3)$$

Where,

$$\Delta T_{LM} = \frac{\Delta T_H - \Delta T_C}{ln\dfrac{\Delta T_H}{\Delta T_C}} = \frac{T_{h1} - T_{c2} - T_{h2} - T_{c1}}{ln\left(\dfrac{T_{h1} - T_{c2}}{T_{h2} - T_{c1}}\right)} \quad (4)$$

Step 8: Construct revised design process flow diagram

This step requires construction of revised design process flow diagram based on outcomes of the application of pinch rules and other predefined constraints after setting optimum energy targets and selecting initial value ΔT_{min}.

Step 9: Construct shifted T-H (Temperature vs. Enthalpy) diagram, feasible cascade diagram, shifted composite curves and grand composite curve

Composite curve (CC): Temperature vs. Enthalpy (T-H) curve is also known as composite curve.

Shifted composite curve: It represents overall energy targets but it does not clearly represent the amount of energy that must be externally supplied by different utility levels. The utility mix can be calculated from grand composite curve.

Shifted grand composite curve (GCC): It can be used for selecting utilities and its requirement. It is also used for determining the temperature of various utilities.

Feasible cascade diagram: It serves the same purpose as GCC does in calculating overall energy target and selecting utilities and its requirement.

GCC and feasible cascade diagram are basic tools. Either of them can be used in pinch analysis for choosing adequate levels of utility and for targeting a given set of multiple utility levels [1,7].

Step 10: Estimate minimum energy cost targets

After selection of ΔT_{min}, determination of minimum cold utility and minimum hot utility from CC, GCC or feasible cascade diagram will be feasible.

If cost of each utility unit, load and levels are known, determine the total energy cost using below energy equation [2]

$$Total\ Energy\ Costs = \sum_{U=1}^{U} Q_U \times C_U \quad (5)$$

Where Q_U -Duty of utility, kW

C_U-Utility unit cost of U, \$/kW per year

U-Total number of utilities used

Step 11: Estimate HEN capital cost targets

HEN capital cost is dependent on the below three factors:

1. Overall network area
2. Total number of exchangers
3. The distribution of area among the exchangers

Heat transfer area is calculated by the following equation [1,4].

$$Area = \frac{Q}{U \times \Delta T_{LM}} \quad (6)$$

$$HEN\ area\ A_{min} = A_1 + A_2 + A_3 + \ldots\ldots\ldots\ldots + A_i \quad (7)$$

$$Exchanger\ cost\ (\$) = a + b(A_{min})^c \quad (8)$$

Where a,b,c are constants

Step 12: Estimate the optimum ΔT_{min} value by energy-capital trade off

To reach at an optimum value of ΔT_{min}, estimate the total annual cost (sum of total annual energy and capital costs) at various values of ΔT_{min}. The following observations will be noticed [8].

1. Reduction in values of ΔT_{min} reduces energy costs but increases capital costs
2. Increase in values of ΔT_{min} increases energy costs but reduces capital costs

Optimum value of ΔT_{min} lies on trend where the total annual cost of energy and capital costs is overlap [5].

Step 13: Assess the optimum design feasibility with regard to engineering design criteria, rules and constraints

This is a decision-making step. The engineer must assess the revised design, achieved from step 12, and study its feasibility with respect to engineering design criteria, standards, and lessons learnt, rules and constraints and check whether the design is practically feasible or is there any further opportunity for further pinch or it is over pinched. If the design is feasible, initiate step 14. Otherwise, repeat steps 7-12 with revised value of ΔT_{min}, revised set of pinch rules and constraints [11].

Step 14: Estimate practical targets for HEN design

The HEN designed on the basis of the estimated optimum values of ΔT_{min} may not always be the most feasible design. Low values of ΔT_{min} may demand very intricate network design with a large total area owing to low driving forces. This may have associated impact on CAPEX, if implemented, owing to complexity of the design. In practice, the designer chooses a high value of ΔT_{min}, perhaps 15°C, and determines the minimal increases in utility duties and required area. If the increase is marginal, the higher value of ΔT_{min} is chosen to get practical pinch point for design of HEN [3].

The following three rules of pinch technology form the basis for practical network design:

1. No external heating below the pinch
2. No external cooling above the pinch
3. No heat transfer across the pinch

Desecration of any of the above rules results in energy penalty than the MER possible [5,6].

Step 15: Finalise the design of heat exchanger network

The systematic application of the pinch analysis in the design phase of HEN allow us to achieve the energy targets within practical limits. The following two fundamental features are key to successful application of this technique:

1. Procedural steps recognize that the pinch region is the most constrained part of the problem
2. Procedural steps allow the designer to choose between match options

Basically, the HEN design inspects which fluid streams can be matched to other fluid [1,2].

Conclusion and Recommendation

The novice algorithm has been developed with an intention to minimise the hurdles of process design engineers and optimisation engineers in applying heat integration to a new or a retrofit design. The algorithm can be followed during development of new soft tool or manual application of pinch analysis. Although it is not mandate, we followed prefixed sequential algorithm follows hierarchical sequential workflow. To eliminate major jargons, flaws in technical integrity of design and minimize the energy penalty, sequential workflow was followed conservatively. Multiple intermittent reiteration check gates are provided by which the designer can optimize the overall energy target by reiteration of intermittent steps at different values of ΔT_{min} till constraint threshold is reached. Designer shall consider all constraints and flexibilities to achieve MER.

There is scope for further refining of the algorithm and incorporation of additional features on future algorithms. The designer should liaise with engineers from different domains to come up with holistic single integrated solution, which takes care of integrity algorithm, carries out coding and testing, and synchronises the code with operating system. Front end graphical templates development should be carried out simultaneously and all parts should be integrated on GUI and validated.

References

1. Chemical Engineers' Resource Page "Pinch Technology: Basics for the Beginners"
2. Linhoff B, Flower JR (1978) Synthesis of heat exchanger networks. AIChEJ
3. Linhoff B, Turner JA (1981) Heat-recovery networks: new insights yield big savings. ChemEng.
4. Linnhoff B, Hindmarsh E (1983) The pinch design method of heat exchanger networks. Chem Eng Sci 38: 745-763.
5. Kemp IC (2007) Pinch Analysis and Process Integration. 2nd edn.
6. Sojitra R, Dwivedi S (2016) Energy Optimisation Of Upstream Separation and Stabilisation Plant Using Pinch Technology. IJSRES.
7. Aspelund A, Berstad DO, Gundersen T (2007) An Extended Pinch Analysis and Design procedure utilizing pressure based exergy for subambient cooling. Appl Therm Eng 27: 2633-2649.
8. Correa DM, Gundersen T (2016) Process design methodology for energy efficient processes operating below and across ambient temperature. AIChE J.
9. Anantharaman R, Abbas OS, Gundersen T (2006) Energy Level Composite Curves- a new graphical methodology for the integration of energy intensive processes. Appl Therm Eng 26: 1378-1384.
10. FU C, Gundersen T (2016) Heat and Work integration: Fundamental insights and applications to carbon dioxide capture processes. Energ Conserv Manag 121: 36-48.
11. Al Moussawi H, Fardoun F, Louahlia-Gualous H (2016) Review of tri-generation technologies: Design evaluation, optimization, decision-making, and selection approach. Energ Conserv Manag 120: 157-196.

Effects of Yeast and Oxygen on Quality Attributes of Wine Produced from Ethiopian Beetroot

Zinabu Hailu* **and Derese Mekonnen**

Department of Chemical Engineering, Adama Science and Technology University, Adama, Ethiopia

***Corresponding author:** Zinabu Hailu, Department of Chemical Engineering, Adama Science and Technology University, Adama, Ethiopia
E-mail: zinisheh@gmail.com

Abstract

The study was focused on the effects of yeast and oxygen on quality parameters of wine produced from Ethiopian beetroot such as Ethanol, Volatile Acidity, Titratable acidity, Specific Gravity and pH. The basic ingredients used for fermentation were water, sugar, yeast, beetroot juice and citric acid. The experimental set up was design through randomized block design with four fermentation treatments such as: F-1: specified beetroot juice mixed with water, sugar, acid and yeast and then placed under anaerobic fermentation. F-2: the conditions were similar with fermentation treatment one, rather yeast was absence. F-3: This also same condition as treatment condition one, but it was put under aerobic conditions and F-4: the conditions were the same as treatment two, but it placed under aerobic. The results obtained showed that treatment F1 of Ethanol- 13% w/v, TA- 0.65% v/v, VA- 0.15% v/v, pH- 3.49, SG- 0.99 and F2, F3 and F4 (15.5, 0.735, 0.11, 2.89, 1.12, 7.15, 0.6, 0.131, 3.31, 0.983 and 11.17, 0.622, 0.121, 3.53, 0.985) respectively. By using statically analysis and sensory analysis treatment F1-A was highly acceptable with ($p<0.02008$) and panelist overall acceptance of 8.8+0.61. Finally, it can be conclude that beetroot juice treated with pure water, yeast, acid and sugar under anaerobic fermentation is produced wonderful wine.

Keywords: Beetroot; Wine; Fermentation; Yeast; Acid

Introduction

Production of wine was first discovered by ancient people using application of fermentation technology of fruits having sugar [1]. Today amateur winemaking is enjoyed by thousands of people throughout the world [2]. Wine consists of flavoring, sugar, acid, tannin and yeast but above all water [3,4]. The whole production of the wine has to be nurtured to perfection and this winemaking process must not be rushed. Wine is an alcoholic beverage produced through the partial or total fermentation of wine producing fruits. It is the fermented product of the fruit of several species of Vitis, mainly *V. vinifera* [5]. Though, the suitability of fruits other than grapes has been investigated all over the world, like sugar beet, banana, pineapple, beet root etc. [3]. Beetroot is a valuable vegetable, which is semi-hardy and biennial. It is grown year-round for its sweet, tender, succulent roots. Beets contain more sugar than any other vegetable, and its earthy taste and aroma comes from an organic compound called geosmin [6,7]. Beets of different colors, sizes and shaped have been grown, ranging from red, yellow, white, multi-colored, round, long cylindrical and huge sugar and mangle beets. Beet root has Red color naturally and its color is mostly stable at pH of 4.5-5.5; it is rather unstable outside this range [7,8]. The composition of beetroot, 100 g Nutritional value per 100 g (3.5 oz) Energy 180 kJ (43 kcal) Carbohydrates 9.56 g, 6.76 g of sugar), Dietary fiber 2.8 g, Fat 0.17 g, Protein 1.61 g, Water 87.58 g and others 0.72 g [5,6,8]. The process begins at the vineyard, where wine beetroot using special techniques for cultivating and maintaining the crop, depending upon the species of Beetroot and type of wine associated [4]. Thereafter, the Beetroot are harvested, upon which they are then either extracted of their juices, called the must, in a process called crushing either mechanically or by treading, the traditional method of basically stepping on the Beetroot in a large container. Beetroot wines, as described in this project, are produced by fermented the raw beetroot juice, from which the alcohol that is produced during the fermenting process will begin to give acceptable wine characterize [9,10]. Fermentation commences by adding the juices, sugar and citric acid and yeast together which then produces alcohol and carbon dioxide [10]. The ethanol content of the beetroot fruit is in between, 7-14% [11]. Thus, the study was focused on the effects of yeast and oxygen in quality attributes of wine produced from Ethiopian beetroot

Experimental Materials and Methods

Aerial Beta vulgar type beetroot was collected from Adama town which is located 98 km far away the capital city of Ethiopia, and then transported to Adama Science and Technology University (ASTU), Chemical Engineering department, Food engineering laboratory. Then, the overall experimental work was set as shown in figure below (Figure 1). About 2.5 kg of beetroot fruit was cleaned or washed thoroughly and sliced thinly using knife after the fresh beetroot selected. The sliced one was placed in juice extractor to be extracted the juice from the beetroot fruit. The chemical composition and quality parameters of wine produced from beetroot were done before and after fermentation. The experimental set up of fermentation treatments were designed in such way: F-1:0.25 kg Sugar, 15 ml of citric acid (to enhance distinct aroma/flavor), 2.25 L of water, 1.25 L of beetroot fruit juice and 0.00017 kg yeast was added and mixed together, then placed under anaerobic fermentation condition. F-2:1.25 L of beetroot fruit juice was mixed with 0.25 kg Sugar, 15 ml of citric acid (to enhance distinct aroma/flavor), 2.25 L of water and no yeast; placed under anaerobic fermentation. F-3:1.25 L of beetroot fruit juice was mixed with 10 ml citric acid, 2 L water and 0.0017 kg yeast; put under aerobic fermentation. F-4:1.25 L of beetroot fruit juice was mixed with 10 ml citric acid, 2 L water and absence of yeast; placed under aerobic fermentation conditions. The chemical analysis of the product was done before (at time of juice) and after 14 days of fermentation time. Total soluble solid (TSS) was measured by using Model-RF18 (India) handheld sugar refractometer. Titrable Acidity (TA) of the fruit was measured by means of an acid base titration method using a juice sample (10 ml juice+10 ml distilled water) and 0.1N Sodium hydroxide (NaOH) with phenolphthalein color indicator. Volatile Acidity (VA) within the distillate samples also expressed in terms of acetic acid mg/100 ml and pH was measured using digital pH meter. The specific gravity (SG) and alcohol content of the product (wine) were measured by hydrometer (AOA 2000 method). For the sensory evaluation tests, fifteen panelists was selected from the university and the tests involved individuals in isolated tasting conditions. All panelists were asked to give their individual ratings on all quality attributes of stored fruits including color, aroma, taste, flavor, firmness and overall acceptability of the wine produced from different fermentation conditions using a 9-point structured hedonic scale to conduct the preference test: 1-dislike extremely and 9- like extremely. The scores marked by panelists were collected and an average was calculated for each parameter. One way Analysis of variance (ANOVA) was performed on the data collected using Microsoft excel and Origin Pro8 softwares.

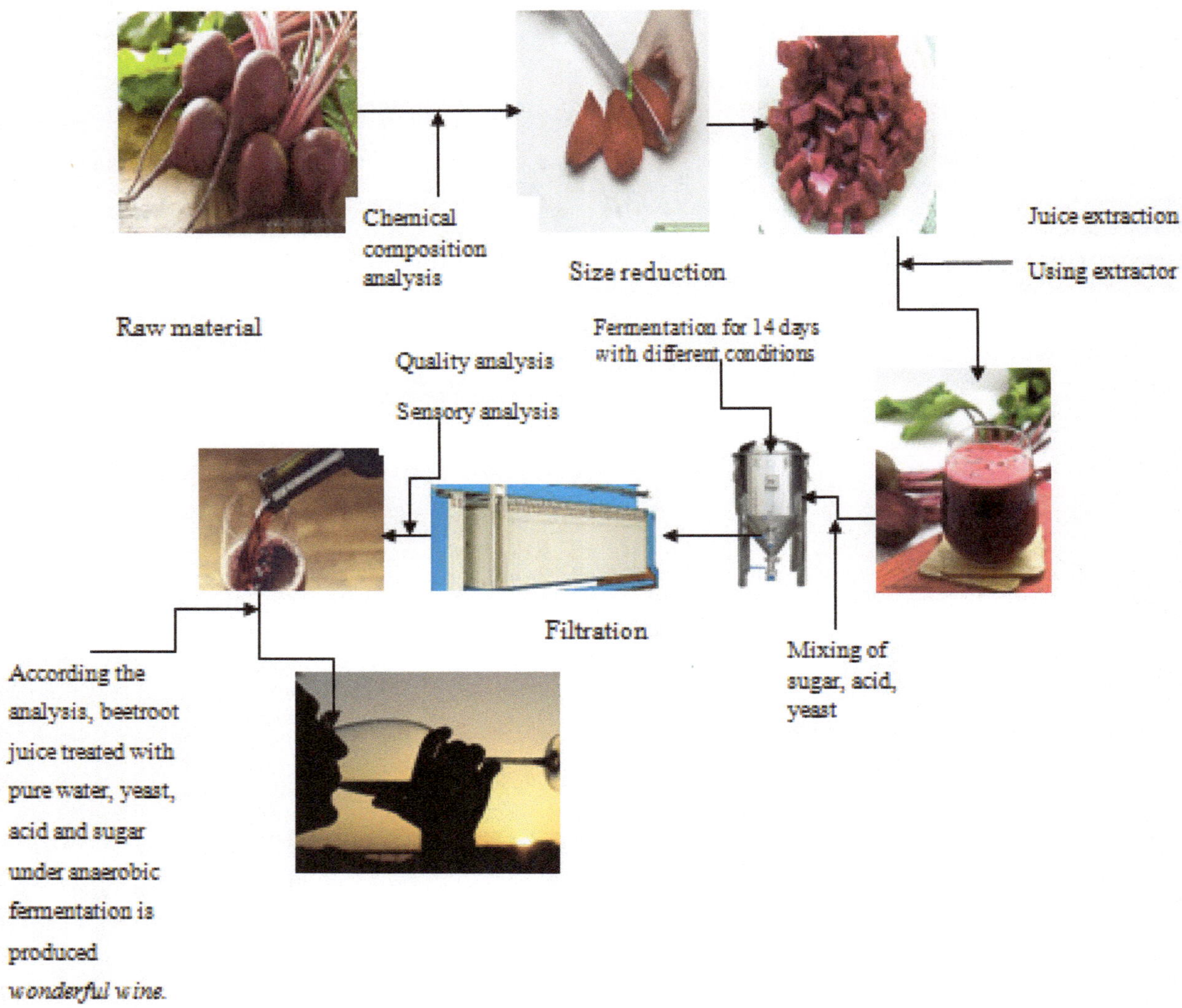

Figure 1: Over all experimental set up to conduct the study.

Results and Discussion

The effects of oxygen and yeast on quality and chemical composition of the product was presented in tables and figures.

Effect of oxygen on quality of wine

Wine produced from anaerobic (with no oxygen in excess) condition (F1-A) had high residual sugar and total soluble solid (24.5% w/v, 18% respectively) before fermentation; F4-D 15% TSS, 21% RS. The highest juice yield was recorded F1-A (640 ml/ kg) followed by 'B', C and D. The main prerequisite character of juice for fermentation is sugar content. Titrable acidity determined as tartaric acid range from

0.33 to 0.45% (w/v). The pH of the musts was 4.2 to 4.8. The low pH (4.8) was recorded in 'D' treatment condition, this is because luck in microbial activities (Table 1).

F-wines	Juice (ml/kg)	RS (% w/v)	TA (%)	pH	TSS* %
A-J1	640 ± 10	24.5 ± 1.22	0.33	4.2 ± 0.43	18.0 ± 1.24
B-J2	510 ± 13	18.5 ± 1.21	0.45	4.6 ± 0.86	16.2 ± 1.87
C-J3	550 ± 17	20.5 ± 1.24	0.36	4.3 ± 0.65	17.5 ± 0.79
D-J4	500 ± 22	21.0 ± 1.15	0.38	4.8 ± 1.10	15.5 ± 1.23

Table 1: Beetroot juice chemical compositions.

After fermentation the total sugar was converting in to 13% ethanol as well as for F4-D 9.7% ethanol (Figure 2), this result was similar with Ref. [12] report, the conversation of reducing sugar in to ethanol and carbon dioxide is due to the activity of microbes.

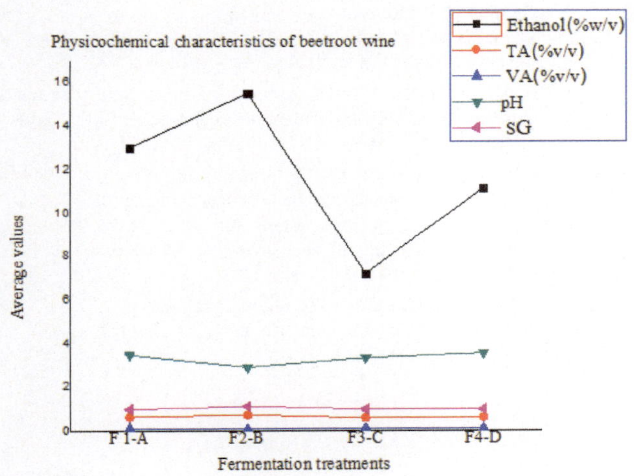

Figure 2: Quality attributes of beetroot wine.

Effect of yeast

The initial environment that affects the microbial makeup of a wine fermentation is that of the vineyard. Although a drastically different environment than juice or wine, the types of microbes present on grapes will have an impact on the ensuing ecology in the wine fermentation, particularly in the early stages. Microorganisms appear to colonize around the grape stomata where small amounts of exudates are secreted [13]. The spontaneous fermentations were shown to contain several different strains of S. cerevisiae that competed within the fermentation, while the inoculated fermentation was dominated by the inoculated strain [14,15]. The result showed that the effect of yeast in conversion of the simple carbohydrate to Ethanol, pH, TA, RS and VA was (13%, 0.65, 3.45, 2.1 and 0.15 respectively) under F1-A fermentation condition and for the other see on Table 2. The pH of the wine produced by aerobic fermentation type is low; this is due to the conversion of all the ingredients (activity of yeast) to the alcohol and has more acidic property. The pH of the wine produced by anaerobic fermentation type is likely moderate; this is due to the reason of fermenting without oxygen that means the nutrients contain in the Fermented cannot be easily converted to alcohol, which shows the wine has weak acid characteristic. The percentage of ethanol percentage was higher with treatment F3 beetroot fermented wines (15.7%w/v) (Table 2) and highest content was in the wine produced from 'F3' and lowest in 'F4', this is because of yeast effect. Preparation of fruit beetroot fruit influenced the synthesis of higher alcohols during fermentation. Higher alcohols may influence certain sensory characteristics although they constitute a relatively lesser quantity of the total Substances. Fermentation changed the aroma of fruit juice, because of the production of yeast volatiles and the metabolism of original fruit volatiles. And it was affected by many factors like variety of fruit, clarification and fermentation conditions [13,16].

Wines	Ethanol (%w/v)	T.A (% v/v)	V.A (%V/V)	PH	SG	RS
F 1-A	13.0	0.65	0.15	3.45	0.99	2.1
F2-B	12.0	0.735	0.11	2.89	1.12	2.4
F3-C	15.7	0.6	0.131	3.31	0.983	2.0
F4-D	9.7	0.622	0.121	3.53	0.985	2.0

Table 2: Chemical compositions of the product/beetroot wine. T.A=Titrable acidity; V.A=Volatile acidity.

Sensory evaluation

Wines produced by cells immobilized on grape skins have a better fruity aroma [16]. Similar results were also reported by Ref. [17] who found a considerable improvement in the wine sensory profile when fermentations were carried out in contact with the skins of Airen white wine grapes because of the transfer of the precursors of volatile compounds like esters, aldhaydes, and alcohols into the wine. The wines produced by immobilized yeast biocatalyst showed fine clarity at the end of fermentation with low free cell concentrations as well as characteristic pleasant soft aroma and fruity taste [6]. The result shoed that, wine produced under anaerobic fermentation with the inoculation of yeast (*S. cerevisiae*) had high over all acceptances (8.8+0.6) than that of produced without inoculating yeast (Figure 3).

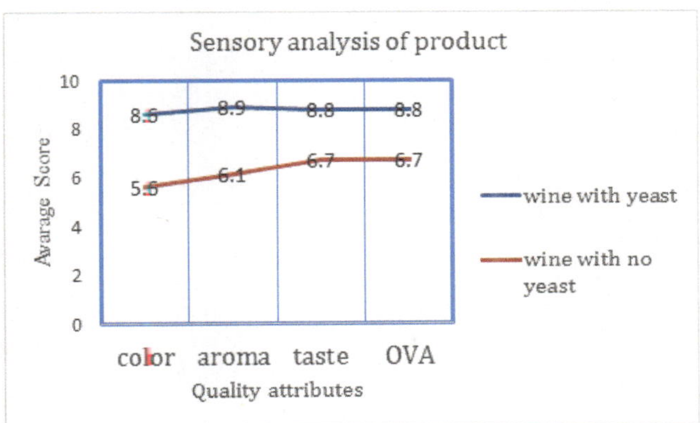

Figure 3: Sensory analysis result of the product.

Conclusions

Beetroot is a valuable vegetable and rich in nutrition, mainly it is rich in carbohydrate. By fermenting the mixed beetroot juice and then purifying the product from by product obtained final analyzed wine. The quality of wine depends up on the compositions of the juice and fermentation conditions. The study was evaluate the quality and compositions of wine produced from beetroot by setting different fermentation treatments; mainly anaerobic condition with and without yeast and aerobic fermentation with and without yeast. Thus, the result showed that, wine produced from fermentation treatment one (F1-A) was best product based on its quality analysis and it was confirmed by sensory analysis. Finally, it can be conclude that beetroot juice treated with pure water, yeast, acid and sugar under anaerobic fermentation is produced wonderful wine.

Acknowledgements

Researchers want to express their thanks to the School of Mechanical, Chemical and Material engineering, Department of Chemical Engineering, Addis Ababa University for providing laboratory facilities. They extend their pleasure to Adama Science and Technology University, Ethiopia, for financial supports for this study.

Competing Interests

The authors declare that they have no competing.

References

1. Aloysius LCP, Lee D (2001) Strategic management of the geography wine region. Tourism Rec Res 26: 81-88.
2. Kunkee R, Amerine MA (1970) Yeast in wine making. The Yeast. J of Science. New York. Academic Press.
3. Beames G (2003) The rock, the reef and the grape: the challenges of developing wine tourism in regional Australia. J Vacation Mark 9: 205-212.
4. Cappuccino JG, Sherman N (2010) Wine Production in Microbiology: A Laboratory Manual. Benjamin Cummings: San Francisco.
5. Millicevic B, Banovic M, Kovacevic-Ganic K, Gracin L (2002) Impact of grape varieties on wine and distillates flavor. Food Technol Biotechnol 40: 227-232.
6. Rapp A (1998) Volatile flavor of wine: Correlation between instrumental analysis and sensory analysis. Nahrung 42: 351-363.
7. Lee WC, Yusuf S, Hamid NS, Baharin BS (2006) Opitimizing Conditions for Enzymatic Clarification of Beetroot juice using Response Surface Methodology. Journal of Food Engineering 73: 53-63.
8. Egan H, Kirk RS, Sawyer R (1987) Pearson's Chemical Analysis of Foods. 8th edn. pp: 347-352.
9. McGregor DR, Bower JF (1975) Home preparation of Juice, Wines and Cider. Canada Department of Agriculture Ottawa. Publication 1406.
10. Son HS, Hwang GS, Park WM, Hong YS, Lee CH (2009) Metabolomiccharacterization of malolactic fermentation and fermentative behaviors of wine yeasts in wine industry. Journal of Agricultural and Food Chemistry 57: 4801-4809.
11. Berruto R, Tortia C, Gay P (2006) Wine bottling scheduling optimization. Transactions of the ASABE 49: 291-295.
12. Sabate J, Cano J, Querol A, Guillamon JM (1998) Diversity of Saccharomyces strains in wine fermentations: analysis for two consecutive years. Lett Appl Microbiol. 26: 452-455.
13. Ribereau-Gayon PD, Dubourdieu B, Doneche A, Lonvaud A (2000) Handbook of Enology. Volume 1: The Microbiology of Wine and Vinifications. John Wiley and Sons, New York, USA.
14. Schutz M, Gafner J (1994) Dynamics of the yeast strain population during spontaneous alcoholic fermentation determined by CHEF gel electrophoresis. Lett Appl Microbiol 19: 253 257.
15. Torulaspora delbrueckii-Saccharomyces cerevisiae culture on high-sugar fermentation. Yeasts. Research Signpost, Kerala, India, pp: 1-17.
16. Bergonzi G, Francolini M, Zaffignani F (1992) Costs of vineyard establishment: comparison of two low-growth systems in Emilia, Romagna. Vignevini 19: 19-22.
17. Amerine MA, Berg HW, Kunkee RE, Ough CS, Singleton VL, et al. (1980) The Technology of Wine Making. 4th edn. Avi Publishing Co. Inc., Westport, CT, USA, pp: 359-380.

Determination of Hydrodynamic Regime of the Reactor Alkylation of Olefins with Paraffin in the Modified Zeolite Catalyst OMNICAT-210P

Khazangul Piriyeva B, Zulfiyya Ismailova R, Azer Gasimov A, Ulviyya Karimova Japharova N*, Sevinj Hajizadeh M, Tunzala Aghayeva F and Rasim Japharov P

Department of Technology of Petrochemical Processes, Institute of Petrochemical Processes of Azerbaijan National Academy of Sciences, Baku, Azerbaijan

Abstract

Calculations for determination of the thermal and diffusive value of Peclet's criteria for establishment of the hydrodynamic mode of the reactor alkylation of hydrocarbons containing in gases of catalytic cracking, over industrial zeolite catalyst (OMNICAT-210P). It has been established that internal diffusion of reagents proceeds much faster than chemical reaction. From which it follows that the chemical reaction is not broken by the speed of diffusion of hydrocarbon into inside of grains of the catalyst. There have been given values of the activation energy and rate constant pre-exponential factors of the principal reactions that occur during the alkylation of hydrocarbons contained in gases of catalytic cracking.

Keywords: Alkylation; Zeolite catalysts; Peclet's criteria

Introduction

As follows from the currently existing structure of oil product consumption, more than half of the funds invested in the development of the oil and oil refining industries, it is necessary for the production of motor fuels for internal combustion engines, as that oil are practically the only source of motor fuels [1]. The development of the motor transport, using oil fuel creates a global environmental problem [2]. With the exhaust gases of vehicles about two hundred different substances, a half of which constitutes a danger to human health and environment are released into atmosphere, and the scale of air pollution increase due to the tendency of increasing the output of cars. In big cities the proportion of air pollution from vehicle exhaust in the total amount of emissions entering the atmosphere, ranges from 80 to 95% [3]. According to the World Health Organization (WHO) three-quarters of human diseases are bound to the ecological reasons caused by anthropogenous activity and motor transport is one of the most serious sources of deterioration in ecology [4].

Very relevant is the involvement in petroleum refining of the alternative sources of carbonaceous materials, such as natural gas, associated petroleum gases, heavy fuel oil, coal and its processing sludge, peat, plant biomass, etc. for the purpose, to produce high-octane gasoline components. At the same time it is very important today to obtain ecologically clean high-octane components of modern gasoline's, complies with Euro-5 and Euro-6 standards [5-8].

The main requirements to change the hydrocarbon composition of motor gasolines consist in increasing of the amount of isoparaffins from 20 to 45, decreasing aromatic hydrocarbons from 42 to 25%, in increase in high-octane additives up to 10-12% [9].

In this regard technologies of producing iso hydrocarbons by isomerization and alkylation of paraffin gain all great value olefins importance [10,11]. The alkylation process is used for recycling C_3-C_4 hydrocarbons obtained from gases of the catalytic cracking by converting their most part to alkylate and also as a possible alternative method for reducing the total sulfur content in gasoline, without losing octane in a product. There have been considered in article technologies of alkylation their prospects for the future have been given.

The impressive pace of scientific discovery in the industry demonstrate how demand alkyl benzene, which proved to the most valuable component of the fuel for carburetor aircraft and automobile engines. Alkyl benzene is one of the most important products of oil processing. For example, in the USA, about 11% of gasoline is currently being produced by an alkylation of isobutene by alkenes C_3-C_5. Alkylate approximately consists of 70% C_8 of alkanes, mainly trimethylpentanes 15% C_5-C_7 and 15% C_9-C_{12} hydrocarbons.

Alkylate meets the technical, operational and environmental requirements of modern European and American standards for motor fuel in internal combustion engines. Currently in the world the gasoline share on the basis of alkylate reaches proportion, about 25% of the total consumption of motor gasoline components, and in aviation gasoline's it is a major component (more than 60%). In abroad the volume of alkylate production exceeds 70 mln. t/year. There are no doubts that as long as the vehicles are operated on a high-octane gasoline's, alkylate will still be a desirable component of automobile gasoline [12,13].

The advantage of the alkylation of paraffin's with olefins process on heterogeneous catalysts is in the stable operation of the catalyst and its multiple use of after regeneration. In this connection with this in Institute of Petrochemical Processes Azerbaijan National Academy of Sciences there have been studied the process of alkylation of paraffin's with olefins on zeolite catalyst OMNICAT-210P modified Ni, Co, Cr.

Experimental Section

Researches on alkylation of hydrocarbons contained in the gases of a catalytic cracking carried out in the reactor in the presence of heterogeneous industrial-zeolite catalyst OMNICAT-210P modified Ni, Co, Cr. Below is a description of the principal technological schemes bigger pilot plant (Figure 1).

***Corresponding author:** Ulviyya Karimova Japharova N, Department of Technology of Petrochemical Processes, Institute of Petrochemical Processes of Azerbaijan National Academy of Sciences, Baku, Azerbaijan
E-mail: u_kerimova@yahoo.com

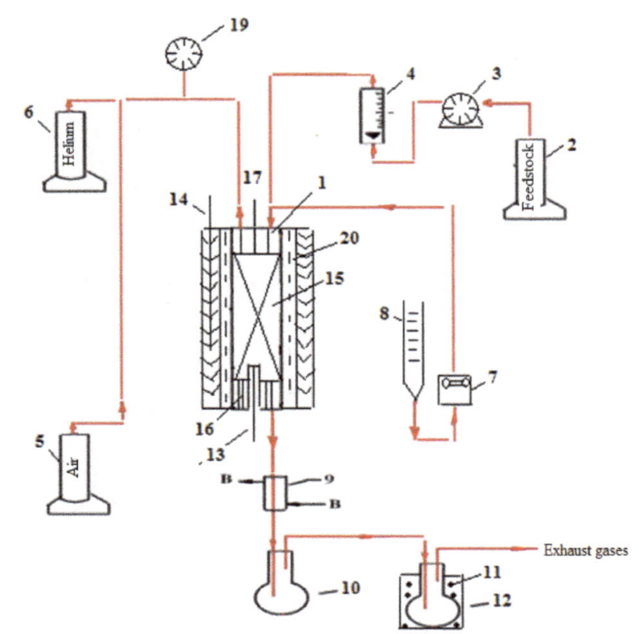

Figure 1: Principal technological scheme bigger pilot installation. 1-reactor (internal diameter 34.2 mm, height 931 mm); 2-feedstock balloon; 3,19-gauges; 4-flowmeter; 5-air ballon; 6-inert gas balloon; 7-water pump; 8- burette with distilled water; 9-fridge; 10,11-receptacles; 12-receptacle with dry ice; 13 thermocouple internal (80 mm depth of immersion in the catalyst bed); 14-external thermocouple (at the height of the catalyst length 360 mm); 15-catalyst zone (volume-270 cm^3, height 437 mm layer); 16-filling (layer height 136 mm); 17-filling (layer height 64 mm); 18 insulation layer; 20-winding.

The raw material from the balloon 2 through the rotameter (4) is fed to the reactor (diameter of reactor -34.2 mm; depth of reaction zone - 931 mm). The catalysts (15) with diameter of pellets 3.2 mm in amount of 270 cm^3 with a height of layer 437 mm are loaded to the reactor (1). Fillings (Raschig rings) have been backfilled on the catalyst layer a 64 mm high (17) and under the catalyst layer 136 mm high (16). Fillings (17) on the catalyst layer are filled with the aim of heating the feed zone to the temperature of 320°C.

The fillings (16), which are loaded into the reactor below the catalyst layer is required with the aim of to equalize the pressure on the cross-section of reactor that will evenly distribute the raw material at the reactor section. The reactor has been isolated with glass wool (18) and heated by a nichrome winding (20). The temperature inside of reaction zone and in isolation layer is controlled by thermocouples (13) and (14). The pressure of raw materials is controlled by the monometer (3). As the reaction temperature in the catalyst zone decreased to 280°C.

The waste gases from reaction zone come to the cooler (9). Liquid reaction products (including the condensed water, which is used for desorbing hydrocarbons from the catalyst surface) are cooled by running water and collected in the reservoir (10) but not condensed reaction products enter into the container (11) cooled by dry ice (12).

For the purpose of a desorption of the reaction products from the catalyst surface the steam is entered on the catalyst layer (15). For this purpose the spout (8) and pump (7) have been provided which supplies water from the dispenser to the reaction zone of the catalyst (15). Previously water evaporates in the filling region (17) and the steam with a temperature of 350-400°C is feed to the catalyst bed. Thus, from the catalyst surface boiling hydrocarbons are desorbed which after passing a cooler (9) is condensed in the receiving tank 10 and 11. After desorption of boiling hydrocarbons at a temperature of 200-400°C the regeneration of oxidizing catalyst is carried out in the presence of air oxygen, which is fed from the balloon (5), having a pressure gauge (19). After regeneration, the reactor is purged with an inert gas in order to release of reactionary volume from air oxygen to its absence in the exhaust gases.

Results and Discussion

The model of industrial reactor calculated taking into account the hydrodynamic mode allows to predict with high accuracy carrying out the studied system at change of the technological regime or composition of raw materials.

The plug flow reactor (PFR) is carried in laminar hydrodynamic regime. In such reactors, the reactants stream flows in one direction longwise of the reactor without hashing.

Hydrodynamic regime in PFR is characterized by the fact that any particle of stream moves in one direction longwise of reactor, the reverse (longitudinal) hashing absent. There is also no hashing on reactor section. It is assumed that the distribution of substances on this section is uniform, i.e., values of parameters of reaction mixture are identical. Each element of volume reaction mass dVr moves along the length of the reactor without mixing up with the previous elements of the volume, and behaves as a piston in the cylinder, foxing out everything that is before it. Therefore, such driving mode of reagents is sometimes called piston or full displacement mode. The composition of each volume element sequentially changes longwise the reactor, as a result of the chemical reaction. So, for example, concentration of initial reagent varies progressively longwise the reactor from initial (C_{A0}) to a final value (C_A). The consequence of this movement mode of the reaction mixture is the fact that the residence time of each particle in the reactor is same.

For the mathematical description of the fixed bed catalyst there have been used quasi-homogeneous model of a layer of the granular catalyst. According to this model, the catalyst layer is presented in the form of a permeable solid medium through which a gas stream flows, and the chemical reaction proceeds with a speed equal to the observed speed of conversion. In this case, the catalyst layer is considered as a homogeneous medium with a continuous distribution of concentration, temperature, pressure and a flow rate.

Reaction rates are averaged by the volume of the layer, but the heat and mass transfer is determined by the effective thermal conductivities and diffusions coefficients, which depend on physical properties of the mixture, a flow rate, grain sizes, shape and the structure of a granular layer. In the adiabatic catalyst bed there is no radial heat-mass transfer. Heat transfer inside of grains takes place, mainly through a solid catalyst bulk which volume heat conductivity on several orders is higher than the thermal conductivity of gas.

As the length of the catalyst layer in industrial reactors is usually large enough, then the effect of axial diffusion is insignificant. While the Reynolds number above 100 heat and mass transfer occurs due to turbulent diffusion. In these conditions, the molecular diffusion can be neglected and the hydrodynamic regime approaches a plug flow regime [14].

There have been studied the influence of the temperature and rate of volume flow of supply of raw materials on liquid yield reaction, there have been installed the mechanism of the process. There have been compiled the kinetic equations describing the process of alkylation of C_3-C_4 paraffins by olefinic hydrocarbons.

The catalyst is the commercial zeolite containing catalyst of modified Ni, Cr, Co in amount of about 1.5-2%. Modification of Ni, Cr and Co allows conducting the dehydrogenation reaction of paraffin's C_3-C_4 to the corresponding olefins. They in their turn together with olefins of feedstock participate in reaction of alkylation of paraffin's.

At volumetric feed rate of 50 h^{-1} and reaction temperature 280-300°C liquid yield of reaction products stored at ~78-80 wt%. Researches of kinetic regularities have been carried out in an isothermal reactor.

The volume of the catalyst was 50 cm^3. For carrying out the process in continuous conditions it is necessary to install three (3) reactors. The first reactor works in an alkylation mode (30-60 min), the second works in the mode of desorption of products from the surface of the catalyst with steam at a temperature of about 400°C and finally the third reactor works in the regeneration mode ($t_{reactor}$=550-580°C) in the presence of atmospheric oxygen.

Then there is a switching of the third reactor in the alkylation mode, the second in the regeneration mode and first in the mode of a desorption of hydrocarbons from the catalyst surface. The process test has been carried out on the enlarged pilot plant with a catalyst volume by ~300 sm^3.

For carrying out the process in industry there is proposed the reactor with the volume of 20 m^3 and with 2.6 m inner diameter (which is a cylindrical device). A possibility of application of model of PRF for the considered process, it is checked by the calculation of diffusion and thermal Peclet's criterion for a stream on an entrance to the alkylation reactor. Taking into account the mean value of a volume flow of a reaction mixture of 1000 m^3/h (V=50 h^{-1}) the peripheral speed of a stream in the reactor makes:

$$w = \frac{Q^t.4}{3600.\pi.d^2} = \frac{1000\times573\times4}{298\times3600\times3.14\times2.6^2} = 0.101 m/s$$

where, w - flow rate in the reactor, taking into account the temperature in the reaction zone (300°C), m/s; Q^t - volume flow of the reaction mixture at a feed rate of 50 h^{-1}, m^3/h; d_p - inner diameter of the reactor, m. Heat Peclet's criterion Pe_T calculated by the formula [15]:

$$Pe_T = w \cdot l \cdot Cp \cdot \rho / \lambda$$

Where, w - in the reactor the flow rate, m/s; l - linear size of the reactor, m; Cp specific heat of the mixture, J/mol · K; ρ - density of the mixture in kg/m^3;

$\lambda_{C_5H_{12}}$ - Thermal conductivity of the mixture, W/m • K [16]

The C_p=59.7 *kcal/g × mol*=249.9 *J/mol × K*

Pe_T=0.101×3.8×249.9×874.7/3.393=24725.6

$$\lambda_{C_5H_{12}} = 0.81\times104 kcal/sm.s = 3.39\times103 Vt/m.s$$

The diffusion criterion of Peclet characterizes the mass transfer in a moving stream and is analogous to the thermal criterion of Peclet characterizing the heat transfer in a moving medium. The ratio of thermal criterion of Peclet Pe_T to the value of the diffusion criterion of Peclet is within the

Pe_T/Pe_D=0.91÷1.2

Pe_D=(ω×1)/D;

Where, ω is the velocity of flow in the reactor, m/s; l- linear size of the reactor, m; D-kinematic diffusion coefficient, m^2/s.

The obtained values of diffusion and thermal criterion of Peclet suggest that the transfer of heat and material flows in the reactor is significantly prevailing over diffusion. In the reactor there is observed hydrodynamic conditions close to the ideal displacement. The kinematic D diffusion coefficient of gas A in the gas B at an absolute temperature T and a pressure P can be calculated by the following formula:

$$D = \frac{0.0043.10^{-4}.T^{1.5}}{P\left(v_A^{1/3} + v_B^{1/3}\right)} \sqrt{\frac{1}{M_A} + \frac{1}{M_B}}; m^2/c$$

Where T, is the temperature, K; P-pressure, atm; M_A and M_B the molar mass of gases A and B, kg/kmol; υ_A and υ_B - molar volume of gases A and B, sm^3/mol [17].

$$v = V_C n_C + V_H n_H$$

where, V_c atomic volume of carbon (14.8), sm^3/atom; n_C-number of carbon atoms (C); V_H - atomic volume of hydrogen (3.7) sm^3/atom; n_H -is the number of hydrogen atoms.

$$D_{C_3H_6/C_3H_8} = \frac{0.0043\times10^{-4}\times573^{1.5}}{(66.6^{1/3}+74^{1/3})^2} \times \sqrt{\frac{1}{42}+\frac{1}{44}} = 0.1894\times10^{-4}, m^2/s$$

$$Pe_T / Pe_D = \frac{24725.6}{20264} = 1.22$$

$$Pe_{DC_4H_8/C_4H_{10}} = \frac{0.101\cdot3.8}{0.1894\cdot10^{-4}} = 27571.84$$

$$Pe_T / Pe_D = \frac{24725.6}{27571.84} = 0.897$$

$$Pe_{DC_3H_6/C_4H_{10}} = \frac{0.101\cdot3.8}{16316\cdot10^{-4}} = 23523$$

The equations for the calculation of comparison values diffusion of paraffinic and C_3-C_4 olefinic hydrocarbons inside catalyst grain with a value of the rate of reaction of an alkylation are given below :

$$\frac{D_{AB}}{\Delta r} \cdot F \square\ K_i \cdot C^{cp}_{C_3-C_4} \Delta V$$

Where, D_{AB} - molecular diffusion coefficient of the reactants in the gas phase, m^2/s; Δr -the radius of the catalyst grains, m; F - the surface area of the catalyst grains, m^2; K_i - reaction rate constant, m^3/mol • sec; $C^{cp}_{C_3-C_4}$ -averaged concentration of C_3-C_4 paraffin's, mol/m^3; V -volume of the catalyst grains, m^3.

The alkylation reaction rate constants are set by solving of the kinetic equations of the process of producing liquid products from the catalytic cracking gas (Table 1).

For the calculations to determine the speed of alkylation reaction there has been chosen VI, X, XIV equations due to the fact that the activation energies of these reactions have a minimum value.

K_i	260°C	300°C	420°C	E_i cal/mol	K_{oi}
$k_1\ c^{-1}$	0.001545	0.00254	0.008027	7556.7	$0.194 \cdot 10^1$
k_2 l/mol·s	0.0151	0.0352	0.2465	12803.5	$0.269 \cdot 10^1$
k_3 l/mol·s	1.1	1.76	5.22	7145.3	$0.937 \cdot 10^0$
$k_4\ c^{-1}$	0.000022	0.000063	0.000693	15767.9	$0.652 \cdot 10^2$
$k_5\ c^{-1}$	0.00000128	0.00000211	0.000006675	7579.2	$0.164 \cdot 10^{-2}$
$k_6\ c^{-1}$	0.00000159	0.00000408	0.00003554	14226.7	$0.109 \cdot 10^1$
k_7 l/mol·s	0.296	0.455	1.23	6542.5	$0.1426 \cdot 10^0$
k_8 l/mol·s	0.0109	0.0249	0.167	12523.6	$0.149 \cdot 10^1$
k_9 l/mol·s	0.348	0.68	3.197	10145.7	$0.518 \cdot 10^1$
K_{10} l/mol·s	22.9	60.6	571.5	14754.2	$0.257 \cdot 10^5$
K_{11} l/mol·s	4.26	7.58	28.6	8725.7	$0.161 \cdot 10^2$
K_{12} l/l/mol·s	1.17	1.845	5.27	6907.6	$0.796 \cdot 10^0$
$K_{13}\ s^{-1}$	0.00567	0.00844	0.0210	6014.1	$0.166 \cdot 10^1$
$K_{14}\ s^{-1}$	0.00001128	0.0000168	0.000042	6044.26	$0.34 \cdot 10^{-2}$
K_{15} l/mol·s	2.42	4.22	15.2	8427.4	$0.69 \cdot 10^1$
K_{16} l/mol·s	894.0	1327.0	3295.0	5980.2	$0.253 \cdot 10^3$
K_{17} l/mol·s	1.156	2.89	23.9	13906	$0.582 \cdot 10^3$
$n_1=0.937$; $n_2=0.817$; $n_3=0.91$; $n_4=0.77$; $n_5=0.98$; $n_6=0.87$; $n_7=0.99$; $n_8=0.93$; $n_9=0.99$					
The value of the adsorption constant k_i^{ads}				Q_i heat of formation cal/mol	
k_1^{ads}	1.359	1.14	0.76	−2658.7	$0.11 \cdot 10^0$
k_2^{ads}	1.33	1.16	0.848	−2063	$0.189 \cdot 10^0$
k_4^{ads}	1.269	1.099	0.79	−2174.5	$0.163 \cdot 10^0$
k_5^{ads}	0.533	0.364	0.151	−5784.3	$0.226 \cdot 10^{-2}$
k_6^{ads}	0.956	0.864	0.684	−1541.3	$0.223 \cdot 10^0$
k_7^{ads}	0.937	0.768	0.486	−3010.5	$0.546 \cdot 10^{-1}$
k_8^{ads}	0.9	0.459	0.09587	−10309.3	$0.537 \cdot 10^{-4}$
k_9^{ads}	0.6934	0.387	0.1009	−8838.2	$0.1647 \cdot 10^{-3}$

Table 1: Kinetic parameters of the process of producing liquid products from a catalytic cracking gas.

$C_3H_8 \underset{k_2}{\overset{k_1}{\leftrightarrow}} C_3H_6 + H_2$	I	$n.C_4H_{10} + i.C_4H_8 \xrightarrow{K_{10}} C_8H_{18}$	IX
$i.C_5H_{12} + C_3H_6 \xrightarrow{K_3} C_8H_{14}$	II	$i.C_4H_{10} + n.C_4H_8 \xrightarrow{K_{11}} C_8H_{18}$	X
$i.C_4H_{10} \xrightarrow{K_4} i.C_4H_8 + H_2$	III	$n.C_4H_{10} + n.C_4H_8 \xrightarrow{K_{12}} C_8H_{18}$	XI
$n.C_4H_{10} \xrightarrow{K_5} n.C_4H_8 + H_2$	IV	$i.C_5H_{12} \xrightarrow{K_{13}} CH_4 + i.C_4H_8$	XI I
$i.C_4H_{10} \xrightarrow{K_6} 2C_2H_4 + H_2$	V	$i.C_4H_8 \xrightarrow{K_{14}} n.C_4H_8$	XI II

Routes, in which are formed connections with same chemical formula but different structure have been combined.

Comparison between the diffusion contribution values and the contribution of chemical reaction to change of reactants concentration in the grain with a diameter of 20 mm indicates that the internal diffusion of reactants proceeds much faster than chemical reactions.

1. Diffusion of C_3H_6 and

i. $C_4H_{10} = \frac{0.163 \cdot 10^{-4}}{0.001} \cdot 12.56\times10^{-6} \square\ 4.5\times10^{-7}\times7.52\times10^{3}\times41.8\times10^{-10} 0.202216 \square\ 0.0000141451$

2. Diffusion C_4H_8 and i.

$C_4H_{10} = \frac{0.137 \cdot 10^{-4}}{0.001} \cdot 12.56\times10^{-3} \square\ 7.58\times10^{-6}\times7.52\times10^{3}\times41.8\times10^{-10} 0.172072 \square\ 0.00023826$

3. Diffusion of C_3H_6 and

C_3H_8=From the above it follows that the rate of chemical reaction is not inhibited by the rate of diffusion of C_3-C_4 hydrocarbons to the inside of the catalyst grains.

Conclusions

As a result of studies established that the alkylation process takes place in a plug flow reactor. The ratio of the thermal criteria of Peclet to the diffusion is within the 0.9-1.22. There have been provided a comparition of the values of contribution of diffusion and chemical reactions while changing of concentration of reactants in catalyst grain. At the same time, it has been found that the inner diffusion of reactants proceeds much faster than a chemical reaction.

It has been established that the internal diffusion of reagents proceeds much faster than the chemical reaction, so for C_4H_8 and C_4H_{10}

is within 0.172072 »0, 00023826. The ratio of thermal Peclet criterion (Pe_T) to diffusion (Pe_D) is in the range 0,91-1,22 that characterizes the process conditions in the reactor in an ideal displacement mode. There have been determined the values of the activation energy and the rate of constants of the principal reactions that occur in an industrial zeolite catalyst OMNICAT-210P. It means that the rate of chemical reaction is not inhibited by the diffusion of hydrocarbon inside the catalyst grains.

References

1. Tretyakov VF, Burdeynaya TN (2003) Motor fuels from non-oil raw materials. Journal of the Russian Academy of Sciences 6: 48-52.
2. Magaril ER (2008) Influence of quality of motor fuels on operational and ecological characteristics of automobiles. M KDU, p: 164.
3. Danilov AM (1996) Additives and additives. Improvement of ecological characteristics of oil fuels. Moscow. Chemistry, p: 232.
4. Trapeznikov NN, Shain AA (1992) Oncology. M Medicine, p: 398.
5. Karpov SA (2007) The quality of automobile gasolines in the light of modern operational requirements. Oil Refining and Petrochemistry 1: 16-19.
6. Lapidus AL, Golubeva IA, Tonkonogov BP (2005) Gas chemistry: problems and prospects. Chemistry and Technology of Fuels and Oils 2: 22-24.
7. Markova NA, Bukina ZM, Kolesnichenko NV (2011) Processing of associated petroleum gases in motor fuels. Petrochemistry and Oil Refining 1: 148-151.
8. Bukina Z, Grafova GM, Ionin DA (2012) Method for obtaining ecologically pure high-octane gasoline. Russian Federation 02: 1095.
9. Gerzeliev IM, Tsodinov MV, Khadzhiev SN (2009) New ways of obtaining isoparaffins-high-octane components of gasoline. Petrochemistry 49: 3-8.
10. Borutsky PN (2010) Catalytic processes for the production of hydrocarbons of a branched structure. Spb NPO Professional, p: 728.
11. Solodova NL, Bashkirtseva NY, Kozin VG, Dullin AI (2009) Modern technologies for the production of motor fuel components. Moscow. KSTU, p: 337.
12. Hommeltoft SI (2001) Isobutane alkylation. Recent developments and future perspectives. Applied Catalysis A: General 221: 421-428.
13. Shiriyazdanov RR (2011) Investigation of the regularities of obtaining and using zeolite-containing catalysts for the alkylation of isobutane with olefins. Dis Cand Those Sciences UFA, p: 145.
14. Slinko MG, Dilman VV, Markeev BM, Kropberg AE (1980) Simulation of fixed-bed reactors. Chemical Industry 11: 22-26.
15. Raseev S (2003) Thermal and catalytic processes in petroleum refining. Marcel Dekker, New York, USA, p: 920.
16. Sherwood T, Reed R (1982) Properties of gases "liquids". Moscow. Chemistry, p: 334.
17. Handbook of the Chemist (1966) Volume V. Moscow. Chemistry, p: 973.

Evaluation of Direct Esterification of Fatty Acid Derivative of Kojic Acid in Co-solvent System: A Statistical Approach

Fatin Amirah Ahmad Norddin[1], Sharifah Nurfadhlin Afifah Syed Azhar[1,2] and Siti Efliza Ashari[1,2*]

[1]Department of Chemistry, Faculty of Science, Universiti Putra Malaysia, 43400 UPM Serdang, Selangor, Malaysia
[2]Process Engineering Laboratory, Enzyme and Microbial Technology Research Center (EMtech), Faculty of Biotechnology and Biomolecular Sciences, Universiti Putra Malaysia, 43400 UPM Serdang, Selangor, Malaysia

Abstract

The present work deals with direct esterification method to synthesize Kojic acid using immobilized lipase as a biocatalyst in acetonitrile with the addition of dimethylsulfoxide (DMSO) as a solubilizing agent Co-solvent respectively. To increase the esterification yield of KMO, modifications of the process were evaluated, including the use of a co-solvent and the use of Novozyme 435 (*Candida antarctica*) as a catalyst. The KMO synthesis has been developed and optimized by using Response Surface Methodology (RSM) with Central Composite Rotatable Design (CCRD). The optimized condition of enzyme was 3.35 wt% and 1:3.64 molar ratio of kojic acid and oleic acid at 82.39°C for 255.24 min of reaction. With these condition, the maximum percentage yield was 42.73% with R2 value of 0.866914 and indicated that 86.69% of the variability in the response could be explained by the model. The model was significant and fitted well with the experimental data and the lack of fit was not significant. The efficacy for cosmetic application was successfully tested and showed as non-irritating with a Human Irritancy Equivalent score between 0.55-0.83 which safe to be applied in cosmetic ingredient.

Keywords: Kojic acid; Kojic ester; Esterification; Co-solvent; Response surface methodology

Introduction

Kojic acid (5-hydroxy-2-(hydroxymethyl)-1, 4-pyrone), is a production of fungal metabolite from many species of *Aspergillus, Acetobacter,* and *Penicilium* [1]. Latterly, the whitening agent in this acid benefits as an essential ingredient in cosmetic cream to block the formation of pigments by the deep cells on the skin [2]. It was studied intensively by several researches due to its potential as a tyrosinase inhibitor to prevent browning in food production as well as a skin lightening agent for cosmetic preparations [3-5]. In Asia, hyroquinone, a well-known whitening agent, was banned for cosmetic usage for its toxicity. Therefore, the cosmetic industry likely opts for kojic acid as an alternative for skin whitening and for its antioxidant properties [6].

However, exposure of sun and air to this compound makes it unstable and reduces its effectiveness. Therefore, to overcome this situation, kojic acid must be modified to improve its properties in terms of storage stability, compatibility, and oil-solubility. Previous studies reported that modifying the esterification method with fatty acid is the best technique to synthesize kojic acid ester [7]. The derivatives have more excellent characteristics than the original starting material, making it more profitable, especially for cosmetic applications. Besides that, solid-phase synthesis of kojic acid-tripeptides and their tyrosinase inhibitory activity, storage stability, and toxicity, concluded that these derivatives were almost 15 times more stable under ambient stored conditions with a lower toxicity than kojic acid itself. Kojic acid derivatives were relatively higher with tyrosinase inhibitory activities than that of kojic acid [4]. Further evidence claimed kojic acid derivatives are major class of many natural and synthetic compounds that own a high activity profile because of their wide range of biological activities [8].

A skin irritancy test and corrosivity potential of an ingredient is necessary for safely assessing cosmetic ingredients. Skin Irritant Assay is one of the methods for assessing chemical irritancy in a product. This assay is a standardized and quantitative *in vitro* test, which works to detect, rank, and predict the dermal irritation potential of product [9]. For the evaluation of skin irritation, physicochemical properties pertinent information that needs to be considered in the assessment. Therefore, the properties of the physicochemical of kojic ester are an important aspect to evaluate their efficacy in the cosmetics field since it is a relatively new compound and research on it is limited.

These esters can be synthesized through chemical or enzyme catalysis. Chemical catalysis might be more economical compared to enzyme catalysis, but this will lead into a complex mixture, thus making it more difficult to purify [10]. Studies from Kobayashi et al. focused on the amino acid derivatives of kojic acid, which showed that the derivatives have stronger activities than kojic acid [11]. In their current study, they attempted to produce a new compound with much stronger inhibitory potency against tyrosinase than that of N-carbobenzoxy-amino acid-kojiate. The synthesis of kojic acid derivatives was performed using DSC (N,N'-disuccinimidyl carbonate) and DMP (4-dimethylaminopyridine). Therefore, ten compounds were obtained by joining the OH group at position C7 of kojic acid with the amino end of an amino acid to form a urethane type bond [11]. This shows that it can be difficult to esterify when a chemical catalyst is used rather than using enzymatic-catalyzed synthesis. Additionally, the high temperature can cause degradation of esters and undesired products, resulting in a high energy cost and low percentage yield [12].

Conversely, enzyme catalysis since is more profitable it requires less amounts of energy, minimum thermal degradation, higher biodegradability, needs a simple step process and therefore, emerges promising substitutes over conventional approaches. Moreover, certain

***Corresponding author:** Siti Efliza Ashari, Department of Chemistry, Faculty of Science, Universiti Putra, 43400 UPM Serdang, Selangor, Malaysia
E-mail: ctefliza@upm.edu.my

aspects such as higher yield, milder reaction conditions, and substrate selectivity can be strong points in favor of biocatalysis production [13]. Raku and Tokiwa, 2003, proposed the regioselective esterification of kojic acid esters using *Bacillus subtilis* protease. They reported that the primary hydroxyl group at the C7 position of kojic acid was easily esterified using a simple one step procedure therefore, giving a higher conversion yield of kojic acid esters. Furthermore, the esters obtained in their study were effective as radical scavenging and improved lipophilicity. It was also indicated that there was no significant change in the biodegradability of kojic acid esters and it can be classified as an environmentally friendly product [14]. In addition, Liu and Shaw improved the lipophilicity of kojic acid by synthesizing lauric and oleic acid esters of kojic acid using nine commercial lipases that were tested [15]. Among all nine lipases, *P. cepacia* and *P. camembertii* showed the best catalytic efficiency and specificity for enzymatic synthesis of kojic acid monolaurate (KAML) and kojic acid monooleate (KAMO). The results show that KAML and KAMO were easily esterified at the hydroxyl group in the C5 position of kojic acid [15].

In enzymatic synthesis, optimization of a reaction is a crucial step in developing an efficient reaction system. The conventional (parameter-at-one-time) method is highly cost and time consuming [16]. For those reasons, optimizing the enzymatic reaction will give the best stage to overcome the lacking of the conventional method. Optimization reaction conditions can be done using Response Surface Methodology (RSM). RSM is defined to be a set of mathematical and statistical methods using a minimal number of experiments for designing experiments, building models, evaluating the effect of variables, and obtaining the optimum condition. The number of experimental trials needed to analyze the interaction between multiple variables can be reduced and thus, save costs and time [7]. This method is more desirable for industrial applications.

The novelty of this research is the usage of Novozym 435 as an enzyme in solvent mixture (acetonitrile and dimethylsulfoxide) for synthesis of kojic ester. The aim of this study is to synthesize, characterize and optimize the reaction condition of kojic ester catalyzed by Novozym 435 in a solvent system. The relationship of reaction variables: enzyme amount, reaction temperature, reaction time, substrate molar ratio and percentage yield, being studied by using Response Surface Methodology (RSM).

Materials and Methods

Materials

Kojic acid (purity 98%) was donated by Tokyo Kasei Kogya Co, Ltd. (Japan). While oleic acid (purity 99%) was purchased from Southern Edible Oil Sdn. Bhd., Selangor, Malaysia. Glycerol tributyrate was used as an internal standard in gas c analysis and was obtained from Sigma-Aldrich (USA). Novozym 435, *Candida antarctica* lipase B immobilized on a macroporous acrylic resin (10,000 propyl laurate units per gram) was purchased from Novo Nordisk A/S (Denmark).

Lipase catalyzed esterification

The esterification was performed by dissolving kojic acid and oleic acid into a solvent mixture of acetonitrile and dimethyl sulfoxide (DMSO) in a 150 ml beaker. The enzyme was added to the mixture and was sealed using aluminium foil. Using a magnetic stirrer, the mixture was stirred at 700 rpm at 80°C for about 360 min. At the end of the reaction, the enzyme was filtered using filter paper (Munktell, 1F) and a rotary evaporator was used to remove undesired solvent. Figure 1 shows the enzymatic synthesis of kojic ester in a solvent system.

Analysis of sample

Isolation and purification of Kojic ester: The reaction was terminated by separating the enzyme from mixture by filtration using filter paper. Then, the solvent was removed by rotary evaporator. Preliminary detection were subjected to column chromatography contain silica gel (Silicycle, 60-200 μm particle size). The sample was eluted using a mixture of hexane and ethyl acetate (75:30, vol/vol) and collected in every 2 ml of flow volume. The purification of product was confirmed by Thin Layer Chromatography (TLC) on pre-coated silica gel plate (Merck, 5 × 10 cm) and developed in hexane:ethyl acetate (75:30, vol/vol).

Characterization: The sample of product was put on the TLC plate using the tip and diluted in a mobile phase which is the mixture of hexane and ethyl acetate (75:30 vol/vol). Then, the TLC plate sheet was dip into the $KMNO_4$ solution and dry to make it clearer to see the spotted obtained.

Experimental design

A central composite rotatable design (CCRD) was employed. The study requires 30 experiments of reactions undergone at various reaction temperatures (70-90°C), reaction time (120-360 min), enzyme amount (1.0-5.0 wt%), and substrate molar ratio (1-5 mmole) for kojic acid and oleic acid. The enzymatic synthesis of kojic ester was performed by dissolving kojic acid with oleic acid in a solvent mixture of acetonitrile and dimethyl sulfoxide (DMSO) in a 150 ml beaker. The enzyme was added to the mixture and was sealed using aluminium foil. The mixture was stirred using a magnetic stirrer at 700 rpm at 80°C for about 360 min. From previous work which showed that an immobilized enzyme like Novozyme 435 has high activities and is stable at a high temperature [17]. Table 1 shows the variables in terms of coded and actual values. The R^2 value obtained indicates that the experiment was significant with the model.

Skin irritancy test

An irritancy test of kojic ester was evaluated to determine its potential to cause dermal irritation. Dermal irritancy was conducted at Malaysian Palm Oil Berhad (MPOB) using the *in vitro* International's Irritection Assay System. Four different amounts (50, 75, 100 and 125 mg) of samples were weighed and placed onto a synthetic membrane disc biobarrier. Then, the protein reagent consisting of oligomeric protein, glycoprotein, lipid constituents and globulins plus blanking buffer were added to the 24-well assay plate. The synthetic disc, which

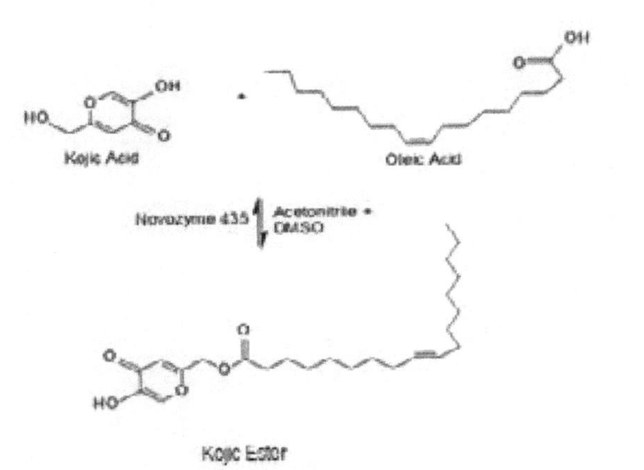

Figure 1: Enzymatic synthesis of Kojic ester in solvent system.

contained various amounts of kojic ester samples, was inserted into the blank and test sample wells of the plate. The assay plate was then incubated at 25°C for 24 h. The optical density (OD) of the protein reagent and blanking buffer were then recorded at 450 nm using an MRX plate reader. The irritancy potential of a test sample was expressed as the Human Irritancy Equivalent (HIE) score. This score is determined by comparing the changes in optical density (OD470) produced by the test material to a standard curve, which is constructed by measuring the increase in OD470 produced by a set of calibration substances. Table 2 shows the relationship of HIE score to the irritancy classification for the Dermal Irritation Test method.

Results and Discussion

Model fitting and Analysis of Variance (ANOVA)

The CCRD design was used in order to achieve a preferable model for the optimization of kojic ester synthesis in solvent system. Table 3 shows the experimental data and the percentage yield obtained.

Fitting of the data to the various models (linear, two factorial interactions quadratic and cubic) and their subsequent ANOVA shows that the reaction of kojic acid and oleic acid in a solvent system is most suitable described with a quadratic polynomial model. The final equation in terms of coded factors is shown in Equation (1):

$$\text{Actual Yield (\%)}=+42.77+1.59A+0.87B+0.12C+3.18D+0.19AB+291AC+6.67AD+4.31BC+3.58BD+5.61CD-7.26A^2-9.36B^2-9.00C^2-7.16D^2 \quad (1)$$

Where A is the reaction temperature, B is the reaction time, C is the enzyme amount, and D is the substrate molar ratio. The positive sign in front of the terms indicates a synergetic effect whereas the negative sign indicates an antagonistic effect. From the ANOVA shown in Table 4, the model of P-values was significant and lack-of-fit with prob>P less than 0.05 values, of which there was only a 0.01% chance that model P-values this large could occur due to noise. Generally, the lack-of-fit test should not significant. Variables in Table 4, state that A, B and C are referred to as the main linear term effects, while AB, AC, AD, BC, BD and CD are the interaction terms and A^2, B^2, C^2, D^2 are the quadratic terms in the responses. In this study, the main linear term (D) is significant ($p<0.05$) whereas interaction term (AD), (CD) and (BC) is significant ($p<0.05$) while for quadratic term A^2, B^2, C^2, D^2 are all significant ($p<0.05$).

According to Bezerra et al. a not significant lack-of-fit indicates the model is good because the model was fitted well to the experimental data [18]. Based on the coefficient value (R^2), the fit between the development model and experimental data can be determined. According to Keng et al. R^2 value >0.900 indicates that the model describes the situation well and indicates that the model has a very high correlation [19]. In this case, the R^2 was equal to 0.866914, and this coefficient value indicated that 86.69% of the variability in the response could be explained by the model. Therefore, the model was preferable to represent the real relationship among the parameters studied.

Effect of reaction parameters

By studying the three-dimensional (3D) response surface plot, the relationship between reaction parameter and responses as well as the optimum level of each variable for production of kojic ester can be determined. Figure 2A illustrates the response surface plots for the interaction between reaction time (B) and reaction temperature (A) with fixed 3.00 wt% of enzyme amount and 1: 3.00 mole substrate molar ratio (KA : OA). According to the Figure 2, the highest yield (43.6%) was obtained at 240 minutes of reaction time and 80°C of reaction temperature. At the beginning, it was seen that when the reaction time increases, the percentage yield also increases. Then, a lower percentage yield is obtained when the reaction time goes further to 250 minutes. Therefore, this result shows that by prolonging the time of reaction (>250) will not increase the production of kojic ester because the esterification reaction will produce water as a side product. The yield of kojic ester was increased by increasing the temperature up to 80°C. When the temperature was raised up to 90°C, the yield becomes decreased which was probably beyond the critical temperature (>90°C), the lipase may have undergone some inactivation and become denatured. These results were similar with the reviewed paper by using the same lipase, which is from *Candida antarctica* (Novozym 435) as optimally used at a temperature of 83°C [19]. Figure 2B represents the effect of the enzyme amount and reaction temperature in the range of 2.00 - 4.00 wt% and 75-85°C by fixing the reaction time 240 minutes and 1 : 3.00 mmole (KA:OA) substrate molar ratio. From the Figure 2C the enzyme amount of 3.00 wt% and a moderate temperature at 80°C appeared to be a favorable condition for the esterification reaction. This phenomenon is related to the active site of enzyme molecules, where the large excess of enzyme cannot be exposed to the substrates due to possible protein aggregation and hence, reducing the conversion of yield [20]. The esterification activity becomes slow as temperature increases more than 80°C at a high amount of enzymes due to the inactivation of the enzyme at a high temperature (Figure 2D). This was proven by Khamarudin et al. in a conventional study of temperature effect on the synthesis of palm-based kojic acid ester [21].

In conclusion, the reaction was not favorable in the highest amount of enzyme. This was proven by Ashari et al. because as the amount of enzyme increases, the production yield decreases and when time increases, the conversion yield also increases [7]. At a certain time, a small decrease of percentage yield was seen as the reaction time goes

Table 1: Range of variables and their levels for the CCRD.

Variable	Level				
	-2	-1	0	+1	+2
Reaction temperature, A (°C)	70	75	80	85	90
Reaction time, B (min)	120	180	240	300	360
Enzyme amount, C (wt %)	1	2	3	4	5
Substrates molar ratio (kojic acid : oleic acid), D (mmole)	1	2	3	4	5

Table 2: Relationship of Human Irritation Equivalent (HIE) Score to Irritancy Classification for the Dermal Irritection Test Method.

Human Irritation Equivalent (HIE)	Predicted Dermal Irritancy Classification
0.00-0.90	Non-irritant
0.91-1.20	Non-irritant/irritant
1.21-5.00	Irritant

Table 3: Central composite rotatable design of kojic ester.

un	A: Reaction Temperature (°C)	B: Reaction Time (min)	C: Enzyme Amount (wt%)	D: Substrate Molar Ratio (mmole)	Actual Yield (%)
1	80 (0)	240 (0)	3 (0)	3 (0)	30.3
2	80 (0)	240 (0)	3 (0)	3 (0)	39.36
3	75 (+1)	180 (+1)	2 (+1)	2 (+1)	28.55
4	75 (+1)	300 (-1)	2 (+1)	2 (+1)	18.4
5	90 (-2)	240 (0)	3 (0)	3 (0)	20.25
7	80 (0)	240 (0)	3 (0)	3 (0)	34.09
8	75 (+1)	180 (+1)	4 (-1)	2 (+1)	8.88
9	75 (+1)	300 (-1)	4 (-1)	2 (+1)	0.08
10	75 (+1)	300 (-1)	4 (-1)	4 (-1)	26.68
11	80 (0)	240 (0)	3 (0)	3 (0)	48.44
12	75 (+1)	300 (-1)	2 (+1)	4 (-1)	2.97
13	85 (-1)	180 (+1)	4 (-1)	4 (-1)	15.91
14	85 (-1)	300 (-1)	4 (-1)	4 (-1)	42.72
15	80 (0)	240 (0)	1 (+2)	3 (0)	7.25
16	85 (-1)	180 (+1)	2 (+1)	4 (-1)	22.02
17	75 (+1)	180 (+1)	2 (+1)	4 (-1)	2.32
18	85 (-1)	300 (-1)	2 (+1)	4 (-1)	6.43
19	80 (0)	240 (0)	5 (-2)	3 (0)	1.48
20	80 (0)	240 (0)	3 (0)	3 (0)	55.98
21	85 (-1)	180 (+1)	2 (+1)	2 (+1)	0.87
22	70 (+2)	240 (0)	3 (0)	3 (0)	2.34
24	80 (0)	240 (0)	3 (0)	1 (+2)	7.67
25	80 (0)	240 (0)	3 (0)	5 (-2)	15.78
26	85 (-1)	300 (-1)	4 (-1)	2 (+1)	0.82
27	80 (0)	240 (0)	3 (0)	3 (0)	48.44
28	80 (0)	120 (+2)	3 (0)	3 (0)	2.32
29	80 (0)	360 (-1)	3 (0)	3 (0)	3.49
30	85 (-1)	300 (-1)	2 (+1)	2 (+1)	0.81

Table 4: ANOVA for the quadratic model developed for synthesis of kojic ester.

Source	Sum of Squares	df	Mean Square	p-value
Model	7459.508	14	532.822	0.0003
A-Reaction temp	60.5155	1	60.5155	0.3874
B-Reaction time	18.2876	1	18.2876	0.6316
C-Enzyme amount	0.357704	1	0.357704	0.9463
D-Substrate molar ratio	242.3797	1	242.3797	0.0950
AB	0.566256	1	0.566256	0.9325
AC	135.1988	1	135.1988	0.2031
AD	712.4896	1	712.4896	0.0080
BC	296.7868	1	296.7868	0.0674
BD	205.4206	1	205.4206	0.1217
CD	504.3393	1	504.3393	0.0213
A^2	1447.147	1	1447.147	0.0006
B^2	2403.595	1	2403.595	< 0.0001
C^2	2219.812	1	2219.812	< 0.0001
D^2	1404.629	1	1404.629	0.0006
Residual	1145.164	15	76.34429	
Lack of Fit	663.891	10	66.3891	0.7119
Pure Error	481.2733	5	96.25466	
Cor Total	8604.672	29		

further [7]. According to Virto et al. and Ismail et al. by prolonging the reaction time, the synthesis of water and concentration of products increases gradually, this promoted the reverse esterification. Therefore, the percentage yield of the product may reduce [22,23].

Optimum conditions

The highest percentage yield was obtained from the optimal combination of parameters by response surface methodology. The predicted highest yield of 44.30% at reaction conditions of 82.39°C,

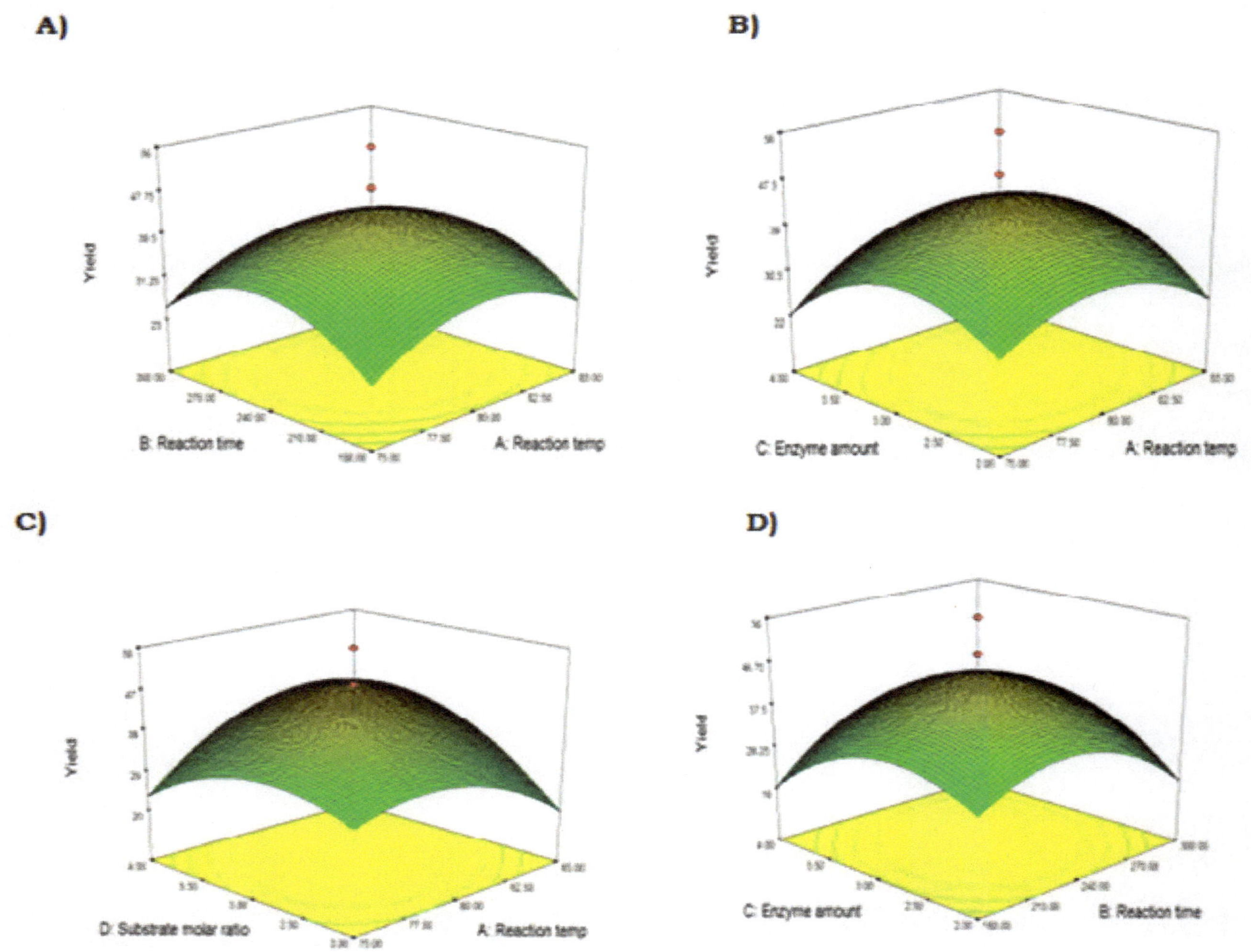

Figure 2: Three-dimentional response surface plot. (A) reaction time (min) versus reaction temperature (°C), (B) enzyme amount (%wt) versus reaction temperature (°C), (C) substrate molar ratio (KA:OA) (mmole) versus reaction temperature (°C) and (D) enzyme amount (%wt) versus reaction time (min) on percentage yield as response.

255.24 min, 3.35 wt% of enzyme and molar ratio of substrate of 1: 3.64 was predicted by using the optimized function of the Design Expert software. As shown in Table 5, the actual experimental value collected was 42.73% with a small difference of 1.57. Hence, these results support the validity of the quadratic model and response surface methodology can be practiced adequately to optimize the lipase-catalyzed synthesis of kojic ester.

Skin irritancy test

A safety evaluation of an ingredient is essential for consumer safety protection. For a novelty ingredient, the toxicity of the ingredients used is of concern. This may give rise to restrictions on their cosmetic applications if the toxicity is highly detrimental and could be eliminated or stopped from further development. In order to assess the safety of cosmetic ingredients, a tiered-approach was suggested [24]. The efficacy of kojic ester in cosmetic formulation was investigated with the use of Irritection Assay System. Table 6 indicates the results of *in vitro* dermal irritancy assay for four samples of kojic ester.

Based on the results, all samples had Human Irritancy Equivalent (HIE) scores of between 0.55-0.83, which is below the non-irritant level. Therefore, the skin compatibility of kojic ester is safe for cosmetic formulation. Previous studies also employed similar dermal irritection Assay methods to test the irritation potential of engkabang wax ester, palm oil esters and dihydroxystearic acid (DHSA) synthesis, respectively [9,18,25]. All new compounds tested were predicted to be non-irritant with the highest HIE scores of 0.31, 0.18, and 0.90, respectively.

Conclusions

Kojic ester was synthesized by enzymatic esterification using Novozym 435 to speed up the reaction and obtained excellent conversion of yield. Optimization was successfully predicted by Response Surface Methodology (RSM) and well-fitting models were established through the quadratic polynomial model. The optimal conditions obtained were 3.35 wt% of enzyme and 1:3.64 molar ratios of kojic acid and oleic acid at 82.39°C for 255.24 min of reaction. The

Table5: Optimum condition for Novozym 435 catalysed synthesis of kojic ester.

Exp	Reaction Temperature (°C)	Reaction Time (min)	Enzyme Amount (wt%)	Substrate Molar Ratio (KA:OA) (mmole)	Yield (%) Actual
1	82.39	255.24	3.35	01:03.6	42.73

Table 6: Irritancy Test of Kojic Ester.

Dose (mg)	Irritancy Score (HIE)	Classifiation
50	0.55	Non-irritant
75	0.63	Non-irritant
100	0.52	Non-irritant
125	0.83	Non-irritant

actual yield obtained was 42.73%. The efficacy for cosmetic application was successfully tested and showed as non-irritating with a Human Irritancy Equivalent score between 0.55-0.83 which safe to be applied in cosmetic ingredient.

Acknowledgements

The authors would like to thank the Department of Chemistry, Faculty of Science, Universiti Putra Malaysia (UPM) for giving opportunity on this project.

References

1. Uher M, Chalabala M, Cizmarik J (200) Kojic acid and its derivatives as potential therapeutic agents. Ceska Slov Farm 49: 288-298.
2. Cabanes J, Chazarra S, Carmona FG (1995) Kojic Acid, a cosmetic skin whitening agent, is a slow-binding inhibitor of catecholase activity of tyrosinase. J Pharm Pharmacol 46: 982-985.
3. Burdock GA, Soni MG, Carabin IG (2001) Evaluation of health aspects of kojic acid in food. Regul Toxicol and Pharmacol 33: 80-101.
4. Kim HCJ, Cho JK, Kim SY, Lee YS (2004) Solid-phase of kojic acid-tripeptides and their tyrosinase inhibitory activity, storage stability, and toxicity. Bioorg Med Chem Lett 14: 2843-2846.
5. Terabayashi Y, Sano M, Yamane N, Marui J, Tamano K, et al. (2010) Identification and characterization of genes responsible for biosynthesis of kojic acid, an industrially important compound from Aspergillus oryzae. Fungal Genetics and Biology 47: 953-961.
6. Gupta AK, Gover MD, Nouri K, Taylor S (2006) The treatment of melasma: A review of clinical trials. J Am Acad Dermatol 55: 1048-1065.
7. Ashari SE, Mohammad R, Ariff A, Basri M, Salleh AB (2009) Optimization of enzymatic synthesis of palm-based kojic acid ester using response surface methodology. J Oleo Sc 58: 503-510.
8. Li Y, Jiang Q, Wang K, Du B, Wang X (2013) A novel and green synthesis of kojic acid derivatives in ionic liquid [bmim] BF_4. Tetrahendron Letters 54: 7147-7150.
9. Abdul RMB, Chaibakhsh N, Basri M, Raja AR, Salleh RNZ, et al. (2008) Modelling and optimization of lipase-catalyzed synthesis of dilauryl adipate ester by response surface methodology. Journal of Chemical Technology and Biotechnology 83: 1534-1540.
10. Chen CS, Liu KJ, Lou YH, Shieh CJ (2002) Optimisation of kojic acid monolaurate synthesis with lipase PS from Pseudomonas cepacia. J Sc of Food and Agric 82: 601-605.
11. Kobayashi T, Adachi S, Nakanishi K, Matsuno R (2001) Semi-continuous production of lauroyl kojic acid through lipase-catalyzed condensation in acetonitrile. Biochemical Engineering Journal 9: 85-89.
12. Awang R, Basri M, Ahmad S, Saleh S (2003) AB Enzyme catalyzed synthesis and characterization of octyl dihydroxystearate acid. J Oleo Sci 52: 7-14.
13. José A, Monica V, Mercedes M, Michel P (2005) Biocatalytic processes for the production of fatty acid esters. Journal of Biotechnology 124: 213-223.
14. Raku T, Tokiwa Y (2003) Regioselective synthesis of kojic acid esters by Bacillus subtilis protease. Biotechnology Letter 25: 969-974.
15. Liu KJ, Shaw JF (1998) Lipase-Catalyzed Synthesis of Kojic Acid Esters in Organic Solvents. JAOCS 75: 1507-1511.
16. Gunawan ER, Basri M, Abdul Rahman MB, Salleh AB, Abdul RNZ (2005) Study on response surface methodology (RSM) of lipase-catalyzed synthesis of palm-based wax ester. Enzyme and Microb Tech 37: 739-744.
17. Noureddin EBN, Ashari SE, Serrano M, Aracil J, Martinez M (2013) Solvent-free lipase-catalyzed synthesis of a novel hydroxyl-fatty acid derivative of kojic acid. Enzyme and microbial Technology 55: 128-132.
18. Bezerra MA, Santelli RE, Oliveira EP, Villar LS, Escaleira LA (2008) Response surface methodology (RSM) as a tool for optimization in analytical chemistry. Talanta 76: 965-977.
19. Keng PS, Basri M, Ariff B, Abdul Rahman MB, Abdul Rahman RNZ, et al. (2008) Scale-up synthesis of lipase-caltalyzed palm ester in stirred-tank reactor. Bioresource Technology 99: 6097-6104.
20. Jumbri K, Rozy MFAH, Ashari SE, Mohamad R, Basri M, et al. (2015) Optimisation and Characterisation of Lipase-Catalysed Synthesis of a Kojic Monooleate Ester in a Solvent-Free System by Response Surface Methodology. PLoS ONE, p: 10.
21. Khamarudin NH, Basri M, Cheng LG, Salleh AB, Abdul Rahman RNZ (2008) Enzymatic synthesis and characterization of palm-based kojic acid ester. J Palm Oil Research 20: 461-469.
22. Virto C, Svensson I, Adlercreuz P (1999) Enzymatic synthesis of Lysophosphtidic acid and phospatidic acid. Enzyme and Microbial Technology 24: 651-658.
23. Ismail A, Soultani S, Ghould M (1999) Enzymatic-catalyzed synthesis of alkylglycosides in monophasic and biophasic systems. The transglycosylation reaction. Journal of Biotechnology 69: 135-143.
24. Salminen WF (2002) Integrating toxicology into cosmetic ingredients research and development. International Journal of Cosmetic Science 24: 217-224.
25. Aldrin Z, Ismail R, Ahmad S (2005) Safety evaluation for dermal and ocular irritation of dihydroxystearic acid as cosmetic ingredient. Journal of Palm Oil Research 17: 160-167.

Chemical Treatment of Petroleum Wastewater and its Effect on the Corrosion Behavior of Steel Pipelines in Sewage Networks

El-Shamy AM[1*], Hanaa A El-Boraey[2] and El-Awdan HF[2]

[1]*Physical Chemistry Department, Electrochemistry and Corrosion Lab, National Research Centre, El-Bohouth, Giza, Egypt*
[2]*Faculty of Science, Menoufia University, Shebin El-Kom, Egypt*

Abstract

The sources of wastewater from municipal and different industries act as a substantial pressure on the surrounding environment, so the waste management is ordinarily required before drainage. The environmental protection agency added specific and limited levels for discharging the pollutants into the municipal networks or into the sea. The treatment process of industrial wastewater became must to remove the contaminants or pollutants, which produced from the production procedure to a maximum, which does not negatively affect the environment. Chemical coagulation process is the highest important technique for industrial wastewater treatment because of its efficiency to minimize the contaminants in wastewater. The coagulants for higher removal of the contaminants and have a low cost is the most important choice of wastewater treatment. In this paper, chemical coagulation process for a petroleum wastewater sample by using ferric chloride, aluminum sulfate and polyacrylamide separately or as a mixture as well as the corrosion behavior of untreated and treated wastewater samples have been investigated. According to the obtained results the best formula, which gives the best result in removal of the organic and inorganic wastes as well as the best corrosion results is the mixture of aluminum sulfate and polyacrylamide. This formula have double functions, decreasing the total dissolved salts of about 21.11% and decreasing the COD and BOD content with about 47.95% and 53.57% respectively. These results proved that this formula is the best in corrosion behavior on mild steel pipelines.

Keywords: Wastewater treatment; Chemical treatment; Corrosion control; Steel pipelines

Introduction

Petroleum industry is important due to its impact on the environment because of the generated huge amount of water during its process. The wastewater from this industry is generally strong and may contain toxic pollutants. Petroleum industrial wastewater usually contains organic and inorganic matters in different concentrations and contains heavy metals, surfactants, grease, oils and different salts. It is well known that the wastewater causes severe problems for the environmental due to their toxicity of coloring materials, huge amount of suspended organic compounds and high chemical oxygen demand (COD). Therefore, they have to treat these pollutants before discharging into the environment [1]. There are many different common and conventional techniques for wastewater treatment; chemical physicochemical and biological methods are used in the treatment of the petroleum wastewater. The most abundant and effective is the chemical treatment [2-8]. Chemical treatment permits reduction of dissolved and suspended pollutants as well as the high concentration of fat, oil and grease from petroleum industrial wastewater. Precipitation process by using coagulants in wastewater treatment includes the addition of organic or inorganic chemicals to alter dissolved and suspended solids and enable their removal by sedimentation. Generally, coagulation is the method in which colloidal particles in water is accumulated together into larger particles, called flocs. Coagulants have been known since early in the twentieth century and they play an essential role in the removal of many impurities from polluted waters [9]. Coagulation is used to achieve maximum removal of COD and TSS [10]. Chemical coagulation [11] is not only the applied technique but also there are many other techniques; adsorption [12], photocatalytic processes [13] and microfiltration [14]. Electrocoagulation is one of the best techniques in petroleum wastewater especially in oil and greasy wastewaters [15,16]. The evaluation of successive technique depends on the removal capacity and reduction of organics, inorganics and toxic materials in oily wastewater. The chemical treatment process depends on the type of coagulants. The most famous and available in the field is inorganic metal based-coagulants. Other examples of inorganic metal coagulants are aluminum sulfate, ferric chloride and ferric sulfate and it is well notorious that the alum is very imperative in water purification for a long time. In the recent years, the pre-polymerized inorganic coagulants have become massive applications due to their availability such as poly aluminum chloride and aluminum chlorohydrate [17]. The poly aluminum chloride has several uses, including the removal of suspended materials and organic matters, metal and heavy metal ions as well as coloring agent [18]. During the last twenty years, the researchers are in a challenge for improving the exclusion of pollutants from the wastewater. The exclusion of organic materials by coagulation process is subjective by many factors such as the conditions and the characteristics of these materials. Consequently, the removal of organic matters by coagulation varies widely beween 10% and 90%. The effectiveness of these coagulants were found to be dependent on the constituents of the wastewater, temperature, dose, mixing rate [19]. The present work aims to evaluate and compare the capability of polyacrylamide or ferric chloride or aluminum sulfate individually and mixed in treatment of the petroleum wastewater. The second task is concerned with the investigation of the corrosion behavior for untreated and treated wastewater.

***Corresponding author:** El-Shamy AM, Physical Chemistry Department, Electrochemistry and Corrosion Lab, National Research Centre, El-Bohouth, Giza, Egypt, E-mail: elshamy10@yahoo.com

Materials and Methods

Materials

The collection of wastewater under study is coming from petroleum company, situated in industrial area, El-Ain El-Sukhna, Suez, Egypt. The coagulant Ferric chloride ($FeCl_3.6H_2O$) and aluminum sulfate ($Al_2(SO_4)_3.18H_2O$) were purchased from El-Nasr Company, Cairo, Egypt. Polyacrylamide, sodium hydroxide (NaOH) and hydrochloric acid (HCl) were come from Ultra Extract Company for water and wastewater treatment.

Methods

Take known weight of the coagulants of ferric chloride, aluminum sulfate and/or polyacrylamide before starting the experiment. We need to filter the petroleum wastewater sample before the treatment as a control. Adjust the pH of the filtered sample by sodium hydroxide to be more than 10. Divide the adjusted sample into seven filtered samples. Chemical treatments were occurred for six of them and the seventh one was remained without any chemical treatment as a blank. On the other hand, one sample was taken before any filtration or pH adjustment or chemical treatment was subjected to full analysis for comparison of the treated samples. A conventional jar test apparatus was used in the experiments to coagulate the six pH adjusted filtered samples of petroleum wastewater by using ferric chloride, aluminum sulfate and polyacrylamide. A batch test was carried out by it, accommodating a series of six beakers together with six-spindle steel paddles; then the samples were mixed homogeneously. After the desired amount of polyacrylamide, aluminum sulfate and ferric chloride was added to the samples, the beakers were agitated at mixing 150 rpm for 10 minutes to coagulation. After stopping the agitation, the suspension was allowed settling for 30 minutes. Then it was filtered to collect 1500 ml where pH is readjusting for them adding HCl or NaOH to be at range 7 to 8 and the gained volume is divided into two portions; one portion is 1400 ml for chemical analytical analysis and another portion is 100 ml for making corrosion experiments.

pH measurements: The efficiency of the wastewater treatment is very much pH sensitive. Monitoring of pH can indicate changes in raw and finished water quality. The efficiency of chlorination and coagulation, and the corrosively of treated water, depends on the pH values. The pH meter is consisted of potentiometer, glass electrode, reference electrode and a temperature in compensating device. Most modern pH meters use a single combined electrode. Set the set points of buffers b1, b2 or b3. Normally the used set points are; 4.0, 7.0 and 10.0, the pH meter will recognize these set points automatically when in calibration mode. The measuring of the pH values are highly required as soon as possible after sample collection, preferably within 15 minutes. If sample must be stored, it should be refrigerated at 4°C with no preservatives added and the test must be performed not later than six hours after collection. If pH values vary widely, standardize for each sample with a buffer solution having a pH within 1 to 2 pH units of the sample.

Alkalinity: The alkalinity of any water represents a measure of its capacity to neutralize an acid. The alkalinity of natural water may exist as carbonates, bicarbonates, hydroxides, silicates, and phosphates. The presence of one or more of these ions depends on the pH of the water. Two titration procedures exist for determination of the alkalinity, the first one is concerned with measuring the pH by using a pH meter and the second one is arisen by using traditional pH indicators. The procedure involves titrating first to a pH of 8.3 using phenolphthalein indicator, the color change of pink to colorless and then continuing the titration to a pH of about 4.5 using methyl orange indicator with a color change from yellow to pink. Calculations are processed form determining carbonate, bicarbonate, and total alkalinity. All values are expressed in terms of equivalent calcium carbonate ($CaCO_3$).

Total residual chlorine: The chlorination of the treated effluent serves primarily to destroy or deactivate disease-causing microorganisms. Chlorination may produce adverse effects due to formation of potentially carcinogenic compounds such as chloroform. Combined chlorine formed on chlorination of ammonia or amine-bearing water adversely affects some aquatic life. To minimize any adverse effects, it is essential that proper testing procedures by using the optimum dose of chlorine. In wastewater, total residual chlorine usually measured while chlorination seldom is carrying far enough to produce free chlorine. Select a sample volume that will require not more than 20 ml 0.01 N $Na_2S_2O_3$ and not less than 0.2 ml for the starch iodide until the end of reaction. For a chlorine range of 1 to 10 mg/l, use a 500 ml sample; above 10 mg/l, use proportionately less sample. Place 5 ml acetic acid, or enough to reduce the pH to between 3.0 and 4.0, in a flask or white porcelain casserole. Add about 1 g KI estimated on a spatula. Pour sample in and mix with a stirring rod. Titrate away from direct sunlight. Add 0.025 N or 0.01 N $Na_2S_2O_3$ from a burette until the yellow color of the liberated iodine is almost discharged. Add 1 ml starch solution and titrate until the disappearance of the blue color.

Total Suspended Solids (TSS): Filter a well-mixed sample through a pre-washed, pre-weighed, glass fiber filter. The suspended solids are retained on the filter then rinsed, dried, and weighed. The increasing in weight represents the suspended solids. TSS samples can be either composite or grab samples and can be collected in either glass or plastic. Analyze the samples as soon as possible or preserve it by ice or refrigeration to reduce microorganism activity. Holding time is no longer than 7 days. Obtain a clean, dry gooch crucible, filtering flask, and vacuum hose then warm the sample to room temperature. Tare weight determination and remove a gooch crucible containing a prewashed filter from the desiccators. Zero the analytical balance, place the gooch crucible containing the filter on the balance using crucible tong. Remove the gooch crucible from the desiccators using the tong and place on the filtering flask. Wet the filter with a small volume of reagent water to seat the filter. Stir the sample on magnetic stirrer with sufficient speed to provide a homogenous solution. Stirring too fast may cause particles to separate by size and weight. Pipet the desired volume from a point mid-depth and mid-way between the wall and the vortex. Exclude large floating particles or submerged agglomerates of nonhomogeneous materials from the sample if it is determined that their inclusion is not representative. Because excessive residue on the filter may form a water-entrapping crust, limit the sample size to that yielding not more than 200 mg residue. For samples of high dissolved solids, thoroughly wash the filter to ensure removal of the dissolved material. Prolonged filtration times resulting from filter clogging may produce high results owing to the increased colloidal materials captured on the clogged filter. Some chloride and nitrate salts may be lost. Choose sample volume to yield between 2.5 and 200 mg dried residue. If volume filtered fails to meet minimum yield, increase sample volume up to one liter. If complete filtration takes more than 10 min, increase filter diameter or decrease sample volume.

Oil and grease: The experiment oil and grease could be defined as any recovered material as a substance soluble in the solvent. It includes other material extracted by the solvent from an acidified sample (such as sulfur compounds, certain organic dyes, and chlorophyll) and not volatilized during the test. Collect a representative grab sample in a wide-mouth glass bottle. Then wash it with soap and rinse it with water. Finally it is rinsed with solvent to remove any residues that may be interfered with the analysis. Do not overfill the sample container and do not subdivide the sample in the laboratory. In case of delaying the analysis for more than 2 h, acidify to pH 2 or lower with either 1:1 HCl or 1:1 H_2SO_4 and refrigerate. Mark the sample bottle at the water meniscus just it is coming into the laboratory. If the sample is not previously acidified, acidify it with either 1:1 HCl or 1:1 H_2SO_4 to pH 2 or lower (generally, 5 ml is sufficient for one liter sample). Using liquid funnel, transfer sample to a separating funnel. Carefully rinse sample bottle with 30 ml extracting solvent (either 100% n-hexane, or solvent mixture) and add the solvent washings to a separating funnel. Shake it vigorously for 2 min. Let layers separate and drain the aqueous layer and small amount of organic layer into original sample container. Drain solvent layer through a funnel containing a filter paper solvent rinsed, into a clean, tared evaporating dish. If a clear solvent layer is not obtained and an emulsion of more than about 5 ml exists, drain emulsion and solvent layers into a glass centrifuge tube and centrifuge for 5 min at approximately 2400 rpm. Transfer centrifuged material to an appropriate separating funnel and drain solvent layer through a funnel with a filter paper into a clean, tared evaporating dish. Recombine aqueous layers and any remaining emulsion or solids in separating funnel. For samples with <5 ml of emulsion, drain only the clear solvent through a funnel with pre-moistened filter paper. Recombine aqueous layers and any remaining emulsion or solids in a separating funnel. Extract twice more with 30 ml solvent each time, but first rinse sample container with each solvent portion. Repeat centrifugation step if emulsion persists in subsequent extraction steps. Combine extracts in tared evaporating dish. Evaporate in a water bath at 85°C. Dry the dish with the oil layer in an oven for at least two hrs. Cool in desiccator for at least 30 min and weigh. To determine initial sample volume, fill sample bottle to mark with water and then pour water into a one liter graduated cylinder.

Nitrates: Nitrate is the highest oxidation state of nitrogen. It may occur in a water supply due to the oxidation of organic nitrogen to ammonia and then to nitrite and then to nitrate. The pollution by organic wastes is identifying by the presence of nitrate in water, since nitrate is presented in trace amounts in surface water, but may reach high levels in ground water. Nitrate at levels above 105 mg/l as N can cause an illness known as Methemoglobinemia in infants. Nitrogen is also one of the fertilizing essential elements of the growth of algae. Samples are collected in glass or plastic bottles with a volume of at least 100 ml. Samples should be analyzed as soon as possible after collection (within l hour), but may be stored at 4°C for up to 24 hours. For longer storage, up to 7 days, preserve the sample with 2 ml concentrated sulfuric acid per liter and store at 4°C. Note, when samples are preserved with acid, the result measurement will include nitrite and nitrate.

Biochemical Oxygen Demand (BOD_5): The BOD test is one of the most difficult tests to run because there are so many variables that affect the outcome. The BOD dilution water must be aerated and non-contaminated and the pH environment must be acceptable (6.5-7.5). The incubation temperature must be controlled and consistent (20 ± 1°C) and the microorganisms must be viable, must be acclimated to the waste material must be active and must be at acceptable concentration. The sample strength must be sufficient to cause a DO depletion of at least 2 mg/l but leave at least 1 mg/l residual and toxic materials must be absent. In summary, the BOD test must create ideal growing conditions, which will encourage microorganisms to affect and efficient utilize the digestible organic materials. In the BOD test, microorganisms are characterized with eating all the organics. In a BOD bottle, organics from a sample are added to dilution water containing nutrients, oxygen, and microorganisms, then capped and incubated at 20°C for 5 days. Initially the microorganism level is low, but the environmental growing conditions are excellent, so the microorganisms quickly enter the logarithmic growth reproduction phase and begin to consume the organics. The large number of microorganisms consumes the organics and because of consumes the oxygen. If the BOD bottle is capped tightly, no external oxygen can enter. The microorganisms can only utilize the oxygen initially available. Over the next couple of days, the microorganisms continue to digest the organics, reproduce logarithmically, and utilize oxygen rapidly. By the 4th day of the BOD test, the organics have begun to become limited, just barely able to maintain the current microorganism population, which is called stationary growth phase. The lack of food causes a slowing in the reproduction rate as well as a decrease for oxygen used. With the high microorganism population, the remaining organics are quickly consuming and the microorganism enters the endogenous phase. During the endogenous phase, the microorganisms utilize internal food reserves, and many die (endogenous phase). The amount of oxygen that has been consumed over the 5 days is proportional to the amount of organics BOD consumed. The BOD represents the total amount of food used up over the 5 days. Complete stabilization of a sample may require a period of incubation too long for practical purposes; therefore, five days have been determined as the standard incubation period.

Chemical Oxygen Demand (COD): The chemical oxygen demand (COD) test represents the amount of chemically digestible organics. COD measures all organics that were biochemically digestible as well as all the organics that can be digested by heat and sulfuric acid. Most permits specify BOD_5 because it gives a better representation of the organics that are affecting the receiving stream. COD has the advantage over BOD_5 in that the analysis is completed within a few hours whereas BOD_5 requires 5 days. The major drawback of the COD test is the presence of hazardous chemicals and toxic waste disposal. COD often is used as a measurement of pollutants in wastewater and natural waters. In domestic wastewater, BOD_5 represents about 50-60% of the COD. Presence of toxic materials or industrial wastewater can alter this ration. Like the BOD test, where, oxygen is used to oxidize the organics to carbon dioxide and water. However, instead of free dissolved oxygen, chemically bound oxygen in potassium dichromate $K_2Cr_2O_7$ is used to oxidize the organics. As the potassium dichromate is used up the Cr^{+3} ion is produced. The amount of dichromate used is proportional to the amount of organics present. Likewise, the amount of Cr^{+3} ion presence is proportional to the amount of organics digested.

Organics+$K_2Cr_2O_7$ Cr^{+3}

(Hexavalent Orange) (Trivalent Green)

Most labs use the Hach method to measure COD. This method uses test tubes with pre-measured amounts of potassium dichromate, sulfuric acid, and catalyst. Be sure the cap is on tight before mixing and mix just prior to place in the digestion reactor. The catalyst is silver. However, silver must be soluble and will precipitate if chlorides are present in the sample. To prevent silver precipitation, mercury is added to the reagents in the tube. The mercury will remove the chloride interference. Once the two hours refluxing period is finished, the remaining potassium dichromate is measured using the spectrophotometer. A reagent blank and standard must be prepared to zero the spectrophotometer. The sample is usually collected the same as BOD_5. The COD sample may be a composite or a grab sample. Because there are no microorganisms involved in this procedure, preservation is usually acidification using sulfuric acid to a pH below two. Refrigeration is acceptable if BOD_5 are also to be running on the sample. The sample holding time is seven days, much longer than the 24 hours allowed for the BOD_5 test. The sample volume is used for the COD test is 2.0 ml so measuring the sample volume is critical. Be sure to mix the sample well and homogenize if necessary. Pipet quickly to avoid settling errors. COD is running on industrial samples that may have high BOD_5. If the COD strength is greater than 1650 mg/l, the sample must be dilute. Make a 1:2 dilution by measuring 50 ml of sample and adding to 50 ml of deionized water, and then add 2.0 ml of the well-mixed dilution to the test tubes.

Electrochemical measurements: Mild steel is used as a working electrode with rod shape was subjected for XRF elemental analysis. The electrode is stacked in a Teflon holder with exposed area of 0.785 cm^2. Platinum wire and SCE electrodes were used as a counter and reference electrodes, respectively. The electrode surface was abraded successively by emery papers of different grades i.e., 150, 320, 400 and 600 then finely polished with a 2400 polishing paper to obtain uniform mirror like finish, degreased with acetone and washed with distilled water before the experiment. The open circuit potential of mild steel electrode was subjected for 30 min in the electrolyte (sediments) in free corrosion conditions before recording the polarization curves. Polarization curves were recorded at scan rates of 1 mV s^{-1} to study the effect of chemical treatment on the corrosion behavior of mild steel. The corrosion behavior is subjected to the untreated and treated wastewater to select the best mixture of chemical treatment on the corrosion results. All the polarization measurements were performed in the potential range from the cathodic to the anodic side (-0.8 to 0.1 V) and at room temperature. The corrosion tendency could be observed by measuring the open circuit potential and by using the Tafel extrapolation method for calculating the corrosion current density (i_{corr}).

pH Experiment	pH before adjustment	pH after adjustment	pH after chemical treatment	pH after Readjustment
Blank	5.57	------	------	------
Filtered blank	6.01	------	------	------
1	5.72	10.6	4.46	7.62
2	5.72	10.60	7.97	7.43
3	5.72	10.60	9.66	7.58
4	5.72	10.60	7.62	7.63
5	5.72	10.60	7.29	7.81
6	5.72	10.60	8.60	7.96

Table 1: pH measurements of wastewater samples before and after treatment.

Experimental Items	Blank	Filtered blank	Standard limit
Temperature	48	46	43
pH	5.57	6.01	Not less 6 and not more 9.5
Biological Oxygen Demand (BOD)	883	790	600 ppm
Chemical Oxygen Demand COD	1633	1560	1100 ppm
Total Suspended Solids (TSS)	926	580	800 ppm
Grease and Oils	170	120	100 ppm
Precipitated materials after 10 min	4.19	2.19	8 cm^3
Precipitated materials after 30 min	6.54	4.54	15 cm^3
Electric Conductivity	100000	100000	Unlimited
Total Dissolved Salts (TDS)	60000	60000	Unlimited
Alkalinity	762	762	Unlimited
Total Organic Carbon (TOC)	1321	1321	Unlimited
Chlorides (Cl^-)	3798.5	3798.5	Unlimited
-2 Nitrates NO3	7.5	7.5	Unlimited
-- Sulfates SO4	131264	131264	Unlimited
-2 Carbonates CO3	77	77	Unlimited
-Bicarbonates HCO3	317.2	317.2	Unlimited
Na^+	54788	54788	Unlimited
K^+	933.48	933.48	Unlimited
Mg^{+2}	960.1	960.1	Unlimited
Ca^{+2}	48033.89	48033.89	Unlimited

Table 2: Analysis of wastewater sample before treatment without and with filtration compared with the permissible limits.

Results and Discussion

Adjustment of pH for wastewater samples

Table 1 represents the pH values before the treatment and at the same time, the pH is measured for filtered sample to compare the wastewater constituents before and after filtration to overcome the effect of the filter paper. It is acknowledging that the chemical treatment is pH dependent reactions; the highest effect for coagulation is obtained at higher pH values. For the Egyptian standard for disposal to the municipal wastewater it is recommended to make the treated wastewater at pH from 6-8, so it is necessary to readjust the pH after treatment.

Chemical analysis of wastewater before treatment

Table 2 presents the chemical analysis of petroleum wastewater before treatment compared with the permissible limits for Egyptian standard.

Treatment of wastewater samples

Treatment of wastewater samples by separate coagulant: Three filtered wastewater samples were used in this experiment, firstly the pH were adjusted, then these samples became ready for the treatment with

Materials Experiment	FeCl3 g/100 ml	Al2(SO4)3 g/100 ml	Polyacrylamide g/100 ml
1	3 gm	----	-----
2	-----	3 gm	-----
3	-----	-----	1.5 gm

Table 3: Treatment of wastewater samples by separate coagulant.

Experimental Items	Treated sample with FeCl3	Treated sample with Al2(SO4)3	Treated sample with polymer	Standard limit
Temperature	33	33	33	43
pH	7.62	7.58	7.97	6-9.5
Biological Oxygen Demand (BOD)	520	430	470	600 ppm
Chemical Oxygen Demand COD	1070	890	980	1100 ppm
Total Suspended Solids (TSS)	352	190	220	800 ppm
Grease and Oils	47	38	45	100 ppm
Precipitated materials after 10 min	1.1	0.85	0.9	8 cm^3
Precipitated materials after 30 min	1.4	1.0	1.2	15 cm^3
Electric Conductivity	286.8	260.2	272	Unlimited
Total Dissolved Salts (TDS)	57000	47644	56887	Unlimited
Alkalinity	703	598	680	Unlimited
Total Organic Carbon (TOC)	877	844	874	Unlimited
Chlorides (Cl^-)	3672.5	2917	3392	Unlimited
-2 Nitrates NO3	6.8	6.3	6.7	Unlimited
-2 Sulfates SO4	122352	86160	89976	Unlimited
-2 Carbonates CO3	60	44	54	Unlimited
- Bicarbonates HCO3	148.8	142.8	144.4	Unlimited
Na^+	48033.89	37659.92	47012	Unlimited
K^+	787.9	735.93	783.96	Unlimited
Mg^{+2}	761.16	361.48	580	Unlimited
Ca^{+2}	46011.2	41241	42723	Unlimited

Table 4: Analysis of wastewater samples after treatment with 3 gm/100 ml of ferric chloride ($FeCl_3$), 3 gm/100 ml aluminum sulfate $Al_2(SO_4)_3$ and 1.5 gm/100 ml polyacrylamide compared with the standard limit.

Materials xperiment	FeCl3 g/100 ml	Al2(SO4)3 g/100 ml	Polyacrylamide g/100 ml
4	1.5 gm	1.5 gm	-----
5	1.5 gm	-----	0.75 gm
6	-----	1.5 gm	0.75 gm

Table 5: Treatment of wastewater samples by mixing the coagulant.

3 g/100 ml ferric chloride, 3 g/100 ml aluminum sulfate and 1.5 g/100 ml polyacrylamide. To 1500 ml of untreated wastewater the coagulant was added and stirred for about 30 min by using magnetic stirrer, then the samples were filtered and the filtrates were collected for the chemical analysis. Table 3 represents the treatment for the selected wastewater by separate coagulant to identify the best one. The chemical analysis after treatment were listed in Table 4a, 4b and 4c respectively. From the gained results, it is found that the chemical analysis depends on the coagulant and the chloride ion is decreased in case of aluminum sulfate treatment if it is compared with ferric chloride or polyacrylamide.

Treatment of wastewater samples by mixing the coagulant: In this experiment, the wastewater samples were treated by mixing the coagulants as mentioned in Table 5. After treatment the water samples were collected and subjected to the chemical analysis and the obtained results were listed in Table 6a, 6b and 6c respectively. The results show that varieties cations and anions values, these values proved that the polyacrylamide when mixed with aluminum sulfate gives lowest concentration of chloride ions. The mixture between the polymer and $Al_2(SO_4)_3$ decreases the salt content, which is preferable in decreasing the corrosion process.

Corrosion experiments

Corrosion behavior of single coagulant: Secondly, 100 ml from each one of six chemically treated samples after adjusting the pH is used to evaluate the corrosion behavior. The 100 ml filtered sample without treatment is used as a blank experiment. All experiments are adjusted as the permissible limit and the recommendation of Egyptian

	Treated sample with mix. FeCl3 and Al2(SO4)3	Treated sample with mix. FeCl3 and polymer	Treated sample with mix. Al2(SO4)3 and polymer	Standard limit
Temperature	33	33	33	43
pH	7.62	7.29	7.96	6 - 9.5
Biological Oxygen Demand (BOD)	460	465	410	600 ppm
Chemical Oxygen Demand (COD)	910	980	850	1100 ppm
Total Suspended Solids (TSS)	230	260	180	800 ppm
Grease and Oils	39	42	31	100 ppm
Precipitated materials after 10 min	0.8	0.88	0.75	8 cm^3
Precipitated materials after 30 min	1.1	1.2	0.9	15 cm^3
Electric Conductivity	263.8	268.2	254.8	Unlimited
Total Dissolved Salts (TDS)	56698	56998	47334	Unlimited
Alkalinity	672	682	586	Unlimited
Total Organic Carbon (TOC)	863	872	825	Unlimited
Chlorides (Cl^-)	2946.5	3337	2692	Unlimited
-2 Nitrates NO3	6.1	6.2	6.0	Unlimited
-2 Sulfates SO4	87816	110768	70160	Unlimited
-2 Carbonates CO3	47	53	35	Unlimited
- Bicarbonates HCO3	140.6	146.6	126.4	Unlimited
Na^+	36791.95	42659.92	30345	Unlimited
K^+	703.95	720.92	671.97	Unlimited
Mg^{+2}	732	744	276.52	Unlimited
Ca^{+2}	33636	36636	30468	Unlimited

Table 6: Analysis of wastewater samples before and after treatment by 1.5 gm/100 ml of ferric chloride ($FeCl_3$) and 1.5 gm/100 ml of aluminum sulfate $Al_2(SO_4)_3$, by 1.5 gm/100 ml of ferric chloride $FeCl_3$ and 0.75 gm/100 ml of polyacrylamide (P) and by 1.5 gm/100 ml of aluminum sulfate $Al_2(SO_4)_3$ and 0.75 gm/100 ml of polyacrylamide (P).

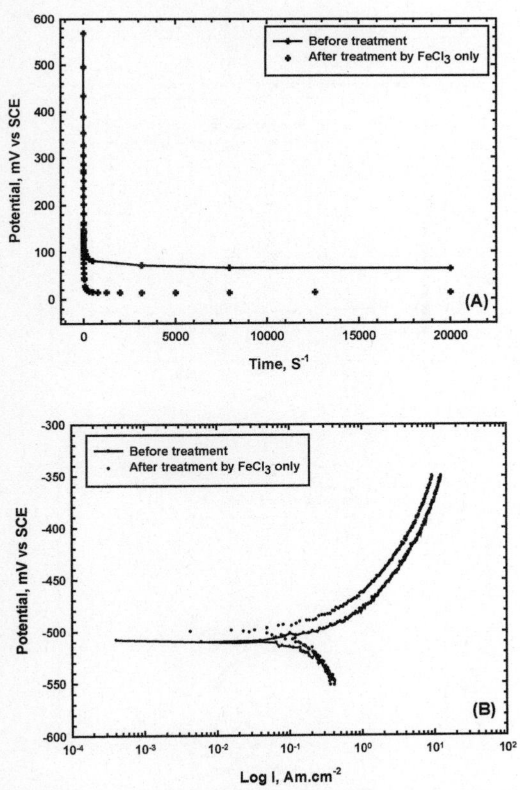

Figure 1: Corrosion curves before and after treatment by 3 gm of $FeCl_3$ (a) Open circuit potential curves (b) Potentiodynamic polarization curves.

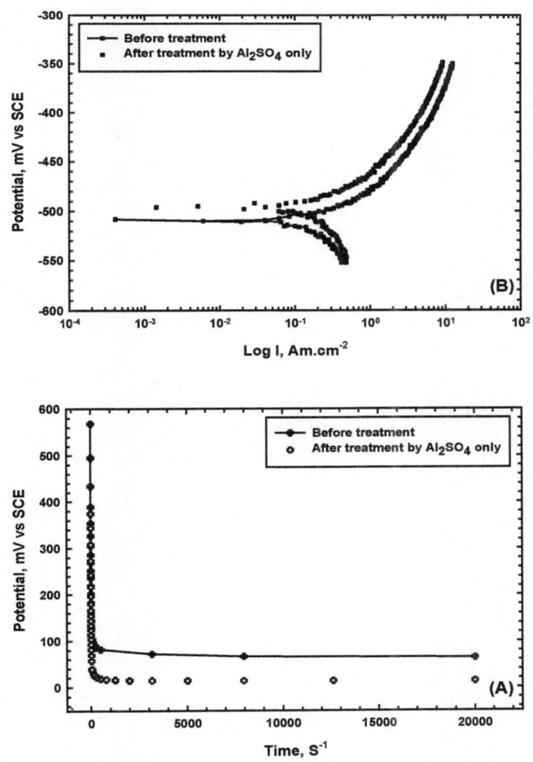

Figure 2: Corrosion curves before and after treatment by 3 gm of $Al_2(SO_4)_3$. (a) Open circuit potential curves (b) Potentiodynamic polarization curves.

Environmental Affairs before carrying out the corrosion experiments. Figure 1 represents the corrosion curves before and after the treatment by 3 gm of $FeCl_3$. Figure 1a is attributing with the relation between open circuit potential and the treated sample with ferric chloride alone. It is clearly noticed that the open circuit potential values are shifted to the lower positive values of potential that means that the medium became corrosive in treated sample. The open circuit potential is decreased from 61.54 mV to 15.67 mV this value prove that the corrosion efficiency reached about 74.5%. The Figure 1b prove that the current density increases continuously with increase the potential, which means the corrosion process is in progress. The results are also showed that the untreated sample is more safe for the steel pipeline although the higher salinity, which is proved that this water was injected with corrosion inhibitor during the production process. After treatment the sample became free of corrosion inhibitor so the corrosion rate increases although the decrease of TDS after treatment.

Figure 2 illustrates the corrosion curves before and after treatment by 3 gm of $Al_2(SO_4)_3$. The Figure 2a shows that the open circuit potential decreases from 61.54 mV to 11.82 mV with corrosion efficiency of about 80.8% and the current density values are also increased with increasing the potential as mentioned in Figure 2b.

The corrosion curves before and after treatment by 1.5 gm/100 ml of polyacrylamide are mentioned in Figure 3. The open circuit potential is shifted from 61.5 mV to 19.53 mV which indicates the increasing the corrosion process with about 68.3% as seen in Figure 3a and the same behavior is clearly noticed in potentiodynamic polarization curves as mentioned in Figure 3b.

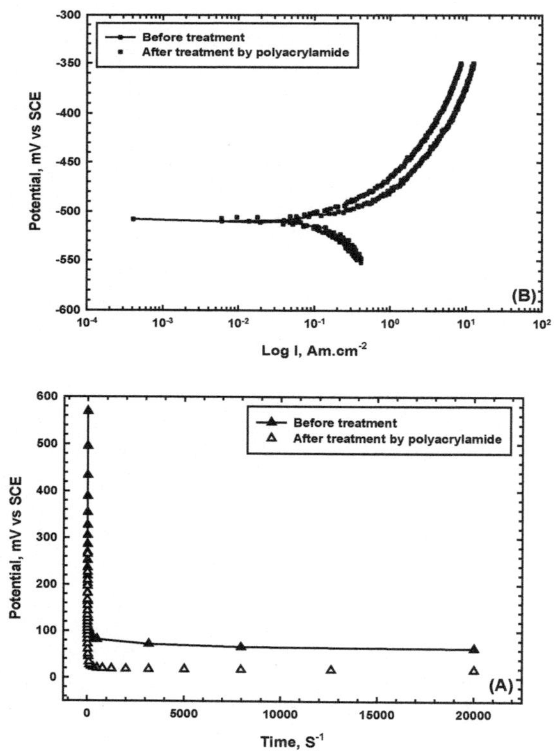

Figure 3: Corrosion curves before and after treatment by 1.5 gm/100 ml of polyacrylamide. (a) Open circuit potential curves (b) Potentiodynamic polarization curves.

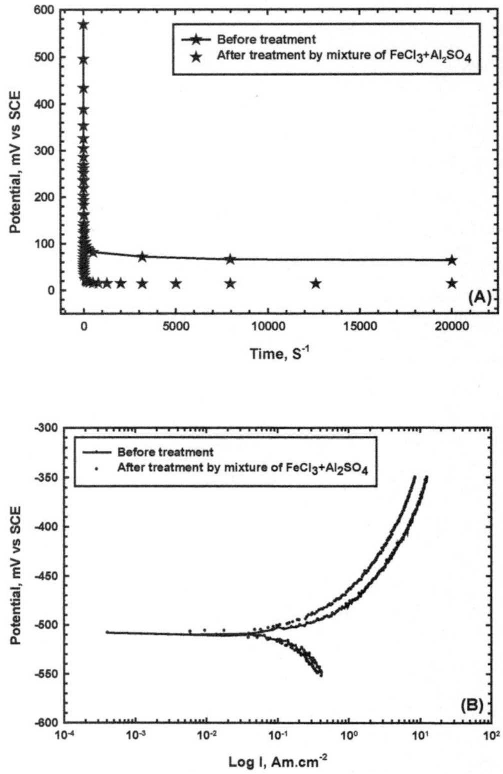

Figure 4: Corrosion curves before and after treatment by mixture of 1.5 gm/100 ml of $FeCl_3$ and 1.5 gm of $Al_2(SO_4)_3$.

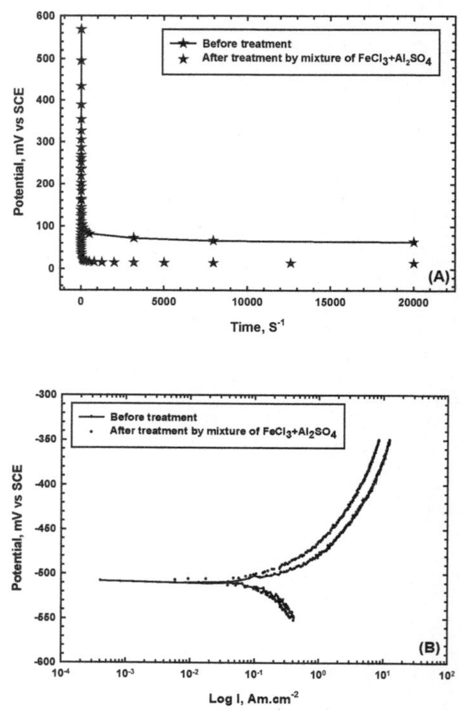

Figure 5: Corrosion curves before and after treatment by mixture of 1.5 gm/100 ml of $FeCl_3$ and 0.75 gm/100 ml of Polyacrylamide.

Corrosion behavior of mixed coagulant: The treated wastewater samples by mixed coagulant are subjected to corrosion behavior as mentioned in Figures 4-6.

The obtained open circuit potential values are 13.66 mV, 13.66 mV and 15.49 mV in case of the mixtures of $FeCl_3$+Al_2SO_4, Al_2SO_4+polyacrylamide and $FeCl_3$+polyacrylamide respectively. The corrosion process efficiencies are 77.80%, 77.80% and 74.8% respectively. These results proved that the best one against corrosion process is polyacrylamide, which register the minimum corrosion efficiency of about 68.30%. The results of open circuit potentials of treated samples with single coagulant comparing with the untreated sample are mentioned in Figure 7a and 7b and the results of corrosion efficiencies are totally mentioned in the collective (Figure 7c).

Conclusions

The chemically treated wastewater is considered the best method to get rid the most pollutants which exist in the petroleum wastewater. The chemical treatment has the capability to reduce the total suspended materials and so it decreases the total residual carbon and nitrogen by removing the most organic compounds. The chemical treatment did not meet the requirements for the permissible limit of heavy metals especially in the water which contain high concentrations of heavy metals. The chemical oxygen demand and biological oxygen demand are reflected the water quality and the chemical precipitation was succeed to meet the standard level for COD and BOD. The reduction efficiency was calculated for COD and BOD and it is found that the percentages of about 47.95% and 53.57% respectively. By comparing the water quality parameters of the chemically treated wastewater we can conclude that the treated wastewater with the mixture of aluminum sulfate and polyacrylamide meets the permissible limit of guidelines of water parameters. The corrosion process is depending on the chloride and sulfide ion concentrations so the reduction of these ions concentration will improve the corrosion resistance for the mild steel pipelines. It found that this mixture have the best corrosion results due to its capability to reduce the total dissolved salts which make reduction of chloride and sulfide ions content.

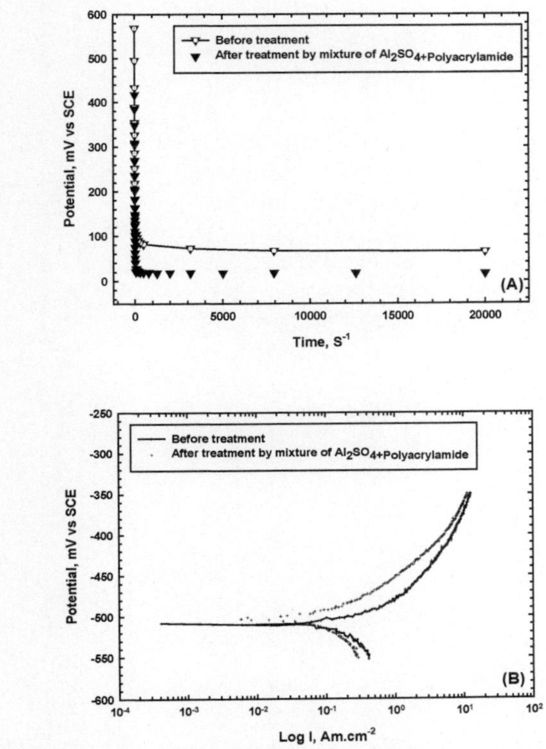

Figure 6: Corrosion curves before and after treatment by mixture of 1.5 gm/100 ml of $Al_2(SO_4)_3$ and 0.75 gm/100 ml of Polyacrylamide.

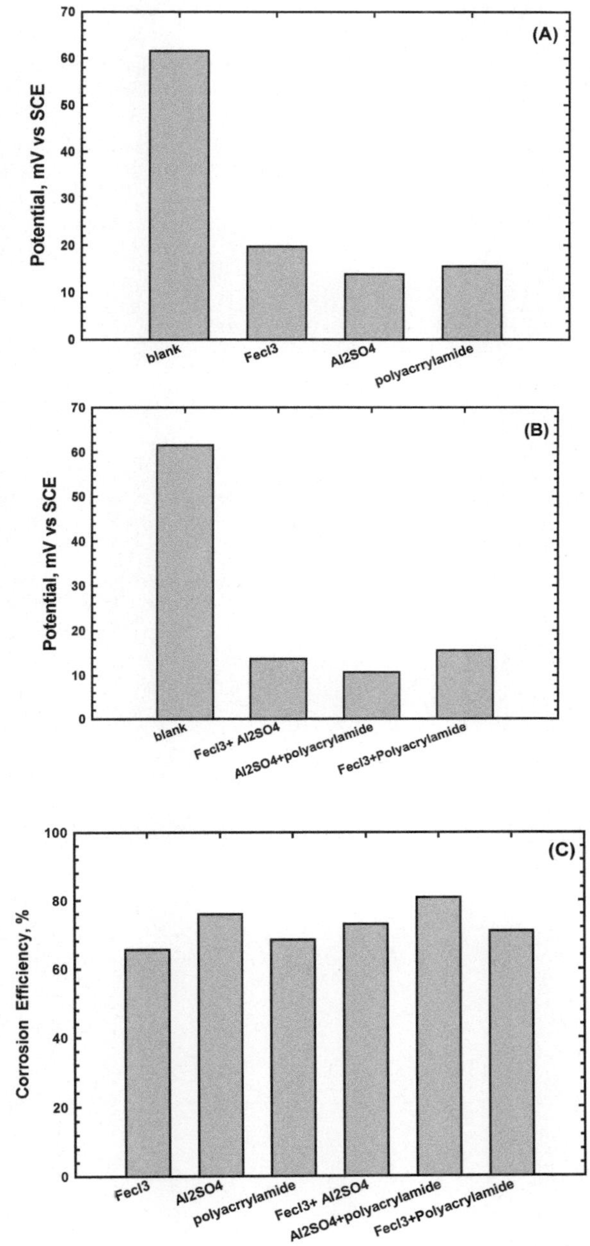

Figure 7: The effect of the treatment method on the calculated corrosion efficiency from the open circuit potential values of mild steel (a) Separate coagulant treatment (b) Mixed coagulant treatment (c) The corrosion efficiency.

References

1. Farajnezhad H, Parvin G (2012) Coagulation treatment of wastewater in petroleum industry using poly aluminum chloride and ferric chloride. IJRRAS 13: 306-310.
2. Pan JR, Huang CP, Chen SC, Chung YC (1999) Evaluation of a Modified Chitosan Biopolymer for Coagulation of Colloidal Particles. Colloids and Surfaces A 147: 359-364.
3. Koohestanian A, Hosseini M, Abbasian Z (2008) The Separation Method for Removing of Colloidal Particles from Raw Water. American-Eurasian J Agric & Environ Sci 4: 266-273.
4. Abu H, Mohd AT, Pei L, Noor ZZ (2009) Coagulation and flocculation treatment of wastewater in textile industry using chitosan. J Chem Nat Resour Eng 4: 43-53.
5. Amokrane A, Comel C, Veron J (1997) Landfill Leachates Pre-Treatment by Coagulation Flocculation. Water Res 31: 2775-2782.
6. Amudaa OS, Amoob IA (2007) Coagulation/Flocculation Process and Sludge Conditioning in Beverage Industrial Wastewater Treatment. J Hazard Mater 141: 778-783.
7. Guibal E, Roussy J (2007) Coagulation and Flocculation of Dye-Containing Solutions Using a Biopolymer Chitosan. React Funct Polym 67: 33-42.
8. Egyptian Environmental Affairs Agency and Egyptian Pollution Abatement Project (2002) Industrial wastewater treatment plants inspection procedures manual, pp: 4-13.
9. Ihsan H (2015) Review of wastewater treatment chemicals and organic chemistry alternatives for abattoir effluent. Australian Meat Processor Corporation.
10. Demirci S, Erdogan B, Ozcimder R (1998) Wastewater Treatment at the Petroleum Refinery, Kirikkale, Turkey Using Some Coagulants and Turkish Clays as Coagulant Aids. Water Res 32: 3495-3499.
11. Koh SM, Dixon JB (2001) Preparation and Application of Organo- Minerals as Sorbents of Phenol, Benzene and Toluene. Appl Clay Sci 18: 111-122.
12. Masschelein WJ, Rice RG (2002) Ultraviolet Light in Water and Wastewater Sanitation. Rice RG (ed.), Lewis Publishers: Boca Raton, Florida, USA, pp: 9-58.
13. Cheryan M, Rajagopalan N (1998) Membrane Processing of Oily, Streams, Wastewater Treatment and Waste Reduction. J Membr Sci 151: 13-28.
14. Chen X, Chen G, Yue PL (2000) Separation of Pollutants from Restaurant Wastewater by Electro coagulation. Sep Purf Technol 19: 65-76.
15. Stephenson R, Tennant B (2003) New Electro coagulation Process Treats Emulsified Oily Wastewater at Vancouver Shipyards. Environ Sci Eng Mag.
16. Al-Shaikhli S (2013) Optimization of coagulation/flocculation process using natural coagulants in raw water treatment. Master Thesis, The University of Southern Queensland, Toowoomba, QLD, Australia.
17. Bazrafshan E, Biglari H, Mostafapour FK (2012) Determination of Hydrophobic and Hydrophilic Fractions of Natural Organic Matter in Raw Water of Zahedan Water Treatment Plant. J Health Scope 1: 25-28.
18. Al-Mutairi NZ (2006) Coagulant toxicity and effectiveness in a slaughterhouse wastewater treatment plant. Ecotoxicol Environ Saf 65: 74-83.
19. Demirci S, Erdogan B, Ozcimder R (1998) Wastewater Treatment at the Petroleum Refinery, Kirikkale, Turkey Using Some Coagulants and Turkish Clays as Coagulant Aids. Water Res 32: 3495-3499.

11

Application of Nano-Fluids to Heat Transfer Enhancement in Double-Walled Reactor

Alireza Mohammadi Ghahdarijani, Faramarz Hormozi* and Ali Haghighi Asl

Faculty of Chemical, Gas and Petroleum Engineering, Semnan University, Semnan, Iran

Abstract

Efficiently cooling of a reactor for an exothermic reaction can lead the efficiency to be increased, while reduces the size of the rector. In the present research, aqueous Nano-fluid of Al_2O_3 and CuO are used as heat transfer media flowing inside the reactor jacket to improve the heat transfer performance of a reacting system. Experiments have been performed to investigate the role of Nano-fluid in cooling of an exothermic reaction. The experimental facility provides conditions to control the generated heat during the reaction. Convective heat transfer coefficient and pressure drop of alumina/water and CuO/water Nano-fluids were experimentally measured. Nano-fluids were using two-step method at mass concentrations of 0.1-0.4%. The reactor is a stainless steel-made annular type, which is cooled by Nano-fluid flowing inside the annular section. For longer stability of Nano-fluids, surfactant and sonication were implemented. Results showed that by increasing the concentration of Nano-fluid and Reynolds number, the heat transfer coefficient increased. For CuO Nano-fluid, rate of enhancement for the convective heat transfer coefficient was 13% at 0.3% in comparison with the base fluid. While for aluminum Oxide at wt. %=0.4, the enhancement rate was 11.7%. Utilizing the Nano-fluid instead of traditional coolants, (e.g., water) resulted in a small penalty for the pressure drop. Higher level of pressure drop was registered for alumina Nano-fluid at wt%=0.4, which was 12% more than that of reported for the distilled water?

Keywords: Convective heat transfer; Pressure drop; Double-walled reactor; Surfactant

Introduction

Cooling performance of a system plays an important role in many sectors such as refineries, power plants, petrochemical, nuclear reactors and process equipment. Therefore, in most of the industrial systems, heat should be transferred as an input energy to the system or should be dissipated from the system. For designing a reactor, thermal performance is a key parameter in terms of both operational and safety criteria. The major chemical reactions are exothermic or endothermic, thus during the reaction, the contents of the reactor may be heated or cooled. Overheating in the reactor can lead to the reactor failure or its control equipment. Most importantly, this can enhance the risk of explosion and leakage. In addition, if the reaction is endothermic, it can be cooled by the progress of the reaction, which leads the reactor to reduce the rate of the reaction even up to the freezing point. Accordingly, Temperature range inside the reactor should be kept in safety region. One of the most common methods to control the reactor temperature is to use a cooling/heating jacket or heat-exchanging coils. In jackets, a layer with higher or lower temperature is wrapped around the reactor to transfer the heat to the reactor indirectly. For such systems, poor thermal conductivity of conventional fluids such as oil, water and ethylene glycol is one of the major challenges of heat transfer fluids. One of the methods to overcome this challenge is to enhance the thermal conductivity and thermal features of coolants using engineered thermal fluids with higher thermal conductivity. For better understanding, following example is explained. Thermal conductivity of water is regarded as reasonable value, which equals to 0.6 W/m. K at temperature of 298 K. However, it is much lower than the thermal conductivity of common metal oxides such as copper, which is usually 700 times greater than water and 3000 times higher than oil. Therefore, a promising passive way to improve the transfer performance of traditional coolant is to use the Nano-fluids.

Nano-fluids are promising materials, which are comprised of the solid particles with nominal size of 0-100 nm dispersing into the traditional coolant. Nano-fluids have some minor issues such as sedimentation, fluid blockage, and erosive behavior in comparison with major challenges belonging to microfluidic working media. However, Nano-fluids are still in lab, although there are slender chances that such coolants can address most of the issues caused by conventional fluids due to the poor thermal conductivity. Following, there are some benefits of Nano-fluid outweighing its drawbacks:

- Higher thermal conductivity in comparison with those of predicted by macroscopic models
- Better stability in comparison with micro-fluids
- less damage caused due to the increase in pressure drop and erosion of pipes

Nano-fluids are extensively used in order to improve the heat transfer performance of heat exchanging media. Some of these efforts have been represented in the following section.

Literature Review

Zarifi et al. [1] used Al_2O_3/water and TiO_2/water in different volume fractions of nanoparticles as cooling fluid in VVER-1000 reactor. Their results showed that with increasing the volume fraction of nanoparticles, the temperature of output cooling fluid increases. Faulkner et al. [2] investigated the fully developed convective heat

***Corresponding author:** Faramarz Hormozi, Faculty of Chemical, Gas and Petroleum Engineering, Semnan University, Semnan, Iran
E-mail: fhormozi@semnan.ac.ir

transfer with Nano-fluids of aqueous carbon nanotube. They observed the significant increase of heat transfer coefficient with increasing the Reynolds number. They also pointed out that in all cases, Nano-fluid was better than base fluid, but Nano-fluids with less volume concentration (1.1%) had better thermal performance rather than Nano-fluids with higher concentration (4.4% and 2.2%). Such findings can be referred to the increase of the interaction between nanoparticles. For increasing the production rate of gasoline in the process of Fischer-Tropsch (FT) synthesis, since the improvement of heat transfer can lead to increase the speed of exothermic reaction, Jayhooni et al. proposed to use the Nano-fluids. Studying the effect of Nano-fluid as coolant in the shell and reaction in the pipe of a tube exchanger showed that increasing the concentration of Nano-fluids could lead to the better heat transfer and higher gasoline production. Xuan and Li [3] performed an experimental study with copper Nano-fluids in a pipe with constant heat flux. They concluded that Nusselt number increased by 12% in the concentration volume of 1%. Chun et al. [4] investigated the heat transfer of Nano-fluids (prepared from oils and solvents and aluminum oxide nanoparticles) at laminar flow condition in a heat exchanger. They found that Nano-fluids containing nanoparticles equivalent to 0.5% by volume have thermal conductivity 10% more than the base fluid. Nazifi et al. [5] investigated the effects of Al_2O_3-water Nano-fluid on thermos-hydraulic parameters to prepare the ground for the application of Nano-fluids in research reactors. Their results showed that use of aqueous Nano-fluids with volume fraction 1%, can enhance the heat transfer coefficient about 4% and pressure drop about 3% approximately. In this paper, application of Nano-fluids in a double-walled reactor is investigated by studying the effects of concentration and flow rate of two copper oxide and aluminum oxide Nano-fluids on the thermal performance and cooling capacity of Nano-fluids. In addition, special emphasis is taken to the pressure drop as a challenging drawback of Nano-fluids.

Experimental Setup

Test rig

Figure 1 schematically demonstrates the experimental setup used in this research. This facility can measure the values of temperature, pressure and flow of fluid in a closed loop to determine the characteristics of heat transfer and pressure drop of Nano-fluid. Fluid enters to the loop from the reservoir, while its conditions (temperature and pressure and flow rate) is measured through the trajectory line of the fluid and then returns back to the reservoir while its temperature and pressure set again to the initial values. The section of fluid flow consists of a storage tank and centrifugal pump (model PKm60) with two valves for precisely regulating the flows, which of them is located on the path of main flow and another one is located on the return path as well. To measure the volumetric flow rate, a turbine flow meter (Model VISION 2000) connecting to a MP5W indicator is used. The Nano-fluid enters to the reactor, heated up and then enters again into the heat exchanger. In heat exchanger, Nano-fluid is cooled by a cooling media flowing inside the heat exchanger. The main aim of using Nano-fluid is to cool the center of the reactor. A detailed schematic diagram of the experimental facility is shown in Figure 2 Different type K thermocouples (connected to indicator Autonics TC4Y) are mounted at different parts of the tank, entrance and outlet of reservoir and the reactor in order to measure the average temperature value at any time during experiments. In order to measure the pressure in the input and output of reactor, two calibrated pressure transmitters (PSCH0.5 SENSYS, VCIA) are implemented as well.

Preparation of nano-fluid

Properties and behaviors of Nano-fluids depend to the type of the base fluid and nanoparticles, size and shape of particulate matter and the quality of particle dispersion within the base fluid. In the present work, CuO nanoparticles with mean size of 40 nm and Al_2O_3 with size of 20 nm are purchased from Research Nanomaterial's, USA. These particles are added to the base fluid to produce Nano-fluid suspensions. Transmission electron microscopy, (TEM) images of nanoparticles was also provided in order to ensure about the size, morphology and represented in Figures 3 and 4. Noticeably, to change the surface properties of suspended particles crack the agglomeration and cluster structures, the ultrasonic vibrator with a surfactant is used to create a stable Nano-fluid. For this purpose, first Sodium dodecyl sulfate, (SDS) is added into the water at concentration of 0.1%, and then it is stirred uniformly. Nano-fluids then are prepared at mass concentrations of 0.1, 0.2, 0.3 and 0.4. For the next step, an ultrasonic is used (20 kHz/400 W), therefore, stable and Nano-fluids are prepared.

Data reduction

In order to investigate the influence of flow rate and Nano-fluid concentration on the heat transfer and Nusselt number, it is necessary to estimate the thermos-physical properties of Nano-fluids accurately. Following equations are used for calculating the thermal properties of Nano-fluids:

$$\rho_{nf} = (1-\phi)\rho_f + \phi\rho_p \quad (1)$$

Figure 1: Image of experimental setup.

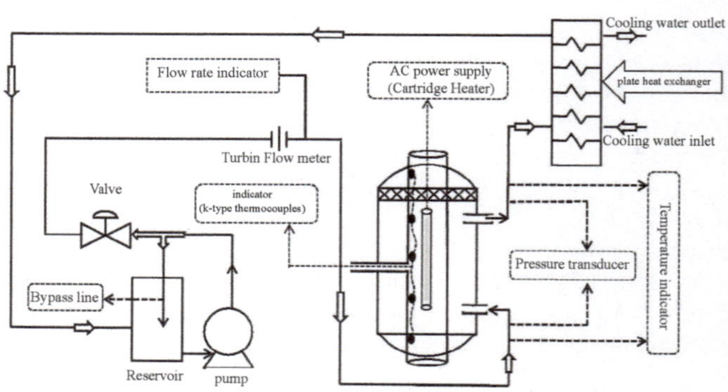

Figure 2: Schematic diagram of experimental setup.

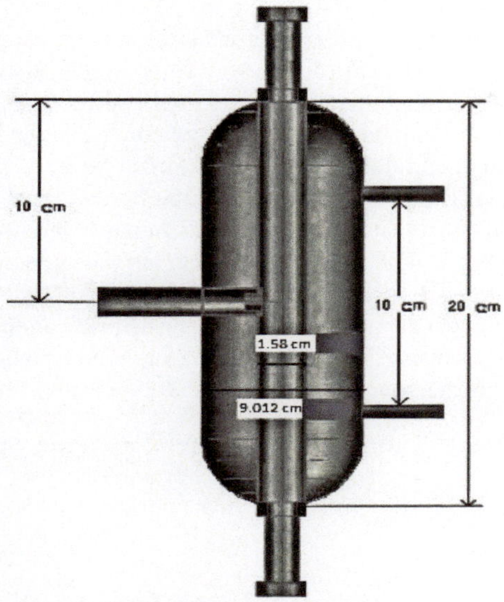

Figure 3: Schematic diagram of the test section.

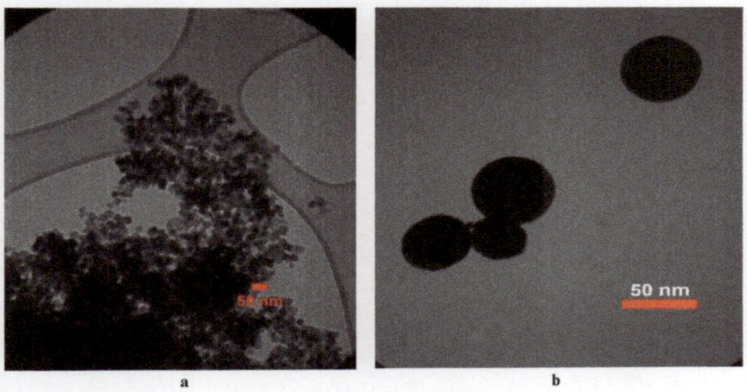

Figure 4: (a) TEM image of Al_2O_3 nanoparticles, **(b)** TEM image of CuO nanoparticles.

$$\left(\rho C_p\right)_{nf} = (1-\phi)\left(\rho C_p\right)_f + \phi\left(\rho C_p\right)_p \quad (2)$$

Corcione [6] has performed studies on dynamic viscosity of Nano-fluids. By using the experimental data available in the literature, he proposed a correlation for dynamic viscosity of Nano-fluids as represented in Equation (3).

$$\frac{\mu_{eff}}{\mu_f} = \frac{1}{1-34.87\left(\frac{d_p}{d_f}\right)^{-0.3} \varnothing^{1.03}} \quad (3)$$

d_f is the average diameter of base fluid molecules, which can be obtained from Equation (4):

$$d_f = 0.1\left(\frac{6M}{N\pi\rho_{f_0}}\right)^{1/3} \quad (4)$$

In this equation, M is the molecular weight of base fluid; N is the Avogadro's number and ρ_{fo} is the density of base fluid at the temperature (T_0=293K). For calculating the thermal conductivity of Nano-fluids, following equation was proposed by Corcione [6-11]:

$$\frac{k_{nf}}{k_f} = 1 + 4.4 Re^{0.4} Pr^{0.66}\left(\frac{T}{T_{fr}}\right)^{10}\left(\frac{k_p}{k_f}\right)^{0.03} \varphi^{0.66} \quad (5)$$

Where P_r is the Prandtl number, T is the temperature of Nano-fluid, T_{fr} is the freezing temperature of base fluid, k_p is the conductive heat transfer coefficient of Nano-particles and Re is the Reynolds number of nanoparticles, which is defined as equation. (6):

$$Re = \frac{2\rho_f k_b T}{\pi\mu_f^2 d_p} \quad (6)$$

Where, k_b is the Boltzmann constant.

It is worth saying that all properties were calculated at the Nano-fluid bulk temperature. The average temperature of Nano-fluid can be obtained from the inlet/outlet thermocouple's reading values. The thermocouples that are installed at the interior wall of reactor can also provide the average internal temperature of reactor. The amount of produced heat in the reactor is calculated by means of Joule's effect, which is the product of voltage and current (see equation. (7)):

$$Q = VI \quad (7)$$

Considering the average temperature of tank, Nano-fluid and the resistor as well (thickness of 2.77 mm), the relation between quantities of heat transfer from the center of the reactor and resistance of chamber and can be obtained using equation. (8):

$$Q = \left\{\overline{T}_r - \overline{T}_{nf}\right\} / \left\{\frac{Ln\left(\frac{r_o}{r_i}\right)}{2\pi k_B L} + \frac{1}{h_o A_o}\right\} \quad (8)$$

By combining equations of (7) and (8), the heat transfer coefficient can be calculated from the following correlation:

$$h_o = \frac{1}{A_o}\left[1 / \left\{\frac{\overline{T}_r - \overline{T}_{nf}}{\gamma VI} - \frac{Ln\left(\frac{r_o}{r_i}\right)}{2\pi kL}\right\}\right] \quad (9)$$

As fluid flows from the annular space between two cylinders, the hydraulic diameter is calculated from the following correlation:

$$D_e = 4 \times \frac{\text{Free cross-sectional area for current passing}}{\text{Wet environment by flow}} = \frac{4\pi\left(D_3^2 - D_2^2\right)}{4\pi\left(D_3 + D_2\right)} = D_3 - D_2 \quad (10)$$

Noticeably, D_e cannot be used to calculate the heat transfer coefficient; because the environment of shell is not considered as heat transfer area (wet area) and should be omitted from the calculations. So following equation is used for estimating the hydraulic diameter:

$$D_e = \frac{4\pi\left(D_3^2 - D_2^2\right)}{4\pi D_2} = \frac{D_3^2 - D_2^2}{D_2} \quad (11)$$

Where D_3 is the inner diameter of wall, D_2 is the external diameter of reactor tank. The amount of Nusslet number is calculated by the hydraulic diameter obtained from equation 11 and by using the following equation.

$$Nu = \frac{h_o D_e}{k} \quad (12)$$

In the present study, different parameters are measured using in-lab instruments with different accuracies. The measurement errors of these parameters may influence on the general accuracy of the experimental results. Specifications of the used devices are presented in Tables 1 and 2.

Physical properties	Diameter of particles (nm)	Specific surface area (m^2/g)	Density (kg/m^3)	Thermal conductivity (W/mK)	Specific Heat
CuO	40	20	6400	69	535.6
Al_2O_3	20	138	3890	46	880

Table 1: Thermo-physical properties of Nano-fluids.

Instrument	Measure	Model	Range	Accuracy
Turbine Flow meter	Flow rate	Vision 2000	0.5-35 l/min	± 0.15 l/min
thermocouples	Temperature	Omega K-type	-73 to 260°C	± 0.2°C
Digital Panel Meter	Current	Ziegler Delta power	0 to 6 A	± 0.001 A
Pressure transmitters	Local pressure	PSCH0.05BCIA	0 to 50 kpa	± 0.01 kpa

Table 2: The model, range, and accuracy of the measuring instruments.

It is necessary to find the influence of the main parameters on the uncertainty of heat transfer coefficient. The Mc- Klintock equation is the implemented to evaluate the uncertainty of the system as equation 13 [12].

$$E_M = \mp\left[\left(\frac{\partial M}{\partial X_1}E_1\right)^2 + \left(\frac{\partial M}{\partial X_2}E_2\right)^2 + \ldots + \left(\frac{\partial M}{\partial X_n}E_n\right)^2\right]^{1/2} \qquad (13)$$

Where X_i is the parameter, M is the calculated from Desired parameter and E_i is measurement error and E_M shows the maximum obtained error in estimating a quantity. Therefore, experimental error of the convective heat transfer coefficient according to Equation 9 and 17 can be obtained as follows:

$$E_h = \mp\left[\left(\frac{\partial h}{\partial \Delta T}E_{\Delta T}\right)^2 + \left(\frac{\partial h}{\partial r_o}E_{r_o}\right)^2 + \left(\frac{\partial h}{\partial r_i}E_{r_i}\right)^2 + \left(\frac{\partial h}{\partial V}E_V\right)^2 + \left(\frac{\partial h}{\partial I}E_I\right)^2\right]^{1/2} \qquad (14)$$

The measurement error of each parameter can be calculated by dividing the maximum error on the measured value. The mean uncertainties of the measured parameters are presented in Table 3 based on the uncertainty equations and Table 2. As can be seen, deviation for heat transfer coefficient is reasonable. Therefore, data represented in this work are reliable.

Results and Discussion

Validation of experimental data

Initial experiments were performed using distilled water to determine the accuracy and reliability of results. In fact, distilled water is regarded as a reference for comparing the results of Nano-fluids.

The equation, which is used for the chamber of a tank to calculate the Nusselt number, is in the form of Eqation. (15) [10]:

$$\mathrm{Nu} = 1.02 Re^{0.45} Pr^{0.33} \left(\frac{D_e}{H}\right)^{0.4} \left(\frac{D_1}{D_2}\right)^{0.8} Gr^{0.05} \qquad (15)$$

Where H is the height of thermal jacket, D_1 is the inner diameter of reactor tank and Gr is the Grashof number, which is calculated by equation (16).

$$Gr = \frac{D_e^3 \rho_f^2 g \beta \left|T_r - T_f\right|}{\mu_f^2} \qquad (16)$$

The experimental results were compared to those of theoretically obtained and represented in Figure 5 for the discharge period of reactor operation. Experimental Nusselt number showed that the obtained results are within the deviation of 1.38-8%, which is a reasonable value.

Heat transfer of nano-fluid

Heat transfer coefficient depends on the thermo-physical properties of the fluid and geometrical characteristics of the system. These quantities include fluid thermal features such as thermal conductivity, specific heat, density and viscosity and system characteristics such as diameter, length and flow speed as well. As can be seen in Figures 5 and 6, by increasing Reynolds number, the heat transfer coefficient increases and clearly, at a given Reynolds number, the heat transfer coefficient of Nano-fluid is relatively higher than water. Because adding the nanoparticles into base fluid changes the flow hydrodynamic properties subsequently, this can increase the thermal conductivity and increases the Brownian motion of particles inside the trajectory line and the heat exchanging media. Such behavior may lead to the local

Parameter	Uncertainty (%)
Convective heat transfer coefficient	3.62
Reynolds number	4.61
Prandtl number	1.43
Nusselt number	3.34

Table 3: The main parameters and their uncertainties.

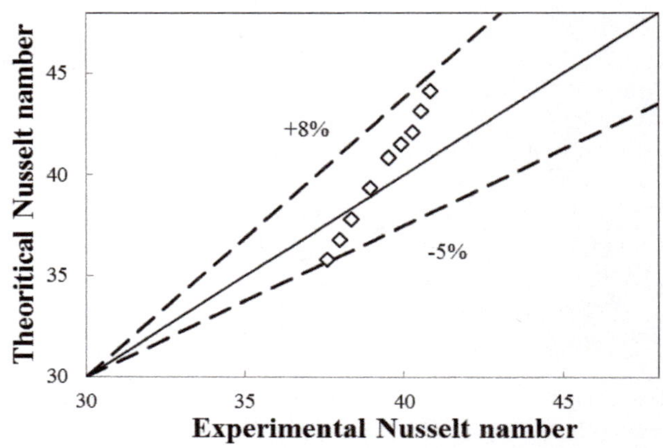

Figure 5: The comparison between experimental Nusselt number and those of predicted by Eq. (15).

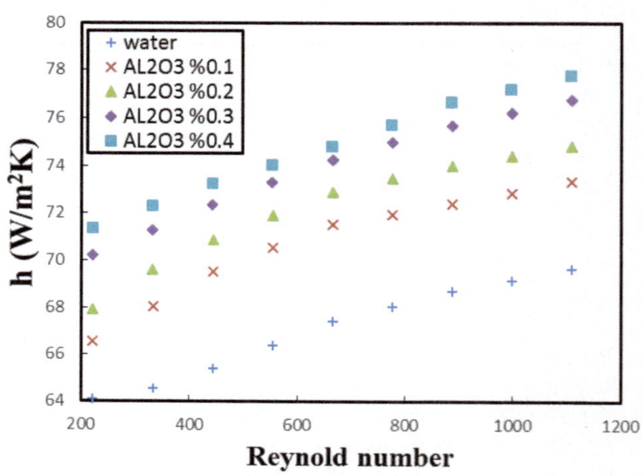

Figure 6: Convective heat transfer coefficient of aluminum oxide nano-fluid versus Reynolds number at different mass concentrations.

turbulent behavior and a slight increase in convective heat transfer coefficient. For the fluid flow with higher discharge rate, the effect of distribution of nanoparticles in the fluid is improved and this may disrupt the formed thermal boundary layer towards the reduction of thickness, which consequently leads to a higher temperature gradient near the wall. All can lead to the increase in heat transfer coefficient and Nusselt number.

For the copper oxide, as can be seen in Figures 7 and 8, the heat transfer coefficient at mass concentration of 0.4 is higher than base fluid; however, it is lower when comparing to mass concentration of 0.3. In fact, the maximum increase of the heat transfer coefficient can be seen for 0.3%. Note that, although with increasing the nanoparticles concentration, thermal conductivity of Nano-fluid can be increased, other parameters may act in a way to reduce the thermal performance of system. Increase in the concentration of nanoparticles results in the enhancement of viscosity and density of Nano-fluid. This can have significant influence on the enhancement of thickness of thermal boundary layer. The Thermophoresis phenomenon is also responsible for the enhancement of heat transfer since it facilitates the heat transfer process by moving nanoparticles from the hot walls towards the cold side. Due to the higher density of copper oxide nanoparticles comparing to aluminum oxide, this phenomenon is less effective for the Nano-fluids of copper oxide-water. Buoyancy force is a function of the specific gravity lead to produce small secondary vortices near the walls, which can be negative parameter for reducing the effective flow rate through the adjacent wall and subsequently leads the heat transfer coefficient to be decreased.

With increasing concentration, the clustering phenomenon occurs and particles may form agglomeration inside the system. Such agglomeration can increase the possibility of sedimentation and decreases the thermal conductivity. It means that there is an optimal value for the concentration of nanoparticles. Since the copper oxide nanoparticles induce larger clustering phenomenon and vast agglomeration rather than aluminum oxide, rate of increase in thermal conductivity coefficient is higher, but its optimum concentration set to the lower concentration value. Figure 9 shows the changes of Nusselt number according to Reynolds number at different mass concentration for copper oxide Nano-fluids.

In Figure 10, the heat transfer coefficient of two Nano-fluids is compared at different weight concentrations. At low concentrations,

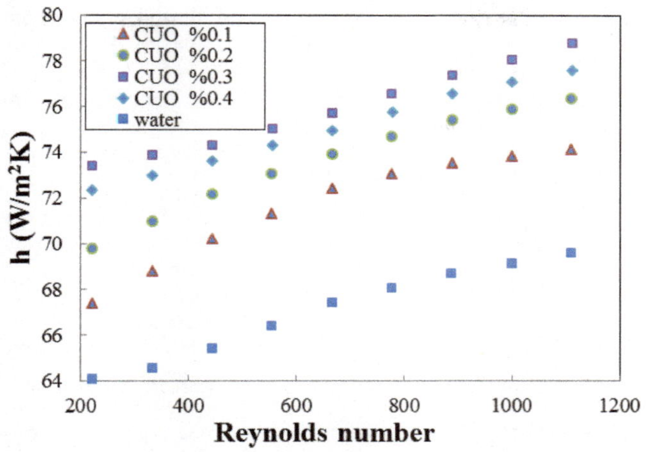

Figure 8: Convective heat transfer coefficient of Cupric Oxide nano-fluid versus Reynolds number at different mass concentrations.

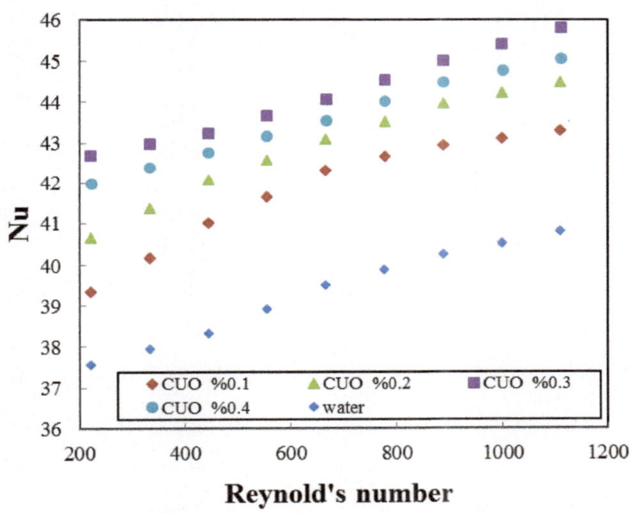

Figure 9: The effect of changes in nanoparticles weight fraction and the Reynolds number on Nusselt number to the aluminum oxide.

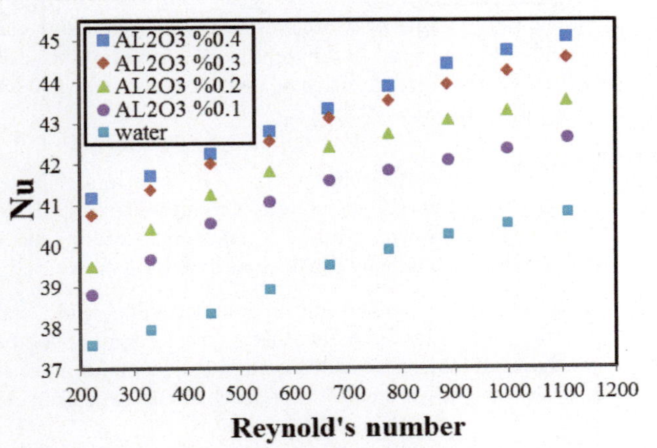

Figure 7: The effect of nano-fluid mass concentration and the Reynolds number on Nusselt number to the aluminum oxide.

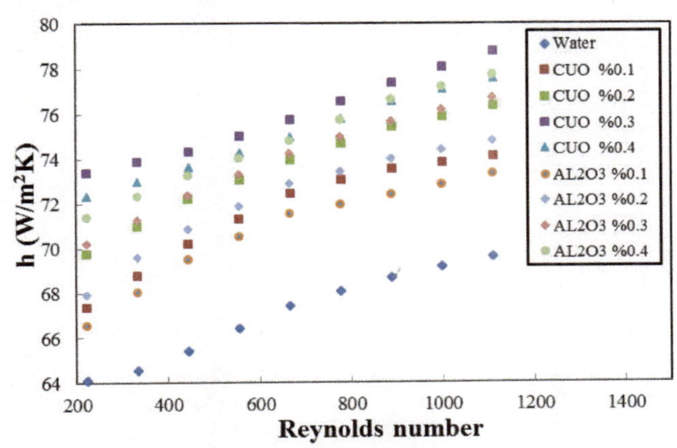

Figure 10: The comparison between heat transfer coefficient of aluminum and copper oxide Nano-fluid versus Reynolds number at different mass concentrations.

due to the higher thermal conductivity of copper oxide, the higher heat transfer coefficient is registered, while based on the above-mentioned point, there is an optimal concentration of copper oxide (0.3% by weight), therefore after this point. The heat transfer coefficient of copper oxide decreases such that the heat transfer coefficient of two Nano-fluids are overlapped at wt. %=0.4.

The effect of nanoparticles on pressure drop

When nanoparticles are added to the base fluid, dynamic viscosity of Nano-fluids increases. Since the pressure drop strongly depends on the viscosity, the higher values for pressure drop are reported. As can be seen in Figures 11 and 12, with increasing the mass concentration of Nano-particles, the viscosity and subsequently, the pressure drop value considerably increases such that at mass concentration of 0.4%, the highest value for pressure drop can be seen. The amount of penalty for the pressure drop (enhancement in pressure drop value) for aluminum oxide Nano-fluid is 12%.

In Figure 13, the changes of pressure drop are shown in terms of mass concentration in volumetric discharge rate of 4 lit/min. By increasing the mass concentration, the Nano-fluid viscosity increases and post-drop pressure represents the increasing upward trend. Because the viscosity of Nano-fluid has a reverse relation with the size of nanoparticles, as a result higher viscosity for aluminum oxide Nano-fluid (particle size=20 nm) is registered, when comparing to the copper oxide (particle size=40 nm). This also leads the higher penalty for pressure drop when aluminum oxide is used in the system.

Thermal performance

For the sake of validation in thermal performance assessment of Nano-fluids, the factor of η is defined as follows [11]:

$$\eta = \left(\frac{Nu_{nf}}{Nu_f}\right)\left(\frac{f_f}{f_{nf}}\right)^{\frac{1}{3}} \quad (16)$$

And, f is the Friction factor, which can be calculated using Equation (17) as follows:

$$f = \frac{\Delta P}{\frac{1}{2}\rho u^2\left(\frac{L}{D_e}\right)} \quad (17)$$

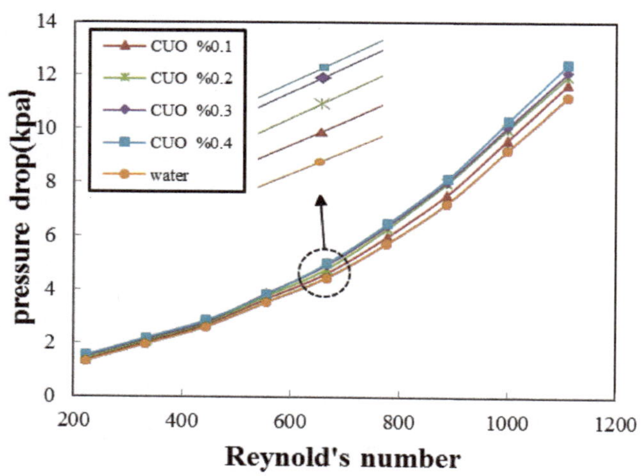

Figure 11: Variations of pressure drop for water and copper oxide Nano-fluid with different mass concentrations versus Reynolds number.

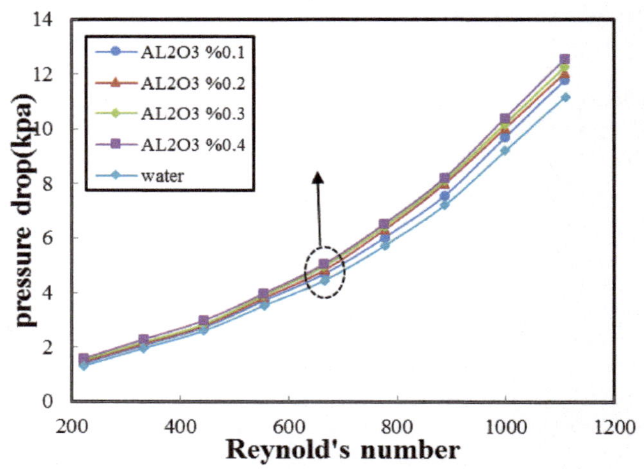

Figure 12: Variation of pressure drop for water and alumina oxide Nano-fluid with different mass concentrations versus Reynolds number.

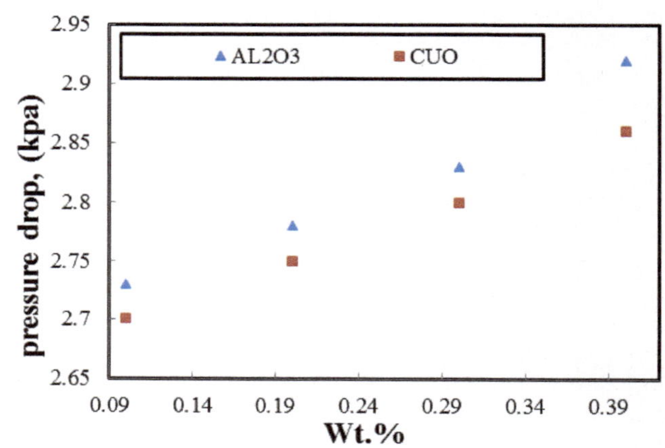

Figure 13: Variation of pressure drop in 4 LPM discharge flow rate.

Variations of thermal performance factor of reactor for the water-CuO and water-Al_2O_3 Nano-fluids are depicted in Figure 14 (A-B). As can be seen, the thermal performance at studied range of Nano-fluid concentration and Reynolds number is higher than unity. In Figure 15, the Thermal performance of two nano-fluids of copper oxide and aluminum oxide are compared in different mass concentrations. The maximum Thermal performance factor related to copper oxide in 0.3% concentration at Re=444 was 1.102.

Conclusions

- The thermal performance of two different Nano-fluids in cooling of a reactor with exothermic reaction is experimentally investigated and following conclusions have been drawn:
- Based on results, two Nano-fluids of copper oxide and aluminum oxide showed higher thermal performance over the base fluid and increase of concentration and Reynolds number led to the increase in the heat transfer coefficient and pressure drop.
- By increasing the mass concentration of aluminum oxide Nano-fluid, the heat transfer coefficient increases up to concentration

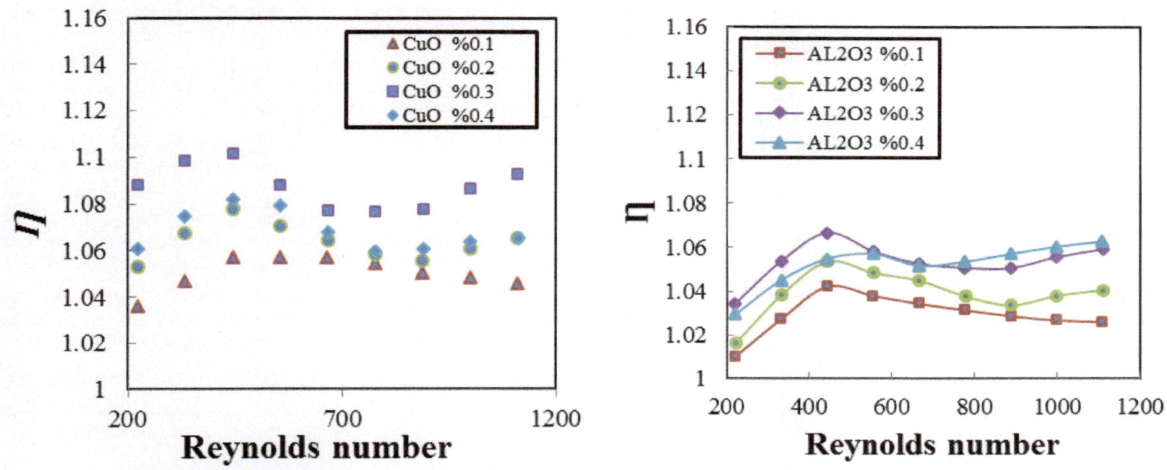

Figure 14: Variations of thermal performance factor of reactor for the (A): water-Cuo (B): water-Al_2O_3 Nanofluids.

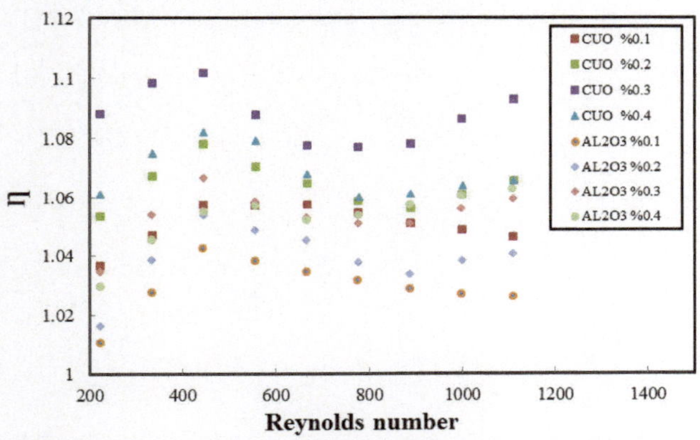

Figure 15: The comparison between thermal performance factor of aluminum and copper Oxide Nano fluid versus Reynolds number in different mass concentrations.

of 0.4%. However, for copper oxide, heat transfer coefficient increased up to concentration of 0.3% and gradually decreased due to the anomalous increase in viscosity and density, which had a negative impact on thermal performance of Nano-fluids. Therefore, there is an optimal concentration for Nano-fluids.

- The maximum heat transfer coefficient for copper oxide was registered at concentration of 0.3% by weight, which reports the 13% enhancement in heat transfer coefficient. This value was 11.7% for alumina Nano-fluid and at concentration of 0.4% by weight.
- Adding the nanoparticles into the base fluid increased the pressure drop as maximum pressure drop was seen at 0.4% by weight for alumina Nano-fluid. The penalty for the pressure drop was 12% higher when comparing to the base fluid.
- A rough comparison between pressure drops of two Nano-fluids was performed. Results demonstrated that in a given volumetric flow rate, the higher pressure drop of Nano-fluid for aluminum oxide was obtained, because particles of aluminum oxide (20 nm) was smaller than copper oxide (40 nm), which caused higher viscosity for aluminum oxide Nano-fluid and subsequently, pressure drop was higher for the aluminum oxide Nano-fluids.

Acknowledgements

Authors of this work appreciate Semnan University for their scientific and financial supports.

Nomenclature

A_o	Outer surface of the inner cylinder (m^2)
C_P	Heat capacity (W/mk)
d_h	Hydraulic diameter (m)
d_p	Nanoparticles Diameter (m)
f	Friction factor
h	convective heat transfer coefficient (W/m^2k)
I	Current strength (A)
K	Thermal conductivity (W/mk)
LPM	liter per minute
Nu	Nusselt number
P	Pressure ($kgm^{-1}S^{-2}$)
Pr	Prandtl number
Re	Reynolds number
r_o	Outer diameter of the central cylindrical (m)
T	Temperature(k)
u	Velocity(m/s)
V	Voltage (v)
Greek symbols	
μ	Dynamic viscosity($kgm^{-1}s^{-1}$)
ρ	Density (kgm^{-3})
φ	nanoparticles volume fraction
γ	Heat dissipation factor
η	thermal performance factor
Subscripts	
f	Base fluid
i	Inner
nf	Nano-fluid
o	outer
p	Particles

References

1. Zarifi E, Jahanfarnia G, Veysi F (2013) Thermal–hydraulic modeling of nano-fluids as the coolant in VVER-1000 reactor core by the porous media approach. Annals of Nuclear Energy 51: 203-212.

2. Faulkner DJ, Rector DR, Davidson JJ, Shekarriz R (2004) Enhanced heat transfer through the use of nano-fluids in forced convection. Paper IMECE2004-62147, presented at the 2004 ASME International Mechanical Engineering Congress and RD&D Expo, Anaheim, CA, USA, pp: 13-19.

3. Jayhooni SMH, Mirvakili A, Rahimpour MR (2012) Nano-fluid concept for enhancement of hydrogen utilization and gasoline production in fixed bed reactor FischereTropsch synthesis of GTL (gas to liquid) technology. Journal of Natural Gas Science and Engineering 9: 172-183.

4. Xuan Y, Li Q (2003) Investigation on convective Heat transfer and flow features a Nano-fluids. J Heat transfer 125: 151-155.

5. Xuan Y, Li Q (2000) Heat transfer enhancement of nano-fluids. Int. J. Heat Fluid Flow 21: 58-64.

6. Xuan YM, Li Q (2003) Investigation on convective heat transfer and flow features of Nano-fluids. ASME J Heat Transfer 125: 151-155.

7. Chun BH, Kang HK, Kim SH (2008) Effect of alumina Nano particles in the fluid on heat transfer in double-pipe heat exchanger system. Korean J Chem Eng 25: 966-971.

8. Nazififard M, Nematollahi MR, Jafarpur K, Suh KY (2012) Numerical Simulation of Water-Based Alumina Nano-fluid in Sub channel Geometry. Science and Technology of Nuclear Installations.

9. Corcione M (2011) Empirical correlating equations for predicting the effective thermal conductivity and dynamic viscosity of nano-fluids. Energ convers Manage 52: 789-793.

10. Bouhenchir H, Cabassud M, le lann MV, Casamatta G (2001) A general simulation model and a heating/cooling strategy to improve controllability of batch reactors. Institution of Chemical Engineers Trans I ChemE 79: 641-654.

11. Rabienataj Darzi AA, Farhadi M, Sedighi K (2013) Heat transfer and flow characteristics of AL_2O_3–water nano-fluid in a double tube heat exchanger. International Communications in Heat and Mass Transfer 47: 105-112.

12. Holman JD (1989) Experimental Models for Engineers. McGraw Hill, p: 5.

Experimental and Numerical Study of Multiphase Mixing Hydrodynamics in Batch Stirred Tank Applied to Ammoniacal Thiosulphate Leaching of Gold

Dmitry Vladimirovich Gradov[1]*, Gerardo González[2], Marko Vauhkonen[2], Arto Laari[1] and Tuomas Koiranen[1]

[1]*School of Engineering Science, Lappeenranta University of Technology, Lappeenranta, Finland*
[2]*Department of Applied Physics, University of Eastern Finland, Kuopio, Finland*

Abstract

Hydrodynamics of aerated slurry is studied experimentally and numerically using the example of thiosulphate leaching of gold concentrate. The studied milled concentrate has shear-thinning fluid rheology and it was imitated by water-based solutions of CMC. Presence of electrolytes, as in the case of the leaching slurry, has great influence on bubble size distribution. Primary phase flow is measured by Particle Image Velocimetry. Local gas hold-up in aerated CMC 50000 (0.15 w%) solution is measured by Electrical Impedance Tomography. Volumetric mass transfer is measured by dynamic method in different CMC solutions over a range of operational conditions in absence and presence of electrolytes. The experimental data was used in CFD modelling of the aerated slurry. Single phase hydrodynamics of shear-thinning fluid (CMC 50000 (0.15 w%)) have been modelled and validated against experimental data with reasonable agreement. Multiphase mixing of the thiosulphate slurry was modelled with the assumption of constant bubble size. The results of the simulations and measurements are presented and discussed.

Keywords: Multiphase flow; Slurry mixing; PIV; EIT; CFD

Introduction

Thiosulphate leaching is a promising alternative to cyanidation, which is frequently used for selective dissolution of gold from ores and concentrates. Slow kinetics of thiosulphate leaching is often boosted by addition of copper (II). The copper oxidizes metallic gold transforming it to ionic form, which reacts with thiosulphate ligands producing stable gold-thiosulphate complexes. Ammoniacal media is required to prevent passivation by sulfur that might appear from the decomposition of thiosulphate when it reacts with ore [1]. It is suggested that ammonia transports gold ions to bulk solution by being more strongly adsorbed on gold surface [2]. Oxygen is needed to recover copper (II) from copper (I) at a little expense of slow thiosulphate oxidation by molecular oxygen [3]. Overall mechanism of gold leaching with thiosulphate can be briefly summarized by the following reaction [4]:

$$4Au + 8S_2O_3^{2-} + O_2 + 2H_2O \rightarrow 4Au\left(S_2O_3\right)_2^{3-} + 4OH^- \quad (1)$$

Selective dissolution of gold with the above-described reaction usually takes place in aerated stirred tanks. Continuous stirred-tank reactors (CSTR) of various types are frequently used in mining, chemical and biochemical industries to promote reaction by providing intensive mixing and high rate of mass transfer. In many cases, mixing affects conversion and yield of product. Poor mixing might result in stagnant zones. It is important to predict locations of stagnant zones. Providing good mixing benefits rate of mass transfer via effective renewal of fluids contact.

As computational power is constantly growing, numerical simulation of mixing tanks may efficiently reveal stagnant zones and help saving resources and time. In the last two decades, Computational Fluid Dynamics (CFD) has been successfully applied for modelling of multiphase hydrodynamics in stirred reactors of different types and shapes. Promising results in resolving flow fields as well as phase interaction parameters are well-described and analyzed in the review article of Joshi et al. [5]. Even though fluid dynamics modelling has been applied successfully to solve engineering tasks, CFD models of multiphase flows involve numerous empirical coefficients caused by spatial discretization and lack of knowledge about phenomena. This leads to uncertainties limiting application area. Tuning of empirical coefficients is a "compulsory measure" that requires validation procedure against experimental data.

There are many methods that have been designed to study fluid flow experimentally. Intrusive techniques such as Hot Wire Anemometry (HWA) are trustworthy and they can produce accurate results of high temporal and spatial resolution. However, they interfere flow and cannot measure many points synchronously [6]. Visual observations and measurements of hydrodynamics has taken significant place in hydrodynamics studies all along. Optical methods such as point-wise Laser Doppler Anemometry (LDA) and field-wise Particle Image Velocimetry (PIV) methods are extensively used in scientific studies dealing with optically transparent flows in stirred tanks [7-10]. For turbid and non-transparent flows Doppler Ultrasonic Velocimetry (DUV) has been developed. The operating principle of DUV is somewhat similar to PIV and it measures flow characteristics by means of ultrasonic beams reflected from seeding particles [11]. Measurements of Bubble Size Distribution (BSD) in mixed multiphase solution by tomographic PIV is challenging since images, taken in two-dimensional space and illuminated by laser pulse, give information on bubble size that is limited to particular area of laser sheet. In addition, local gas hold-up remains unknown in whole reactor. Application of Electrical Impedance Tomography (EIT), which is based on voltage measurements, has shown great potential in studies of multiphase flows for phase visualization in 3D [12-14]. Since conductivity is a function of gas hold-up, it is possible to obtain local void fraction of secondary phase by measuring voltages difference between primary phase alone and multiphase flow. Measuring voltage at several electrodes simultaneously

***Corresponding author:** Dmitry Vladimirovich Gradov, School of Engineering Science, Lappeenranta University of Technology, Lappeenranta, Finland
E-mail: dmitry.gradov@lut.fi

reconstruction of conductivity difference distribution and hence local gas hold-up is possible [15,16].

Studying of mixing hydrodynamics facilitates finding local k_L, which is needed to evaluate specific volumetric interfacial mass transfer rate in CSTR. Based on Higbie's penetration theory Kawase et al. [17,18] showed that mass transfer coefficient in liquid film for viscous fluids can be expressed through local energy dissipation:

$$k_L = C\sqrt{D_L}\left(\frac{\varepsilon\rho}{K}\right)^{\frac{1}{2(1+n)}}, \tag{2}$$

where ε is the turbulence energy dissipation rate, m^2/s^3, D_L is the gas diffusivity in liquid, m^2/s, ρ is the liquid density, kg/m^3, K is the consistency index, $Pa{\cdot}s^n$, n is the dimensionless flow behavior index, C is the proportionality coefficient. The effect of gas bubbles to the mass transfer coefficient is taken into account with the proportionality coefficient. The value of 0.301 for the coefficient, proposed by Kawase [18], has been tested by Garcia-Ochoa and Gomez [19] and was found to produce accurate results for gas-liquid mass transfer in stirred reactors.

In addition to local mass transfer coefficient, local interfacial area is also needed to evaluate local k_La. Measured locally distributions of bubble sizes and gas void fraction can provide information about local interfacial area between gas and liquid.

The idea of the current work is to acquire sufficient experimental data in order to be able to connect hydrodynamics of stirred multiphase flow to reaction kinetics. Thiosulphate leaching of gold in ammoniacal solution with the presence of copper (II) in stirred reactor was studied experimentally first and then simulated by means of CFD in order to get distribution of local mass transfer rate of oxygen into solution. The data presented in this study are useful for multiphase flows of finely-milled concentrate as well as for shear-thinning fluids agitated by Rushton turbine in aerated baffled reactors.

Materials and Methods

Object of study

The process of gold leaching with thiosulphate in ammoniacal solution, studied earlier [20] in a lab-scale batch STR, was taken as an example of typical multiphase pulp mixing process frequently encountered in mining and pharmaceuticals industries. The reaction proceeds under almost ambient conditions (atmospheric pressure and 30°C).

A lab-scale baffled stirred reactor, equipped with standard six-blade Rushton turbine, was investigated in the current research. A schematic representation of the vessel geometry and its dimensional parameters are presented in Figure 1. The impeller clearance was set at the distance equal to its diameter from the bottom of the tank. The gas sparger is a perforated metal pipe mounted at the bottom of the tank. The tank, made out of plexiglass, was designed and constructed in two versions. EIT uses controlled current injections to measure responding voltages at electrodes placed around the flow. Therefore, the vessel, meant to be used in EIT experiments, was equipped with electrodes on the walls arranged in rows. In particular, an arrangement of 4 rows consisting of 16 electrodes, made out of stainless steel, was mounted on perimeter of the reactor wall. Thus, approximate resolution of the spatial reconstruction is roughly 2 cm^3. Another tank of the same size was encased inside a rectangular plexiglass container to be filled with liquid of the same refractive index as the studied solution in order to eliminate visual distortions.

The power input per volume of fluid that was used in the reaction kinetics research was scaled via the known power number of the dual impeller to the reactor scale used in the current work. The scaled power input corresponds to the mixing rate of around 400 rpm.

Experimental techniques

Rheology study: The mean diameter of the gold concentrate particles used by Lampinen et al. [20] was 7 μm. [21] Having measured the total heat production in calorimeter (universal Calvet calorimeter C80 CS Evolution of SETARAM Instrumentation with reversing mechanism) as the function of solids content, and the maximum value corresponded to 30% of solids content, which was suggested to be optimal in the leaching solution. The range of the shear rate in the pulp mixing with the four different impeller speeds was roughly estimated using the PIV data (Figure 6) and the CFD model. Single phase mixing of the above-mentioned reactor was simulated in Fluent 16 and the results are summarized in Table 1. Looking ahead, rheological measurements of the gold concentrate suspended in water revealed shear-thinning behavior (Figure 4). The suspension used in the leaching experiments is non-transparent. Therefore, it is not suitable for hydrodynamics study by PIV. Suspended solids may also cause some bias when measured by

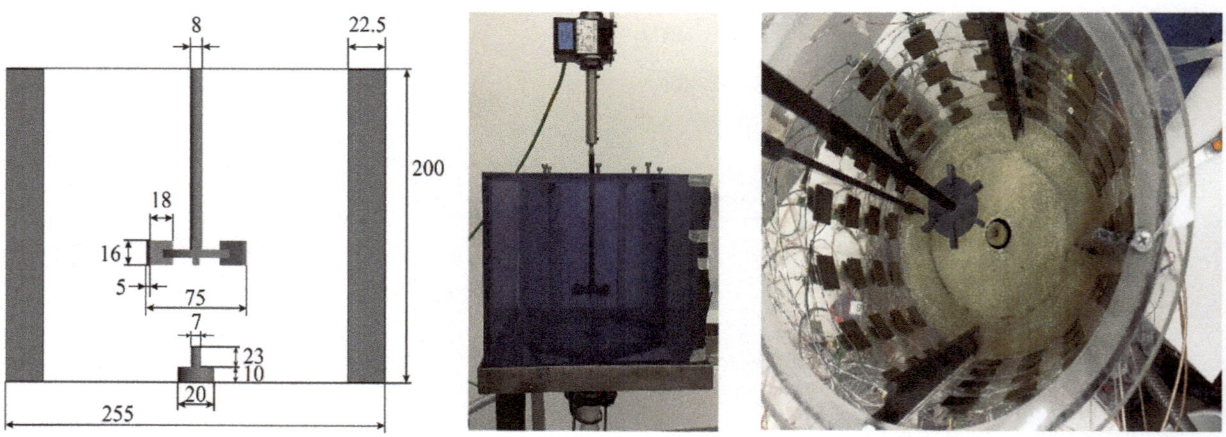

Figure 1: Reactor illustrations. a – scheme and sizing in mm, b – photo of the tank encased inside a plexiglass prism for PIV experiments, c – photo of the vessel equipped with electrodes for EIT experiments.

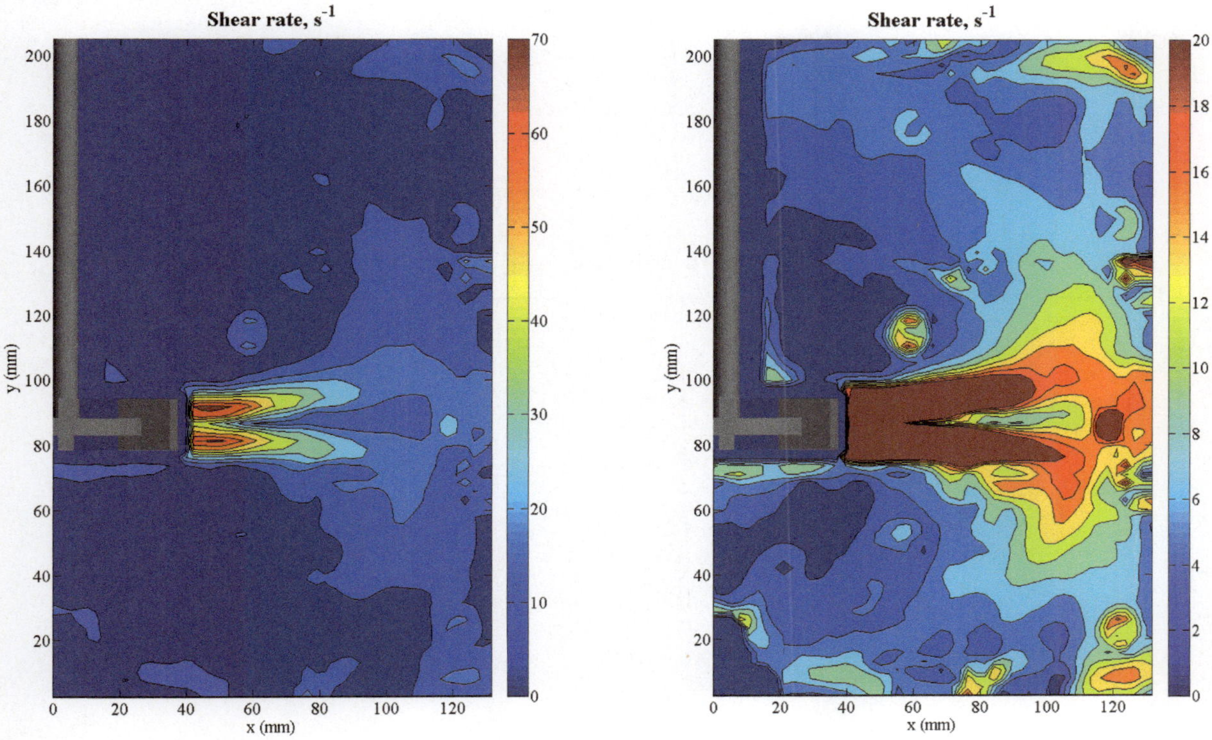

Figure 2: Contours of shear rate of CMC 50000 (0.15 w%) water-based solution at 400 rpm.

EIT due to conductivity difference between solids and liquid. For that reason, it was decided to imitate the solution experimentally by using Sodium Carboxy Methyl Cellulose (CMC) dissolved in water. Water solutions of CMC also have shear-thinning behavior and effect of CMC to rheological properties depends on molecular length. Two CMC powders were tested: CMC 30 and CMC 50000. The number going after CMC denotes viscosity of 2 % water-based solution measured at 25°C [22].

Dynamic Mechanical Analysis of the leaching pulp and water-based CMC solutions was performed using Anton Paar Modular Compact Rheometer 302. By means of swirling cone-analyzer (standard measuring cylinder CC27/T200/AL) the effective viscosity was measured in the shear rate range from 1 to 200 s^{-1}. Suspended solids in the leaching pulp had a tendency to sedimentate. Therefore, the pulp viscosity was measured additionally with a stirrer-sensor (ST24-2D/2V/2V-30/109). Measurements made by the cone and the blade analyzers produced similar results meaning that sedimentation for such tiny particles is a relatively slow process. Thus, the cone mixer-sensor was used alone in all the experiments in further.

Overall gas hold-up measurements: Gas hold-up measurements were made based on visual observation of liquid surface level of the agitated solutions. The reliability of the measurements was compromised by surface fluctuations and appearance of foam. In order to reduce uncertainties, the rise of the surface level Δh was measured at two positions: between the baffles next to the vessel wall. In each experiment, the agitated solution was allowed to reach equilibrium and then 10 readings from each point were made and averaged. The overall gas hold-up was computed based on the following equation:

$$\varepsilon_{\Delta h} = \frac{h_{gassed} - h_{ungassed}}{h_{gassed}}, \tag{3}$$

where $\varepsilon_{\Delta h}$ the overall gas hold-up, h_{gassed} and $h_{ungassed}$ are the heights of the mixed solutions with and without gas introduction correspondin gly.

EIT: The EIT set-up used in the experiments involved a signal generator, receiver, controller, and computer. The purpose of the signal generator is to produce current pulses of constant voltage to the solution according to a predefined scheme of pulses order. There are two different scenarios for the order of current injections that can be applied. Voltage injections can be sent pair-wise e.g., adjacent pairs of electrodes or multiple electrodes simultaneously. In both scenarios, voltages are measured at all electrodes. Then, obtained signal goes to demodulation process, where input voltage is decomposed into real and imaginary parts that are needed to get phase and amplitude according to the following formulations [23]:

$$v_{real} = \frac{2}{N}\sum_{i=0}^{N-1} v(i)\sin\frac{2\pi i}{N} \tag{4}$$

$$v_{image} = \frac{2}{N}\sum_{i=0}^{N-1} v(i)\cos\frac{2\pi i}{N} \tag{5}$$

$$A = \frac{2}{N}\sqrt{v^2{}_{real} + v^2{}_{image}} \tag{6}$$

$$\varphi = \tan^{-1} v_{image} / v_{\mathrm{real}}, \tag{7}$$

where N is the number of electrodes, v(i) is the measured voltage

Impeller speed, rpm	γ s⁻¹			Method of determination
	Minimum	Maximum	Volume averaged	
100	0.002	350	2	CFD
200	0.005	708	5	CFD
400	0.012	1443	12	CFD
400	0	70	8	2D PIV plane (Figure 2)
600	0.024	2186	19	CFD

Table 1: Range of shear rate in the reactor.

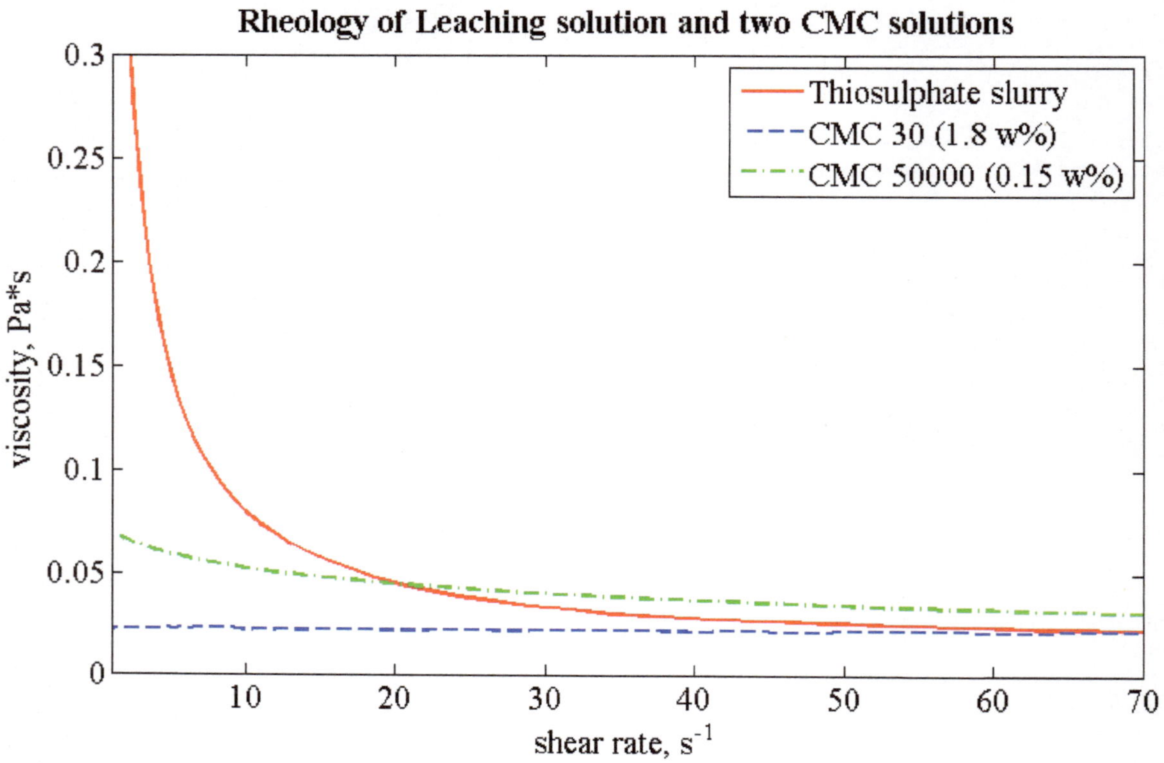

Figure 3: Rheological analysis of the leaching suspension and the two CMC water-based solutions of similar viscosity.

at i-th electrode, V, V_{real} is the real part of measured voltage, V, V_{image} is the imaginary part of measured voltage, V, ϕ and A are the signal phase and amplitude correspondingly.

Finding a conductivity distribution out of measured voltage outcomes of preset current injections represents itself an inverse problem. Computation of conductivity distribution is a challenging task to solve, since even a slight shift of voltages may be a result of significant change of conductivity distribution in solution, meaning that the "equation" might have several solutions. Therefore, some additional information about object of study is needed to solve the inverse problem. Prior information on conductivity distribution may be achieved statistically and used as a reference [23,24].

The difference method exploits measurements of non-gassed mixed solution as a reference in order to get gas void fraction distribution in the reactor. The conductivity distribution is the function of fluid resistance, and hence, local gas hold-up could be found as follows [23]:

$$\varepsilon_{EIT} = \frac{2(\sigma 1 - \sigma 2)}{2\sigma 1 + \sigma 2}, \quad (8)$$

where σ_1 is the conductivity distribution of liquid phase, S/m, σ_2 is the conductivity distribution of dispersed gas phase, S/m. Thereby, the difference method automatically removes the influence of non-conductive parts of the vessel such as baffles and impeller.

Many researches describe successful applications of the technique for stirred tanks to measure local gas hold-up [13,14,25-28]. Value of current injections value is dictated by conductivity of the studied solution. The higher the solution conductivity, the higher the current injections are required in order to increase sensitivity of the system.

PIV: Particle Image Velocimetry was used to study the hydrodynamics of the liquid phase agitated in multiphase mixing. The PIV technique measures velocity flow fields by capturing the movement of tracing particles, which are small enough not to interfere the flow. By slicing the flow with a laser sheet, PIV can produce a two-dimensional picture of the fluid motion within the laser sheet of illumination. A vector velocity field is constructed based on spatial shift of tracer-particles between a pair of laser pulses emitted consecutively in accordance with a preset time delay. The PIV set-up included a double-pulsed Nd:Yag laser and two Charge-Coupled Device (CCD) cameras of the following resolution 1600 × 1200 pixels. The laser and the cameras were managed by Davis 7.2 PIV software. The imaging of tracers shift was carried out by two CCD cameras which were synchronized with laser pulsing. Thus, velocity vector fields of axial and radial components were measured in 2D space.

In order to measure liquid phase flow, fluorescent particles of size 20 – 50 µm, made of polyamide, were used as tracers. The fluorescent tracers emit light at different wavelength than the laser. This property facilitates getting images with uniformly illuminated background. High bandpass filters (Light red bandpass filter: BP635) were installed on the CCD cameras to separate light within the wavelength range 615 – 645 nm to filter out the green light coming from light-reflecting surfaces of the reactor [29]. Gas bubbles pictures were made using the light reflected from bubbles. In this case, a camera was capped with a grey filter (B+W, 52 103 ND 0.9 8x) only to reduce intensity of the reflected light.

The laser sheet in the reactor was positioned vertically near the baffle and 2 mm behind the vertical axe of the shaft. The reason for this positioning was to remove as many visual obstacles and strong reflections as possible. Size of single frame taken (measurement window) depends on camera resolution and focal length of objective. A typical size of the measurement frame was 127 × 100 mm. Such frames were processed twice via cross-correlation algorithm. At first, the frame square is split by the PIV software into interrogation areas of 128 × 128 pixels to search vectors. Then, interrogation areas were decreased to 64 × 64 pixels with 50% overlap and the frame was treated in similar way to search for tracers shift with higher precision. The final PIV results have resolution of 2.2 mm. A sample of the taken frames and the resulting vector field of instantaneous and average velocities are presented in Figure 3. In order to get a full picture of the reactor, it was necessary to take three frames at different heights along the vertical axe with 50 mm overlapping (Figure 5). Thus, when the captured frames were assembled, a full image of the reactor, from the tank wall to the shaft, of size 127 × 200 mm could be obtained. The resolution of post-processed and vertically combined frames was 51 × 79 vectors. Totally 1000 frames were taken and processed in order to obtain time-averaged results.

k_La measurements: Volumetric mass transfer coefficient was measured via "gassing-out" dynamic method by using dissolved oxygen meter MARVET BASIC. All the experiments were performed at 20°C. The reactor was at first bubbled with nitrogen until dissolved oxygen concentration reached its minimum. Then a valve of nitrogen pipe line was closed and air was introduced. The evolution of the changing oxygen concentration was recorded until it reached its maximum value. The data was analyzed by parameter estimation software MODEST [30] and the volumetric mass transfer rate was calculated using using Eq. 9.

$$\frac{c_{sat} - c_{liquid}(t_1)}{c_{sat} - c_{liquid}(t_2)} = \frac{e^{(-k_L a t)} - k_L a \tau_p e^{-t/\tau_p}}{1 - k_L a \tau_p}, \quad (9)$$

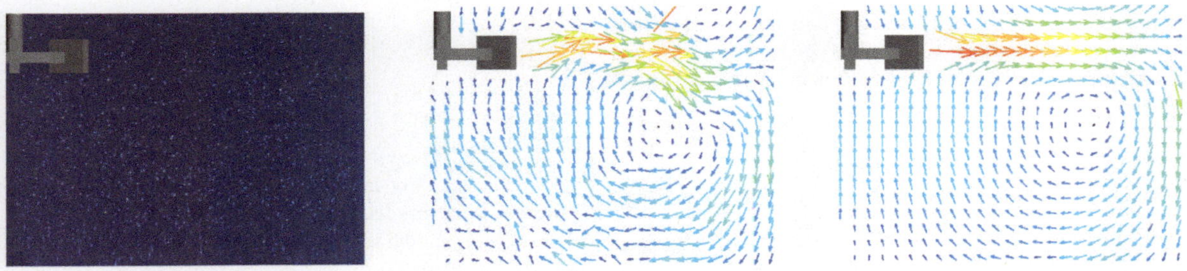

Figure 4: Measurement window and the calculated velocity vector fields near the impeller. a – PIV frame, b – instant velocity vector field, c – averaged velocity vector field.

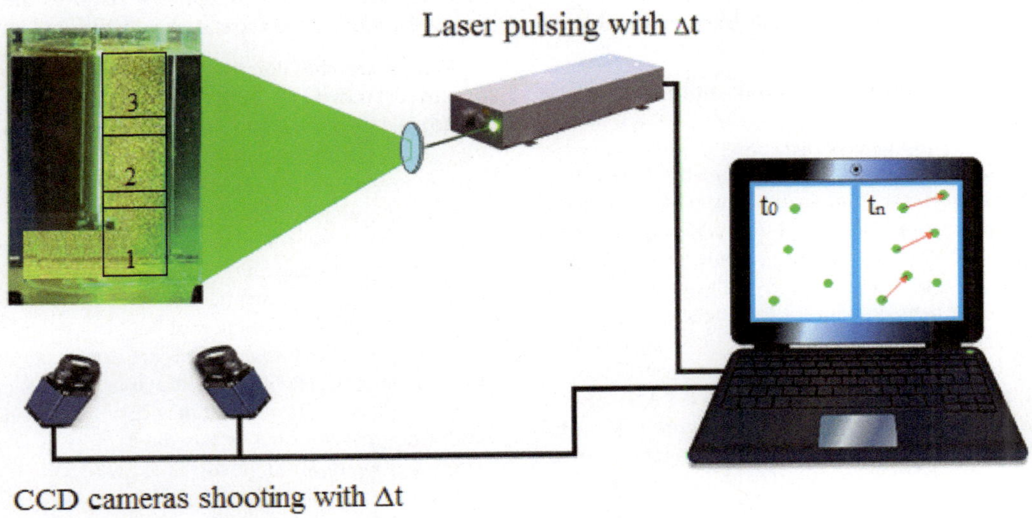

Figure 5: Schematic representation of PIV set-up

where c_{sat} is the saturated oxygen concentration in water, mol/m³, c_{liquid} is the oxygen concentration in liquid, mol/m³, at t_1 and t_2.

Time constant shows how quickly a measuring probes responses to a sudden step change of oxygen concentration. In order to expect accurate results the value of the time constant should be lower than $k_L a^{-1}$ [31,32]. The response time constant τ of the probe was measured separately for each solution studied. In the experiments, the oxygen probe was placed between baffles where the mixing is vigorous and the bulk solution around the probe is constantly renewed by the flow. Otherwise, the probe might consume all the oxygen presented in the bulk around at some moment and further measurements become inaccurate.

Modelling theory

Multiphase fluid flow: The Stoke's number (Eq. 10) for the studied system is more than 1, meaning that the secondary phase influence onto the primary phase is not negligible and it has to be taken into account.

$$St = \frac{\tau_p}{\tau_q} = \frac{d^2{}_p\varepsilon^{0.5}}{36\nu^{1.5}}, \tag{10}$$

Where τ_p and τ_q are the relaxation time of primary phase and bubble, s, d_p is the bubble diameter, m, ε is the energy dissipation, m²/s³, ν is the kinematic viscosity, m²/s.

Therefore, in the gas-liquid modelling, Eulerian-Eulerian multiphase approach was used, in which gas and liquid phases are considered as interpenetrating continua. A set of conservation equation is solved for each phase (Eq. 12 and 13). Additionally, volume fraction is introduced (Eq. 11).

$$\sum_i a_i = 1 \tag{11}$$

$$\nabla \overline{u} = 0 \tag{12}$$

$$\overline{u}.\nabla\overline{u} = -\nabla\overline{p} + \nabla.[(\mu + \mu_{turb})\nabla\overline{u}], \tag{13}$$

where α_i is the phase volume fraction, $\overline{u}$ - phase velocity, m/s, $\overline{p}$ is the pressure, Pa, μ_{turb} is the turbulent viscosity, Pa·s.

Suspended solids: The tiny milled concentrate was used in the original leaching tests with mean size of 7 µm. Pulp flow can be modeled via effective viscosity [33]. Rheology of the leaching slurry has been measured experimentally and modelled via power-law model (Eq. 23).

Turbulence: In CFD modelling, available computing power and acceptable simulation time were limited, for which reason it was decided to apply a Reynolds-Averaged Navier-Stokes (RANS) turbulence model in the current simulations. Turbulence formulation in RANS models are based on statistical analysis rather than on actual physical phenomena. By contrast, the Large Eddy Simulation (LES) model is based, at least partially, on resolving turbulence fluctuations. Approximations such as the Boussinesq hypothesis can lead to mismatches with real flow behavior due to turbulence isotropy and homogeneity. As a result, empirical constants are introduced to tune the model (Eq. 14 and 15). In the current work, it was decided to evaluate the suitability of the most frequently applied RANS models rather than attempt to implement a novel turbulence model by tuning the model constants for the hydrodynamics of the specific stirred tank. In our previous work [34], three popular models in the RANS family were tested, namely the Realizable k-ε model, the Shear Stress Transport k-ω model, and the Reynolds Stress Model (RSM). The Realizable k-ε model produce the most accurate results for single and multiphase mixing in round bottom STR. The Realizable k-ε model is a recent model with an improved computational mechanism preventing turbulence energy from being negative. Such energy is not physical but can occur using the Standard k-ε model.

$$\overline{u}\cdot\nabla k = P - \varepsilon + \nabla\cdot[\left(\mu + \frac{\mu_{turb}}{\sigma_k}\right)\nabla k] \tag{14}$$

$$\overline{u}\cdot\nabla\varepsilon = C_{\varepsilon 1}\frac{\varepsilon}{k}P - C_{\varepsilon 2}\frac{\varepsilon^2}{k}P + \nabla\cdot[\left(\mu + \frac{\mu_{turb}}{\sigma\varepsilon}\right)\nabla\varepsilon], \tag{15}$$

where k is the turbulence kinetic energy, m²/s². Production P includes one more constant.

Dispersed turbulence formulation model was used since the secondary phase is dilute and fluctuating quantities of the gaseous phase may be found as the function of the mean terms of the liquid phase. At first, turbulence terms k and ε are first computed for primary phase. Then, secondary phase turbulence terms are found and exchange of turbulence momentum between phases is computed based on Tchen-theory [35].

Phase interaction: Momentum conservation equation (Eq. 16) includes cumulative force $\vec{F}_q$ (Eq. 17) acting on primary phase. The cumulative force comprises several forces occurring in multiphase interaction, including drag, virtual mass, lift, wall lubrication, turbulence diffusion forces etc.

$$\frac{\partial}{\partial t}(\mathrm{a}\rho\bar{\mathrm{u}})_q + \nabla\cdot(\mathrm{a}\rho\bar{\mathrm{u}}\bar{\mathrm{u}})_q = \nabla\cdot(\mathrm{a}\,\sigma)_q - \mathrm{a}_q\nabla p + \vec{F}_q + a_q\rho_q\vec{g} \tag{16}$$

$$\vec{F}_q = \sum_{p=1}^{n}(\mathrm{k}_{pq}(\bar{\mathrm{u}}_p - \bar{\mathrm{u}}_q) + \dot{\mathrm{m}}_{pq}\bar{\mathrm{u}}_{qp} - \dot{m}_{qp}\bar{\mathrm{u}}_{qp}) + (\vec{F}_q + \vec{F}_{lift,q} + \vec{F}_{wl,q} + \vec{F}_{vm,q} + \vec{F}_{td,q}), \tag{17}$$

where $\vec{F}_q$ is the external body force, $\vec{F}_{wl,q}$ the lift force, $\vec{F}_{wl,q}$ is the wall lubrication force $\vec{F}_{vm,q}$ is the virtual mass force, and $\vec{F}_{td,q}$ is the turbulence diffusion force.

Different drag models can be found in published literature. The most popular model, proposed by Schiller and Naumann [36] in Eq. 19, is a drag model suitable for rigid spherical particles.

$$kpq = \frac{\rho_p C_D \mathrm{Re}}{144\tau_p} \mathrm{d_p A_i}, \tag{18}$$

where ρ_p is the particle density, kg/m³ C_D is the drag coefficient, A_i is the relative Reynolds number, and A_i is the interfacial area, m². The drag force coefficient is the function of Re according to the following formulation:

$$CD = \begin{cases} 24(1+0.15\,\mathrm{Re}^{0.687}) & \mathrm{Re} \le 1000 \\ 0.44 & \mathrm{Re} > 1000 \end{cases} \tag{19}$$

The drag force model is based on bubble rise velocity measured in stagnant fluids, which is higher than rise velocity in turbulent flow. A model correcting drag force coefficient for turbulence was proposed by Brucato et al. [37]. This turbulence modification factor concept (Eq. 20) changes stagnant fluid drag force to make it suitable for adoption in turbulent multiphase flow simulation Later, Lane et al. [24] critically considered Brucato's formulation and found that ratio of slip and terminal velocity as a function of bubble size and Kolmogorov length scale does not take into account particle density or density difference between phases. Having pointed out this deficiency in Brucato's model, Lane et al. suggested a new correlation (Eq. 21) for turbulence modification factor that is based on ratio of stagnant to turbulent terminal velocity $\frac{U_s}{U_T}$ to be correlated with ratio of particle relaxation time to integral time scale of turbulence (Eq. 22).

$$K'_{pq} = \eta K_{pq} = \left(1 + K\left(\frac{d_p}{\lambda}\right)^3\right)K_{pq}, \tag{20}$$

Where η is the turbulence modification factor, K= 6.5·10-6, λ is the Kolmogorov length scale, m.

$$\eta = \left(1 - 1.4\left(\frac{\tau_p}{T_L}\right)^{0.7} e^{\left(\frac{0.6\tau_p}{T_L}\right)}\right)^{-2} \quad (21)$$

$$T_L = 0.135\frac{k}{\varepsilon}, \quad (22)$$

Gradov et al. showed that the Schiller-Naumann drag force model in conjunction with Lane's turbulence modification factor gave the most accurate results for air-water mixing in stirred tank at the assumption of constant bubble diameter equal 1 mm. Also, the effect of non-drag forces was found to be insignificant air-water mixing simulations. Therefore, in order to reduce computational complexity and promote solution stability the above mentioned combination of models was used to simulate drag force, while other forces models were ignored [34].

Modelling strategy: In the gas-liquid flow simulations, the gas was introduced into converged liquid flow field in order to reduce computational instabilities. In the CFD simulations, the stirred tank can be considered as a semi-batch system with gas is supplied via sparger and exit through liquid surface. The gas phase was introduced in a transient manner at 0.02 s time step until mass flow rate balance was reached and stabilize within 5% deviation. Under relaxation factors for pressure and velocity were reduced up to 0.15 and 0.35 correspondingly to facilitate convergence. Second order discretization scheme was applied to all the parameters. Convergence criterion was set at 10^{-4} for all the variables to be simulated. When multiphase flow hydrodynamics have been computed, mass transfer was calculated on top of multiphase results according to Eq. 2. The operational conditions simulated and models used in single phase and multiphase mixing simulations are summarized in Tables 2 and 3.

Results and Discussion

Experimental results

Rheology: The rheology of the leaching suspension as well as the water solutions of CMC 50000 (0.15 w%) and CMC 30 (1.8 w%) were measured and shown in Figure 3. The presented data was plotted in log-log scale, and the rheological parameters were found from the Ostwald-de Waele power law [38]:

$$\mu_{eff} = K\left(\dot{\gamma}\right)^n, \quad (23)$$

where μ_{eff} is the effective viscosity, Pa·s, K is the consistency index, Pa·s^n, n is the flow behavior index.

The viscosity of all the solutions approach constant level as the shear rate is increased. However, only the solution of short molecular chain CMC 30 shows insignificant shear-thinning rheology. Therefore, only CMC 50000 has been used in further as an imitation model of leaching suspension. It is advisable to set limits for the effective viscosity when modelling non-Newtonian hydrodynamics in order to avoid numerical errors. Calculated power law parameters are summarized in Table 4.

Gas hold-up: Measurement of gas hold-up in STR by level difference has better accuracy at high gas hold-ups. Gas hold-up has linear dependence on superficial velocity as described by Rewatkar et al. [39] numerically and approved experimentally experimentally later by Kraume and Zehner [40] for glass beds suspended in water and by Laakkonen [41] for Xanthan solutions. In the current study, gas hold-up was measured over the range of gas flow rates from 1 to 10 L/min at constant impeller speed of 400 rpm. The results are plotted in Figure 6.

The obtained results show almost linear dependence of gas hold-up from gas flow rate. It is necessary to use relatively low gas flow rate in PIV measurements to ensure good optical transparency of solution and reduce impact of gas bubbles onto laser intensity. Therefore, gas flow rate could be safely reduced to 1.25 L/min.

EIT: Increased viscosity promotes growth of bubble size due to increased coalescence rate. However, it is also known that presence of electrolytes in solution inhibits bubble coalescence [42]. At sufficiently high salt concentrations, bubbles, covered by ion layer, repel each other. The total optimal concentration of salts for thiosulphate leaching was found to be 0.4 mol/L [20]. Unfortunately, the EIT set-up loses its sensitivity at high conductivity of solution since the maximum current was limited to 6.5 mA. For that reason, the salt concentration had to be reduced to lower value, where a compromise between the similarity in the multiphase solution properties compared to the original solution and the set-up limits could be met. It can be concluded, that significant amount of ions in liquid suppresses the bubble-growing effect generated by higher viscosity. The influence of the salt concentration on gas phase can also be tested indirectly by measuring the power draw of the impeller. Impeller torque was measured as salts were added step-wise (0.05 mol/L per step). The results (Figure 7) show that 0.05 mol/L of salts was enough to effectively cover the major part of the bubbles surface in the solution. Further addition of salts has insignificant effect on the mixing power and hence on the gas hold-up. However, the power draw is manly driven by the average density of the multiphase solution, and it is little influenced by the distribution of bubble size which has, however, significant influence on the contact surface area between the phases. Therefore, the overall k_La was also measured at different salt concentrations, as depicted in Figure 9 (left plot) as depicted in Figure 7 (left plot) (left plot). The slow increase of the overall mass transfer is an evidence of the ongoing shift in the BSD towards small bubbles. At concentration 0.1 mol/L, with a conductivity of 7.9 mS/m (which is within the maximum conductivity limit of the EIT), the k_La value is 91% of the maximum k_La value at concentration 0.4 mol/L. Therefore, this concentration was selected as a reasonable compromise and was used in all the experiments.

In order to evaluate the performance of the EIT set-up, several experiments were made at three gas flow rates without mixing. The

Operational conditions	Modelled phenomenon	Model used
400 rpm	Turbulence	k-ε Realizable
	Shear-thinning rheology of mixed fluid	Power-law model
	Flow near the solid walls	Standard wall function

Table 2: Summarized conditions for single phase mixing simulations.

Operational conditions	Modelled phenomenon	Model used
400 rpm	Turbulence	k-ε Realizable
	Shear-thinning rheology of mixed fluid	Power-law model
1.25 L/min of air flow rate	Flow near the solid walls	Standard wall function
	Bubble size	Constant d_b=1.5 mm

Table 3: Summarized conditions for multiphase mixing simulations.

Solution	n	K, Pa·s^n	μ_{min}, Pa·s	μ_{max}, Pa·s
Thiosulphate slurry	0.200	0.044	0.02	0.40
CMC 50000 (0.15 w%)	0.746	0.086	0.02	0.09

Table 4: Power law model parameters.

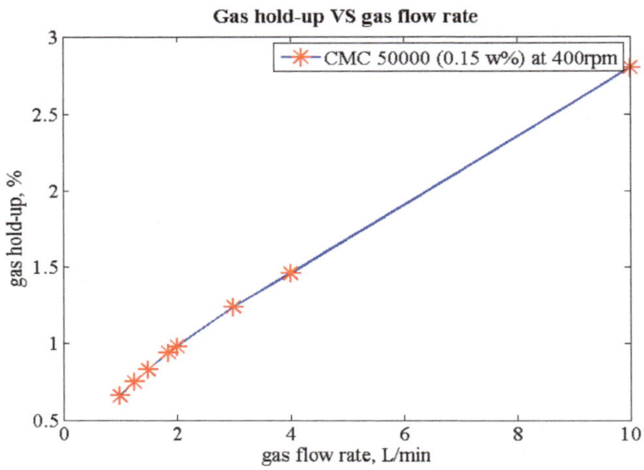

Figure 6: Gas hold-up vs superficial gas velocity in CMC 50000 (0.15 %) water-based solution at 400 rpm.

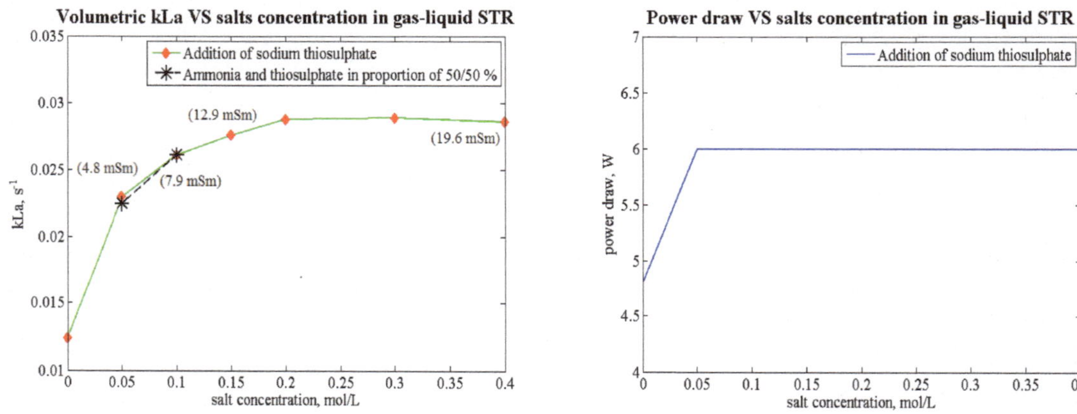

Figure 7: Overall kLa (right) and power draw (left) with conductivity of CMC 50000 (0.15 %) water-based solution at 600 rpm and 2 L/min gas flowrate vs salt concentration

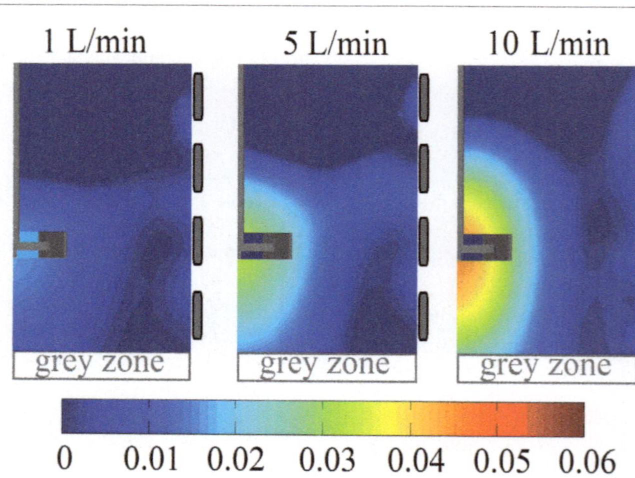

Figure 8: Local gas hold-up in CMC 50000 (0.15 %) and 0.1 mol/L of sodium thiosulphate in 2D plane (middle plane positioned between the baffles) measured by EIT without mixing.

obtained local gas hold-ups are illustrated in Figure 10. The resolution of the resulting images was around 2 cm^3. It is clear that at large flow rates, the contrast in gas-liquid conductivity is high and the EIT results of gas-liquid mixing become more reliable.

The local gas hold-ups of the water-based solution of CMC 50000 mixed at 200, 400 and 600 rpm over the range of gas flow rates from 1 to 2 L/min are presented in Figures 10-11.

The overall gas hold-ups increase as the gas flow rate is increased with all the impeller speeds tested. Unfortunately, the poor resolution of the EIT set-up does not capture small bubbles present in high quantities in the most agitated region around the impeller. At low gas flow rates and high impeller speeds conditions are challenging when bubbles are smaller and distributed more uniformly in the volume of the reactor. Decreased and evenly distributed gas void fraction allows electrical current bypassing small bubbles especially at high loads of electrolytes in the solution. The measured zero value of gas hold-up around the impeller indicates narrow size distribution of small bubbles. The bubbles become traceable only in the region of low turbulence intensity where bubble coalescence takes place. The regions of bubble growth are most distinct at low ratio of impeller speed to gas superficial velocity. It can be noticed that bubble size grows in radial direction towards the wall. Below the impeller, a down pumping circulation loop generates a stagnant zone where the retention time of bubbles is bigger. Meanwhile, bubbles, escaping the loop, gravitate to the free surface and move into less turbulent region where they coalesce. In addition, intensive foaming takes place when the CMC-water-thiosulphate solution is mixed which facilitates accumulation and growth of bubbles near the free surface. As a result, high local gas hold-up next to the surface is formed which is also presented in all the figures.

PIV: PIV results of CMC 50000 (0.15 w%), showing contours of the averaged velocity vector field and the r.m.s instantaneous velocity field, are shown in Figure 12. The averaged velocity vector field is a great source of information. The spatial velocity gradients may give information about the range of shear rate in the 2D slice of the flow (Figure 2). That information was used to define the range of shear rate in the rheology study of the suspension and the CMC water-based solutions. When studying the shear rate distribution in the PIV plane more closely one may notice that the agitated flow is rapidly attenuated in the viscous solution. It can be clearly seen that the visible stable streams which have shear rates of 70 s^{-1} are pushed by the agitator and dissipated into the bulk solution where the initial shear rate is reduced

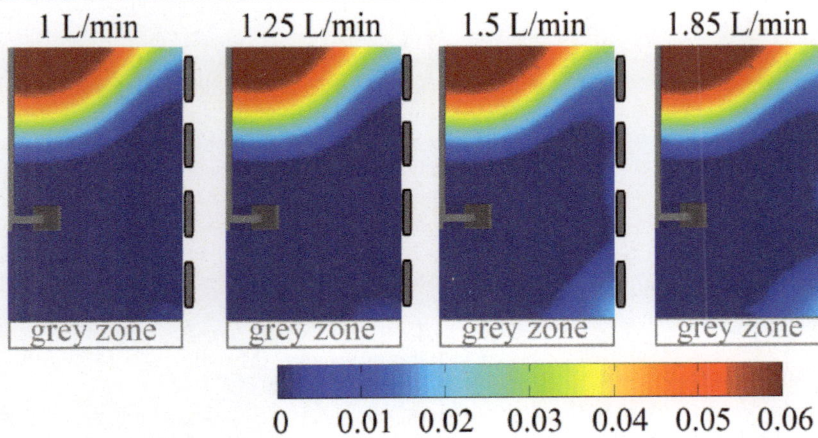

Figure 9: Local gas hold-up in water solution of CMC 50000 (0.15 %) and 0.1 mol/L of sodium thiosulphate in 2D plane (middle plane positioned between baffles) measured by EIT at 600 rpm and the range of gas flow rate.

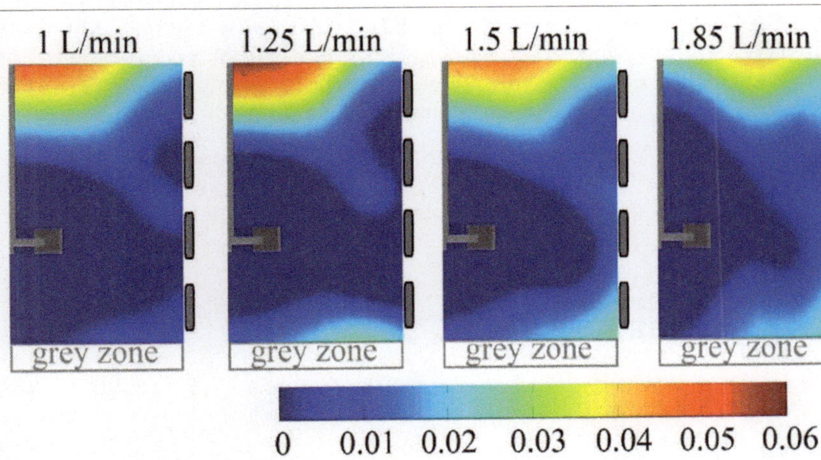

Figure 10: Local gas hold-up in water solution of CMC 50000 (0.15 %) and 0.1 mol/L of sodium thiosulphate) in 2D plane (middle plane positioned between the baffles) measured by EIT at 200 rpm and the range of gas flow rate.

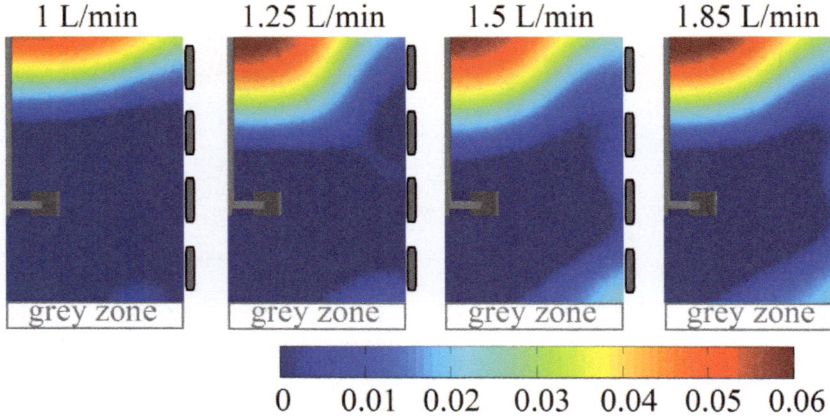

Figure 11: Local gas hold-up in water solution of CMC 50000 (0.15 %) and 0.1 mol/L of sodium thiosulphate in 2D plane (middle plane positioned between baffles) measured by EIT at 400 rpm and the range of gas flow rate.

to around 10 s^{-1}.

Salts free single phase mixing of the CMC solution at 400 rpm (Figures 2 and 6) at 1.25 L/min were measured by the PIV and used for validation of CFD results.

As it was stated earlier, the leaching solution used in the original tests of thiosulphate leaching of gold [20] had high salts concentration. In the PIV tests, CMC 50000 (0.15 w%) solution with 0.1 mol/L concentration of sodium thiosulphate was used to study gas bubbles. Presence of salts has significant impact onto BSD. Ions of the same sign cover surface of gas bubbles and block coalescence rate by repelling mechanism. As a result, one may notice narrow distribution of gas bubbles in Figure 13.

Bubble size distribution is almost uniform in the most turbulent region next to the impeller as well as also elsewhere in the image. Bubble size varies from 1 mm to 1.8 mm as it is seen in the zoomed picture. The bubbles are densely populated in the plane of measurement. However, it should be mentioned that only those bubbles that are in focus were considered when evaluating the bubble size.

Such a narrow range of BSD may be approximated as a constant bubble size (1.5 mm) throughout the reactor under the current conditions. Considering CFD modelling this approximation greatly simplifies modelling and use of Population Balance Model (PBM) can be avoided.

Overall k_La: Two water-based solutions of CMC 30 (1.8%) and CMC 50000 (0.15%) were used to measure the volumetric mass transfer coefficient at three impeller speeds and the range of gas flow-rates with and without addition of salts. The results are presented in Figure 14.

The volumetric mass transfer rate is proportional to interfacial contact area and mixing intensity. At mixing speed 600 rpm k_La grows as gas flow rate is increased. However, at lower mixing speeds k_La increases up to some point and then goes down with the increase of gas input. The reason for this behavior is the contact area that drops if mixing intensity is not enough to keep the bubble size small. Without salts both CMC solutions have similar mass transfer due to similar apparent viscosity. Presence of ions in water-based solution of CMC 30 does not lead to significant drop of viscosity and the similar mass transfer can be observed. However, ion presence affects CMC 50000 so that the long chain molecules are broken and the apparent viscosity of solution is decreased. Low viscosity and high content of ions in the solution give nearly constant and small bubble size that results in high values for k_La.

Numerical results

Grid independence test: Spatial discretization influences results of numerical simulation. Five different grids have been tested. Water mixing simulations were used in grid independence study. Reactor geometry was meshed with almost doubling factor relatively one another and scaling uniformity was maintained as well. Re number differs significantly within the vessel therefore 14 sampling points of velocity have been defined in the reactor domain. A normalized root mean square error index (NRMSE) was employed as a criterion for grid independency validation [43]. Its formulation is as follows:

$$\text{NRMSE} = \frac{3}{\left(\frac{N_{fine}}{N_{coarse}}\right)^{p^*} - 1}\sqrt{\frac{\sum_{i=1}^{n}\left(\varphi_1(i) - \varphi_2(i)\right)^2}{\sum_{i=1}^{n}(\varphi_2)^2}} \cdot 100\% \tag{24}$$

where N is number of cells in a grid (it is advised to keep grid size ratio above 1.3 corresponds to order of discretization, n is number of samples, φ_1 and φ_2 are predictions on coarse and fine grids accordingly.

According to the authors, grid difference might be considered negligible if NRMSE value is less than 10%. During mesh sizing independence tests simulations were performed with second order discretization scheme. Results presented in Table 5 clearly indicate grid 4 to be optimal one for the needs of the current study.

Shear-thinning fluid mixing: To reduce numerical instability the gas-liquid multiphase simulation strategy included two steps. The gas-liquid simulations were made in two-way coupling. At first,

Grid name	Number of elements	Torque, mNm	∫kdv $m^2/s^2 \cdot 10^{-5}$	∫εdV, $m^2/s^3 \cdot 10^{-5}$	NRMSE $\bar{u}$ %
1	108 276	54.0	9.9	139.2	20
2	269 654	56.0	10.8	161.2	13
3	464 532	56.6	10.7	166.4	15
4	754 056	65.6	13.0	209.1	9
5	1 969 376	64.4	12.6	208.5	NaN
Measured experimentally		65 ± 3			

Table 5: Grid test results of water mixing at 400 rpm in STR.

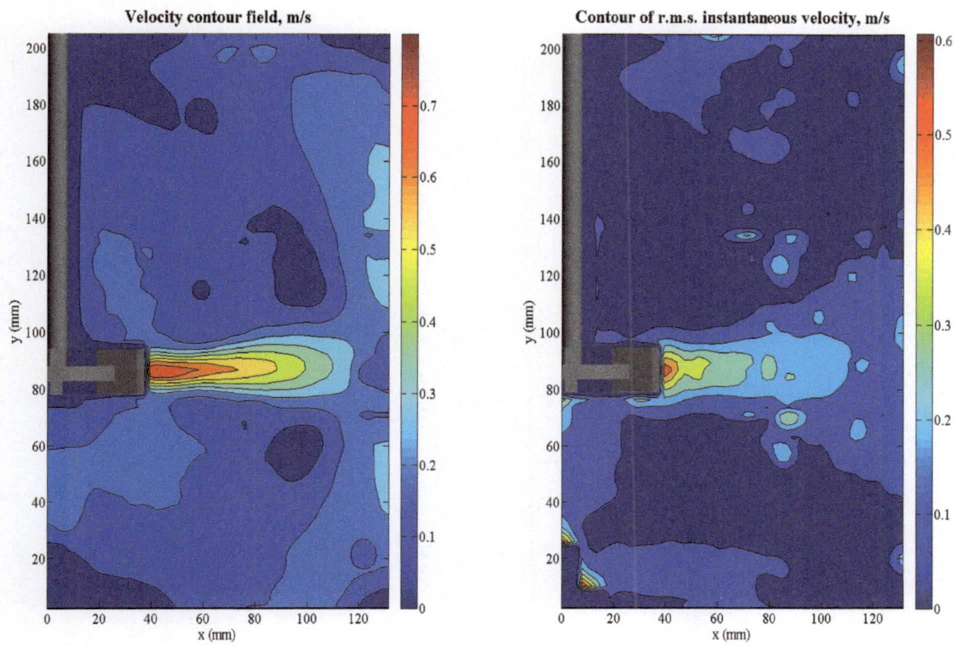

Figure 12: Velocity field contour (left) and contour of r.m.s. instantaneous velocity (right) for CMC 50000 (0.15 w%) water-based solution at 400 rpm.

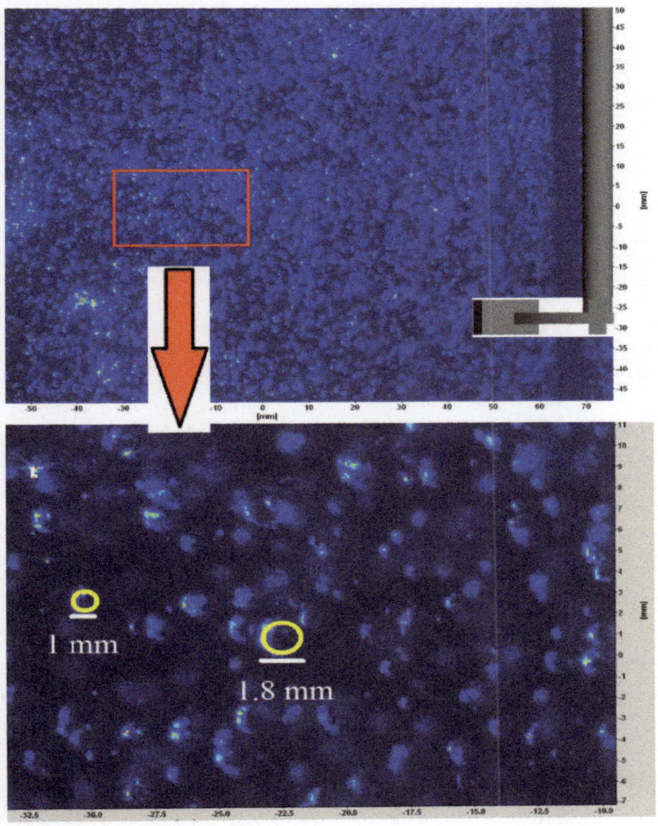

Figure 13: PIV frame with measured bubble sizes in CMC 50000 (0.15 %) water-based solution with salt addition (1 mol/L) at 400 rpm and 1.25 L/min gas flow rate.

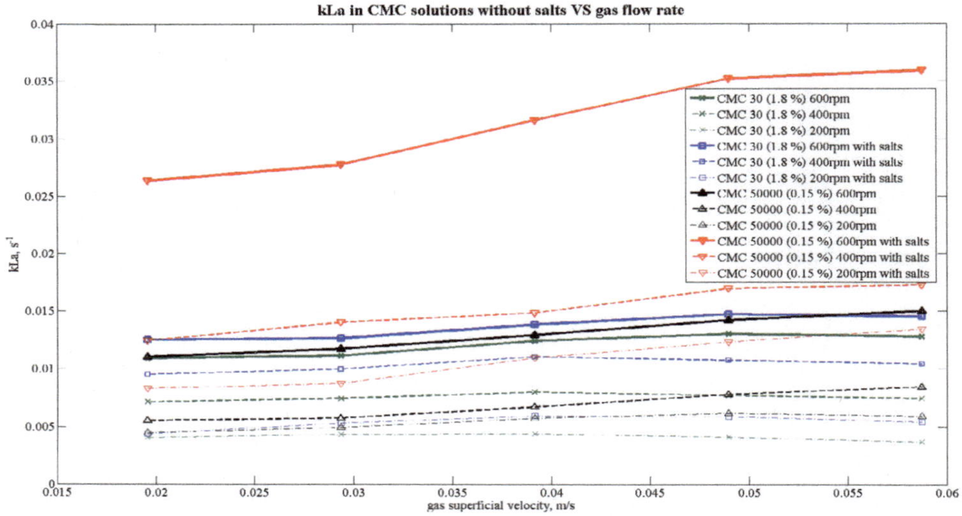

Figure 14: Measured overall kLa via dynamic method for water-based solutions of CMC 50000 (0.15 %) and CMC 30 (1.8 w%) in gas-liquid STR alone and in presence of salts (1 mol/L).

primary phase flow has been resolved, converged solution results of which were used as initial data before activation of secondary phase. It is important to resolve primary phase accurately because primary phase hydrodynamics has great impact onto secondary phase. Before gas phase introduction in a transient manner, the simulation results of mixed non-Newtonian solution were validated against experimental results.

Qualitative analysis was performed by comparison of the velocity contour fields in vertical 2D plane sampled at the identical spatial position and it is shown in Figure 15. Particle Image Velocimetry used in this work measured only two velocity components out of three. Therefore, the velocity contour of CFD simulation results was adjusted for qualitative comparison and includes only radial and axial components. The tip speed of the Rushton turbine rotating at 400 rpm is 1.57 m/s. Radial and tangential at the tip of impeller can be equaled. However, accurate measurements of impeller jet-stream are complicated by rotating blades of the turbine and slight swinging of the shaft. That makes slight underprediction of maximum radial velocity component in the impeller region in PIV measurements.

Overall velocity distribution profiles look quite similar between CFD and PIV results. Both circulations taking place above and below jet-stream are captured at the same locations and in similar manner. However, jet-stream velocity field profile is a bit thicker in CFD contour. It could be explained by isotropic turbulence being used by the turbulence model, while the flow turbulence produced by a Rushton turbine is anisotropic [44,45].

Since velocity filed contours comparison might be biased, the velocity components normalized with the tip speed were validated against experimental results in radial profiles (Figure 16). In order to sample hydrodynamics at different regions of the reactor, the profiles were taken at three vertical positions 30, 80 and 150 mm starting from the bottom. The simulation of shear-thinning flow agitated by the Rushton turbine describes the flow accurately.

Power draw was found experimentally by measuring torque via on-shaft torque transducer. The measured and simulated torque of single phase mixing of CMC 50000 (0.15 w%) and the leaching slurry are presented in Table 6. As expected, the reduction of power draw is observed in case of gas-liquid mixing, because average density of multiphase solution is lower than of single phase. The torque of mixed CMC 50000 was simulated accurately.

Aerated slurry mixing: When the solution of the leaching slurry mixing has reached convergence, air has been introduced via velocity inlet in transient manner with 0.02 s as time step. The simulation lasted until the balance of air mass flow rate was observed. Average density of the suspension (30 w%) is 1570 kg/m³.

Average velocity field of the mixed slurry is presented in Figure 17. Maximum value of velocity magnitude is around tip speed. One may notice velocity contour of the air jet supplied via sparger.

As it was found during the gas bubbles study in PIV section, the BSD is very narrow in presence of electrolytes (0.1 mol/L). Therefore, 1.5 mm bubble size was used in the simulations. Specific volumetric surface area of secondary phase can be found as follows:

$$A_p = \frac{6\varepsilon_p}{d_p} \quad (25)$$

where ε_p is the volume fraction of secondary phase, fraction.

The volume fraction and specific surface area of gas phase in the mixed slurry are shown in Figure 18. On one hand, high apparent viscosity of the leaching slurry increases dissipation rate of energy and power draw growth as a result. On other hand, higher viscosity hinders rise velocity of gas bubbles that leads to higher gas hold up almost twice up to 1.35% (Figure 8). Being small the bubbles follow the flow of the primary phase. The flow pattern of the radial impeller has two loops

Flow type		Torque, mN·m	
		Experimental	CFD
Single phase	CMC 50000 (0.15 w %)	68 ± 3	65.6
	Thiosulphate slurry	-	103.0
Multiphase	Thiosulphate slurry	-	101.2

Table 6: Simulated VS experimental torque.

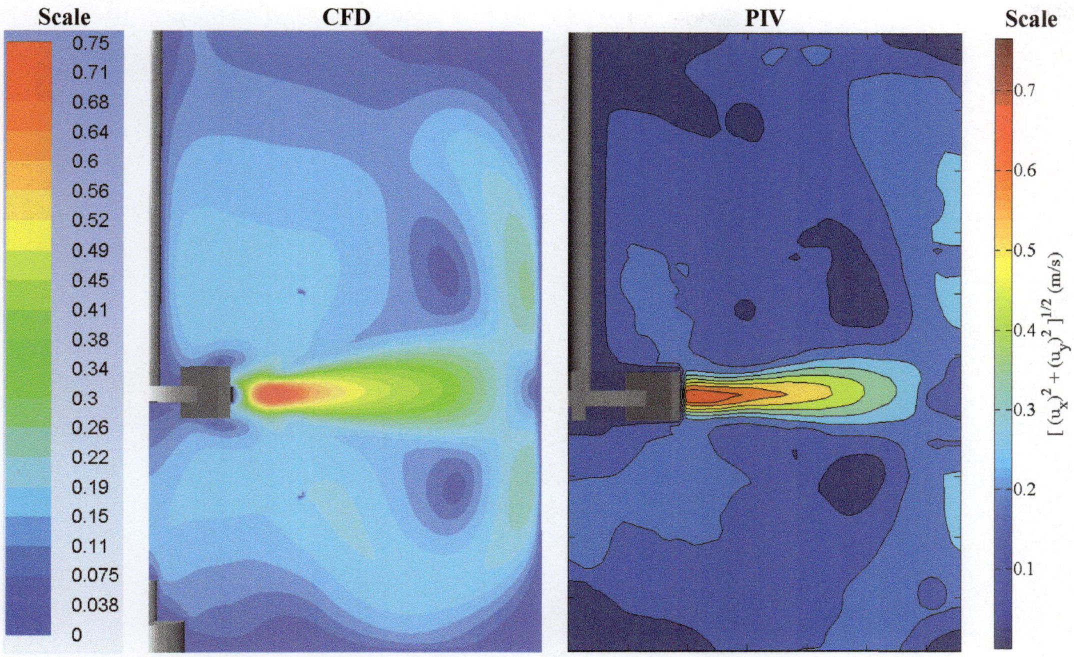

Figure 15: Average velocity field contours of water-based CMC 50000 (0.15 %) mixing at 6.67 rps modeled by CFD and measured by PIV.

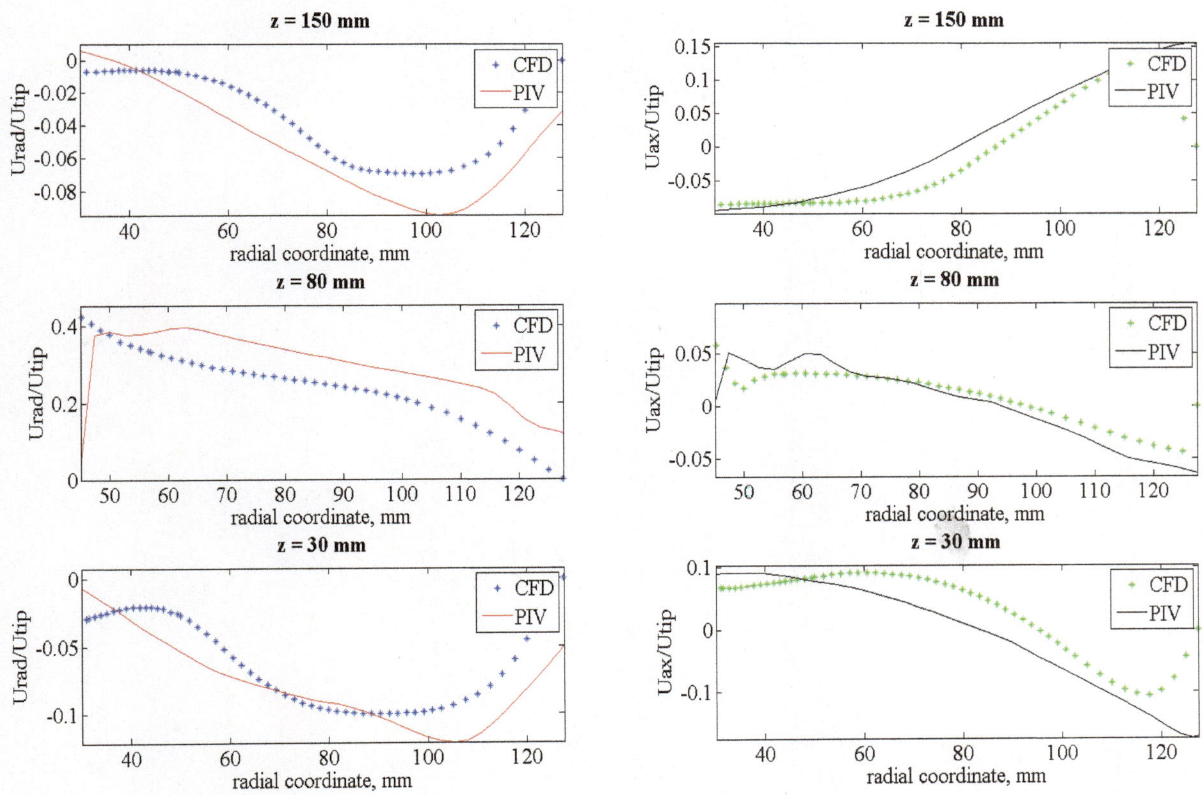

Figure 16: Comparison of normalized mean velocity components, measured by PIV and simulated by CFD of water-based CMC 50000 (0.15 %) mixing at 400 rpm, sampled at different vertical positions.

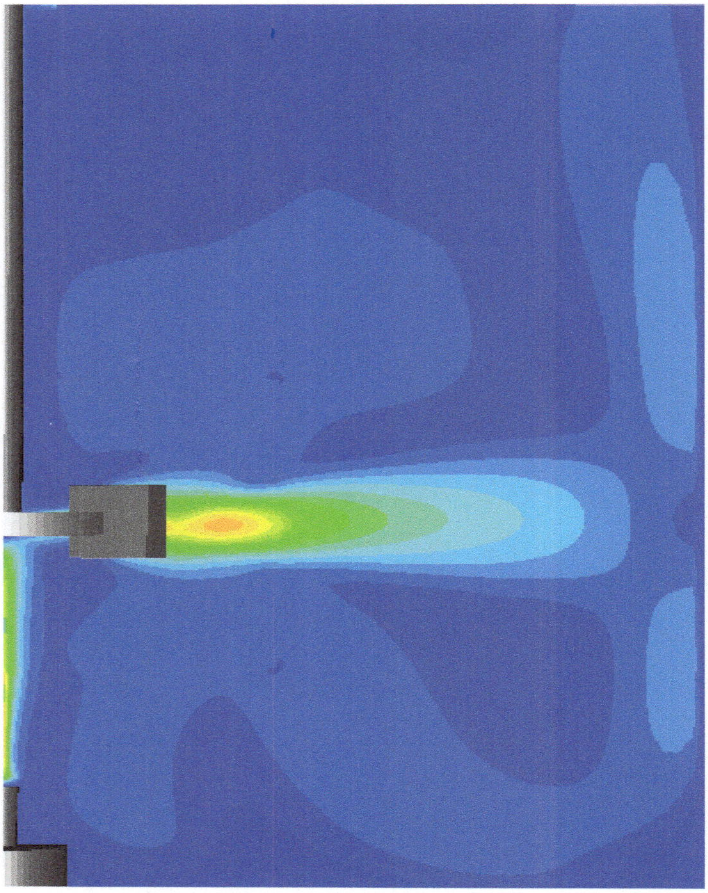

Figure 17: Average velocity contour fields of the slurry mixed at 400 rpm and aerated at 1.25 L/min (d_b=1.5 mm).

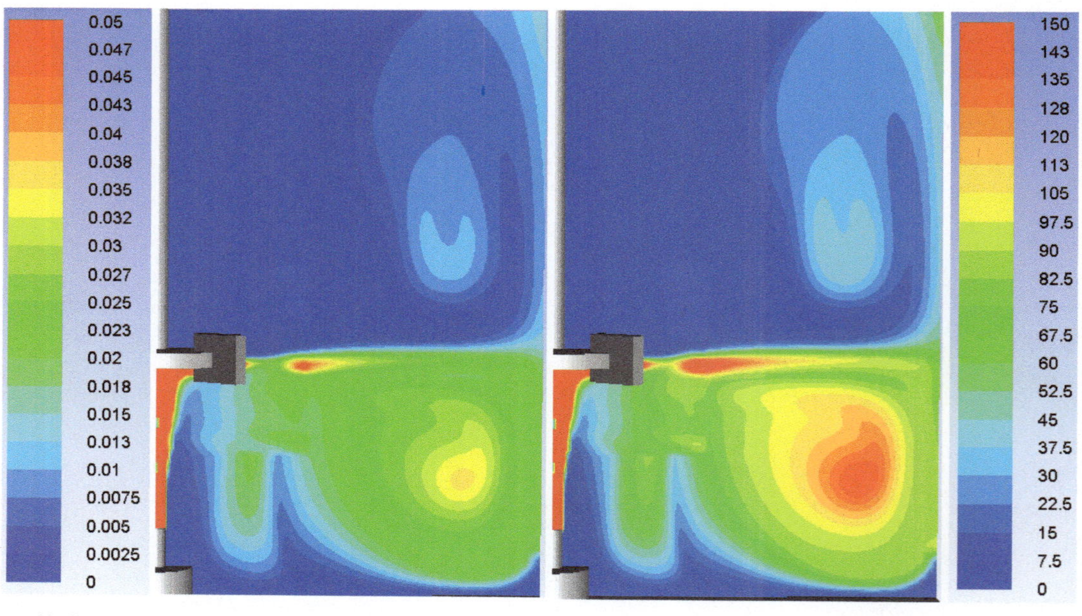

Figure 18: Contour of air volume fraction and specific surface area of the slurry mixed at 400 rpm and aerated at 1.25 L/min (d_b = 1.5mm).

of circulation. Following radial jet-flow of Rushton turbine they are dragged into down circulation while only small portion of gas goes up. The upper loop can also be distinguished since some amount of air bubbles are accumulated there. Total interfacial area is 0.26 m^2.

Mass transfer: Mass transfer has been modeled using Eq. 2. Diffusivity of oxygen in liquid depends on physical properties of solution. According to Hung and Dinius [46] increase of electrolytes concentration drops oxygen diffusivity. Diffusion coefficient of oxygen was calculated based on measured values of $D_L(O_2)$ in water with presence of sodium sulphate. Linear interpolation from the above mentioned work resulted in $D_L(O_2)=1.65 \times 10^{-9}$ m^2/s. Influence of suspended solids interaction with gas bubbles onto diffusion coefficient is unknown and neglected in this work since particles are finely milled. Theoretically, tiny particles can partially cover gas-liquid interface and cause some decrease of oxygen mass transfer into the liquid phase. However, overall k_La reduction should not be significant since interfacial area was increased drastically due to small bubbles. The distributions of specific volumetric mass transfer coefficient and mass transfer rate of oxygen are presented in Figure 19.

As it can be noticed from the distribution of specific mass transfer, the rate of mass transfer is higher in the regions where gas hold-up and energy dissipation rate are high. Therefore, the most intensive mass transfer is in the radial jet created by Rushton turbine. However, high gas hold-up of small bubbles, which do not coalesce, in the lower loop promotes mass transfer. Also, the region of moderate k_La value, caused by bubble retention in the upper circulating loop can be seen. Overall simulated mass transfer in mixed thiosulphate slurry is 0.0183 s^{-1}, which is 30% higher compared to the mass transfer measured in the mixed CMC 50000 (0.15%) solution under analogous operational conditions [47,48].

Conclusions

During this work mixing of aerated slurry on the example of thiosulphate leaching has been studied experimentally and simulated numerically. Rheology of the suspended solids was measured and found similar to those of shear-thinning fluids, which can be described by power-law model. Therefore, the slurry was imitated with water-based Carboxy Methyl Cellulose solutions (CMC 30 and CMC 50000) in order to study mixing hydrodynamics experimentally. Velocity vectors and shear rate fields of CMC 50000 (0.15 w%), mixed at 400 rpm were measured by stereoscopic PIV in vertical 2D plane.

Volumetric mass transfer was measured by dynamic method in the CMC solutions over the range of operational conditions in absence and presence of electrolytes. Presence of electrolytes promotes mass transfer of oxygen. Electrolytes decrease viscosity by breaking molecular chains of CMC that increases mass transfer coefficient. Bubbles coalescence blockage causes significant growth of surface area. Measurements of volumetric mass transfer at constant operational conditions were made with stepwise addition of electrolytes. The study revealed that decrease of bubble size is dominating factor upon k_La because viscosity drop takes place at lower salt amount. When level of electrolytes saturation is reached slight decrease of mass transfer was found at stable power draw. It could be due to contamination of the gas-liquid interfacial area.

Local gas hold-up in the aerated CMC 50000 (0.15 w%) solution was measured by EIT. Potentially the technique is capable to measure local gas hold-up in slurry. However, the sensitivity of the set-up was not enough to measure gas hold-up distribution of dispersed phase in electrolyte solution accurately.

The achieved experimental data was used in CFD modelling of the aerated slurry. Single phase non-Newtonian fluid mixing, modelled via effective viscosity, was validated against the PIV measurements of velocity components in vertical plane. Simulated and measured power consumptions of the CMC mixing at 400 rpm were in reasonable agreement. In the presence of electrolytes (0.1 mol/L) the bubble size in the CMC solution (400 rpm and 1.25 L/min) was found to be almost constant (1.5 mm). Multiphase mixing hydrodynamics of the thiosulphate slurry was modeled at the assumption of constant bubble size. Results of the simulation were presented and discussed. The simulated k_La in the leaching pulp was compared to the measured one

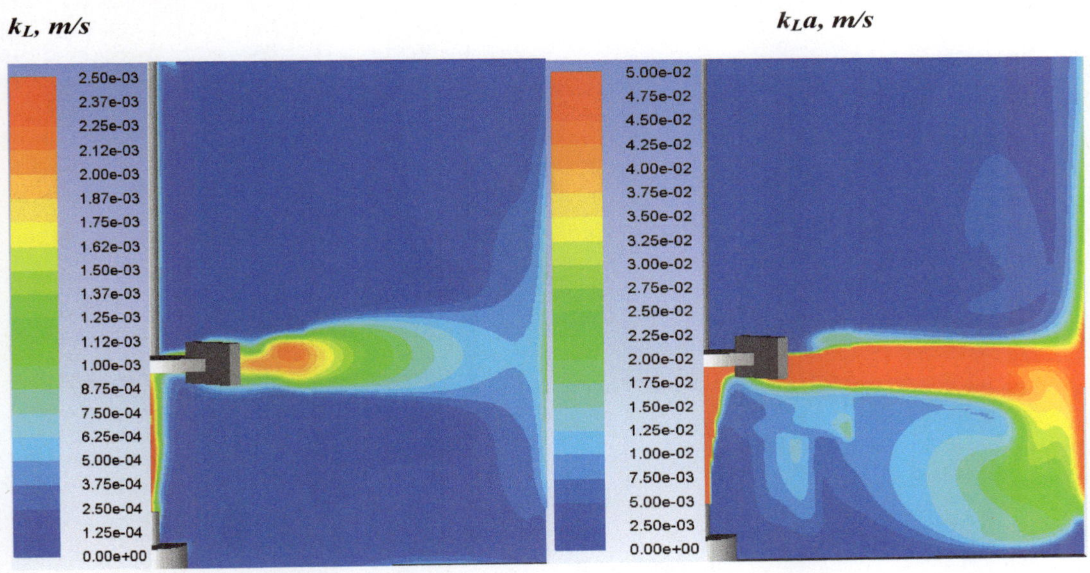

Figure19: Mass transfer coefficient and mass transfer rate of the slurry mixed at 400 rpm and aerated at 1.25 L/min (d_b = 1.5mm).

in CMC 50000 (0.15 w%) under analogous operating conditions. It was found that bigger density of the leaching slurry caused higher global gas hold-up and hence oxygen mass transfer rate [49].

Acknowledgements

The authors are grateful to the Finnish Funding Agency TEKES and, in particular, to Outotec Finland Oy collaborating under FERMATRA project (908/31/2016 and 958/31/2016), for active supervision and financial support. Also, the work of Marko Vauhkonen and Gerardo González was supported by the Academy of Finland (Project 250215 Finnish Center of Excellence in Inverse Problems Research).

Nomenclature

Symbols	Meaning	Units
A	Signal amplitude	-
A_i	Interphase surface area	m^2
C_i	Empirical coefficient	-
C_D	Drag force coefficient	-
c_{gas}	Oxygen concentration in gas phase	mol/m^3
c_{liquid}	Oxygen concentration in liquid phase	mol/m^3
d_b	Bubble diameter	m
D_L	Diffusivity in liquid phase	m^2/s
$\vec{F}_i$	Force	N
h_{gassed}	Surface level in gassed reactor	m
$h_{ungassed}$	Surface level in non-gassed reactor	m
K	Consistency index	$Pa \cdot s^n$
K_{pq}	Interphase exchange coefficient	-
k	Turbulence kinetic energy	m^2/s^2
k_L	Mass transfer coefficient in liquid phase	m/s
k_La	Volumetric mass transfer coefficient	s^{-1}
N_{fine}	Number of cells in fine grid	-
N_{coarse}	Number of cells in coarse grid	-
n	Dimensionless flow behavior index	-
NRMSE	Normalized root mean square error	%
$\overline{p}$	Pressure	Pa
P^*	Order of discretization	-
Re	Reynold's number	-
St	Stoke's number	-
T_L	Turbulence length scale	m
t	Time	s
U_T	Particle terminal velocity in turbulent flow	m/s
U_S	Particle terminal velocity in stagnant flow	m/s
$\overline{u}$	Mean velocity	m/s
u_r	Radial velocity component	m/s
u_z	Axial velocity component	m/s
V	Volume of reactor	m^3
V_{image}	Imaginary part of measured voltage	V
V_{real}	Real part of measured voltage	V

Greek letters	Meaning	Units
α_i	Phase volume fraction	-
ε	Turbulence kinetic energy dissipation rate	m^2/s^3
ε_{EIT}	Global gas hold-up measured by EIT	-
$\varepsilon_{\Delta h}$	Global gas hold-up measured by level differencing	-
$\varepsilon_{i,EIT}$	Local gas hold-up in element i, measured by EIT	-
μ	Dynamic viscosity	$Pa \cdot s$
μ_{eff}	Effective viscosity	$Pa \cdot s$
μ_{turb}	Turbulent viscosity	$Pa \cdot s$
η	Turbulence modification factor for drag force	-
ρ	Density	kg/m^3
σ_1	Conductivity distribution of non-gassed mixing	S/m
σ_2	Conductivity distribution of gas-liquid mixing	S/m
τ	Time constant	s
τ_b	Relaxation time of bubble	s
τ_q	Relaxation time of primary phase	s
ν	Kinematic viscosity	m^2/s
γ	Shear rate	s^{-1}
λ	Kolmogorov length scale	m
φ	Signal phase	-
ϕ_i	Prediction of property value on grid i	-

Acronyms	Transcript
CCD	Charged-Couple Device
CFD	Computational Fluid Dynamics
CISTR	Continuous Ideally Stirred-Tank Reactor
CMC	Carboxy Methyl Cellulose
FEM	Finite Element Method
HWA	Hot Wire Anemometry
LDA	Laser Doppler Anemometry
MCR	Modular Compact Rheometer
Nd:Yag	Neodymium-doped yttrium aluminum garnet
PIV	Particle Image Velocimetry
UDV	Ultrasonic Doppler Anemometry
vvm	Volume of gas per volume of mixed solution per minute

References

1. Jiang T, Chen J, Xu S (1993) A kinetic study of gold leaching with thiosulfate. Hydrometallurgy, Fundamentals, Technology and Innovations. AIME, pp: 119-126.
2. Chen J, Deng T, Zhu G, Zhao J (1996) Leaching and recovery of gold in thiosulfate based system research summary at ICM. Transactions of Indian Institute of Metallurgy 49: 841-884.
3. Naito K, Sheh C, Okabe T (1970) The chemical behaviour of low valence sulfur compounds. Decomposition and oxidation of tetrathionate in aqueous ammonia solution. Bulletin of the Chemical Society of Japan 43: 1372-1376.
4. Groose AC, Dicinoski GW, Shaw MJ, Haddad PR (2003) Leaching and recovery of gold using ammoniacal thiosulfate leach liquors (a review). Hydrometallurgy 69: 1-21.
5. Joshi JB, Nere NK, Rane CV, Murthy BN, Mathpati CS, et al. (2013). CFD simulation of stirred tanks: comparison of turbulence models. Part II: axial flow impellers, multiple impellers and multiphase dispersions. Canadian Journal of Chemical Engineering 89: 754-816.
6. Jensen K (2004) Flow measurements. 10th Brazilian Congress of Thermal Sciences and Engineering 26: 400-419.
7. Khopkar A, Aubin J, Rubio-Atoche C, Xuereb C, Le Sauze N, et. al. (2004) Flow Generated by Radial Flow Impellers: PIV Measurements and CFD Simulations. International Journal of Chemical Reactor Engineering 2: 1-14.
8. Moilanen P (2009) Modelling gas-liquid flow and local mass transfer in stirred tanks. Helsinki, PhD, Finland: Helsinki University of Technology.
9. Xinhong L, Yuyun B, Zhipeng LI, Zhengming G (2010) Analysis of turbulence structure in the stirred tank with a deep hollow blade disc turbine by time-resolved PIV. Fluid Flow and Transport Phenomena 18: 588-599.
10. Busciglio A, Grisafi F, Scargiali F, Brucato A (2013) On the measurement of local gas hold-up, interfacial area and bubble size distribution in gas-liquid contactors via light sheet and image analysis: Imaging technique and experimental results. Chemical Engineering Science 102: 551-566.
11. Takeda Y (1986) Velocity Measurement by Ultrasonic Doppler Shift Method. International Journal Heat and Fluid Flow 7: 313-318.
12. Mann R, Dickin FJ, Wang M, Dyakowski T, Williams RA, et. al. (1997) Application of electrical resistance tomography to interrogate mixing process at plant scale. Chemical Engineering Science 52: 2087-2097.

13. Wang M, Dorward A, Vlaev D, Mann R (2000) Measurements of gas-liquid mixing in a stirred vessel using electrical resistance tomography (EIT). Chemical Engineering Journal 46: 93-98.

14. Kourunen J, Niiti T, Heikkinen LM (2011) Application of three-dimensional electrical resistance tomography to characterize gas hold-up distribution in laboratory flotation cell. Minerals Engineering 24: 1677-1686.

15. Arizmendi-Morquecho AM, Perez-Garibay R, Uribe-Salas A, Nava-Alonso F (2002) On-line solids hold-up measurement in mineral slurries by the standard addition method. Minerals Engineering 15: 61-64.

16. Gomez CO, Finch JA (2007) Gas dispersion measurements in flotation cells. International Journal of Mineral Processing 84: 51-58.

17. Higbie R (1935) The rate of absorption of a pure gas into a still liquid during short periods of exposure. Institution of Chemical Engineers 35: 36-60.

18. Kawase Y, Halard B, Moo-Young M (1992) Liquid-phase mass transfer coefficients in bioreactors. Biotechnology and Bioengineering 93: 1133-1140.

19. Garcia-Ochoa F, Gomez E (2004) Theoretical prediction of gas-liquid mass transfer coefficient, specific area and hold-up in sparged stirred tanks. Chemical Engineering Science 59: 2489-2501.

20. Lampinen M, Laari A, Turunen I (2015) Ammoniacal thiosulfate leaching of pressure oxidized sulfide gold concentrate with low reagent consumption. Hydrometallurgy 151: 1-9.

21. Sharikov FY, Zhukov VV, Lampinen M (2012) Investigation of Gold Oxidized Leaching Process from Concentrate of Base Ore by Utilization Calve Calorimetry. Problem Definition and First Results, Material Investigation by Utilization Methods of Thermal Analysis, Calorimetry and Gas Trapping. International Conference Reports. Saint-Petersburg: Poltorak: 59-61.

22. Wuestenberg T (2014) Cellulose and Cellulose Derivatives in the Food Industry: Fundamentals and Applications. Germany: Wiley-VCH.

23. Kourunen J (2009) Imaging of Mixing in Selelcted Industrial Processes Using Electrical Resistance Tomography. Kuopio, PhD Thesis, Finland: University of Eastern Finland.

24. Heikkinen L (2005) Statistical Estimation Methods for Electrical Process Tomography. Kuopio, PhD Thesis, Finland: University of Kuopio.

25. Holden PJ, Wang M, Mann R, Dicki FJ, Edwards RB (1999) On Detecting Mixing Pathologies Inside a Stirred Vessel Using Electrical Resistance Tomography. Chemical Engineering Research and Design 77: 709-712.

26. Tarkiff MS, Hamzah AA, Karnarudin SK, Abdullah J (2009) Electrical Resistance Tomography Investigation of Gas Dispersion in Gas-Liquid Mixing in an Agitated Vessel. Journal of Applied Science 9: 3110-3115.

27. Bawadi A, Chirag D, Nguyen T, Cooper CG, Adesoji AA (2011) Electrical resistance tomography-assisted analysis of dispersed phase hold-up in a gas-inducing mechanically stirred vessel. Chemical Engineering Science 66: 5648-5662.

28. Montante G, Paglianti A (2015) Gas hold-up distribution and mixing time in gas-liquid stirred tanks. Chemical Engineering Journal 279: 648-658.

29. Adrian R (2011) Bibliography of particle image velocimetry using imaging methods: 1917 - 1995. University of Illinois. Illinois: TAM Report 817.

30. Haario H (1994) MODEST User's guide. ProfMath Oy, Helsinki, Finland.

31. Van't Riet K (1979) Review of measuring methods and results in non-viscous gas-liquid mass transfer in stirred tanks. Industrial and Engineering Chemistry Process Design and Development 18: 357-364.

32. Ruchti G, Dunn IJ, Bourne JR (1981) Comparison of dynamic oxygen electrode methods for the measurement of kLa. Biotechnology and Bioengineering 23: 277-290.

33. Bhutani G, Cilliers JJ, Brito-Parada PR (2017) A three-dimensional numerical framework for the modeling of the pulp phase in froth flotation Computational Modeling 17: 1-8.

34. Gradov DV, Laari A, Turunen I, Koiranen T (2016) Experimentally Validated CFD Model for Gas-Liquid Flow in a Round-Bottom Stirred Tank Equipped with Rushton Turbine. International Journal of Chemical Reactor Engineering 15: 1-20.

35. Holden J (1955) Fundamentals of the hydrodynamics mechanism of splitting in dispersion processes. American Institute of Chemical Engineers Journal 289-295.

36. Schiller L, Naumann Z (1935) A Drag Coefficient Correlation. Z. Ver. Deutsch. Ing 77: 318-320.

37. Brucato A, Grisafi F, Montante G (1998) Particle drag coefficients in turbulent fluids. Chemical Engineering Science 53: 3295-3314.

38. Scot Blair GW (1939) An Introduction to Industrial Rheology. Journal of Physical Chemistry 43: 853 - 864.

39. Rewatkar VB, Deshpande AJ, Pandit AB, Joshi JB (1993) Gas hold-up behaviour of mechanically agitated gas-liquid reactors using pitched blade down flow turbines. Canadian Journal of Chemical Engineering 71: 226-237.

40. Kraume M, Zehner P (2001) Experience with Experimental Standards for Measurements of Various Parameters in Stirred Tanks. Chemical Engineering Research and Design 79: 811-818.

41. Laakkonen M (2006) Development and Validation of Mass Transfer Models for the Design of Agitated Gas-Liquid Reactors. Helsinki, PhD, Finland: Helsinki University of Technology.

42. Prince MJ, Blanch HW (1990) Transition Electrolyte Concentration for Bubble Coalescence. American Institute of Chemical Engineers Journal 36: 1425-1429.

43. Wang H, Zhai Z (2012) Analyzing grid independency and numerical viscosity of computational fluid dynamics for indoor environment applications. Building and Environment 52: 107-118.

44. Lee KC, Yianneskis M (1998) Turbulence Properties of the Impeller Stream of a Rushton Turbine. American Institute of Chemical Engineering 44: 13-24.

45. Fentiman NJ, Lee KC, Paul GR, Yianneskis M (1999) On the trailing vortices around hydrofoil impeller blades. Chemical Engineering Research and Design 77: 731-740.

46. Hung GW, Dinius RH (1972) Diffusivity of oxygen in electrolyte solutions. Journal of Chemical & Engineering Data 17: 449-451.

47. Busciglion A, Grisafi F, Montante G (1998) Particle drag coefficients in turbulent fluids. Chemical Engineering Science 53: 3295-3314.

48. Hugo P, Steinbach J (1986) A comparison of the limits of safe operation of a SBR and a CSTR. Chemical Engineering Science 41: 1081-1087.

49. Lasa HI (1986) Chemical Reactor Design and Technology. Ontario, Canada: University of Western Ontario.

Crystallization of Kainite from Solutions in System K^+, Mg^{2+}, Na^+ // Cl^-, SO_4^{2-} - H_2O

Kostiv IY[1] and Ya I Basystiuk[2]*

[1]State Enterprise Scientific - Research Gallurgi Institute 5a, Fabrichna Str., 76000 Kalush, Ukraine
[2]Precarpathian National University named after V. Stephanyk 57, Shevchenko Str., 76025 Ivano-Frankivsk, Ukraine

Abstract

In isothermal conditions have been studied the influence of system solution sea salts during their evaporation and crystallization to composition of received liquid and solid phases. It is shown that evaporation of the solution leading to its supersaturation by sulfate salts. In the liquid phase before crystallization of kainite concentration of SO_4^{2-} ions increases to 10% and above. Through this probability of formation of crystals increases and the result of evaporation is the formation of finely dispersed kainite. During crystallization of kainite concentration of SO_4^{2-} in the liquid phase decreases rapidly. Decreasing its intensity increases with increasing value k=E SO_4^{2-} : E Mg^{2+} of initial solution. For the degree of evaporation of 20.0% and 31.0% concentration of SO_4^{2-} in evaporated solution on the value of k does not change. Most forms of potassium salts in the solid phase for the value k=0.7573 and reaches a maximum value at 82% evaporation rate of 31%. With the increasing degree of evaporation of the solution the degree of separate of SO_4^{2-} in solid phase also increases. Most of the value achieved the degree of evaporation of the solution is 31% and about 90% for all values of k. This is an increase of salting action of $MgCl_2$ of liquid phase at kainite.

Keywords: Kainite; Schenite; Crystallization; Evaporation; Magnesium chloride solution; Kieserite; Langbeinite; Potassium chloride; Sodium chloride

Introduction

Crystallization of kainite in technology of polymineral potassium processing raw of materials chloride sulfate type and natural salt solutions in nochloride potassium and potassium-magnesium fertilizers is a major operations of process. This is because during crystallization schenite of saturated solutions obtained by dissolving natural raw in the liquid phase as a result of reaction conversion chloride minerals (kainite $KCl{\cdot}MgSO_4{\cdot}3H_2O$, sylvine KCl) in sulfate (schenite $K_2SO_4{\cdot}MgSO_4{\cdot}6H_2O$) increases the concentration of $MgCl_2$. So part of the solution after crystallization of schenite separated from the sulfate process. With this solution according to the potash production separated to 40% of potassium and sulfate on their number of processed raw materials. Therefore, it is sent for multistage evaporation to separate of sulfate salts and return them to the production of sulfate potassium fertilizers. During evaporation in industrial conditions is not achieved a high degree of extraction of potassium and sulfate from excessive schenite solutions in kainite, high dispersion of the solid phase and therefore the difficulty of filtering and significant losses of liquid phase, which adversely affects to the work stages of dissolution of ore and crystallization schenite. This causes the low efficiency of polymineral potash ore processing in general. To increase the size of crystals kainite suggest reducing the intensity of evaporation from 450-600™ 50-250 kg/m³· hour [1], and mixing the initial solution with evaporated to reduce the correlation concentrations of $MgSO_4$ and $MgCl_2$ [2]. Previously [3], we have found that the change of composition of system solution sea salts for constant degree and intensity of evaporation has little effect on the size of the crystals kainite. Evaporation of the addition of solid phase kainite at constant composition of the liquid phase makes it possible to increase the average size of the crystals kainite only 11.9%. Increasing the depth of evaporation of the solution leads to a substantial increase in the average size of the crystals kainite, but it increases the content of solid phase and the density of the suspension, which hinders evaporation. So evaporated suspension should be diluted by concentrated on the final $MgCl_2$ solution. After separation of crystalline magnesium chloride solution which containing sulfates [4] the past remain in the liquid phase and finely dispersed solid phase. Return them to the process of crystallization of kainite leads to the binding coarse grained potassium chloride in finely dispersed kainite. The result is a suspension of the solid phase, which can easily be cleaned of impurities coarse grained sodium chloride. Evaporation of sea system solutions of salts is accompanied by movement of figurative point of the system from one field to another crystallization. So for a solution with index x=20 ÷ 30 and above 368 K first crystallized mixture of NaCl+KCl. After reaching the limit saturation by sulfates crystallizing kainite, while previously separated KC1 dissolved [5]. With the concentration of Mg^{2+} 7.95 starts crystallization of kieserite and carnallite, accompanied dissolution of kainite reaction by the equation:

$$KCl{\cdot}MgSO_4{\cdot}3H_2O+MgCl_2=KCl{\cdot}MgCl_2{\cdot}6H_2O+MgSO_4$$

The above suggests that, being in the field of crystallization kainite, all other minerals are not stable. Similarly in other fields, crystallized mineral that corresponds this field, while the others are dissolved, increasing its output. This explains the fact that during the change of temperature changes border fields crystallization and system with constant composition of the liquid phase moves from one field to another crystallization. Therefore, during the temperature increasing chlorides salting sulfates, but with decreasing - sulfates salting chlorides [6]. Evaporation is carried out in vacuum evaporators, the temperature in corps which is 323 to 388 K. It is known [7] that the maximum size of the crystals kainite without impurities langbeinite is achieved by evaporation of the solution after separation sodium chloride from

***Corresponding author:** Ya I Basystiuk, Precarpathian National University named after V. Stephanyk 57, Shevchenko Str., 76025 Ivano-Frankivsk, Ukraine
E-mail: yaroslava_artus@ukr.net

the addition solid phase of kainite at a temperature of 363-368 K. The composition of the initial solution significantly affects the coefficient supersaturation of solution in the crystallization process at the rate of formation crystals and granulometric composition of sediment.

Experimental Section

Research of processes of evaporation and crystallization kainite was carried out on a laboratory vacuum evaporating installation and thermostatted glass reactor with paddle mixer. The first included throat round bottom flask with a capacity of 2500 cm^3, mechanical paddle mixer with electric meter and velocity, heater, glass refrigerator of vapors, calibrated collection of condensate, vacuum pump and system manually adjust the depth of vacuum, by which the temperature of evaporation supported in the temperature range 323-388 K with precision of ± 0.5 K. Thermostatted reactor includes a rubber cover closed double-walled glass vessel with a capacity of 1200 cm^3 elliptical bottom, glass paddle mixer with electric drive and electric thermostat which is supported temperature with a precision ± 0.1 K pumping water between the walls of the reactor.

Initial for researches was polymineral processing solution of potassium ores after evaporation and crystallization of sodium chloride composition by weight.%: 5.73 K^+, 4.33 Mg^{2+}, 2.82 Na^+, 16.76 Cl^-, 7.37 SO_4^{2-}, 63.00 H_2O, previously diluted with water to 30% to density of 1246 kg/m^3. In the flask for evaporation poured 2 dm^3 diluted solution. And added to change the composition of the solution different quantity of crystalline KCl, $MgSO_4{\cdot}7H_2O$, $MgCl_2{\cdot}6H_2O$, Na_2SO_4. This solution is heated to temperature of experiment and evaporated water, added earlier to dilution. Obtained initial solution. Thereafter evaporated the different quantity of water which corresponded to the specified degree of evaporation (α), which represented a ratio of the mass of condensate to initial solution. The temperature of evaporation supported constant and was 368 K. The intensity of evaporation in all experiments was constant and was equal to 0.45 h^{-1} (1 kg evaporated water with 1 kg of starting solution for 1 hour). After evaporation turned off heater and vacuum pump, open the flask and the sample of suspensions were selected for analysis by special sampler, and then again turn on the installation work. A sample of the suspension was filtered under vacuum in a heated crucible Schott № 16. The filtrate flow down in previously weighed test tube and solid phase after filtration were washed on the filter from intercrystalline solution first purified vegetable oil, and with organic solvent and dried at 323 K in an air thermostat for 30 min. The obtained liquid and solid phases were analyzed using tetraphenylborate weight method (for K^+ ions), complexometric method (for Mg^{2+} and Ca^{2+} ions), mercurimetry (for Cl^- ions), weight method (for SO_4^{2-} ions), as difference between the sum of equivalents of anions and cations (for Na^+ ions), H_2O – by difference. A result of analysis in a special program material balances were calculated and determined the degree of extracted components in the solid phase.

Results and Discussion

We have been studied the influence of composition and degree of evaporation to concentration in the solution evaporated SO_4^{2-} in solution. The results are shown in Figures 1 and 2. The composition of the solution given as equivalent indices X=E SO_4^{2-} : (E $2K^+$+E $2Mg^{2+}$); Y=E Mg^{2+}: (E $2K^+$+E $2Mg^{2+}$). Changing value X:Y of the initial solution and the ratio k=X:Y or, respectively, k=E SO_4^{2-}:E Mg^{2+}. The purpose of evaporating the solution at the stage of separating kainite is the maximum excretion of ions K^+ and SO_4^{2-} in the solid phase and obtaining a solution with a low concentration of SO_4^{2-}. The equivalent ratio E SO_4^{2-} i E Mg^{2+} corresponds to the ratio molar concentration of magnesium sulfate to the sum molar concentrations of magnesium chloride and magnesium sulfate.

We investigated the value of the maximum concentration of SO_4^{2-} ion of evaporation process at a temperature of 368 K. The results are shown in Figure 1.

As follows from the data obtained (Figure 1), evaporation of the solution leads to a gradual supersaturation for its sulfate salts. In this time there is crystallization of potassium chloride and sodium [6]. In solutions with a high index value X before crystallization kainite SO_4^{2-} ion concentration is achieved 10% and above. This suggests that with increasing X index increasing quantity of kainite which crystallized. Due to this, the probability of formation of new crystals increases and the result of evaporation is the formation of a large number of finely dispersed crystals of kainite. After achieving a critical supersaturation begins intensive crystallization kainite and SO_4^{2-} concentration in the liquid phase decreases rapidly. Decreasing its intensity increases with increasing value of k=E SO_4^{2-} : E Mg^{2+} the initial solution. During the dilution of initial solution by evaporated final solution is not enough crystallized material to increase the size of the crystals. Slowest decrease concentrations of SO_4^{2-} in the liquid phase is for

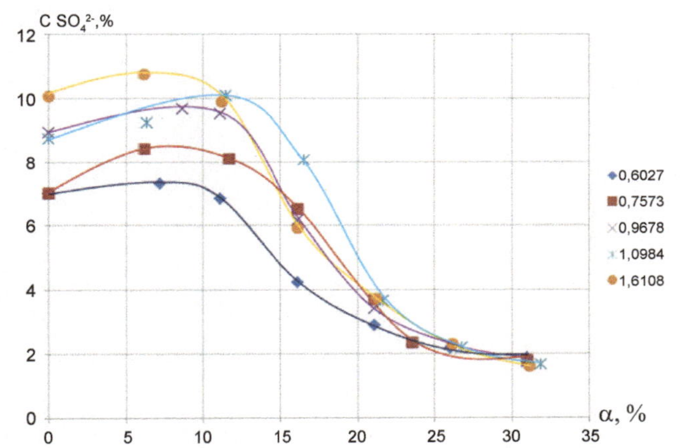

Figure 1: The dependence of the concentration (C) of sulfate ions in the solution *vs* the degree of evaporation (α) ratio for various equivalents $MgSO_4$ and $MgCl_2$ in the initial solution.

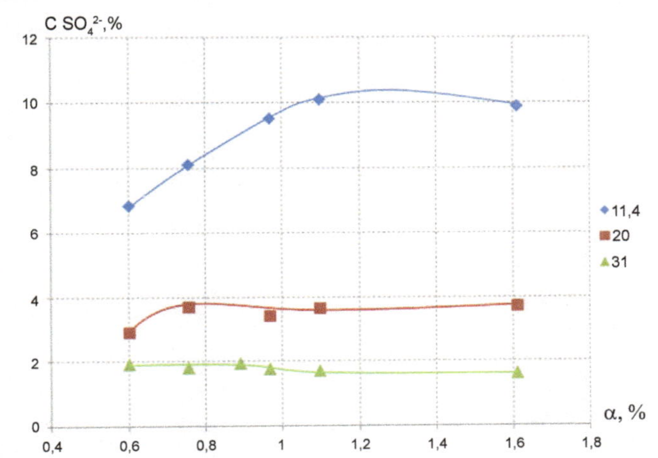

Figure 2: The dependence of the concentration (C) of sulfate ions in the evaporated liquid phase *vs* ratio of equivalents (k) $MgSO_4$ and $MgCl_2$ in the initial solution at different degrees of evaporation: 11.4; 20.0; 31.0%.

the value of k=0.6027, namely the highest concentration of magnesium chloride. This is caused by the lower number of ions SO_4^{2-} to form a solid phase kainite. Figure 2 shows the dependence of the concentration of SO_4^{2-} in evaporated solution under different degree of evaporation solution of the value k of initial solution. As can be seen from Figure 2, SO_4^{2-} concentration in solution evaporated at the degree of evaporation of a solution is 11.4% increasing with the value of k of initial solution to 1.09. For the degree of evaporation 20.0 and 31.0% SO_4^{2-} concentration in solution evaporated from the value of k does not change. So to achieve lowest value of concentrations of SO_4^{2-} in the evaporated solution can be by evaporation depth and addition part of evaporated solution to the original. Return part of the solution evaporated to evaporation causes reducing the intensity formation of crystals kainite and increase the size of its crystals. Figure 3 shows the results of calculation of the degree of extracted of K^+ from the degree of evaporation and the initial value of k solution.

As seen from the above data, the most of potassium salts separated in the solid phase for the value k=0.7573 and achieves maximum value for the 82% degree of evaporation of 31%. With the increase in the value of k degree of extracted potassium in the solid phase decreases. Thus, the return of the evaporated solution on evaporation promotes the separation of potassium salts in the solid phase by salting action on $MgCl_2$ kainite. Figure 4 shows the results of calculation of the degree of extracted SO_4^{2-} ions solid phase of the degree of evaporation for different values of k in the initial solution.

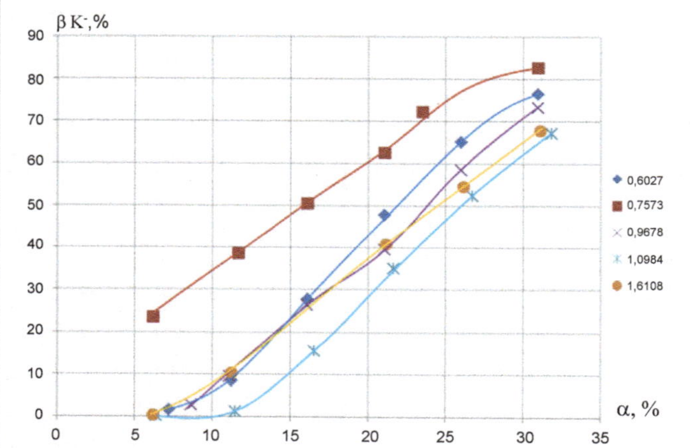

Figure 3: The dependence of the degree of extracted ions K^+ in the solid phase (β) *vs* the degree of evaporation (α) and the ratio E $MgSO_4$/E $MgCl_2$ in solution.

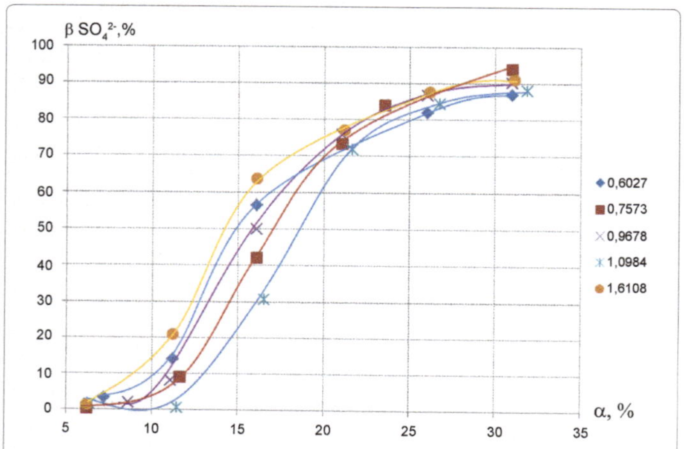

Figure 4: The dependence of the degree of extracted ions SO_4^{2-} in the solid phase (β) *vs* the degree of evaporation (α) and the ratio $MgSO_4$/E $MgCl_2$ in solution.

As can be seen from the data, with increasing degree of evaporation of the solution degree of extracted SO_4^{2-} in solid phase increased for all values of magnitude k. Most of the value is achieved by the degree of evaporation of the solution is 31% and about 90% for all values of k. This is explained by the increase concentration and salting action $MgCl_2$ liquid phase at kainite.

Conclusions

Consequently to achieve a high degree of excretion of salts K^+ and SO_4^{2-} in the solid phase can be due to evaporation of a solution to the high concentration of $MgCl_2$. This increases the number of extraction of salts and content of solid phase in evaporated suspension, which limits the depth of evaporation. Return part of the evaporated solution contributes to increases salting action of $MgCl_2$ on potassium and sulfate, and also reduction of solid phase in suspension during evaporation. Results of laboratory researches have shown that for practical realization of the process of evaporation potassium chloride solutions which containing sulfate ions on stage of crystallization kainite rationally choose three stages vacuum evaporators with direct flow movement of the heating vapor and evaporation of the solution initially in the first corps at a temperature of 383-388 K, then in the third at a temperature 333-335 K for formation crystals of kainite and then pump over the suspension into a second corps for evaporation and crystallization kainite at a temperature of 363-368 K. Evaporated solution expedient partially to return in the third corps for formation of crystals kainite. Evaporation at a temperature of 363-368 K without a solid phase kainite leads to partial extraction except kainite and langbeinite, which complicates the subsequent conversion of selected minerals in schenite. Suspension of kainite which containing coarse grained sodium chloride crystals, which are separated by hydroseparation and clean suspension of kainite thickened and sent to the stage of crystallization schenite.

References

1. AC 882915 (USSR), cl. C01 D 3/08. The method for allocating potassium and magnesium salts of schenite solutions // E.S Dubiley et al. – B.I. 1981, № 43.
2. AC 1049427 (USSR), cl. C01 D 3/08, 5/12. The method for allocating salt and potassium-magnesium salts of schenite solution // YG Goroshchenko et al. – B.I. 1983, № 39.
3. Kostiv I (1999) Untersuchung der Kristallisationskinetik von Kainit und Carnallit aus Losungen der Verarbeitung von polymineralischen Kalirohsalzen. Freiberger Forschungshefte A 853: 70-81.
4. Basystiuk YI, Kostiv IY (2016) Getting Hydrated Magnesium Chloride from Magnesium Chloride Solutions of Potassium Sulfate Fertilizers Production. J Chem Eng Process Technol 7: 2.
5. Kostiv IY (1995) Research of technology separate kainite from processing solutions of polymineral potassium ores. Ukraine Chemical Industry 2: 50-56.
6. Orlova VT, Yanateva OK (1971) Journal of Inorganic Chemistry 16: 2779-2781.
7. Kostiv IY (1990) Research process of evaporation solutions in system K^+, Mg^{2+}, Na^+|| Cl^-, SO_4^{2-} - H_2O at temperature 323-383 K. Journal of Applied Chemistry 1: 3-8.

Assessment of Ultrafiltration Cleaning Protocols in Drinking Water Treatment Processes

Vera M[1,2*], Martín-Alonso J[2], Mesa J[2], González S[3], Lefèvre B[3], Gibert O[1,3] and Cortina JL[1,3]

[1]*Barcelona Research Center in Multiscale Science and Engineering, EEBE, Universitat Politècnica de Catalunya (UPC) – Barcelona TECH, c/Eduard Maristany, Barcelona, Spain*
[2]*Aigües de Barcelona, Empresa Metropolitana de Gestió del Cicle Integral de l'Aigua, Gral. Barcelona, Spain*
[3]*CETAQUA, Water Technology Center, Ctra. D'Esplugues Cornellà de Llobregat, Spain*

Abstract

Membrane-based processes efforts are focused on mitigating membrane fouling by optimizing operating conditions, especially when low quality surface waters are treated. The study conducted here was driven to evaluate the current measures applied to mitigate fouling in Ultrafiltration (UF) stages when integrated as a pre-treatment of the Reverse Osmosis (RO) in a Drinking Water Treatment Plant (DWTP). First, the effect of coagulants in fouling and water quality was evaluated by using synthetic and surface waters in bench-scale. Both aluminum and iron-based coagulants rejected the organic compounds when dosed above 1 mg·L^{-1}. Also, micro-coagulation with $FeCl_3$ previous to the UF impacted positively the membrane performance. Then, two modes of operation were proposed and tested in bench-scale to improve the cleaning strategies when filtering waters from different sources. The results highlighted the feasibility to apply a different cleaning protocol when groundwater was filtered as it would improve the water yield and production by elongating the filtration times. To conclude, an exploratory analysis reviewed the membrane performance and cleaning strategies in the Ultrafiltration stage at full-scale during 1 year of operation. The results showed high variations in Specific flux (Js) throughout the year and, also, suggested that chemical cleaning procedures were being applied in excess.

Keywords: Water treatment; Ultrafiltration; Fouling indices; Cleaning strategies; Micro-coagulation; Water blends; Surface and ground water

Introduction

The decrease of water availability in many regions suffering water scarcity is forcing the need of using surface waters of poor quality. To achieve the water quality standards in such scenarios, it is necessary to integrate pressure driven membrane technologies. Reverse Osmosis (RO) and Nanofiltration (NF) are common solutions to remove both excess of salinity and undesired levels of microbiological pollutants. These membrane treatments are usually linked to a pre-treatment stage incorporating Ultrafiltration (UF) membranes to ensure the safety of RO/NF membranes by accomplishing low levels of particulate matter (SDI<4) and low residual levels of Al-based coagulants (below 100 μg/L). Successful operation of the UF treatment stage depends upon a proper membrane fouling control, which is especially complex in treatment trains fed with waters of different quality, such as sand filtered water (SFW) and groundwater (GW) blends. This is the case presented here in which river water and groundwater sources are exploited depending on the low river quality associated to water droughts and rainfall events.

Membrane fouling control is especially critical in Drinking Water Treatment Plants (DWTPs) located in semiarid climate regions where production of tap water depends upon the resources availability as water composition, especially in terms of dissolved organic matter (DOM), changes substantially.

Monitoring fouling is essential to properly control filtration performance as fouling is the main obstacle during filtration. Fouling leads to a partial or total blocking of the membrane pores negatively affecting its operation either by increasing transmembrane pressure (TMP) and/or reducing the permeate flux [1]. Foulants can deposit at the top of the surface layer, or they can also get trapped inside the pores, hindering partially or totally the pass of water. The latter case is usually more problematic and can lead to irreversible fouling [2].

The potential of water compounds to foul the membranes also depend on the synergistic effects of the organic and inorganic fractions. Hao et al. studied the effect of humic substances in combination with metal ions in UF membranes [3]. Also, Qin et al. studied the contribution that humic substances in combination with silica particles had on fouling [4].

In order to mitigate fouling, membranes need to be cleaned periodically. In hollow fiber UF membranes, current efforts focus on optimizing the cleaning protocols. Hydraulic cleaning is applied periodically and consists in circulating clean water in reverse mode during a specific period of time. By doing so, foulants detach from the membranes and the water solution is discarded. This procedure is commonly known as backwash (BW) and the fouling removed by BW is known as hydraulically reversible fouling [5-7]. Fane et al. studied the effect of the operational conditions (BW duration, BW strength and air scouring) when using low pressure hollow fiber PVDF membranes BW [8,9].

Since the membrane permeability does not fully recover in every BW, there is a moment in which the operation of the membrane becomes unfeasible [10]. When this occurs, a chemical cleaning (CC) must be applied. The CC strategies can include the chemical enhanced backwash (CEB), in which a BW is performed with chemical agents which

***Corresponding author:** Vera M, Barcelona Research Center in Multiscale Science and Engineering, EEBE, Universitat Politècnica de Catalunya (UPC) – Barcelona TECH, c/Eduard Maristany, Spain
E-mail: marc.vera.canudas@estudiant.upc.edu

enhance removal of foulants, and the cleaning-in-place (CIP), in which membranes are soaked in a chemical solution for a specific period of time [11]. When this period of time is relatively short, the CIP can also be called as maintenance cleaning (MC). Some chemicals are preferred over others depending upon the type of fouling to be removed. The most commonly used are sodium hypochlorite (NaOCl), citric acid ($C_6H_8O_7$), phosphoric acid (H_3PO_4) and hydrogen peroxide (H_2O_2) [12]. Among these chemicals, NaOCl is the agent most used and is applied primarily when organic fouling occurs. When utilized in concentrations around 1%, it has been found to reduce the presence of undesired compounds adhered to the membrane with an efficiency close to 95% [13,14].

The frequencies at which membranes need to be cleaned depend, among other factors, on the quality of water to be treated. Because of that, efforts have focused on developing methodologies to predict and quantify how a specific water would affect membrane performance. In order to measure the potential of water to foul membranes, different fouling indices have been developed. They are used by membrane suppliers to fulfill the membrane specifications. Water streams with index values above the specified ones are not suitable to be treated by the considered membrane. Among the different indices, the silt density index (SDI) and the modified fouling index (MFI) are the most used (mostly in RO systems). In these tests, fouling is quantified by filtering water through 0.45 micrometer membranes at a constant pressure and are widely used because of their simplicity [15]. However, they are not very accurate and their applications can lead to overestimations. For instance, the methods ignore the fact that, when tested in real membranes, the diameter of pores is much smaller than 0.45 micrometers [16]. Regarding UF membranes, more appropriate indices have been developed. Boerlage et al. improved the modified fouling index (MFI) procedure by using a membrane with a lower molecular weight cut off (MWCO) that better mimicked full-scale UF systems [17]. Sim et al. introduced the Cross flow Sample Modified Fouling Index Ultrafiltration (CFS-MFI-UF), which accounts for the hydrodynamics of the system [18]. Further modifications include the normalized silt density index (SDI+) developed by Alhadidi et al., which introduces corrections for temperature, pressure and membrane resistance [19].

The unified membrane-fouling index (UMFI) developed by Huang et al. presents the advantage of accounting for the membrane-specificity and for being independent of filtration scale. Huang et al. demonstrated the significance of developing indices to quantify the fouling potential of different waters to assess membrane performance in bench-scale and pilot plant experiments [20,21]. Also, Nguyen et al. developed the same mathematical expression although they approached it using a resistance in-series method [22]. They also differentiated the fouling indices into total fouling index (TFI), hydraulic-irreversible fouling index (HIFI) and chemical irreversible fouling index (CIFI).

The present work was conducted (i) to evaluate the MCs performed in a full-scale DWTP by identifying Specific flux improvements before and after cleaning the membranes and measuring the frequencies in which MCs were applied, (ii) to quantify the fouling potential of two water sources feeding the UF stage of a full-scale DWTP by using the fouling indices developed by Nguyen et al. and (iii) to quantify, in bench-scale, the benefits and disadvantages of coagulation with Fe(III) and Al(III) in terms of rejections and fouling rates regarding water quality and operation, respectively. As a model system, a DWTP incorporating a UF stage as a pre-treatment of a RO stage feeding surface and groundwater resources has been used.

Materials and Methods

Full-scale DWTP description

The Drinking water treatment plant (DWTP) of study is located in the lowest part of the Llobregat River basin (Sant Joan Despí, NE Spain) and has a nominal capacity of approximately 5.3 m^3/s. The raw water used by the DWTP comes from the Llobregat River and, also, from its aquifer. The proportion of both source waters in the UF is ruled mainly by the Llobregat River water quality. Since the plant is located in a European semi-arid region, the river suffers from water draughts and poor quality. On the one hand, the average flow rate during the year is 20 m^3/s with episodes of dryness in which the flow rate can go down to 5 m^3/s. Moreover, the river is subject to high changes in quality due to heavy rainfall events (turbidity values above 500 NTUs), problems of dilution of large inputs of industrial and urban origins and episodes of pollution.

The whole treatment process of the DWTP is displayed in Figure 1. It includes a conventional pre-treatment comprised of preliminary screening, pre-chlorination with ClO_2, coagulation/flocculation by the addition of aluminium coagulant, subsequent sedimentation and sand filtration. It is at this stage where groundwater, when required, is incorporated. From this point on, water flow is split into two lines: (i) the conventional line undergoes ozonation and granular activated carbon (GAC) filtration, while the (ii) membrane-based line (implemented in 2009) undergoes in-line coagulation with $FeCl_3$, ultrafiltration (UF), UV irradiation, reverse osmosis (RO) filtration and mineralization. The use of the micro-coagulation stage with Fe(III) is justified by the need to achieve residual levels of Aluminum below 100 µl Al/L. Both treated streams are blended and the resulting stream is post-chlorinated prior to distribution.

The membranes in the UF stage are subject to cleaning protocols to sustain permeability values. A hydraulic BW is applied approximately every 50 minutes to remove the reversible fouling. The chemical cleaning (CC) procedures are applied in a lower frequency to mitigate the hydraulic irreversible fouling. The chemical agents used to conduct the CC are sodium hypochlorite, citric acid and phosphoric acid. There are two types of CC procedures conducted in the UF: a Maintenance cleaning (MC) is applied in which the membranes are soaked in solution for approximately 45 minutes. A more aggressive cleaning procedure, known as Recovery cleaning (RC), must be applied when episodes of sever fouling occurs.

Bench-scale membrane ultrafiltration system

A bench-scale UF setup was used to conduct the experiments. The configuration was in dead-end mode and Zenon/GE provided PVDF hollow-fiber membranes in order to use the same configuration and membranes as in the DWTP. The membranes had a nominal pore size of 20 nm and an effective membrane area of 0.042 m^2. The system operated under vacuum with a pressure ranging from 0 to 0.5 bar. Pressure was monitored using a micro-processor based pressure transducer (Model LEO3 from KELLER). The permeate flow was controlled with a peristaltic pump (Heidolph, Pump drive 5201) and was set constant during each run; the flux was measured by collecting and weighting water at the outlet over a known period of time. The membrane was immersed in a 15 L tank with a thermostat to set the temperature at 10°C. A second tank of 50 L was used to continuously recirculate water to the first in order to sustain the water level and keep the concentration of feed water constant. Data from the manometer and the balance was continuously logged over time via a data logger (Pico log 1216) and loaded using an in-house made program (created with MATLAB 2013). Figure 2 shows a scheme of the bench scale set-up.

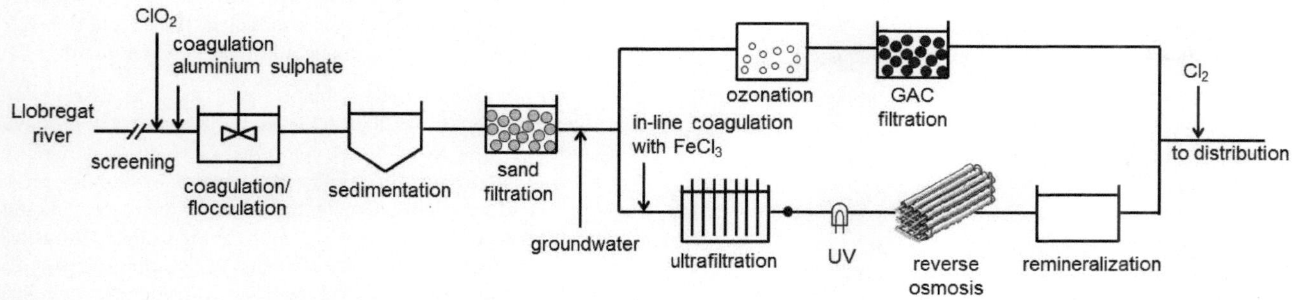

Figure 1: Schematic representation of the DWTP in SJD.

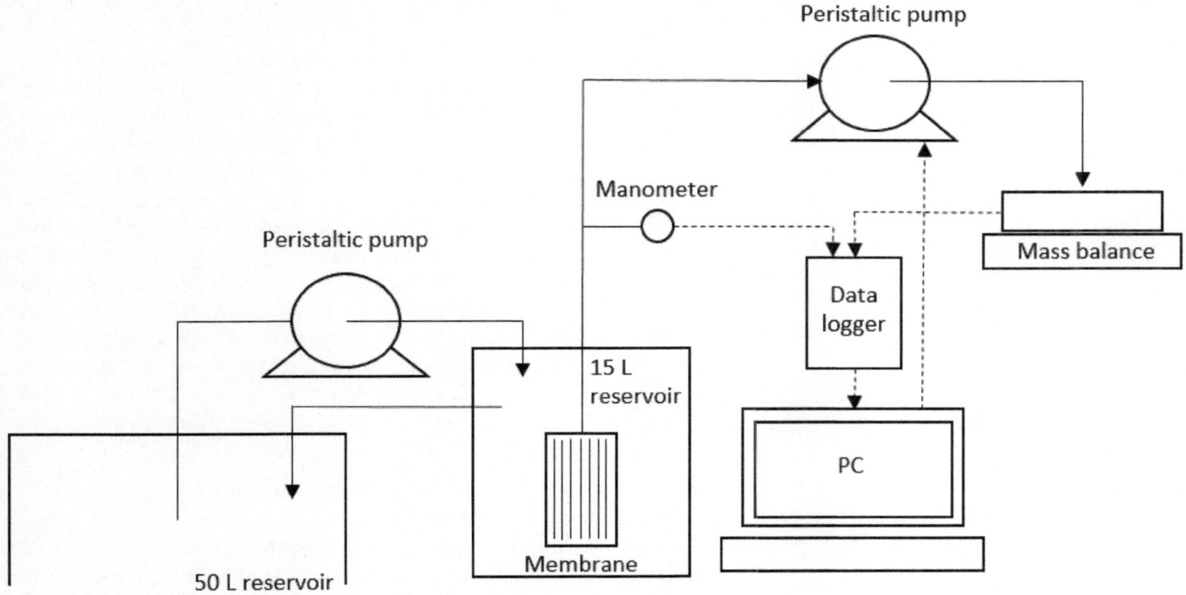

Figure 2: Scheme of the bench-scale UF dead-end set-up.

In order to properly study fouling for extended periods of time, it was necessary to automatize the filtration sequences. This was achieved by using an in-house program (created with MATLAB 2013) designed to mimic the operating conditions used in the DWTP. The filtration process was constituted by the following steps: (i) filtration for 60 minutes, (ii) 2 minutes relaxation (iii), 2 minutes backwash filtration (iv) 6 minutes relaxation. From experiment to experiment, membranes were chemically cleaned in place (CIP) either with NaOCl (400 $mg{\cdot}L^{-1}$) or citric acid (1000 $mg{\cdot}L^{-1}$) and the permeability was checked at the beginning of every experiment to ensure proper cleaning.

Analytical techniques

Absorbance spectroscopy at a wavelength of 254 nm was used to track variations in DOM. A DR5000 spectrophotometer (Hach Lange) equipped with 1 cm cuvettes was used for absorbance measurements (5 cm cuvettes were used in solutions of BSA). To measure particulate matter, the 1720E Low Range Process Turbidimeter and sc100 Controller (Hach Lange) were used. Also, a Particle counter (Particle Sense, Izasa Scientific) to measure particle ranges from 2 μm up to 100 μm.

Data treatment: Principal component analysis

A Principal Component Analysis (PCA) was conducted to compare the water quality originating from surface and ground sources. The analysis reduces the complexity of a system by projecting objects and variables in a lower dimensional space. It applies restrictions of orthogonally and maximum variability, which makes data easier to interpret.

The PCA results are usually discussed in terms of component scores and loadings. The scores are projections of the samples in the new space defined by principal components (PCs). Regarding the loadings values, they indicate the magnitude of the contribution of every original variable to every principal component (the weight by which each standardized original variable should be multiplied to get the component score). Variables with large values of loadings on the same component are assumed to correlate. If they present the same sign, they present a positive correlation whereas if they present opposite signs, they correlate inversely [23].

Parameters used to evaluate membrane performance

In this study, operation data (transmembrane pressure, TMP, and specific Flux, Js) were treated to calculate the total fouling index (TFI), the hydraulically irreversible fouling index (HIFI), the BW efficiency, the permeate flux capacity and the percentage of water wasted from BW.

The pressure driven flux through the membrane (J, flow per unit area) can be described by Eq. 1:

$$J = \frac{TMP}{\mu R} \quad (1)$$

Where TMP is the transmembrane pressure through the membrane, μ is the viscosity of the water and R is the coefficient of resistance that accounts for the resistance of the membrane and the resistance associated to fouling. In this model, it is assumed that fouling increases linearly with the specific volume (Vs, volume filtered per unit area). The Specific flux can be defined by Eq. 2:

$$J_S = \frac{J}{TMP} = \frac{1}{\mu(R_{mem} + \Upsilon_{fouling} V_s)} \quad (2)$$

The Specific flux can also be normalized by the Specific flux at the beginning of the operation when there is no fouling (Vs=0) as indicated by Eq. 3:

$$\frac{J_S}{J_S(vs=0)} = \frac{R_{mem}}{R_{mem} + \Upsilon_{fouling} V_s)} \quad (3)$$

Thus, the expression can be rearranged as shown by Eq. 4:

$$\frac{J_S}{J'_S} = 1 + FI.Vs \quad (4)$$

Where J's is the normalized permeability and FI accounts for the fouling index. Depending upon the interval of data used to calculate the FI, the index provides different information. For instance, if the FI is calculated from one chemical cleaning operation to the next one the HIFI is obtained, whereas if the range evaluated is only between two hydraulic cleanings, the TFI is obtained [21,22].

The BW efficiency was calculated by Equation 5:

$$BW_{eff}(\%) = \frac{JS_{afterBWi}}{JS_{afterBWi-1}}.100 \quad (5)$$

The water wasted due to BW was also calculated by Eq. 6:

$$BW_{ww}(\%) = (1 - \frac{J_f.t_f - J_{bw}.t_{bw}}{J_f.t_f}).100 \quad (6)$$

Finally, the MC improvements (%) were also calculated by Eq. 7:

$$MC_{improvement}(\%) = (\frac{Js_{bMC} - Js_{aMC}}{Js_{bMC}}).100 \quad (7)$$

Where Js_{bMC} and Js_{aMC} are the Specific fluxes before and after the MC were conducted [24].

Results and Discussion

Evaluation of micro-coagulation on natural organic matter removal by bench-scale UF Experiments

Groundwater (GW) and Sand filtered water (SFW) differ in terms of natural organic matter (NOM) but also by the presence of coagulants; SFW contains residual aluminum coagulants dosed in the pre-treatment as well as iron coagulant dosed previous to the UF stage. In this section, a two-step method consisting in a jar test and a filtration using 0.45 μm membranes was conducted to evaluate the effect that dosing aluminum and iron coagulants have in water quality.

First of all, synthetic foulants were used to assess selectivity between humic-like and protein-like substances as they have been identified as the main NOM components. Next, SFW replaced the synthetic compounds to estimate the benefits of filtering pre-treated water regarding quality. Finally, the effect of micro-coagulating Sand filtered water (doses below 1 mg Fe/L) prior to the UF was assessed not only from a quality stand point but also by conducting bench-scale experiments in the UF. This will help assess the benefits of coagulating pre-treated water (at such low doses).

Evaluation of model NOM foulants removal by micro-coagulation: First, synthetic model compounds typically found in water were studied; humic acid (HA) and bovine serum albumin (BSA) were selected because they resemble the humic-like and protein-like fractions found in water and have been previously used to evaluate fouling in membranes [25,26]. The solutions were prepared using ultrapure water in which known amounts of salts were added to account for the effect of ions on coagulation and removal of dissolved organic matter (DOM) (277 $mg{\cdot}L^{-1}$ $MgCl_2$, 300 $mg{\cdot}L^{-1}$ Na_2SO_4 and 300 $mg{\cdot}L^{-1}$ Na_2CO_3 was added). In every jar test run, six different one litter beakers were used and concentrations of coagulant ranging from 0 to 10 $mg{\cdot}L^{-1}$ were dosed during constant agitation. The coagulants used to conduct the experiments were Fe(III) (PIX-511, Kemira Water Solutions) and Al(III) (PAX18, Kemira Water Solutions). After fifteen minutes, agitation was stopped and the beakers were left to settle for another 15 minutes. Water from each beaker was then filtered using 0.45 μm PVDF membranes and the resulting solution was analyzed.

Figure 3a and 3b exhibit UV_{254} and rejection values measured for HA and BSA when coagulated with Fe(III) and Al(III), respectively. Both coagulants showed no DOM rejection when dosed at 1 $mg{\cdot}L^{-1}$. From 1 to 3 $mg{\cdot}L^{-1}$ the rejection of HA and BSA increased significantly. Then, the rates at which the coagulants initially increased were reduced. Also, based on the results obtained, there were almost no differences between coagulants and within the rejections obtained for both model compounds.

As it could be seen, improvement on the removal of NOM in the pre-coagulation stage were only efficient at doses of Fe above those coagulated in the UF pre-treatment stage. Next, Sand filtered water (SFW) was coagulated following the same procedures as with synthetic compounds. The objective was to quantify the effect of aluminum and $FeCl_3$ when using SFW. Figure 4 shows no DOM rejections below 1 $mg{\cdot}L^{-1}$ as when model compounds were used. Unlike the previous case study, lower rejections were found when Al(III) was used.

Evaluation of NOM removal by UF bench experiments using Fe(III) micro-coagulation: As stated earlier, there is a micro-coagulation chamber in the full-scale DWTP in which iron coagulant ($FeCl_3$) is dosed at low concentrations (below 1 $mgFe{\cdot}L^{-1}$). A battery of bench-scale UF assays, using 1 mgFe/L, were conducted to evaluate its impact on membrane performance. Figure 5a and 5b show fouling indices obtained from the different runs. Both total fouling index (TFI) and hydraulic irreversible fouling index (HIFI) showed slight improvements in membrane filtration upon coagulation. Rojas-Serrano et al. also found similar results when testing the coagulants at a larger scale. Higher coagulant doses decreased its beneficial effect. Zupančič et al. also studied the effects of coagulation with $FeCl_3$ prior to the UF stage using TFI and HIFI values. They found that coagulation had a

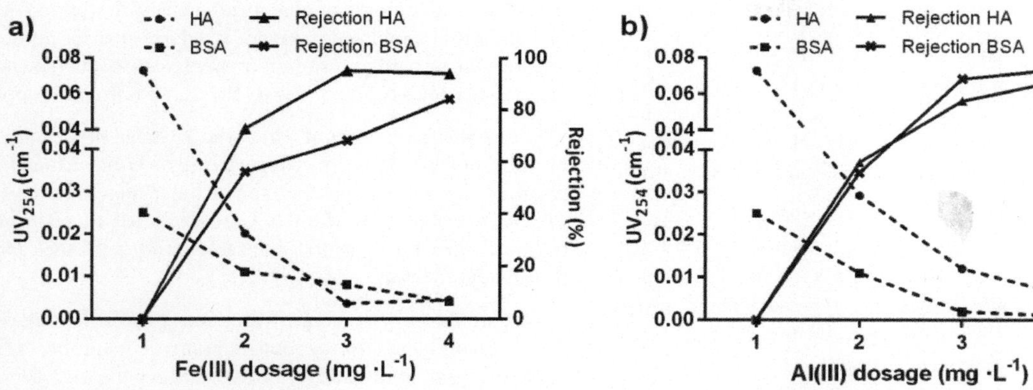

Figure 3: Results from jar tests evaluating synthetic solutions of HA and BSA. The dashed lines exhibit the absorbance values after coagulation whereas the regular lines show rejections of HA and BSA upon coagulation of a) Fe(III) and b) Al(III).

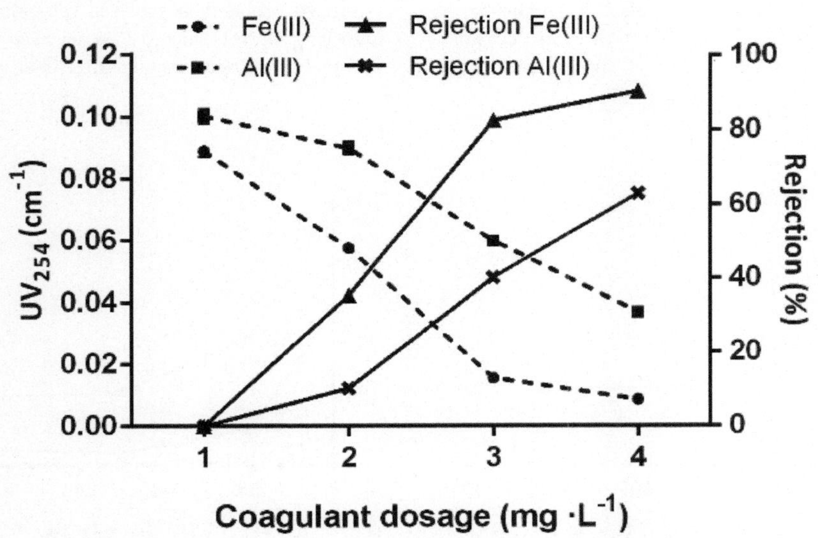

Figure 4: Variation of NOM removal efficiency by following changes of the molar UV absorption at 254 nm as a function of the coagulant dose for Al(III) and Fe(III) salt by using jar tests.

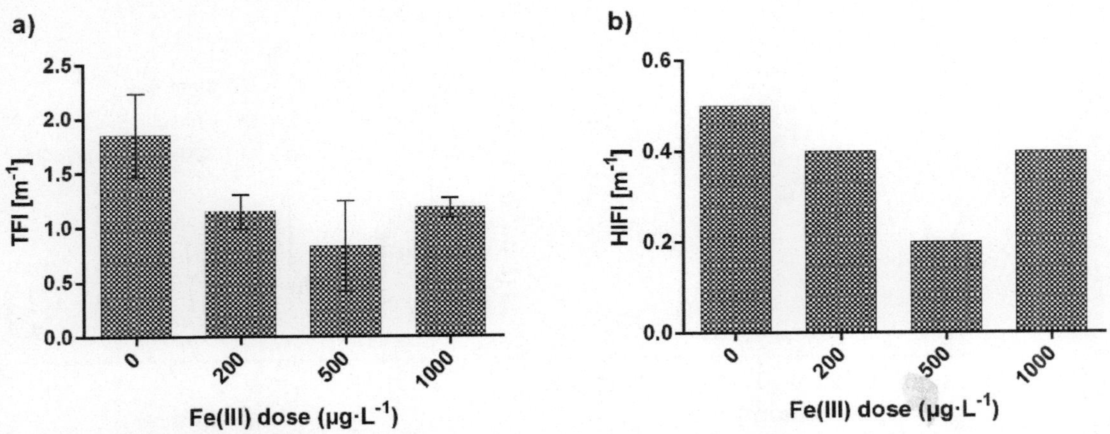

Figure 5: TFI values and b) HIFI values obtained from bench-scale UF experiments with type SW water at different coagulant doses. The error bars indicate the standard deviations observed.

positive impact on both reversible and irreversible fouling. However, in their system the turbidity was much higher (>5 NTUs) [27-30].

Overall, the experiments conducted showed no improvements in terms of membrane performance when dosing $FeCl_3$ at concentrations higher than 500 µg Fe·L^{-1}. In fact, $FeCl_3$ may act as a foulant compound itself. The results are in agreement with the study conducted by Wray et al. found an optimal $FeCl_3$ dose with a beneficial impact on membrane filtration and water quality [31].

The water quality changes due to micro-coagulation were also considered. Figure 6a and 6b show an increase in rejection of particulate matter (analyzed as turbidity and particle counts) associated to micro-coagulation dosage. Permeate was also analyzed for UV254 but no differences in absorbance were appreciated at different $FeCl_3$ doses.

Exploratory analysis of the cleaning protocols and membrane performance in the full-scale UF stage

Figure 7a shows the dynamics of the Specific flux (Js) in the UF stage of the DWTP of SJD measured during 1 year of operation. The Js experienced high oscillations throughout the year with occasional episodes in which Js dropped drastically. Figure 7b shows Js values and highlight the maintenance cleanings (MCs) conducted over a shorter period of time (1 month of operation). During this period, it is observed that the MCs conducted were not regularly spaced. Instead, the MCs for phosphoric acid and sodium hypochlorite occurred closely spaced whereas the MCs conducted with citric acid were more spread out.

Since the frequencies at which the MCs are applied are fixed either by time or by volume of permeate, there is no actual frequency or protocol that account for the type of water being filtered (GW/SFW). Finally, the chemical cleans (CCs) conducted in plant alternate the chemical agents used so that the same number of MCs are conducted throughout the year.

Figure 8a shows averaged values of Specific flux measured before each MC conducted throughout 10 month of operation: Js exhibits high variability among the different MCs conducted which denotes that the cleaning protocols were not optimally applied. Figure 8b exhibits the improvements (%) among the MCs conducted: citric acid, phosphoric acid and sodium hypochlorite improved Js by 11 ± 6%, 9 ± 6% and 10 ± 5% respectively. Also, the values were especially low during months 9 and 10 probably because Js values before the MC were especially high (Figure 8a). Martín-Pascual et al. also found high variations in permeability improvements after chemically cleaning the membranes

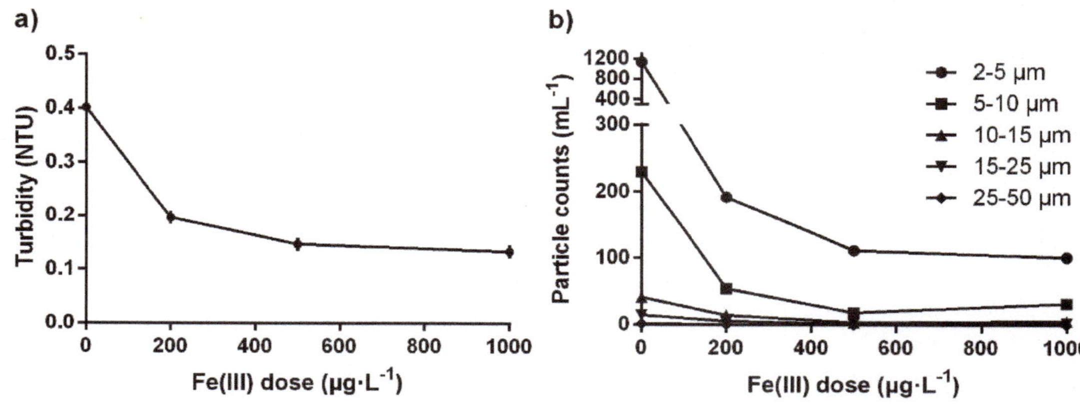

Figure 6: a) Turbidity values and b) particle counts analyzed from water samples after micro-coagulation and bench scale ultrafiltration.

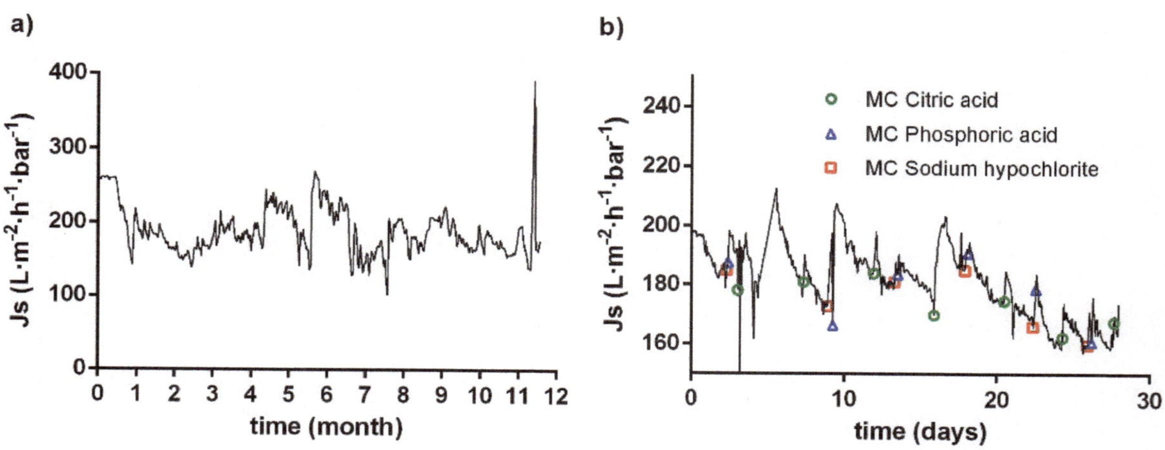

Figure 7: Full-scale membrane specific flux values during a) 1 year of operation and b) 1 month of operation.

although they obtained higher improvements (20-68%) [32]. Based on the analysis conducted, the MCs evaluated here may be ineffective due to a slightly high cleaning frequency.

During this period, a MC was performed approximately every 1200 L/m^2; which corresponds to approximately 2.5-3 MCs per week. Porcelli et al. reviewed 85 MF/UF plants and showed that less than half of the plants carried more than 1 MC per week [33]. An excess of MCs was found to affect the membrane integrity as the fibers suffer from breakage and degradation. Akhangelsky et al., in their study regarding the effect of NaOCl in polymeric membranes, showed a diminution in the mechanical strength of fibers (rupture stress and Young modulus) associated to an increase in the concentration of NaOCl and exposure time [34]. Also, Childress et al., when studying the effect of MCs in the lifetime of PVDF membranes, detected a total degradation of the superficial layer translated in an increase in their hydrophilic properties. An optimal NaOCl concentration between 0.5% and 1.0%, depending upon the operational conditions was suggested [35].

Accordingly, it is concluded that MC may be over-applied in the DWTP. Future work should focus on reducing the number of MC performed by: (i) being selective with the chemical reagents [36] and (ii) determining the frequency of the MCs based on the water quality.

Influence of water source and quality on membrane fouling

The physicochemical properties of water feeding the UF stage were analyzed through a Principal component analysis (PCA) by filtering either mostly SFW or mostly GW in full-scale. Figure 9 shows a bi-plot in which the scores (observations) and loadings (physicochemical parameters within water) are plotted. The abscissa axis provides information regarding the first component (PC1): it segregates the water sources completely. Also, the position of the physicochemical variables can be used to depict correlations. Thus SFW exhibited higher TOC, UV_{254} and turbidity values whereas the correlation was inversed for GW. Also, it is observed that SFW shows much higher variability than GW throughout the year.

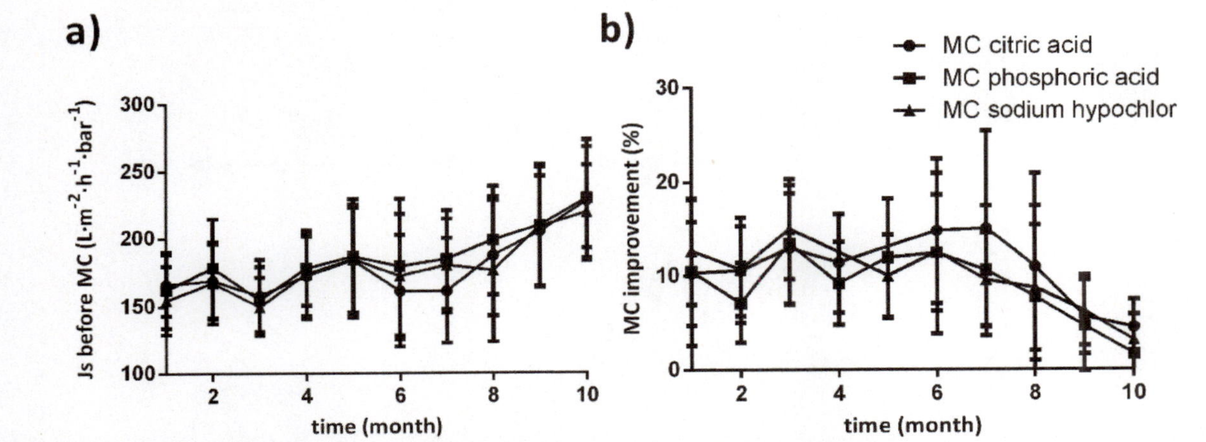

Figure 8: a) Averaged permeability values measured before a MC and b) averaged permeability improvements (%) measured during 2014. The error bars indicate the standard deviations observed.

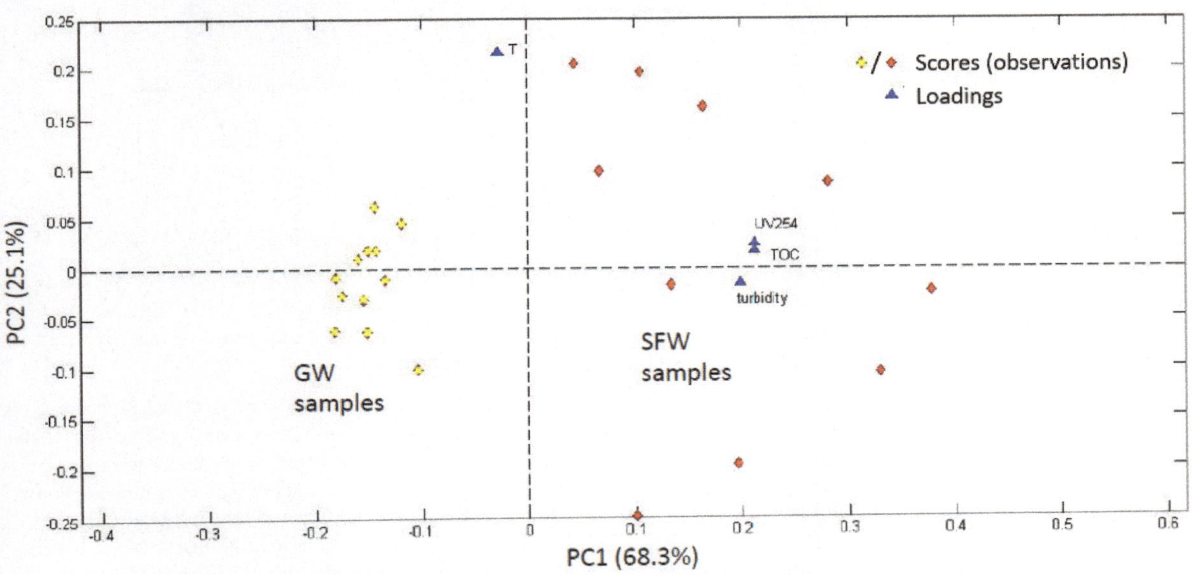

Figure 9: Principal component analysis (PCA) conducted on physicochemical parameters from two water sources.

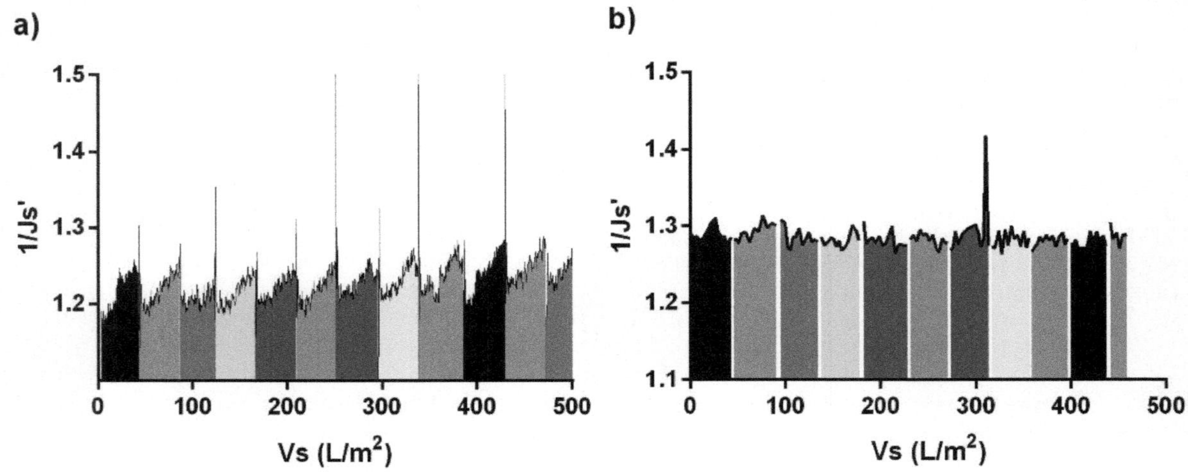

Figure 10: Full-scale fouling data of a) type SW and b) type GW water during 1 day of operation.

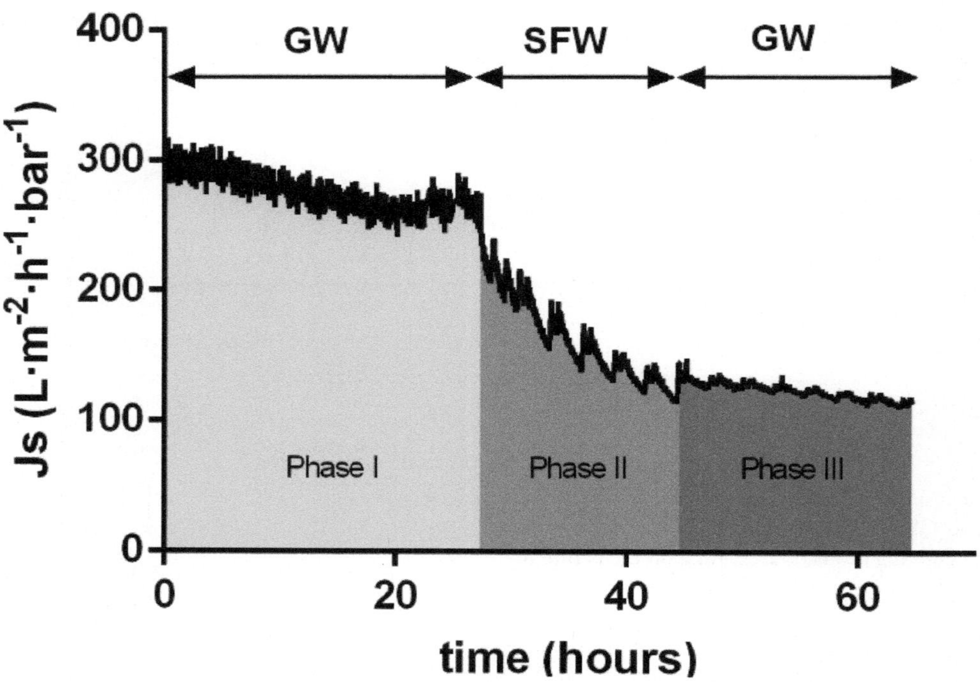

Figure 11: Js values from the Bench-scale UF setup obtained from alternating filtration of type SW and type GW water during 65 hours.

Finally, PC2 (axis of ordinates) was influenced by the temperature variations observed during the year but do not segregate the samples between SFW and GW. Again, the variability shown by SFW samples is much higher (the temperature ranges from 7-27 °C throughout the year).

Additionally, fouling arising from both water sources was evaluated at full-scale during one day of operation. Figure 10a and 10b show the variation of inverse of the flux normalized (1/Js') upon filtration of water (expressed as specific volume) which depict the rates at which membranes were fouled upon filtration of water. In comparison, SFW fouled the membranes at a much higher rate. The averaged TFI value for SFW water was 1.4 ± 0.5 m^{-1} and the HIFI value was $9{\cdot}10^{-4}$ m^{-1}. On the other hand, fouling upon filtration of GW was irrelevant (considering only the one-day period).

This operation behavior was analyzed and reproduced using bench experiments as it is shown in Figures 6 and 11 shows the Js values obtained upon filtration of water in bench-scale filtration. First, GW was filtered during 27 hours with BW intervals of 60 minutes (Phase I). During this period, the TFI was $3.0{\cdot}10^{-1} \pm 1.9{\cdot}10^{-1}$ m^{-1} and the HIFI was $1.4{\cdot}10^{-1}$ m^{-1}. Then, the membrane started filtering SFW for which a significant permeability drop was observed; the TFI and HIFI values were 5.5 ± 1.6 m^{-1} and $1.6{\cdot}$ m^{-1}, respectively (Phase II). Finally, the membrane started

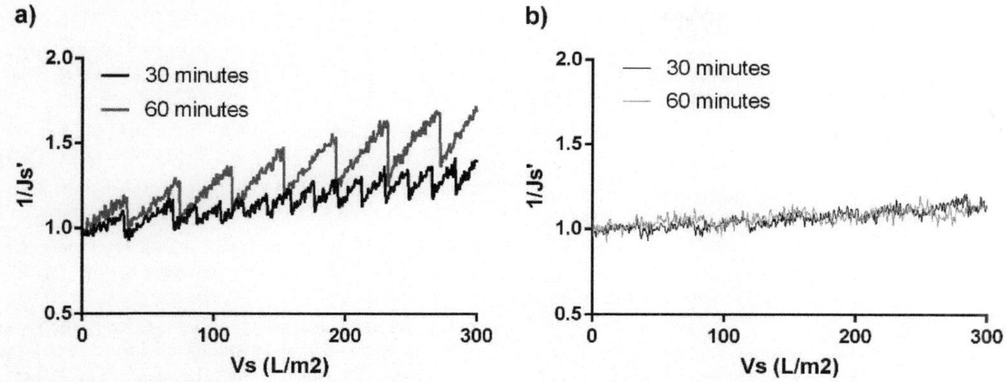

Figure 12: Operating conditions obtained during filtration of a) SFW and b) GW working at different frequencies.

SFW					
BW_{freq} (min)	TFI (m^{-1})	HIFI (m^{-1})	BW_{eff} (%)	BW_{WW} (%)	Water flux ($L \cdot m^{-2} \cdot h^{-1}$)
30	7.4 ± 1.4	1.1	94 ± 3	14	43
60	8.6 ± 1.3	2.1	98 ± 3	7	36

Table 1: Operating performance values obtained from filtration of type SW water for filtration cycles of 30 and 60 minutes.

filtering GW again during 20 hours (Phase III). During this period, the TFI and HIFI values dropped again ($5.1 \cdot 10^{-1} \pm 1.4 \cdot 10^{-1}$ m^{-1} and $5.7 \cdot 10^{-1}$ m^{-1}, respectively). Even though GW in Phase III was visibly "cleaner" than SFW, it was not possible to recover the irreversible fouling inflicted in Phase II. In fact, the irreversible fouling increased as compared to the beginning of the experiments in which the same water was used, which could be related to the membrane being already partially blocked up. To conclude, when the full-scale data analysis and the bench-scale experiments conducted were compared, the latter results showed slightly higher overall TFI and HIFI values both when filtering GW and SFW.

Influence of BW frequencies on membrane filtration performance

Membrane performance and water quality properties behaved different when SFW and GW were filtered which made evident the need to use a different operating protocol depending on the nature of the feed water.

Figure 12a and 12b show the variation of inverse of the flux normalized (1/Js') upon filtration of water (expressed as specific volume) for two different BW frequencies when tested in bench-scale for SFW and GW, respectively.

Figure 12a shows a higher increase in 1/J' values upon BW conducted every 60 minutes than when BW was applied every 30 minutes. The fouling indices calculated for TFI_{30} and TFI_{60} were 7.4 ± 1.4 m^{-1} and 8.6 ± 1.3 m^{-1}, respectively. Higher differences were found regarding the irreversible fouling, in which the $HIFI_{30}$ and $HIFI_{60}$ were 1.1 m^{-1} and 2.0 m^{-1}, respectively. Also, the BW efficiencies dropped from 98 ± 3% to 94 ± 3% during filtration at 30 and 60 minutes, respectively. Moreover, the average TMP values were 0.21 ± 0.02 bar and 0.25 ± 0.04 bar for filtration cycles of 30 and 60 minutes, respectively (the values are summarized in Table 1. Raffin et al. also found that the BW frequency had a major effect on irreversible fouling [37].

Overall, the results showed significant improvements when working with 30 minutes filtration cycles as the irreversible fouling was reduced and the average TMP also decreased. However, this was in detriment of the production yield (Table 1). Thus, picking the optimal BW frequency is a tradeoff in which it is necessary to prioritize within different aspects.

On the contrary, Figure 12b show no differences between the two frequencies used as the data showed mild fouling upon filtration.

Conclusion

In this study, the cleaning procedures performed in the UF were reviewed. Also, the influence of groundwater and sand filtered water on the membranes as well as the role they play on the cleaning procedures was studied. The work included full-scale data, bench-scale tests and laboratory experiments.

The following conclusions can be drawn from the results:

- Membrane-based systems operating in water scarcity regions require the attention of special protocols to improve operation due to the strong changes on water quality streams (GW and SW) especially in terms of organic matter composition and content.
- The study conducted to review the maintenance cleaning in the DWTP concluded that there is room for optimization by (i) improving the selectivity between chemical reagents and (ii) modifying the frequencies based upon water quality information.
- Based on the significant differences observed when filtering GW and SFW, a bimodal system has been suggested to work with two different BW frequencies: filtering GW during longer times was feasible and could improve the water yield and production.
- The coagulants did not show significant selectivity in removing HA over BSA and both coagulants showed rejections when dosed above 1 $mg \cdot L^{-1}$ but higher uses of coagulant doses will reduce water production yield due to the increase of cleaning stages.
- Micro-coagulation with $FeCl_3$ previous to the UF exhibited a positive impact in membrane performance and particulate matter but did not improve DOM rejections.

Acknowledgements

This research was supported by the Exima project (Industrial PhD Program through Agaur and Aigües de Barcelona 2013-DI-026) and the Waste2Product project (CTM2014-57302-R) financed by the Ministerio de Economíay Competitividad (MINECO) and the Catalan Government (Project Ref. 2014SGR50), Spain.

References

1. Xiao P, Xiao F, Wang D, Qin T, He S (2012) Investigation of organic foulants behavior on hollowfiber UF membranes in a drinking water treatment plant. Sep. Purif. Technol 95: 109-117.
2. Cuartas UB, Álvarez BS, Lora GJ (2011) Analysis of fouling resistances under dynamic membrane filtration. Chem. Eng. Process. Process Intensif 50: 404-408.
3. Hao Y, Moriya A, Maruyama T, Ohmukai Y, Matsuyama H (2011) Effect of metal ions on humic acid fouling of hollow fiber ultrafiltration membrane. J. Memb. Sci 376: 247-253.
4. Qin JJ, Li Y (2005) Hollow fiber ultrafiltration membranes with enhanced flux for humic acid removal. Memb. Sci. 247: 119-125.
5. Li Q, Elimelech M (2006) Synergistic effects in combined fouling of a loose nanofiltration membrane by colloidal materials and natural organic matter. Memb. Sci 278: 72-82.
6. Peiris RH, Jaklewicz M, Budman H, Legge RL, Moresoli C (2013) Assessing the role of feed water constituents in irreversible membrane fouling of pilot-scale ultrafiltration drinking water treatment systems. Water Res 47: 3364-3374.
7. Peldszus S, Hallé C, Peiris RH, Hamouda M, Jin X, et al. (2011) Reversible and irreversible low-pressure membrane foulants in drinking water treatment: Identification by principal component analysis of fluorescence EEM and mitigation by biofiltration pretreatment. Water Res 45: 5161-5170.
8. Ye Y, Sim LN, Herulah B, Chen V, Fane AG (2010) Effects of operating conditions on submerged hollow fibre membrane systems used as pre-treatment for seawater reverse osmosis. J. Memb. Sci 365: 78-88.
9. Akhondi E, Wicaksana F, Gordon A (2014) Evaluation of fouling deposition, fouling reversibility and energy consumption of submerged hollow fiber membrane systems with periodic backwash. J. Memb. Sci 452: 319-331.
10. Delaunay D, Rabiller-baudry M, Lilian B, Paugam L, Chaufer B (2008) A dual approach of membrane cleaning based on physico-chemistry and hydrodynamics Application to PES membrane of dairy industry. Chem. Eng. Process. Process Intensif 47: 267-275.
11. Regula C, Carretier E, Wyart Y, Gésan-Guiziou G, Vincent A, et al. (2014) Chemical cleaning/disinfection and ageing of organic UF membranes: a review. Water Res 56: 325-65.
12. Zondervan E, Betlem BHL, Blankert B, Roffel B (2008) Modeling and optimization of a sequence of chemical cleaning cycles in dead-end ultrafiltration. J. Memb. Sci 308: 207-217.
13. Puspitasari V, Granville AM, Le-Clech P, Chen V (2010) Cleaning and ageing effect of sodium hypochlorite on polyvinylidene fluoride (PVDF) membrane. Sep. Purif. Technol. 72: 301-308.
14. Strugholtz S, Sundaramoorthy K, Panglisch S, Lerch A, Brügger A, et al. (2005) Evaluation of the performance of different chemicals for cleaning capillary membranes. Desalination 179: 191-202.
15. Chuang SH, Chang WC, Chang MC, Sung MA (2009) The effects of soluble organic matters on membrane fouling index. Bioresour Technol 100: 1875-1877.
16. Alhadidi AMM, Kemperman AJ, Blankert B, Schippers JC, Wessling M, et al. (2011) Silt Density Index and Modified Fouling Index relation, and effect of pressure, temperature and membrane resistance. Desalination 273: 48-56.
17. Boerlage SFE, Kennedy MD, Petros M, Abogrean E, Tarawneh ZS, et al. (2003) The MFI-UF as a water quality test and monitor. J. Membr. Sci 211: 271-289.
18. Sim LN, Ye Y, Chen V, Fane AG (2010) Crossflow Sampler Modified Fouling Index Ultrafiltration (CFS-MFIUF)-An alternative Fouling Index. J. Memb. Sci 360: 174-184.
19. Alhadidi A, Kemperman AJB, Schippers JC, Blankert B, Wessling M, et al. (2011) SDI normalization and alternatives. Desalination 279: 390-403.
20. Jacangelo JG (2008) Low Pressure Membrane Filtration of Natural Waters: Principles and Methodology. Environ Sci Technol 42: 714-720.
21. Huang H, Young T, Jacangelo JG (2009) Novel approach for the analysis of bench-scale, low pressure membrane fouling in water treatment. J. Memb. Sci 334: 1-8.
22. Nguyen AH, Tobiason JE, Howe KJ (2011) Fouling indices for low pressure hollow fiber membrane performance assessment. Water Res 45: 2627-2637.
23. Hwang K, Chan C, Tung K (2009) Effect of backwash on the performance of submerged membrane filtration. Journal of Membrane Science 330: 349-356.
24. Jolliffe IT (1986) Principal Component Analysis. Springer, New York, USA.
25. Kim YJ, Yun T, Lee S, Kim D, Kim J (2014) Accelerated testing for fouling of microfiltration membranes using model foulants. Desalination 343: 113-119.
26. Shutova Y, Karna BL, Hambly AC, Lau B, Henderson RK, et al. (2016) Enhancing organic matter removal in desalination pretreatment systems by application of dissolved air flotation. Desalination 383: 12-21.
27. Citulski J, Farahbakhsh K, Kent F, Zhou H (2008) The impact of in-line coagulant addition on fouling potential of secondary effluent at a pilot-scale immersed ultrafiltration plant. J. Memb. Sci 325: 311-318.
28. Dong BZ, Chen Y, Gao NY, Fan JC (2007) Effect of coagulation pretreatment on the fouling of ultrafiltration membrane. J. Environ. Sci 19: 278-283.
29. Rojas-serrano F, Pérez JI, Gómez MÁ (2016) Comparative study of in-line coagulation and / or ozonization pre-treatment for drinking-water production with spiral-wound ultrafiltration membranes. Chem. Eng. Process. Process Intensi 105: 21-29.
30. Zupančič M, Novak D, Diaci J, Golobič I (2014) An evaluation of industrial ultrafiltration systems for surface water using fouling indices as a performance indicator. Desalination 344: 321-328.
31. Wray HE, Andrews RC (2014) Optimization of coagulant dose for biopolymer removal : Impact on ultrafiltration fouling and retention of organic micropollutants. J. Water Process Eng 1: 74-83.
32. Martín-pascual J, Rivas PR, Muñio MM, González-lópez J, Poyatos JM (2016) Membrane fouling of a hybrid moving bed membrane bioreactor plant to treat real urban wastewater. Chem. Eng. Process. Process Intensif 104: 112-119.
33. Porcelli N, Judd S (2010) Chemical cleaning of potable water membranes: A review. Sep Purif. Technol. 71: 137-143.
34. Arkhangelsky E, Kuzmenko D, Gitis NV, Vinogradov M, Kuiry S, et al. (2007) Hypochlorite Cleaning Causes Degradation of Polymer Membranes. Tribol. Lett 28: 109-116.
35. Childress AE, Le-Clech P, Daugherty JL, Chen C, Leslie GL (2005) Mechanical analysis of hollow fiber membrane integrity in water reuse applications. Desalination 180: 5-14.
36. Porcelli N, Judd S (2010) Chemical cleaning of potable water membranes: The cost benefit of optimization. Water Res 44: 1389-1398.
37. Raffin M, Germain E, Judd SJ (2012) Influence of backwashing, flux and temperature on microfiltration for wastewater reuse. Sep. Purif. Technol 96: 147-153.

Effect of the Pre-treatment on the Performance of MBR, Al-Kut WWTP, Wasit Governorate

Sadik Hameed*

Department of Chemistry, College of Science, Wasit University, Kut, Wasit, Iraq

Abstract

Pilot scale experiments were carried out to examine the effect of the pre-treatment methods on the performance of MBR. The TITAN MBRTM module was used in this study. In order to investigate the effect of pre-treatment on the behavior of membrane, samples were collecting at different locations in Al-Kut WWTP, Wasite. The first samples group has been collected directly from the main source as raw sewage to determine its main characteristics of wastewater. The second group samples have been collected from outlet of Al-Kut WWTP (conventional treatment) and the third group and fourth group samples have collected from outlet of pretreatment of MBR system and finally treated wastewater from MBR system. The study showed that the membrane bio-reactor filters out nearly all solids, the pre-treatment has a positive effect on the MBR performance, and the pre-sedimentation is more effective than fine screening. Moreover, aeration is considered as one of the intrinsic parameters in both hydraulic and biological process performances because of its ability to maintain solids in suspension, scours the membrane surface, limits fouling, and provide oxygen to the biomass, which results in a better biodegradability.

Keywords: Flux; Membrane bioreactor; Membrane fouling; TITAN™ MBR module

Introduction

The membrane bioreactor is an activated sludge process coupled with membrane filtration instead of the settling process for liquid-solid separation. Because of the membrane filtration, the suspended solids are completely removed from the treated water to the extent that the effluent contains no bacteria in microfiltration or no virus in nanofiltration [1]. This very compact arrangement produces a MF/UF quality effluent that is suitable for reuse applications. Depending upon the membrane nominal pore size, the virus removal can consequently be attained by providing a barrier to certain chlorine resistant pathogens as *GIARDIA* [2]. The present study discusses the performance of the membrane bioreactor (MBR), which, in recent, is widely used for the municipal and industrial wastewater treatment and the effect of the pre-treatment on its performance. Main variable parameters as COD-pH-DO-MLSS-SVI-TKN-PO_4-P-Vacuum and flux obtained are accurately monitored and recorded.

Membrane Bioreactors (MBRs) have become a standard technology for wastewater treatment. However, understanding how to ensure sustained membrane permeability remains a key operational consideration. Several approaches have been introduced in order to reduce fouling including intermittent suction, back-flushing, module design improvement, or the optimization of aeration such as with the MBR aeration technology which has enabled dramatic reductions in operational energy cost to be realized. Nevertheless, membrane fouling remains important, particularly understanding how to respond to system perturbation, and as such it is suggested that MBR cost could be further reduced by better control of membrane fouling [3].

Membrane fouling is a consequence of the interaction between the membrane and a complex mixture comprising colloids, bacterial flocs and dissolved macromolecules. It has been widely reported that soluble microbial products which comprise mainly of proteins and polysaccharides are primarily responsible for the clogging and blocking of membrane pores. As these compounds are generally colloidal in nature, their transport is primarily controlled by Brownian motion and as such the shear forces applied either through pumping (in side stream) or aeration (in immersed) are insufficient to provide back-transport toward the bulk which results in preferential colloidal deposition at the membrane. Deposition of these high molecular weight compounds result in formation of a highly-hydrated gel matrix into which microorganisms are embedded, resulting in a significant resistance to permeate flow during membrane operation [4].

To ameliorate the potential impact of SMP on membrane fouling, several authors have introduced the use of coagulants, which enable the assimilation of SMP compounds into aggregates (or flocs), that are then more strongly influenced by inertial lift and shear induced diffusion, thereby promoting back-transport away from the membrane surface into the bulk [5].

Numerous chemical compounds have been trialed including metal salts, biopolymers, starch and organic polymers and have been considered particularly pertinent for reactive dosing of the coagulant to limit the impact of sudden SMP release in response to process perturbation, such as saline shock. Crawford [6] described the use of MPE50 which is a modified biopolymer with a net cationic charge that has been demonstrated to effectively reduce membrane fouling both at laboratory and full scale [7] suggested that the cationic structure of MPE50 enabled the neutralisation of the negatively charged colloidal biopolymers, thereby encouraging floc growth, which can be considered analogous to conventional coagulation-flocculation. However, in conventional application, coagulation and flocculation processes are generally configured in series. In the coagulation step,

***Corresponding author:** Sadik Hameed, MSc, Department of Chemistry, College of Science, Wasite University, Kut, Wasit, Iraq
E-mail: Sadikhameed15@yahoo.com

coagulant is added within a high shear zone with an average velocity gradient (G) exceeding 1500 s^{-1} to ensure the coagulant is distributed homogeneously to achieve particle destabilization and charge neutralization. Flocculation then proceeds in a secondary shear zone with a gentler average velocity gradient of around 100-200 s^{-1} to encourage successful collision of destabilized particles and is provided with a retention time of up to 60 min to ensure subsequent floc growth is complete.

For comparison, there is no high shear zone specified within present MBR design to incorporate coagulant dosing. Furthermore, the average velocity gradient imposed by coarse bubble aeration in MBR is around 150-200 s^{-1}. Consequently, without a clearly delineated high shear zone to provide homogeneous distribution of the coagulant, and with gas mixing in MBR providing a velocity gradient which is equivalent to flocculation, it is suggested that coagulation-flocculation will be rate limited within MBR which will cause a delay in the time from which coagulant is reactively dosed to when colloids destabilize and form aggregates. As flocculators are generally designed with residence times of up to 60 min to enable sufficient time for aggregate growth, it is posited that with direct dosing of MPE50 into the MBR, the time at which coagulant is added during the filtration cycle could therefore be of significance to sustaining membrane permeability during filtration [8].

Organic matter in raw wastewater is partially converted to biomass, which is eventually used for biogas production (energy recovery) in some cases. However, the conversion rate to biomass is not high [9,10]. At present, a large portion of organic matter in wastewater is not recovered but is degraded into carbon dioxide and water with external energy input. A paradigm shift is necessary: organic matter in wastewater should not be degraded but recovered for energy production. Wastewater from domestic usage contains a significant amount of potential energy. Capturing this organic matter as a renewable energy source will be an attractive process. Wastewater treatment plants can be net energy producers by utilizing organic matter in municipal wastewater that is currently degraded with external energy input [11].

Anaerobic treatment processes are viable options for producing energy from organic matter in wastewater. Concentration of COD in municipal wastewater is in the range of 250-800 mg/L [12], whereas it is difficult to apply anaerobic processes to wastewater with COD concentrations of <1500-2000 mg/L. Concentrating raw wastewater can therefore facilitate application of anaerobic processes to produce biogas. Analysis of the particle size distribution of organic matter in raw municipal wastewater showed that 63-70% of total organic carbon (TOC) was associated with particles that were larger than 0.1 μm. Loose membranes such as microfiltration (MF) membranes can efficiently retain particulate and colloidal organic matter in wastewater. Concentrating organic matter in wastewater can therefore be carried out by membrane processes, facilitating recovery of energy from wastewater via anaerobic digestion. Direct Membrane Filtration (DMF) of wastewater has advantages including simplicity of design and maintenance. However, in DMF of municipal wastewater, severe membrane fouling is very likely to occur. Although several attempts have been made to concentrate organic matter in wastewater by using membranes, few studies have focused on membrane fouling in DMF. It is not clear whether long-term operation of a membrane process for recovery of organic matter from wastewater is feasible [13].

Materials and Methods

Lab-scale submerged membrane bioreactor

A cylindrical lab scale reactor of 25 cm diameter and 55 cm height with open top has been installed in the laboratory (Al-kut WWTP laboratory, Wasite). The capacity of the reactor is 25l, in which a TITAN membrane module is submerged. The reactor is fed with wastewater from a 50 L polyethylene feeding tank via a dosing pump with feeding rate varies from zero to 100 mL/min. For generating permeate, filtration is induced by the vacuum pump. Thus, the trans-membrane pressure (TMP) of the membrane is monitored as a negative value ranging from 20 to 50 kPa. The system is aerated by diffused air. The HRT of the bioreactor is 6 h while the SRT was 8 days. The Figure 1 illustrates the schematic diagram of the bioreactor. Raw wastewater collected daily from the inlet of the primary sedimentation tank of Al-kut Wastewater Treatment Plant, Wasite, Iraq was supplied as feed water in this study. The sewer system connected to the treatment plant is a combined system and covers a population of 500,000.

Membrane filtration was carried out in the outside-in mode at constant membrane fluxes. Resultant trans-membrane pressure (TMP) was recorded automatically by digital pressure meters. Intermittent filtration (3 min pause for every 12 min of filtration) was performed with peristaltic pumps. To mix the tanks, aeration was also provided in both tanks at the flow rate of 4 L/min throughout the experiments.

Assessment of the distribution of fouling resistance

At the termination of each filtration run, fouled membranes were taken out from the tanks and subjected to dead-end filtration tests using

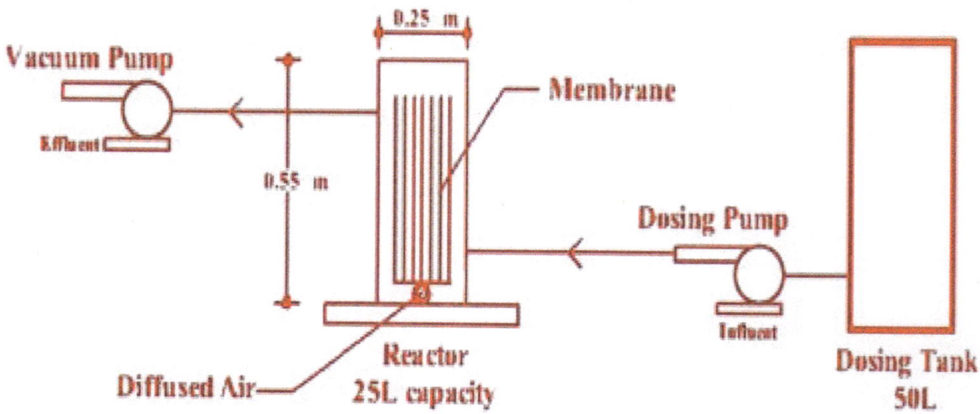

Figure 1: Schematic diagram of the MBR.

pure water. Pure water permeabilities of the fouled membranes were measured to investigate the distribution of fouling (i.e., reversible or irreversible fouling). The fouled membranes were gently wiped with a soft sponge to remove deposits from the membrane surface. Pure water permeability of the wiped membrane was then measured. Restoration of pure water permeabilities achieved by wiping represents the degree of reversible fouling, whereas filtration resistance that remained after wiping represents the degree of irreversible fouling [14]. Conversion from pure water permeability to filtration resistance was made based on Darcy's law:

$$J = \frac{\Delta P}{\mu R_t}$$

Turn Math Jaxon; where Rt is total filtration resistance (m^{-1}), J is permeate flux ($m^3/m^2/s$), ΔP is TMP difference (Pa), and μ is viscosity of water (Pa s).

Air diffusion

In a submerged MBR, Cross-Flow Velocity (CFV) is created by aeration, which not only provides oxygen to biomass, resulting in a better biodegradability and synthesis of the cells, but also, maintains the solid in suspension and controls the membrane fouling by hydraulic shear force and agitation. The air-induced cross-flow can efficiently remove or at least reduce the fouling layer on the membrane surface. The rate of air diffusion in this research was 15 l/h.

Start up

The reactor has been seeded with an activated sludge from the plant of Al-Kut WWTP for two weeks. The sludge is continuously aerated by diffused air. The concentration of the dissolved oxygen in the reactor was 5.8 mg/L at the ambient temperature (26.4°C).

Analytical methods

Analysis of MLSS, TSS, TKN, COD, DO, TP and NO_3-N for the influent and effluent was conducted using the procedures described in standard methods. These parameters were monitored and recorded continuously, all tests methods were conducted according to APHA [15].

Experimental evaluation of membrane fouling

The Flux is the quantity of material passing through the unit area of membrane per unit time $m^3/m^2/h$ (LMH) and is occasionally referred to as the permeate velocity m/h. The driving force for the process may be a trans-membrane pressure gradient. Since the flux and driving force are interrelated, either one can be fixed. In order to observe membrane fouling; the driving force was fixed according to the required vacuum by membrane manufacturer at the range between 20 and 50 kPa.

Results and Discussion

Al-Kut wastewater treatment plant treats about 65000 m^3/d. The primary treatment contains screens, aerated grit chamber and primary settling tank. The biological treatment consists of four activated sludge basins and final two sedimentation tanks. To start up, the reactor was seeded by an activated sludge which is withdrawn from the return sludge tank. The MLSS was 3600 mg/l and the SVI was 110 mg/l. Before studying the effect of the pre-treatment on the performance of the membrane, raw sewage samples were withdrawn and analyzed. The main characteristics of the raw sewage are determined and tabulated in Table 1.

The raw sewage has a ratio of C:N:P of approximately 139:4.1:1.0 so there is a surplus of nitrogen and phosphorus. Dissolved oxygen concentration indicates that the influent is in aerobic conditions. To investigate the performance of the MBR as compared to the activated sludge system, a comparison between the characteristics of the effluents of the membrane bioreactor and activated sludge system in Al-Kut WWTP should be carried out. For six months (Feb. to Jul. 2016) a complete analysis of the influent and the effluent was carried out in Al-Kut WWTP (Figure 2). The Table 2 summarizes the main characteristics of the influent and effluent in Al-Kut wastewater treatment plant. The given values are the mean values during the six months of measurements. The obtained results of the effluent indicate that the performance of the activated sludge system inefficient and the results didn`t comply with the environmental laws in Iraq (TSS, COD, TKN and PO_4-P values were exceeding the Iraqian permissible limit (Figures 3 and 4).

Effect of pre-treatment on the performance of the membrane bioreactor

As mentioned before, the aim of this study is to investigate the effect of the pre-treatment on the performance of the membrane. Fine screening and pre-settling are proposed as the methods of pre-treatment. Fine screening screening must be in place to remove larger materials, but most waste products can be decomposed with the right bacterial composition. Effluent water is almost free of solids and larger particles. The raw sewage was screened via stainless steel screen meshes 1 and 2 mm size. Although a 1 mm screen is standard, a finer screen is required for some types of hollow fiber membrane modules. On the other hand, a coarser screen will suffice for some flat sheet membrane modules. Pre- settling tank Al-Kut WWTP contains a primary settling tank with a volume of 5500 m^3. It removes approximately 36% COD, 26% PO_4-P, 11% TKN and 52% TSS. The MBR is fed with the pre-settled wastewater from the effluent of the primary sedimentation tank.

Three groups of the experimental work are carried out. The first and second groups are to study the performance of the MBR for the fine screened influent while the third group studies the performance of the MBR for the pre-settled influent. Each set of experiments and analysis was carried out for two weeks. The average of the measured values is recorded and tabulated. The Tables 3 and 4 summarizes the main characteristics of the influent and effluent during the application of MBR in the laboratory after the pre-treatment. The MBR results show that there is no significant difference between the obtained results using 1 and 2 mm fine screening. The pre-settling has a higher effect than the fine screening on the performance of the MBR. This seems to be attributed to the high removal of the suspended solids in the primary settling tanks due to the HRT.

Parameter	Average value (six months data)
pH	8.15
D.O	1.1 mg/L
TSS	316 mg/L
COD	1640 mg/L
TKN	48.9 mg/L
NO_3-N	11.8 mg/L
PO_4-P	6.4 mg/L
Conductivity	1930 S/cm

Table 1: Results of raw sewage analysis.

Parameter	Raw sewage	After primary treatment	After biological treatment
pH	8.15	7.74	7.38
TSS (mg/l)	316	112	44.8
COD (mg/l)	1640	844	82.6
TKN (mg/l)	48.9	46.1	12.6
PO_4-P (mg/l)	6.4	4.52	1.32

Table 2: The main characteristics of the influent and effluent in Al-Kut WWTP.

Parameter	Raw sewage	After fine screening (1 mm)	After primary settling (2 mm)	MBR effluent		
				(1 mm)	(2 mm)	Pre-settled
pH	8.12	8.06	7.98	7.56	7.44	7.36
TSS (mg/l)	265	213	117	2.12	0.92	0.77
COD (mg/l)	1560	1420	786	26.10	17.0	13.6
TKN (mg/l)	49.70	48.10	45.3	6.20	5.34	4.2
PO_4-P (mg/l)	11.2	11.1	6.34	0.84	0.43	0.31

Table 3: Wastewater quality of the influent and effluent during the application of MBR.

Parameter	Raw sewage	Removal (%)				
		After fine screening (1 mm)	After primary settling (2 mm)	MBR effluent		
				(1 mm)	(2 mm)	Pre-settled
pH	8.12	0.74	1.72	6.90	8.37	9.36
TSS (mg/l)	265	19.62	55.85	99.20	99.65	99.71
COD (mg/l)	1560	8.97	49.62	98.33	98.91	99.13
TKN (mg/l)	49.70	3.22	8.85	87.53	89.26	91.55
PO_4-P (mg/l)	11.2	0.89	43.39	92.50	96.16	97.23

Table 4: The removal percentages after MBR treatment.

Figure 2: Al-kut WWTP.

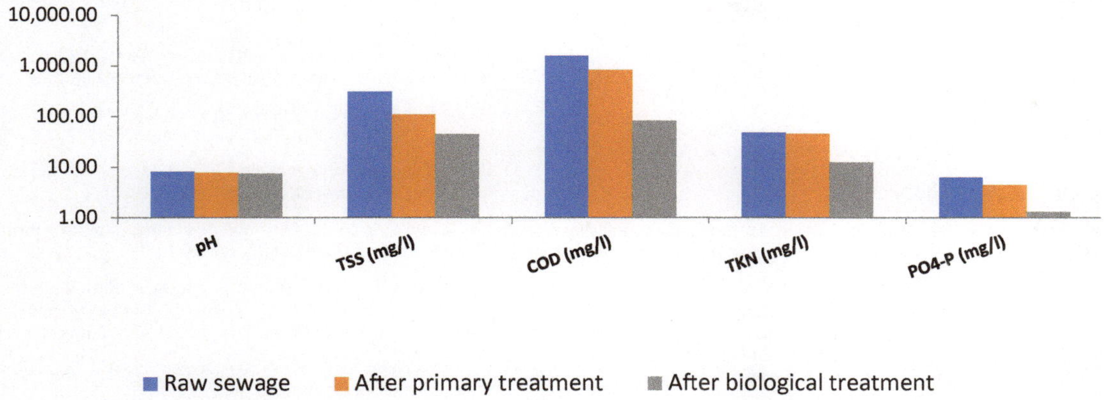

Figure 3: The main characteristics of the influent and effluent in Al-kut WWTP.

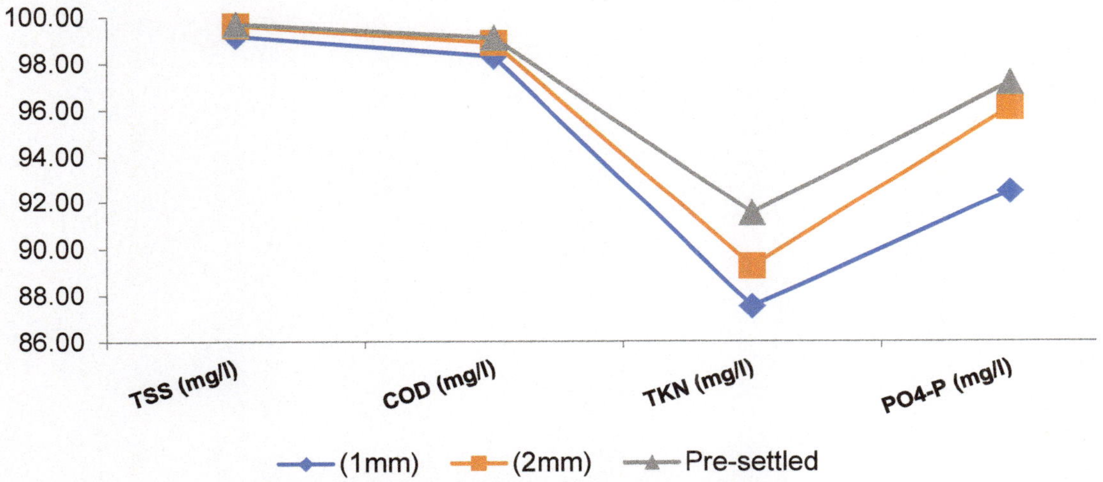

Figure 4: The removal percentages after MBR treatment.

Comparison between the effluent of the activated sludge in Al-Kut WWTP and the effluent of the membrane bioreactor

COD, TKN, PO_4-P and TSS of the effluent of the activated sludge system and membrane bioreactor. As a consequence, the comparison between the activated sludge system and membrane bioreactor is established. It can be seen that the difference between the characteristics of the effluent resulting from activated sludge treatment in Al-Kut WWTP and the effluent resulting from membrane bioreactor is significant due to suspended solids concentration, COD and PO_4-P exceed the Egyptian permissible limits (40, 80 and 1 mg/l, respectively) while the effluent treated water from MBR is complying with Iraq regulation. The obtained results of the two pre-treatment methods (1 & 2 mm pre-screened and pre-settled) in the MBR system meet the requirements of the environmental legislations for reuse [16].

Since the membrane acts as a barrier to bio solids and microorganisms, the effluent quality is much better than that produced by a conventional plant. Also, the membrane barrier eliminates the secondary clarifier and allows the activated sludge to be more highly concentrated. This reduces the capacity needed for biological tanks, saving space and money.

Conclusion

This study aims at investigating of the effect of the pre-treatment on the performance of the membrane biological reactor for six months. A TITAN MBR module from PALL Membrane Systems was used in this study. The obtained results reveal that the pre-sedimentation is more effective than the fine screening (1 & 2 mm fine screening), in this case study, as a pre-treatment method. Additionally, the membrane bio-reactor filters out nearly all solids and satisfies the environmental requirements of reuse. The TITAN MBR module employs a unique air scour nozzle, delivering air in the center of each membrane fiber bundle to effectively shake the membrane and scour the entire fiber length. The advanced air scour nozzle design minimizes power consumption by delivering air efficiently in a cyclical fashion while eliminating sludging. Moreover, aeration maintains solids in suspension, controls the membrane fouling by hydraulic shear force and agitation and provides oxygen to the biomass, leading to a better biodegradability and synthesis of the cell.

Acknowledgements

The author wishes to express his appreciation to Al-Kut WWTP staff for his assistance during the research and for supplying with samples and analysis.

References

1. Jefferson JB (2003) Membranes for Industrial Wastewater Recovery and Re-use.
2. Wiesmann U, Choi S, Dombrowski EM (2007) Fundamentals of Biological Wastewater treatment. Wiley-VCH Velag GmbH & Co. KGaA, Weinheim, pp: 34-36 & 291-233.
3. Myung SW, Choi IH, Lee SH, Kim IC, Lee KH (2005) Use of fouling resistant nanofiltration and reverse osmosis membranes for dyeing wastewater effluent treatment. Water Sci Technol 51: 159-164.
4. Judd S (2004) Fouling control in submerged MBR, in: Proceedings of IWA Specialized Conference on Water Environment-Membrane Technology, Seoul, Korea.
5. Guo WS, Vigneswaran S, Ngo H (2004) A rational approach in controlling membrane fouling problems: pretreatments to a submerged hollow fiber membrane system. In: Proceedings of IWA Specialized Conference on Water Environment-Membrane Technology 2004, Seoul, Korea.
6. Crawford G (2001) The Evolution of Membrane Bioreactor System Design for Wastewater Treatment. IWA.
7. Judd S (2006) The MBR Book: Principles and Application of Membrane Bioreactors in Wastewater Treatment (2nd edn.). Elsevier Science Ltd., USA.
8. Le-clech P, Jefferson B, Judd S (2003) Impact of aeration, solids concentration and membrane characteristics on the hydraulic performance of a MBR. J Membr Sci 218: 117-129.
9. Ognier C, Wisniewski A (2004) Grasmick MBR fouling in subcritical filtration conditions: a local critical flux concept. J Membr Sci 229: 171-177.
10. Wozniak T (2012) Comparison of a conventional municipal plant and an MBR plant with and without MPE. Desalin Water Treat 47: 341-352.
11. Water Environment Federation (2006) Membrane Systems for Wastewater Treatment, McGraw-Hill, NY, USA, pp: 202-207.
12. Isik M, Sponza DT (2008) Anaerobic/aerobic treatment of a simulated textile wastewater. Sep Purif Technol 60: 64-72.
13. Koseoglu H, Yigit NO, Iversen V, Drews A, Kitis M (2008) Effects of several different flux enhancing chemicals on filterability and fouling reduction of membrane bioreactor (MBR) mixed liquors. J Membr Sci 320: 57-64
14. Cui ZF, Chang SAG, Fane AG (2003) The use of gas bubbling to enhance membrane process. J Memb Sci 2211: 1-35.
15. APHA (2012) Standard Methods for the Examinations of Water and Wastewater, 22th edn., American Public Health Association, Washington DC, USA.
16. Law 48 (2008) and amendments article 92, 2013, Amyria press, Egypt.

CFD Validation for Forces on Immersed Tubes in Fluidized Bed

Sushil Kumar*

Corporate R&D, BHEL, Hyderabad, Telangana, India

Abstract

Fluidized bed is widely used in chemical processing industries and heat transfer operations because of high heat transfer and mass transfer rates due to intermixing. This work deals with estimation of the force imparted by the bed material on the tube assemblies in a fluidized bed by CFD technique. The gas-solid flow was simulated by means of a multi-fluid Eulerian model incorporating the kinetic theory for solid particles. The pulsating forces were compared to experimental data of Kennedy et al. A good approximation was observed in force exerted over different tubes. However minor differences were observed in interval duration of the peak forces.

Keywords: Fluidized bed; CFD; Gas-solid flow; Pulsating forces

Introduction

Fluidized beds with immersed tubes are extensively used in industries. These tubes provide large surface area for heat transfer with the bed materials. Fluidized bed encounters time varying forces of an irregular nature due to the action of gas bubbles and turbulence within the bed. These forces contribute to vibrations which can lead to failures of the tube and their support system. To have a basis for structural design, a definition of load environment is required. Information on the magnitude and frequency composition of the applied forces is necessary in order to predict the fatigue life of a structure. The objective of this investigation is to validate the fluent capability of predicting the forces of the tubes. Forces predicted are compared with the experiment performed by Kenndy et al. [1]. In the experiment the forces were measured on various length of tube and ranges of fluidization conditions. Three beds considered had following dimension 0.3 × 0.3 m (1 ft by 1 ft), 0.91 × 0.91 m (3 ft × 3 ft) and 2.4 m × 0.3 m (8 ft × 1 ft). Out of these 3 bed, a bed with dimensions of 0.91 m × 0.91 m was chosen for our investigation as it has aspect ratio of industrial fluidized bed. Also for this case there were more experimental data available to compare with CFD results.

Experimental Setup

The fluidized bed test facility bed in experiment by Kennedy et al. [1] had various cross-sectional dimensions. This fluidized in the present discussion had a cross section of 0.91 m × 0.91 m (3 ft × 3 ft) and height of 1.4 m (4.5 ft). Here fluidizing air is supplied by a rotary, positive displacement blower with maximum capacity of 3.78 m^3/sec at 51.7 kPa (8000 cfm at 7.5 psi). Air from blower enters a plenum below the air distributor which consists of perforated plate. Air flow is monitored with a venturimeter and superficial gas velocity is controlled by speed of the blower. Figure 1 shows the arrangement of tube array used in the experiment. Tubes are numbered in order 1 to 8 from the top where instruments were fixed in the experiment to capture the force data. Tubes of 5 cm OD were fixed in fluidized bed with fixed pitch as shown in Figure 1. In the experiment forces on the individual tubes were measured by supporting each end of a 5 cm (2 inch) diameter tube with strain gauze that were designed and built for this specific application. These load cells can measure both the vertical and horizontal components of force transmitted by tubes to its end support.

CFD Modeling

Traditionally, two approaches are followed for the study of multiphase flow phenomena in CFD: Eulerian–Lagrangian and the Eulerian–Eulerian. The first method is a fundamental approach which involves the balance of forces that act upon each of the particles and requires considerable computational effort. This approach also takes into account a collision model for commending the energy dissipation caused by the non-ideal particle-particle interactions. Thus this approach is limited to only fluidized bed of small in size and with very few particles. The Euler–Euler approach considers the dispersed phase (bed particles) as a continuous phase and is based on the Navier–Stokes equations applied to each phase. It is a more realistic approach for investigation of fluidized bed of Industrial scale [2,3]. An important aspect which influences the accuracy of CFD results for fluidized beds is the methodology used to extract the data. The data extraction methodology and the subsequent results, such as bed properties, bubble characteristics and bed expansion, suffers from great variation between studies reported in the literature [4]. Also, according to Asegehegn et al. [4] data extraction can have as much influence as the use of different constitutive relationships. For instance, Hulme et al. [5] shown that different volumetric solid fraction at the inlet led to different bubble average parameters (Table 1).

Modelling and meshing

A geometrical CAD model of 3 × 3 ft. fluidized bed with height of 2.8 m is imported in Ansys ICEM CFD. The height of the bed is doubled here in the model so that fluidized particles does not come out of the top section of the bed during start of the fluidization process in CFD. Tubes with diameter of 5 cm are laid as per experimental setup show in Figure 1. Meshing of the fluid zone was carried out with hexahedral elements. To capture the boundary layers around the tubes, fine prism mesh layers were created surrounding the tubes as shown in Figure 2. Total mesh elements for the fluid zone was approximated at 0.3 million. Mesh was imported in the Ansys fluent software for defining the problem setup. The bed was initially patched with 0.8 mm spherical particles with density of 2700 kg/m^3 such that all the tubes are submerged. A uniform velocity of 1.5 m/s is applied at the bottom of the fluidized bed as inlet

***Corresponding author:** Sushil Kumar, Corporate R&D, BHEL, Vikasnagar, Hyderabad-500 093, Telangana, India, E-mail: sushilkumar@bhelrnd.co.in

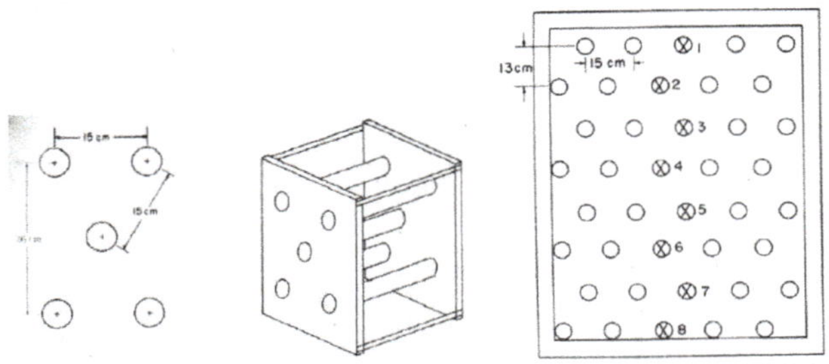

Figure 1: Tube array and location of tubes in the experimental setup by Kennedy [1].

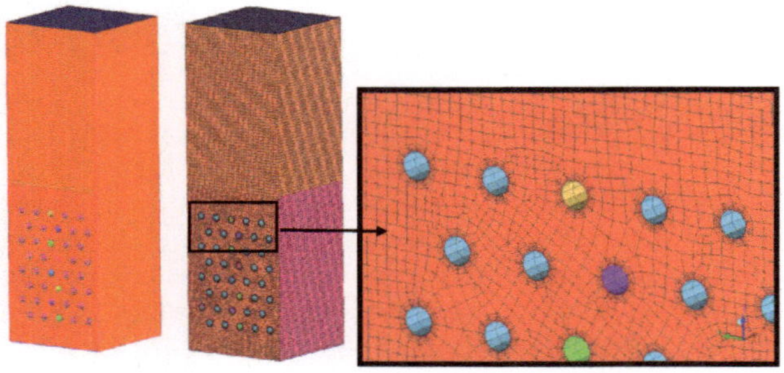

Figure 2: Geometrical model and mesh in ICEM CFD.

Parameter	Model (Ansys Fluent)	Reference
Granular viscosity	Syamlal-Obrien	[6]
Granular bulk viscosity	Lun et al.	[7]
Frictional viscosity	Schaeffer	[8]
Frictional pressure	Based-ktgf	[7]
Solid pressure	Lun et al.	[7]
Radial distn function	Lun et al.	[7]
Drag law	Syamlal-Obrien	[6]

Table 1: Closure equations.

boundary condition. The fluidized bed system was isothermal at 300K and initial pressure was set at 1 bar. The minimum void fraction of bed packing was set to 0.45. Top of the fluidized chamber was considered as pressure outlet boundary condition. The particle gas interaction was characterized by Syamlals- Obrien Drag law [6]. The coefficient for the drag law are obtained from the literature and Fluent theory guide based on particle size and phase. The particle- particle and particle-wall interaction was considered to be elastic in nature with coefficient of elasticity as 0.9. The simulation is initiated with time step of 1e-4 s and is simulated for 5 sec of real fluidization time. The QUICK and second order upwind were used for the spatial discretization of the continuity and momentum equations respectively while time was discretized using first order implicit. The Phase-Coupled SIMPLE algorithm was used for the pressure-velocity coupling.

Results and Discussion

This simulation was performed for 4 seconds of real flow time. The initial 1.5 sec were neglected to reduce the effect of divergence in some of the cell zones during the bed expansion. Once the bed is fully fluidized. The properties like forces, pressures and volume fraction of solids are produced in Ansys Fluent 15. In literature there is no distinct definition for bubble boundaries, but many investigators have considered solid volume fraction equal to 0.2 [7-9] and same has been adopted here too. From the solid volume fraction contour plots as shown in Figure 3 it was observed that small bubble forms when air enters into the bed though distributor plates. These bubble grows in size while rising up in the bed and encounters tubes above it. There were two major observations regarding the bubble rise. First, the bubble is hitting the tubes encounter its path while rising and breaks into small bubbles; second it elongates vertically between the tubes and passes between the tubes without breaking. Sizes of some bubble increase also by coalescence with other bubbles. Finally, when bubble crosses all the rows of tubes it erupts at the top of the bed. The solid volume fraction contour below shows the same phenomenon happening inside the bed at any time t. Forces were monitored on each of the eight tubes marked in the Figure 1. Vertical forces for tube 6 and tube 8 were measured and is shown in Figures 4 and 5. The force time history here consists of pulses with durations of approximately of 0.2 s which closely matches with the pulses in experimental force data. In the Figure below it can be observed that forces on tube 6 and 8 are almost similar in nature. The force appears as pulses occurring at a rate of 2-3 per second with magnitude of around 250-350 N. In experimental data shown the peak forces vary from 200-350 N. Pulses on tube 8 (lower tube) precede those on the tube 6 (upper tube) by 0.1- 0.2 s. The distance between these tubes is 26.4 cm. Thus it can be concluded that pulse is propagating upwards with a velocity of 1.3-2.6 m/s. This velocity corresponds to bubble rise velocity predicted by CFD, while those by experiment is in the range of 1-1.7 m/s. The total vertical force predicted by CFD on the eight tubes is shown in Figures 6 and 7 below. The pulses were of magnitude 700 to 900 N with duration of around 0.8-1 sec. whereas in the experiment, forces were of magnitude 600 N to 800 N with a duration of 0.5-0.8 s. This gives a near match of CFD predicted results with the experimental data.

Conclusion

Hydrodynamics simulation were carried out to validate the experimentally determined forces on the tube to with the CFD simulated forces. The two fluid Eularian-Eularian model available in Ansys Fluent were able to simulate the bubble behavior in the fluidized bed. Tubes immersed in the bed significantly affect the bubble shapes and sizes which eventually decreases the mean diameter of bubble and bubble rise velocity. Forces on the tubes predicted through CFD matches trend of the experimental determined forces. Although there is reverse trend in the simulated forces pattern, as tubes at bottom of bed is exerting slightly lesser force than those above it. A complete information of experimental test facility bed is not available, which could have given a more accurate modelling of the test bed in CFD. However, the forces predicted in CFD occurs in pulses of 2-3 per second with magnitude of 250-350 N matches nearly with the experiment. This complex mechanism of bubble movement around the tubes, which alters the bubble aspect ratio can be the main reason for deviation of force predicted by the simulation and experiment. Deviation could also be due to the approximation of uniform velocity at the bottom of bed as nozzle details of distributor plate were not available. An intensive investigation of fluidized bed is needed to verify the mechanism of bubble rise in presence of tubes and forces acting upon the tubes. Further, a close experimental and numerical studies with dense tube arrangements can reveal more insight about the bed. Moreover, CFD modelling was capable of predicting the forces on the tubes and thus provides a basis in structural design of the fluidized bed tube bundles.

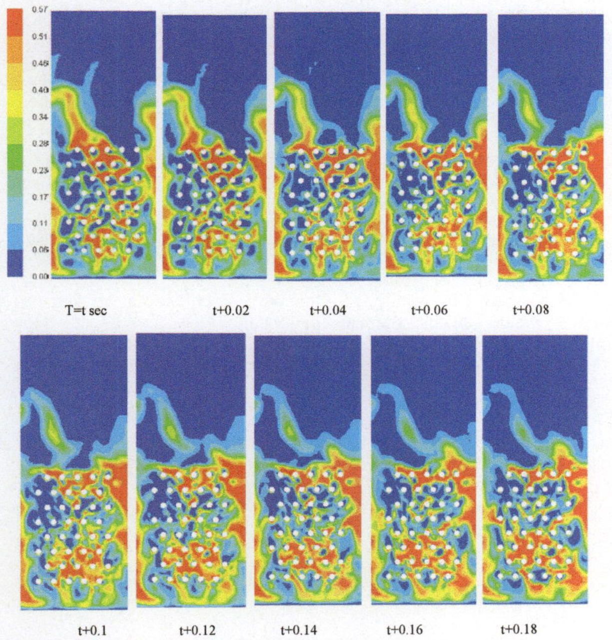

Figure 3: Volume fraction contours for solids in a mid-plane of fluidized bed.

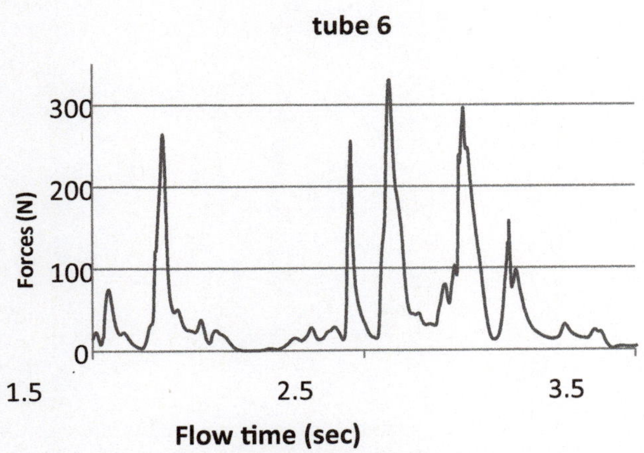

Figure 4a: CFD predicted forces on tube 6.

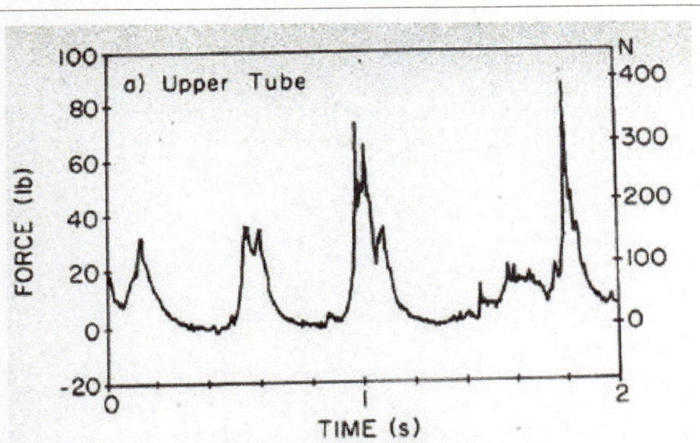

Figure 4b: Experimental forces on tube 6.

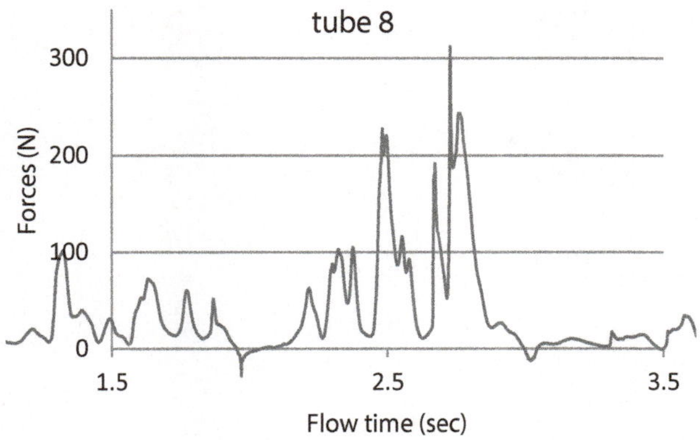

Figure 5a: CFD predicted forces on tube 8.

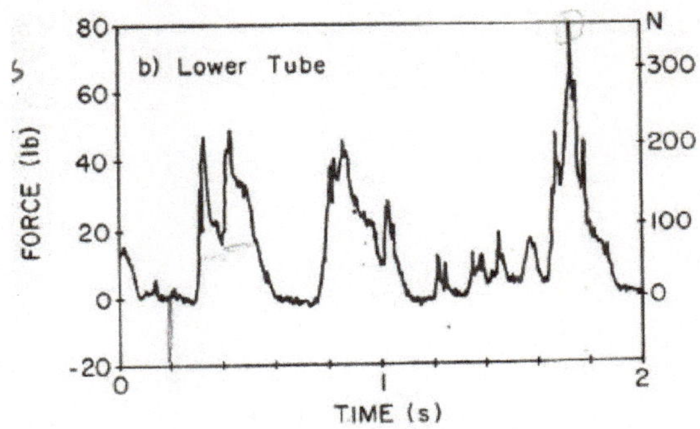

Figure 5b: Experimental forces on tube 8.

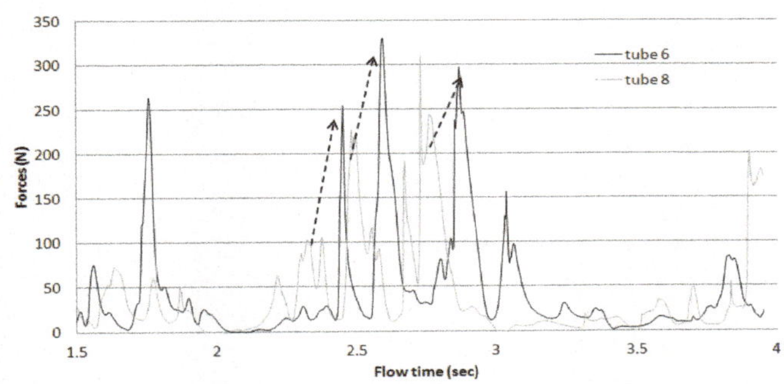

Figure 6: Force comparison between tube no 6 and tube no 8 in CFD.

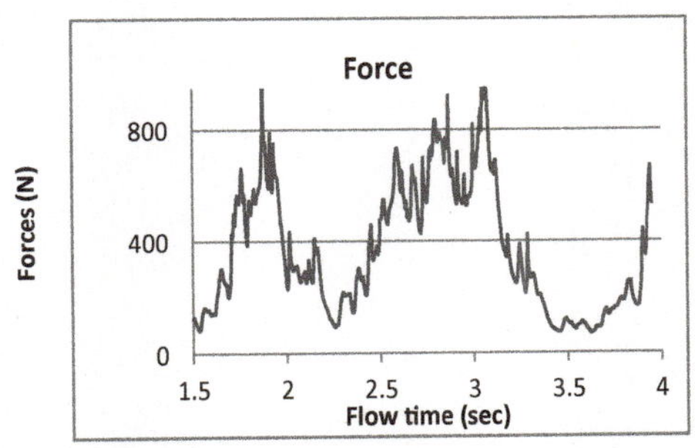

Figure 7a: CFD predicted forces summed all 8 tubes.

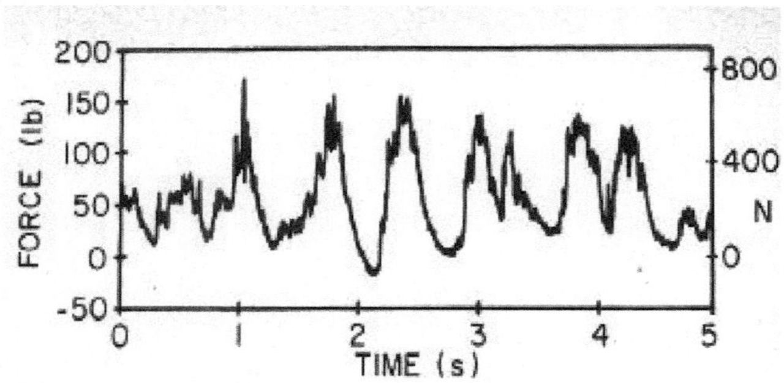

Figure 7b: Experimental forces summed on all 8 tubes.

References

1. Kennedy TC, Donovan JE, Trigas A (1981) Forces on immersed tubes in Fluidized bed. AIChE Journal 27: 31-357.
2. Van Wachem BG, Schouten JC, Van den Bleek CM, Krishna R, Sinclair JL (2001) Comparative analysis of CFD models of dense gas–solid systems. AIChE Journal 47: 1035-1051.
3. Van der Hoef MA, van Sint Annaland M, Deen NG, Kuipers JA (2008) Numerical simulation of dense gas-solid fluidized beds: a multiscale modeling strategy. Annu Rev Fluid Mech 40: 47-70.
4. Asegehegn TW, Schreiber M, Krautz HJ (2011) Numerical Study of a Gás-Sólid Fluidized Beds Hydrodynamics: Influence of Immersed Horizontal Tubes and Data Analysis. International Journal of Chemical Reactor Engineering 9: 1-39.
5. Hulme I, Clavelle E, van der Lee L, Kantzas A (2005) CFD Modeling and Validation of Bubble properties for a Bubbling Fluidized Bed. Industrial e Engineering Chemistry Research 44: 4254 4266.
6. Syamlal M, O'Brien TJ (1989) Computer simulation of bubbles in a fluidized bed. AIChE Symposium Series 85: 22-31.
7. Lun CK, Savage SB, Jeffrey DJ, Chepurniy N (1984) Kinetic Theories for Granular Flow: Inelastic Particles in Couette Flow and Slightly Inelastic Particles in a General Flow Field. Fluid Mech J 140: 223-256.
8. Schaeffer DG (1987) Instability in the Evolution Equations Describing Incompressible Granular Flow. Journal of Differential Equations 66: 19-50.
9. Hulme I, Clavelle E, van der Lee L, Kantzas A (2005) CFD modeling and validation of bubble properties for a bubbling fluidized bed. Ind Eng Chem Res 44: 4254-4266.

How are Waste Entirely Avoided in Solid-State Productions?

Gerd Kaupp*
Institute of Applied Chemistry, Oldenburg University, Oldenburg, Germany

Abstract

The mechanism, synthetic application, scaling, and industrial production with gas-solid and solid-solid reactions without producing wastes is outlined. These reactions profit from lowered activation energy (with respect to melt reactions), favorable kinetics with definite end of reaction, and mostly reaction specificity giving only one product in pure form without workup necessity. They proceed thus solventless (not only "solvent free") and wasteless. The bargain of the solid-state should not be given up by heating to melt (e.g., milling without temperature control) or by "adding a little solvent". Molecular (or ionic) solid-state reactions, be they gas-solid or solid-solid, profit from the presence of cleavage-planes, channels or voids for intracrystalline long-range anisotropic molecular migrations. Such release of internal local pressure upon chemical change is essential. This is experimentally verified. In the solid-solid case mixing and contacting leads to the thermodynamically more stable product, but this does not exclude kinetic control. It is of utmost importance to control the temperature so that eutectic temperatures (melts) are not reached during all reaction stages. Typical examples out of more than already thousand 100% yield reactions from 26 reaction types for particularly delicate (not otherwise available) or otherwise interesting products are presented, and the scaling up to continuous industrial production is exemplified. Also, mechanochemical reactions (for example counter-thermodynamic mechanical bond breaking of infinite covalent materials) can be waste-free, ending with the most stable possibility when the bond-breaking ceases at the nanoscale. The proper experimentation (most important is strict temperature control) is outlined and easily and safely achieved.

Keywords: Activation energy decrease; Gas-solid production; Scaling up to industrial; Solid-solid production; Wasteless

Introduction

Since as early as 1985 it is known that molecular solid-state reactions (gas-solid and stoichiometric solid-solid reactions proceed to completion. Pure solid products are obtained without waste forming solvent use for purification from side products, starting materials, catalysts and other auxiliaries. AFM-measurements proved with molecular precision that no melts are involved and that surface-specific anisotropic molecular migrations occur beyond these surfaces. Clearly, avoiding (intermediate) melts is the requirement for making profit from the solid-state, be it crystalline or amorphous. Thus, temperature control is essential.

Roughly, these solid-state reactions proceed rapidly at about 100 K lower temperatures than melt reactions, and they stop when eutectic temperatures are reached, if that is not avoided by proper cooling. Nevertheless, these very profitable bargains of the solid-state (more than 1000 quantitative wasteless syntheses due to favorable kinetics and decrease of activation energy) have not widely been appreciated, because it is at variance with a hypothesis of Schmidt, Tel-Aviv, Israel, of 1964 (called "Schmidt's topochemistry), "forbidding" long-range molecular migrations within crystals. Since such disproved thinking is still widely in use, people tried with molecular milling without temperature control via melts, or even with adding "a little solvent" [1] later called LAG (liquid assisted grinding) and claimed this to be an advancement [2]. However, this is at best applicable to recrystallizations with formation of cocrystals and supramolecular objects by kneading (not grinding), and the here mechanistically incorrect term "mechanochemistry" should not be applied. When LAG is applied to molecular reactions involving covalent bond formation (rather than mechanical bond breaking, cf. the Review [3]) it is a severe retrogression by losing the bargain of the solid state. Typically, such procedures, are incomplete (e.g., 59% yield, not citing the review with more than 1000 waste-free solid-state reactions in at least 26 different reaction types all with 100% yield [4], and wasteful with producing side products requiring chromatography. These could be wastefree syntheses when liquid phases are avoided. A typical example (34 cases) for the latter is the authors' temperature controlled solventless milling of aldehydes/ketones with $NaBH_4$ (4:1 ratio) to obtain quantitatively the elusive high melting sodium-tetraalkoxyborates in pure form even with liquid carbonyls, from where the free alcohols can be quantitatively obtained with water vapor, leaving pure sodium borate for reuse [5]. Nevertheless, the journal "Green Chemistry" had published the wasteful reduction by "high speed ball milling" of four arylaldehydes and four acetylbenzenes with $NaBH_4$ (all 1:1) but without temperature control at 95% conversion and yields from 65-70% [6]. It followed a paper of the carbonyl reductions with NH_3-BH_3 (explosive borazane from $NaBH_4$+$(NH_4)_2SO_4$ in THF+gaseous H_2), using only one of the three B-H bonds in water, giving poor yields (85%) after chromatographic workup and production of huge amounts of waste. Remarkably, this paper [7] claims to be "environmentally benign" and does not even cite the unbeatably sustainable paper [5]. Or Fiscic's group, mills carbodiimides, isocyanates, and isothiocyanates as long as 2 h to add amine derivatives according to the "LAG technique" requiring copperchloride-catalysis with poor yield (e.g., claimed 59% or 74%) and chromatographic workup [8], but not citing [4,9], where more than 18 (hetero)cumulenes, provide waste free solid-state syntheses without added solvent and thus without catalyst giving 100% yield without producing wastes. Milling of melts without temperature control in planetary mills with too much impact (clogging) require exceedingly long reaction times. For example, 2.2 4-phenyl-cyclohexanone + 2.0

***Corresponding author:** Gerd Kaupp, Institute of Applied Chemistry, Oldenburg University, Oldenburg, Germany
E-mail: gerd.kaupp@uni-oldenburg.de

4-nitro-benzaldehyde+0.2 (S)proline give 42% aldol condensate and 29% recovered aldehyde after 48 h planetary milling at 250-400 rpm [10].

Using mills as heating source for milling melts, or LAG for bond-forming reactions (not only recrystallization) are severe retrogressions. They do not consider proper experimentation, but are apparently appreciated in "green chemistry", despite their vast inferiority in terms of cost and sustainability. It appears therefore timely to point out the much superior solid-state techniques in both the mechanistic and scale-up sense as a viable supplement to the review in [4].

Mechanism of solid-state molecular and salt reactions

Solid-state reactions are kinetically favorable and experience lowered activation energy when compared to melt or solution reactions. The benefits are shown in Figure 1. These do not cover mechanochemical reactions that require mechanical breakage of covalent bonds.

The solid-state mechanism was exhaustively studied with atomic force microscopy since the 1980s. Figure 2 shows that unlike the hypothesis of "Schmidt's topochemistry" of 1964 molecules migrate upon chemical reaction anisotropically within crystals along channels or cleavage planes in the phase-rebuilding stage and at a certain conversion sudden transformation occurs into the product phase in the reaction zone, followed by disintegration, creating fresh surface for further reaction. This has been called the "phase-rebuilding mechanism". The formed surface features are face-selective depending on the orientation of channels and/or cleavage-planes. An exhaustive review of submicroscopic features classifies about 8 basic types with various characteristic appearances [11]. These provide also proof at the molecular scale that liquid phases are excluded. If these occur the scanning contact-AFM tip would plough and smear materials along its path [11]. The reason for the molecular migrations is the local internal pressure by the geometrical change upon the chemical reaction that must be immediately released. The phase transformation occurs when the product lattice becomes energetically more favorable. Such features occur equally in solid-state photoreactions, thermal reactions, gas-solid and solid-solid reactions.

When the mechanistic requirements are obeyed in the experimentation, complete reactions (close to zero order reactions) with 100% yield are obtained at decreased activation energy (roughly 100 K temperature decrease with respect to melt reactions is possible [12]). This has been verified with more than 1000 reactions all across molecular chemistry in 26 reaction types (Table 1) [4], mostly with highest specificity (one product), and that represents unbeatable "atom economy". The importance of the results as listed in Table 1 appears to urgently list the major requirements for proper application of the mechanistic requirements in Table 2.

How are wastes entirely avoided in solid-state productions?

Solid-state syntheses: complete reactions without producing wastes

No solvents, mostly no catalysts, no supports, new selectivities

Gas-Solid Reactions

Solid-Solid Reactions

Melt Reactions (second choice, less reactive, less selective)

temperature control is essential

100% yield, no workup no removal of catalysts/supports, no solvents, more than 1000 examples in >26 reaction types, several on the kg scale

modern techniques of analyses:

AFM (atomic force microscopy), SNOM (scanning near-field optical microscopy), Nanoscratching

Figure 1: Achievable benefits of solid-state molecular syntheses.

Gas-solid reactions

Figure 3 shows the easy lab-scale equipment for gas-solid reactions. Temperature control is primarily obtained by slow addition of the reactive gas to the evacuated vessels that may be cooled or heated when required [13]. Vacuum tightness is essential. The flow-systems must use inert gas mixtures to calm down the reaction, because cooling from outside may not be sufficient. Care has to be taken for extra free volume if the powder volume increases.

Some examples for the particular wealth of gas-solid reactions, for example completely anhydrous conditions allowing for the synthesis of otherwise inaccessible new products, shall be exemplified [14]. For example, elusive aminomethylene hydrochlorides are only stable as solids at low temperatures (here -18°C) (Figure 4). The now available aminomethylene-hydrochlorides are important highly reactive building blocks.

A long-standing problem has been solved by gas-solid addition of HCl gas to obtain the crystalline unrearranged camphene hydrochloride in pure form. There are not the moisture problems with that technique [15]. In solution, this addition product undergoes Wagner-Meerwein rearrangement and no pure product could ever be obtained. Now the solid-state thermal rearrangement of camphene hydrochloride to isobornylchloride can be studied kinetically and with synchrotron X-ray (the within-crystal transformation is complete after 3 years at room temperature and 6 h at 80°C without melting), which might give a clue for a revival of the non-classical carbocation question (Figure 5).

The quantitative diazepine syntheses in Figure 6 proceed only with gaseous acetone [16]. If liquid phase acetone is used one needs expensive catalysts, supports and microwave in 80 times smaller runs with far from quantitative yield and chromatographic purification in six following papers of others. These are, of course, neither green nor sustainable as claimed, and surprisingly, they do not cite our early work. Nobody should create enormous amounts of dangerous wastes for these reactions any more [17].

Large scale gas-solid exhaust-gas purification of acetone vapor from exhaust gas atmospheres has been achieved in a two-column system below detection limit in a fluidized bed. At well-adjusted acidity, the volatile products are expelled at 80°C according to $(H_2NOH)_3 \cdot H_3PO_4 + K_2HPO_4 \cdot xH_2O + 3Me_2C{=}O \rightarrow 3\ Me_2C{=}NOH + 2\ KH_2PO_4 + (x+3)\ H_2O$. The acetone-oxime is condensed and separated from the water of reaction, KH_2PO_4 is a value added couple product.

The gas-solid addition of bromine gas to cholesterol [18] (Figure 7) provides directly a better quality intermediate dibromide for hormone production of technical interest. If necessary, the hydroxyl group of cholesterol can be protected by solid-solid esterification with oxalic acid.

Gas-solid additions of HCl to oxiranes a low temperatures (where these are solidified) occur stereoselectively (Figure 8). Others occur stereospecifically in the synthesis of contraceptive steroid hormones [4,19]. More complicated gas-solid reactions with elimination of water and rearrangements (not available in solution) in steroid-hormone syntheses are found in [20].

Carbonic acid halfester salts are highly versatile merchandisable reagents. They are synthesized in suspension bed columns from commercial alcoholates with carbon dioxide that is initially strongly diluted with nitrogen, to remove the enormous heat that is produced

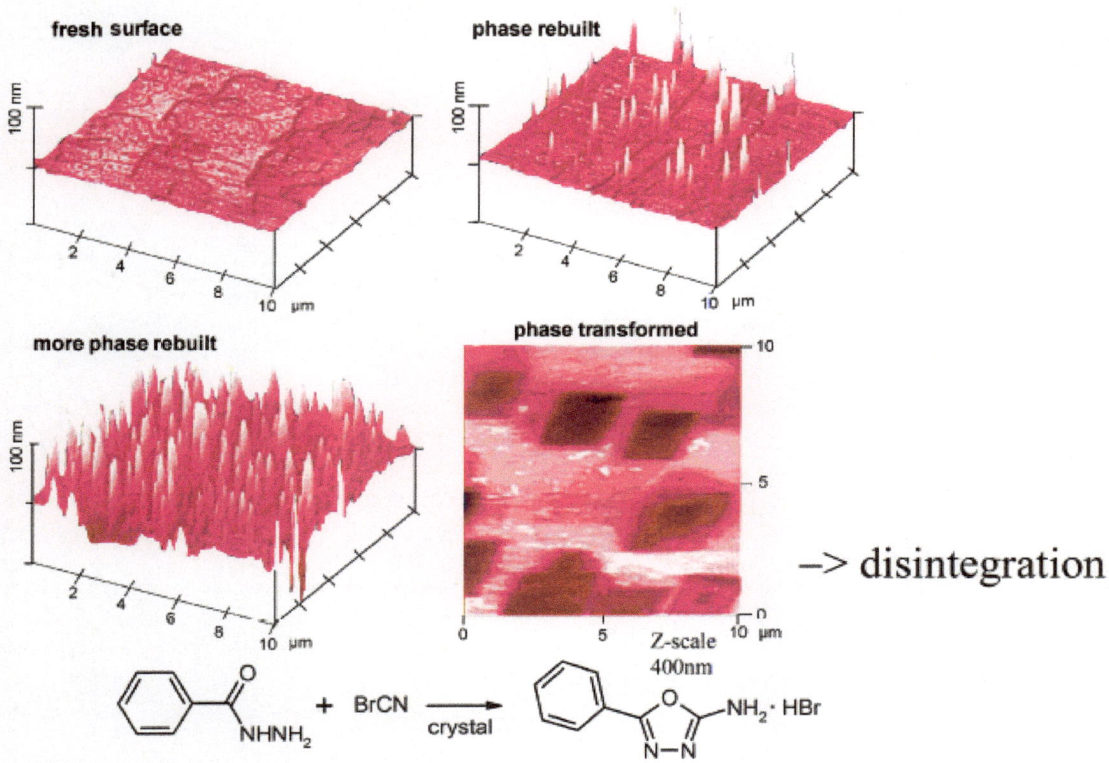

Figure 2: The three stages in the solid-state phase-rebuilding mechanism as revealed by contact-AFM.

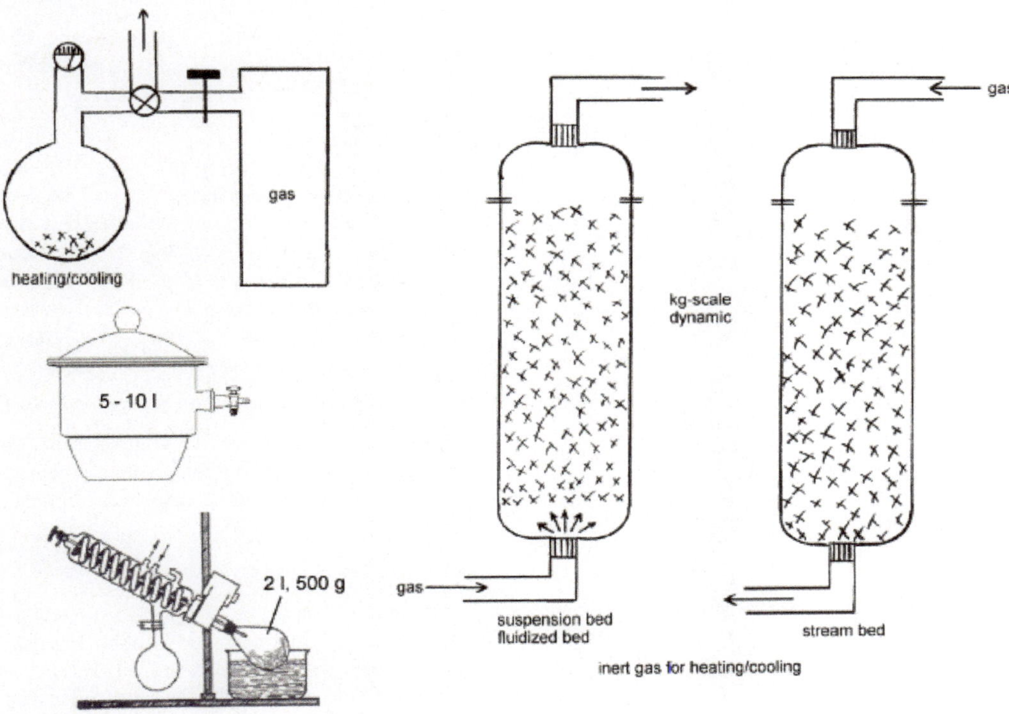

Figure 3: Lab-scale equipment for gas-solid reactions [13].

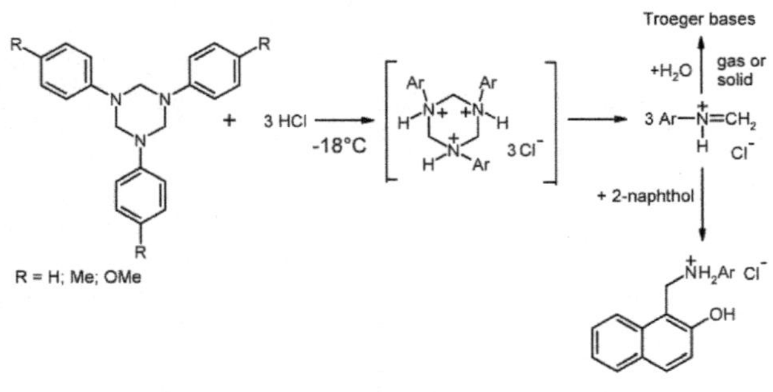

Figure 4: Synthesis of aminomethylene hydrochorides by quantitative gas-solid reaction at -18°C [14].

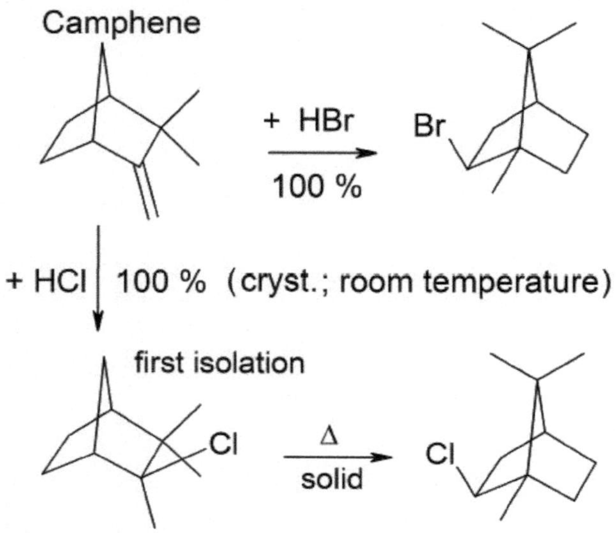

Figure 5: Gas-solid synthesis of unrearranged camphene hydrochloride and its thermal rearrangement to isobornylchloride, similar to the directly obtained isobornylbromide.

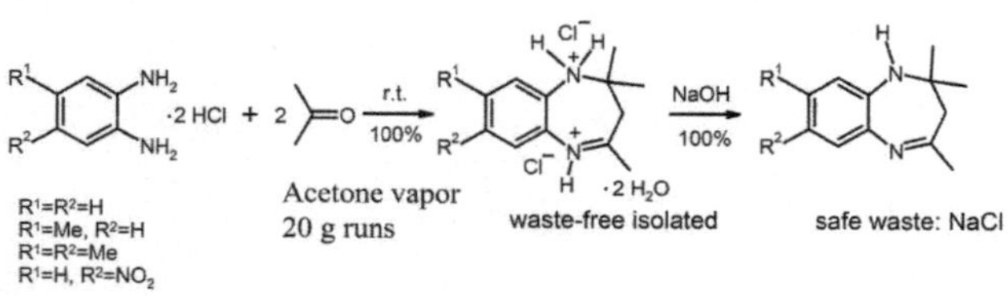

Figure 6: Gas-solid syntheses of 1,4-diazepines from o-phenylenediamine dihydrochlorides with acetone vapor [16,17].

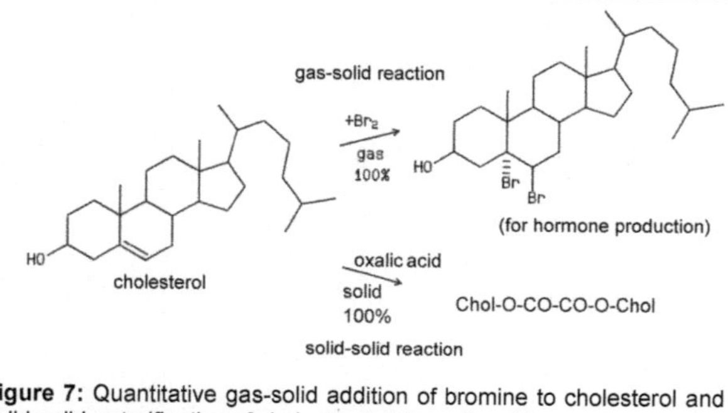

Figure 7: Quantitative gas-solid addition of bromine to cholesterol and the solid-solid esterification of cholesterol with oxalic acid.

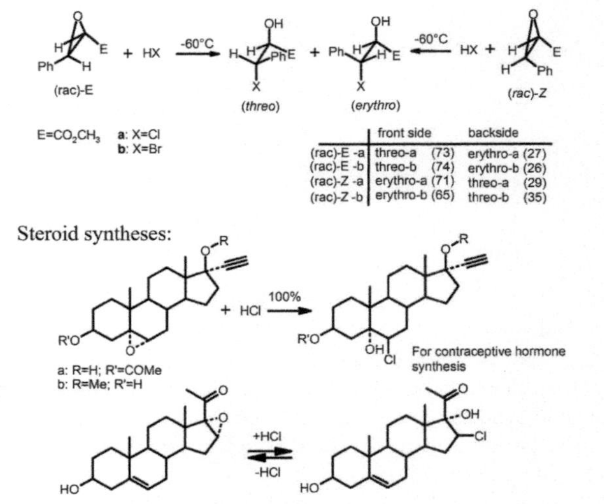

Figure 8: Stereoselective and stereospecific gas-solid additions of HCl to oxiranes.

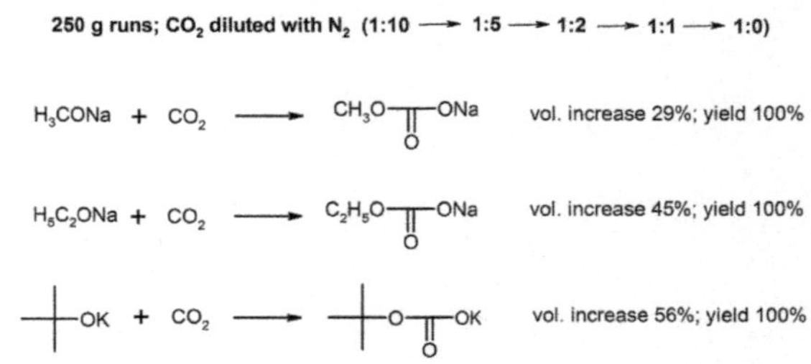

Figure 9: Gas-solid synthesis of carbonic acid halfester salts [13,21,22].

Inclusion reaction	Condensation of carbonyls
Electron transfers	Cycloadditions
Proton transfers	Cycloreversions
Oxygen transfers	Substitutions with RX
Oxygenations	Aromatic substations
Hydrogenations	Cyclization's
Additions of AR'NH.H_2O.ROH	Methylations
Additions of halogens and HX	Nitration's at N and C
Eliminations	Diazotations
Rearrangements	Azocouplings
C-C bond formations	Sandmeyer reactions
Carboxylation's with CO_2	Diazo transfer
Catalysed reactions	Cascade reactions
More than 1000 rapid 100% yield one-product reactions in these types	

Table 1: Optimized organic gas-solid and solid-solid reactions.

(Figure 9). At the end, pure CO_2 completes the reaction [21]. Free space has to be provided due to strong volume increases. Numerous further gas-solid syntheses including large and technical ones [13] are to be found in [4].

Solid-solid reactions

Quantitative wasteless solid-solid reactions require mills. Grinding in mortar are for initial tests, and with shock-sensitive materials like solid diazonium salts that must not be milled. Figure 10 depicts a temperature controlled vibration ball-mill MM_2 (Retsch Ltd, Germany) with double-walled beaker for milling up to about 200 mg organic stoichiometric mixtures, and for finding the temperatures that are below any eutectics for large-scale milling [22].

A larger scale 2 L horizontal rotor ball-mill and semibatch powder feeding and powder collection equipment is shown in Figure 11. In the absence of continuous production one does not need the feeding/collection equipment but leaves about 10% of the material in the mill during milling-out through a cyclone and bypass for protection of the balls. Furthermore, one determines the yield with the following batches. Such mills can be obtained up to 900 L size with fully continuous operation in industry. The 2 L version can be run in academic labs due to low trembling of the balls-driving rotor and moderate noise [23].

Figure 12 depicts a 400 L Simoloyer® mill, Figure 13 illustrates the blueprint of the feeding/collection device and Figure 14 shows it at work. Temperature control of the larger mills (≥ 2 L) is by cooling with cooling mantels and automatic alternating milling with cooling cycles using the different processing state. Since several hundreds of waste-free solid-solid syntheses have been reported in the exhaustive review [4] we select here the already realized large-scale examples of solid-solid molecular reactions [24].

Figure 15 contains two examples of unbeatable stereo specificity. The six OH-groups of D-mannitol (r.t.) or meso-inositol (95°C) are completely regio- and stereo-specifically condensed with stoichiometric phenylboronic acid upon milling in the temperature controlled MM_2 vibration mill [25] and in the 2 L Simoloyer [13]. The water of reaction is taken up by the product crystals and can be removed by evaporation in a vacuum. This more than underlines the profit of solid-solid reactions, all starting materials exhibiting proper channels for the necessary molecular migrations. As expected, the corresponding solution reactions lead to intractable mixtures of stereoisomers.

Less spectacular are formations of molecular complexes by stoichiometric milling of the components, which is also possible at different molecular ratios. Figure 16 gives typical examples with caffeine. This is certainly an interesting and easy processing for drug formulations and many others expected; also different ratios.

It is also possible to achieve the synthesis of pharmaceuticals. For example, the antibacterial sulfonamide phthalazole is obtained free of imide and bisamide side products (that occur upon reaction in solution or in the melt), if stoichiometric solid-state milling of the reactants is performed for the acylation reaction as in Figure 17 [13].

Two of the early quantitative 2 mmol Knoevenagel solid-solid syntheses (in Figure 18a and 18b) between the aldehyde and barbituric acid [9,12] were waste-free scaled to 200 g in the 2 L Simoloyer [13] at typical out-milling times of 10 min. The reaction with vanillin (in Figure 18c) occurs already as melt reaction at 50°C. It was directly scaled 2500-fold from 200 mg at 0°C in the cooled MM_2 vibration mill to 500 g in the 8 L Simoloyer at ≤ 10°C without any problems [26]. The water of reaction was taken up by the product crystals and did thus not impede, and the reaction time is very short.

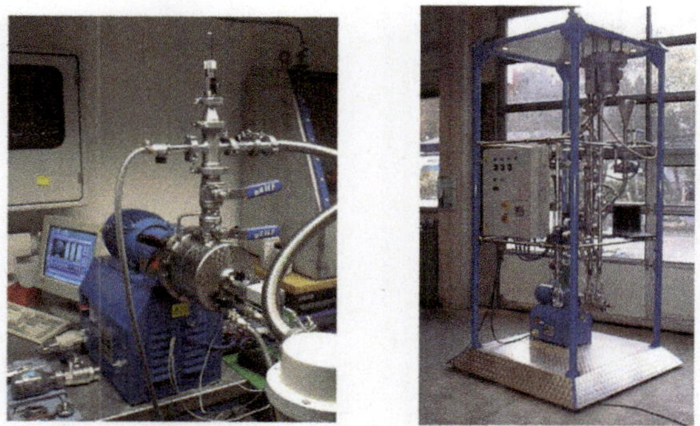

2 L horizontal rotor ball-mill Powder charging and collection

Figure 11: Technical double-walled ball mill Simoloyer® C01, 2 L of Zoz Group Ltd, Wenden, Germany with temperature control and feeding/collecting equipment for semibatch operation [23,24].

Figure 12: 400 L Simoloyer(R) ball-mill (extensions up to 900 L); with kind permission of the Zoz-group Ltd, Wenden, Germany.

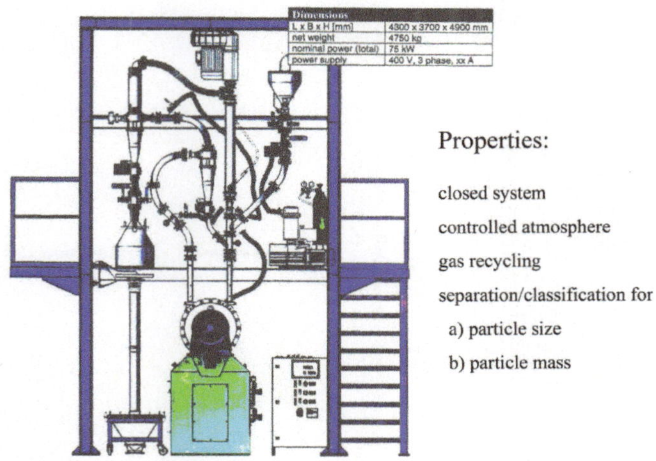

Figure 13: Scheme of the large scale semicontinuous feeding/collecting equipment. With kind permission of Zoz-Group Ltd, Wenden, Germany.

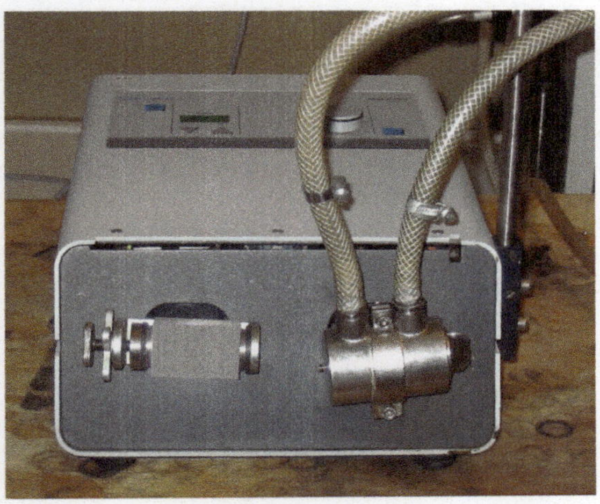

Figure 10: Lab-size vibration ball-mill: 20 ml beakers with temperature control (-78 to 100°C) for initial tests (do not use planetary mills) [23].

Figure 14: Different 100 L Simoloyer® ball-mills and Auto-batch-operation (continuous production). We thank the Zoz-Group Ltd. Wenden, Germany for permission to present these images.

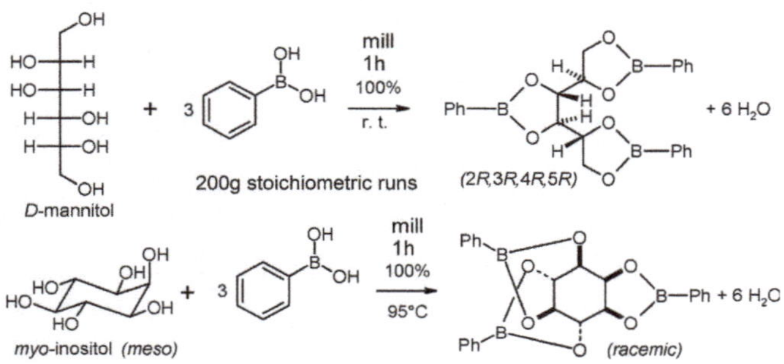

Figure 15: Waste-free multi cascades of sugar alcohols upon stoichiometric molecular milling [13,25].

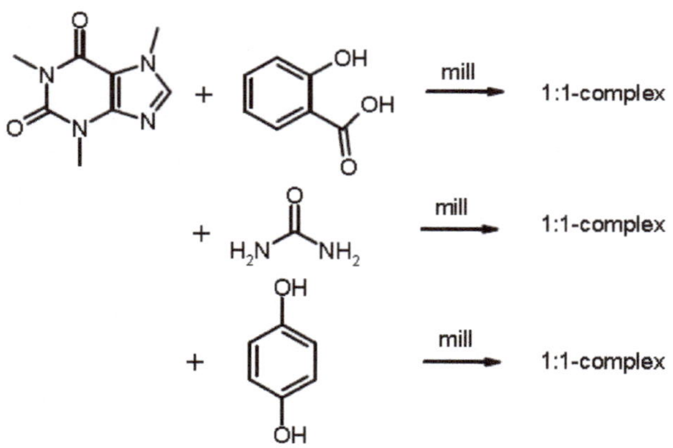

(and many others expected; also different ratios)

Figure 16: Waste-free 200 g batches at 25-30°C in the 2 L Simoloyer; milling time 20 min at 1000 rpm with 100% yield.

Scaling in the ≥ 2 kg ranges requires the 20 L Simoloyer. The 1:1-milling of glucose with urea for 5 min gave the glucose-urea complex, providing new reactivity to the glucose [23]. This complex cannot be obtained from solution. This technique shall also enable formulation of urea complexes in cosmetics and to synthesize ureidoglucoses. 2 kg of a stoichiometric mixture were milled and out-milled at a maximum

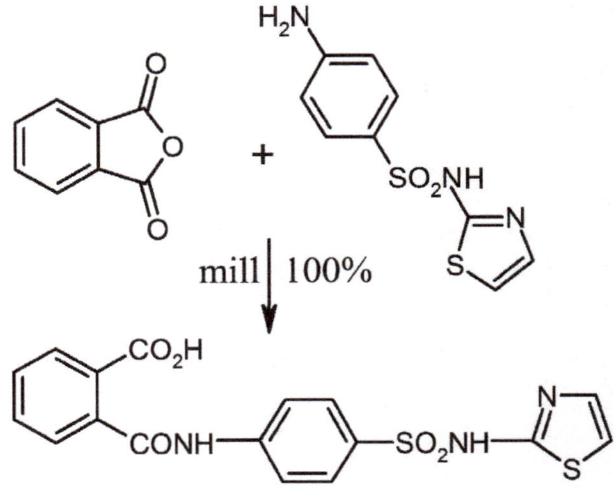

Figure 17: Solid-solid synthesis of the directly pure pharmaceutical sulfonamide by milling of 200 g stoichiometric batches in the 2 L Simoloyer ball-mill at 25-30°C.

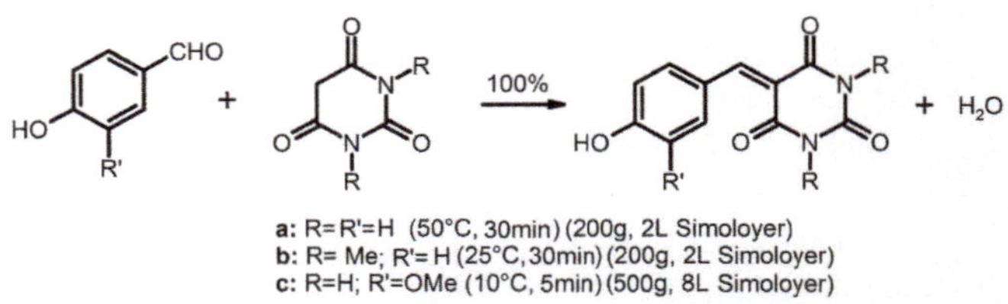

Figure 18: Quantitative large-scale solid-solid Knoevenagel condensations.

of 25°C using the automatically intermitting cooling times at lower rotation speed. The complexation (Figure 19) that had previously succeeded at the 200 g scale (2 L mill [13]) is complete within 5 min, so this process could easily be continuously performed.

Also salt formations with CO_2 release succeeded at the 2 kg scale in the 20 L Simoloyer at 25-30°C. 185 L of CO_2 had been manually released [26]. The quantitative reactions in Figure 20 that had previously succeeded at the 200 g scale proceeded fully selectively, depending on the stoichiometric sodium carbonate or sodium hydrogen carbonate [13]. The tartaric salts are widely used food additives. No detour via sodium hydroxide is required and costly evaporation of water is avoided.

Mechanochemical reactions

Also, genuine mechanochemical (with covalent bond-braking at 3D-covalent networks) syntheses (Review [3]) are waste-free by proper milling at stoichiometric ratios of the components. Figure 21 contains four 200 g batches in the 2 L Simoloyer [13] and various general reaction types. A particular characteristic is the exclusive formation of the most stable product that survives despite further bond breaking and recombination until no further breakage can occur at the nanoscale.

It is seen in Figure 2 that a wide variation of successful oxide combinations reacts quantitatively. In the literature, most of these have only been performed small-scale in planetary mills. But our 200 g batches in the C01Simoloyer indicate, that much shorter milling times (30-60 min) would be possible in temperature controlled small vibrating or large rotor ball-mills. Of industrial interest is the continuous milling of aluminum with multi-walled technical carbon nanotubes (Baytubes®),

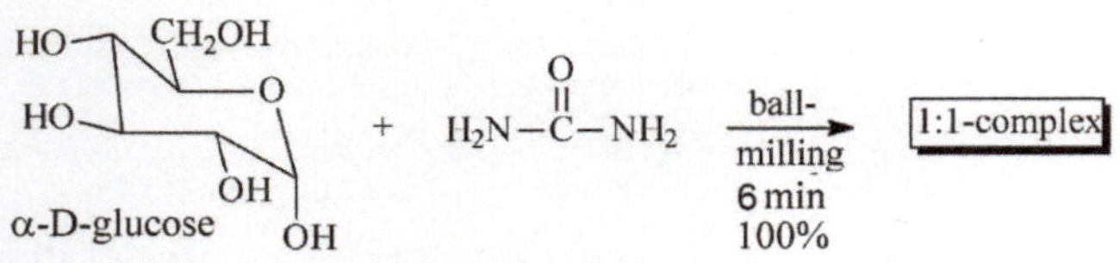

Figure 19: Rapid solid-solid synthesis of an otherwise not available glucose-urea complex at the 2 kg scale in the 20 L Simoloyer at 25-30°C [26].

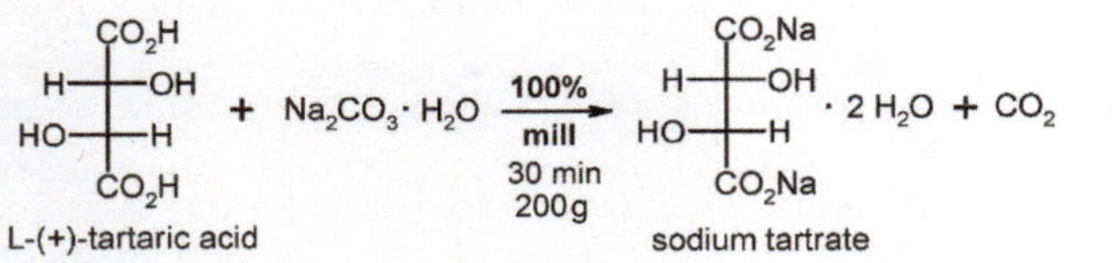

Figure 20: Quantitative stoichiometric solid-solid syntheses of sodium tartrates at the 200 g and 2 kg scale in the 20 L Simoloyer ball-mill with CO_2 release.

$PbO + CrO_3 \longrightarrow PbCrO_4$

$PbO + PbCrO_4 \longrightarrow PbO \cdot PbCrO_4$ — **200g batches**

$PbO + TiO_2 \longrightarrow PbTiO_3$

$ZnO + \alpha\text{-}Fe_2O_3 \longrightarrow ZnFe_2O_4$

$M_2O_3 + M'_2O_3 \longrightarrow 2\,MM'O_3$ — all valences are possible

$3\,Y_2O_3 + 5\,Al_2O_3 \longrightarrow 2\,Y_3Al_5O_{12}$ — numerous combinations

$MO_2 + M'O \longrightarrow MM'O_3$ — syntheses of ceramics at rt

$\alpha\text{-}Fe_2O_3 + MO \longrightarrow MFe_2O_4$ — similarely other infinite 3D-networks: Si, BN, WC, etc.

$2\,Bi_2O_3 + 3\,TiO_2 \longrightarrow Bi_4Ti_3O_{12}$

$M_2O_3 + M'_2O_5 \longrightarrow 2\,MM'O_4$

Figure 21: Mechanochemical cold syntheses of ceramics by reactive milling of oxides (always 3D-covalent networks).

giving Zentallium® (grain-size controlled aluminum by milling at 700 MPa) [27]. This material after passivation and hot extrusion is lighter than aluminum and as strong as steel. It can be used for helicopter shafts, threads, rod/billet, fasteners, etc. Its density is 2.63 g/cm³, specific strength 270 MPa/g/cm³. The C20 Simoloyer ball mill at autobatch processing provides 120 kg/d. The safety of the process has been officially checked by BGIA, Germany, excluding aerosol emission of the industrial process (Figure 22).

Still another waste-free industrial process is the production of nano-cement, simply by continuous milling of dried blast slag, a genuine mechanochemical processing. The obtained high performance cement is called "Futurcement". It provides futurconcrete at high strength and super durability by the sub-micronizing of the blast furnace slag. It is 3 times stronger (140 MPa) and of superior durability (95% GP) than Portland cement concrete. Its production saves 20% of the latter's CO_2-emission. The milling consumes 347 kWh/ton of futurcement. At a production of 43.000 tons p.a. the remarkable advantages are obtained for 7,00 €/ton extra costs. But that is superceded by the saved CO_2-emission costs. Figure 23 shows the construction of a public traffic bridge in Germany out of futurconcrete.

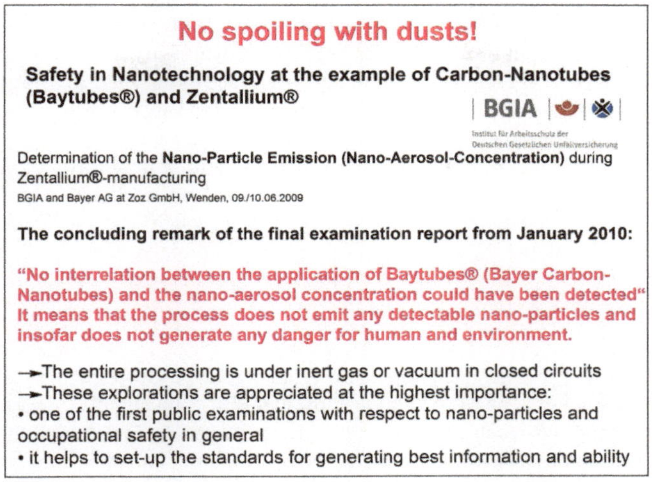

Figure 22: Exclusion of spoiling by aerosols from the industrial Zentallium® production.

Figure 23: A public bridge built out of futurcement/futurconcrete.

1	Obey the consequences of the experiences of the experimental three-stage solid state mechanism (avoid thinking in terms of misleading Schmidt's topochemistry hypothesis)
2	Avoid solvents after reaction for collection of directly pure product
3	Do not add "a little solvent" (a misleading concept, relict "topochemistry")
4	Use the favourable bargain of the solid state that decreases the activation energy and serves with kinetics close to zero order, which provides complete conversion
5	Care for temperature control from about -78 to +12°C (also frozen liquids work)
6	Catalysis should be avoided, due to reaction at much lower temperature than in liquid
7	If rarely necessary use only gaseous catalysts that can be waste less recovered
8	Use precise stoichiometric ratio of pure starting materials to avoid workup
9	When producing gaseous couple products collect them for further use.
10	Adjust milling impact, do not exaggerate (e.g., in planetary mills)
All of that is straight forward and published since the 1980s	

Table 2: Easy essentials for reaching waste-free productions.

Conclusions

Wasteless syntheses and productions requires gas-solid stoichiometric solid-solid processing without (intermediate) liquid phases. Thus, temperature control and cooling below eutectic temperatures is essential. This is not possible in planetary mills, but no problem with small vibration and large rotor ball-mills. Clearly, ball-mills should never be used as heating devices for generating melt reactions. Only the bargain of the solid-state leads to decrease of activation energy, to mostly complete specificity when also further paths would appear feasible, and to favorable kinetics with definite completion of the reaction (close to zero order). It is thus counterproductive to add "a little solvent", and the so-called liquid assisted grinding (LAG) loses all benefits of the solid-state, being a severe retrogression. Even when recrystallization of cocrystals might occur with formation of supramolecular objects by kneading (not grinding) it is not profiting from the solid-state and thus no advancement.

Solid-state syntheses make available otherwise not accessible delicate products at proper experimentation with respect to the long-known "phase-rebuilding mechanism", involving pressure relieving far-reaching anisotropic molecular migrations. Molecular (also ionic) solid-state reactions require mixing/contacting and lead to more stable product. They are thermodynamically governed. Mechanochemical reactions (mechanical bond breaking) start counter-thermodynamically. Proper experimentation is easy, and the scaling up to industrial production has been pointed out with striking examples from more than 1000 waste less ones not requiring workup with solvents. Numerous further applications are envisioned and should be used for a cleaner environment.

References

1. Braga F, Grepioni G, Lampronti I (2011) Supramolecular metathesis: co-former exchange in co-crystals of pyrazine with (R, R)-,(S, S)-,(R, S)-and (S, S/R, R)-tartaric acid. Cryst Eng Comm 13: 3122-3124.
2. Friscic T (2012) Supramolecular concepts and new techniques in mechanochemistry: cocrystals, cages, rotaxanes, open metal–organic frameworks. Chem Soc Rev 41: 3493-3510.
3. Kaupp G (2009) Mechanochemistry: the varied applications of mechanical bond-breaking. Cryst Eng Chem 11: 388-403.
4. Kaupp G (2005) Organic Solid-state reactions with 100% yield. Top Curr Chem 254: 95-183.
5. Naimi-Jamal MR, Mokhtari J, Dekamin MC, Kaupp G (2009) Sodium tetraalkoxyborates: intermediates for the quantitative reduction of aldehydes and ketones to alcohols through ball milling with NaBH4. Eur J Org Chem pp: 3567-3572.
6. Mack J, Fulmer D, Stofel S, Santos N (2007) The first solvent-free method for the reduction of esters. Green Chemistry 9: 1041-1043.
7. Lei S, et al. (2012) Selective reduction of aldehydes and ketones to alcohols with ammonia borane in neat water. Green Chemistry 14: 1372-1375.
8. Friscic T (2015) lecture at the ICCOSS XXII, Niigata, Japan.
9. Kaupp G, Schmeyers J, Boy J (2000) Quantitative solid-state reactions of amines with carbonyl compounds and isothiocyanates. Tetrahedron 56: 6899-6911.
10. Rodriguez B, Bruckman A, Bolm C (2007) A highly efficient asymmetric organocatalytic aldol reaction in a ball mill. Chem Eur J 13: 4710-4722.
11. Kaupp G (2006) Atomic Force Microscopy, Scanning Nearfield Optical Microscopy and Nanoscratching - Application to Rough and Natural Surfaces. Springer Series NanoScience and Technology 1-292.
12. Kaupp G, Naimi-Jamal MR, Schmeyers J (2003) Solvent-free Knoevenagel condensations and Michael additions in the solid state and in the melt with quantitative yield. Tetrahedron 59: 3753-3760.
13. Kaupp G (2006) Waste-free large-scale syntheses without auxiliaries for sustainable production omitting purifying workup. Cryst Eng Comm 8: 794-804.
14. Kaupp G, Schmeyers J, Boy J, prakt J (2000) Iminium salts in solid-state syntheses giving 100% yield. J Prakt Chem 342:269-280.
15. Kaupp G, Schmeyers J, Boy J (2001) Waste-free solid-state syntheses with quantitative yield. Chemosphere 43: 55-61.
16. Kaupp G, Pogodda U, Schmeyers J (1994) Gas/solid reactions with acetone. Chem Ber 127: 2249-2261.
17. Kaupp G, Boy J, Schmeyers J, prakt J (1998) Iminium salts in quantitative gas/solid and solid/solid reactions. J Prakt Chem 340: 346-355.
18. Kaupp G, Seep C (1988) Diastereospecific gas/solid additions with cholesterol derivatives. Angew Chem Int Ed Engl 27: 1511-1512.
19. Kaupp G, Ulrich A, Sauer G (1992) Selective additions of gaseous hydrochloric acid to crystalline epoxides and steroid epoxides. Prakt J Chem-Chem Ztg 334: 383-390.
20. Kaupp G, Jaskulska E, Sauer G, Michl G (1994) New products by ring expansion of solid 17-hydroxy-20-ketosteroids with gaseous hydrochloric acid J prakt Chem-Chem Ztg 336: 686-689.
21. Kaupp G, Matthies D, de Vrese C (1989) J prakt Chem-Chem Ztg 113: 219-220.
22. Kaupp G (1991) Carbonic acid halfester salts - versatile building blocks. Merck Spectrum 3: 42-45.
23. Kaupp G (2003) Solid-state molecular syntheses: complete reactions without auxiliaries based on the new solid-state mechanism. Cryst Eng Comm 5: 117-133.
24. Kaupp G, Naimi-Jamal MR, Ren H, Zoz H (2003) Reactive dry-milling for environmental protection encouraging industrial applications for high kinetic processing. Worldwide PROCESS Germany 4: 24-27
25. Kaupp G, Naimi-Jamal MR, Stepanenko V (2003) Waste-Free and Facile Solid-State Protection of Diamines, Anthranilic Acid, Diols, and Polyols with Phenylboronic Acid. Chem Eur J 9: 4156-4161.
26. Kaupp G, Funk B, Zoz H Publication in preparation. information available from the Zoz Group Ltd, Wenden, Germany

Soret and Dufour Effects on Unsteady Boundary Layer Flow and Heat Transfer of Nanofluid Over a Stretching/Shrinking Sheet: A Stability Analysis

Nor Fadhilah Dzulkifli[1,2], Norfifah Bachok[2], Ioan Pop[3*], Nor Azizah Yacob[1], Norihan Md Arifin[2] and Haliza Rosali[2]

[1]*Department of Mathematics, Faculty of Computer and Mathematics Sciences, Universiti Teknologi MARA Pahang, Bandar Pusat Jengka, Pahang, Malaysia*
[2]*Department of Mathematics and Institute for Mathematical Research, Universiti Putra Malaysia, UPM Serdang, Selangor, Malaysia*
[3]*Department of Mathematics, Babes-Bolyai University, Cluj-Napoca, Romania*

Abstract

The effects of Soret and Dufour parameters on the boundary layer flow in nanofluid over stretching/ shrinking with time dependent is studied using Buongiorno model. The system of partial differential equations is transformed to the system of ordinary differential equations by applying similarity transformation. The results are obtained numerically using bvp4c in Matlab. The reduced skin friction coefficient reduced Nusselt number, velocity, temperature and concentration profiles are shown graphically with different values of Soret effect, Dufour effect, mass flux parameter, unsteadiness parameter, thermophoresis as well as Brownian motion parameter where the dual solutions are obtained. The unsteadiness parameter and mass flux parameter expand the range of solution for stretching/ shrinking parameter. Meanwhile, the Soret and Dufour parameters are found to affect the heat transfer rate at the surface. In order to determine the stability of the solutions, stability analysis is performed.

Keywords: Unsteady boundary layer; Nanofluid; Soret effect; Dufour effect; Heat transfer; Stability analysis

Nomenclature

A	unsteadiness parameter
C	nanoparticle volume fraction
C_f	skin friction coefficient
C_∞	ambient fluid concentration
C_s	concentration susceptibility
D_B	Brownian diffusion coefficient
D_f	Dufour number
D_m	coefficient of mass diffusivity
f	dimensionless stream function
k	thermal conductivity
Le	Lewis number
Nb	Brownian motion parameter
Nt	thermophoresis parameter
Nu_x	local Nusselt number
p	fluid pressure
Pr	Prandtl number
q_w	surface heat flux
Re_x	local Reynolds number
s	mass flux parameter
Sr	Soret number
t	time
T	fluid temperature
T_w	plate temperature
T_∞	ambient temperature
T_m	mean fluid temperature
u,v	velocity components along the x and y directions, respectively
u_w	velocity of the plate
v_w	velocity of mass flux
x,y	Cartesian coordinates along the surface and normal to it, respectively

Greek symbols

α	thermal diffusivity
γ	eigenvalue
η	similarity variable
θ	dimensionless temperature
ε	stretching/ shrinking parameter
μ	dynamic viscosity
ν	kinematic viscosity
ϕ	nanoparticle volume fraction parameter
ρ	fluid density

***Corresponding author:** Ioan Pop, Department of Mathematics, Babes-Bolyai University, Cluj-Napoca, Romania
E-mail: popm.ioan@yahoo.co.uk

$(\rho c_p)_{nf}$ heat capacity of the nanofluid

$(\rho c_p)_p$ heat capacity of the nanoparticle

τ_w surface shear stress

ψ stream function

Subscripts

w condition at the surface of the surface

∞ ambient condition

nf nanofluid

p nanoparticle

f basefluid

Superscript

Differentiation with respect to η.

Introduction

Generally, it was known that heat and mass fluxes were created from temperature and concentration gradient, respectively. However, heat flux is actually can existed due to the concentration gradient which is known as Soret effect. Same goes to the mass flux where the flux occurred by the temperature gradient and is called Dufour effect. Configuration involving the heat and mass transfer with Soret and Dufour effects is an important subject due to a wide range of applications such as the solidification of binary alloys, groundwater pollutant migration, chemical reactors, geosciences multi-component melts, oil-reservoirs, isotope separation, and in mixture between gases. Generally, the effects of diffusion of matter caused by temperature gradients (Soret effect) and diffusion of heat caused by concentration gradients (Dufour effect) can be become influential when the temperature and concentration gradients are very large. Joly et al. [1] analyzed the thermal and solutal effects on natural convection in a vertical enclosure. Mansour et al. [2] have obtained a multiplicity of solutions induced by thermosolutal convection in a square porous cavity with horizontal concentration gradient in the presence of Soret effect. Chamkha and Rashad [3] studied the Soret and Dufour effects on unsteady double-diffusive convection flow from a rotating vertical cone with chemical reaction effects. Kafoussias and Williams [4] investigated the effects of Soret and Duffour on mixed convection and mass transfer laminar boundary layer flow over a vertical flat plate. Besides Soret and Dufour effects, Alam et al. [5] has considered suction variable on mixed convection flow over a semi-infinite vertical porous flat plat and they found that the wall suction stabilized the boundary layer growth of velocity, temperature and concentration. Similar to Alam et al. [5], El-Kabeir [6] also considered the porous medium in the study but he considered stretching cylinder rather than flat plate. In addition, the effect of chemical reaction was also investigated in the work.

Low thermal conductivity of working fluids such as water, mineral oils, ethylene glycol is a primary limitations of enhancing the heat transfer performance. To avoid this situation to add high thermal conductivity of nanoparticles in the working fluid. Due to the excellent thermal property, nanofluids are used in the areas of building heating, heat exchangers, and the automotive cooling process. The term of nanofluid first introduced by Choi [7] in 1995 is one of the mean to improve or enhance the thermal conductivity, which is proportional to heat transfer of the conventional regular fluid such as water, ethylene glycol and mineral oil. Basically, these regular fluids have low thermal conductivity and by dispersing nanoparticles in the base fluid, the thermal conductivity of the fluid can be enhanced. Several works are concerned with natural convection filled with nanoparticles, Khanafer et al. [8], Tiwari and Das [9], Oztop and Abu-Nada [10], who concluded that the Cu-water based nanofluid has the higher heat transfer rate compared with other nanofluids. Lai and Yang [11], Aminossadati and Ghasemi [12] reported that increase the solid volume fraction of the nanoparticles yields a increasing value of heat transfer.

Buongiorno [13] proposed a model for momentum, heat and mass transfer in nanofluid where this model focussed on two slip mechanism that can produced relative velocity between nanoparticles and regular fluid which were Brownion diffusion and thermophoresis. Buongiorno model has been used by Kuznetsov and Nield [14] to study the natural convective boundary layer flow passing through a vertical plate in nanofluid. Meanwhile, different boundary layer condition has been applied in the model by Mansur and Ishak [15], Kuznetsov and Nield [16]. Some of the authors studied the effect of nanouids for the case of mixed convection Nemati et al. [17], Cimpean and Pop [18], etc. Very good literature reviews on convective flow and applications of nanofluids have been done by Jou and Tzeng [19], Das et al. [20], Kleinstreuer et al. [21], Kakaç and Pramuanjaroenkij [22], Wong and Leon [23], Wen et al. [24], Jaluria et al. [25], Mahian et al. [26], Nield and Bejan [27], Sheikholeslami and Ganjii [28], and Shenoy et al. [29] have given reviews of nanofluid on natural convection heat transfer.

Recently, the stability analysis is important to be performed when there are more than one solution exists in order to determine the stability of the solutions. The analysis can be obtained by adopting the work proposed by Merkin [30]. According to According to Ishak [31] the implementation of the analysis was to investigate the growth of disturbances for first and second solutions where the solution with initial decay of disturbance represented the stable solution, while the solution with initial growth of disturbance indicated the unstable solution. There were some authors who performed the analysis in their studies over various cases such as Weidman et al. [32]. Aleng et al. [33], Nazar et al. [34], Hafidzuddin et al. [35], Bachok et al. [36] and Najib et al. [37] studied several problems on stretching/shrinking sheet in a viscous and incompressible fluids and they found that the first solution was stable while the second solution was unstable. It is worth mentioning that the flow caused by a stretching/shrinking surface gets up mostly in the field of chemical engineering and metallurgy such as aerodynamic and polymer extrusion, cooling of metallic plate, drawing of paper films, glass blowing, and paper production.

The present study is an extension of Bachok et al. [38] with new boundary condition proposed by Kuznetsov and Nield [16]. The main purpose of this present work is to investigate the boundary layer flow and heat transfer characteristics over a stretching/ shrinking sheet in nanofluid when the Soret and Dufour effects are taken into consideration for unsteady problem. The governing equations are transformed to ordinary differential equations using dimensionless similarity transformation parameter and are solved numerically by Matlab. The stability analysis is performed in order to determine the stability of the numerical solutions.

Problem Formulation

Unsteady boundary layer flow over a stretching/ shrinking surface immersed in nanofluid is considered. At $t<0$, it is assumed that the surface is in stationary state with velocity $\mu_{w=0}$. As $t>0$, the surface begin to stretch or shrink where the velocity of the sheet is $\mu_{w=}Ax / t$ which $A>0$ is dimensionless acceleration parameter. The velocity of mass flux is represented by v_w where $v_w>0$ is for injection and $v_w<0$ is for suction. Let the uniform temperature of the plate is T_w T_∞ and C_∞ are the temperature and the nanoparticle volume fraction of the ambient fluid. Following the assumptions above, the governing equations of the problem are, see Bachok et al. [38], and Alam et al. [5],

$$\frac{\partial u}{\partial x}+\frac{\partial v}{\partial y}=0, \tag{1}$$

$$\frac{\partial u}{\partial t}+u\frac{\partial u}{\partial x}+v\frac{\partial u}{\partial y}=-\frac{1}{\rho}\frac{\partial p}{\partial x}+v\frac{\partial^2 u}{\partial y^2}, \tag{2}$$

$$\frac{\partial v}{\partial t}+u\frac{\partial v}{\partial x}+v\frac{\partial v}{\partial y}=-\frac{1}{\rho}\frac{\partial p}{\partial y}+v\frac{\partial^2 v}{\partial y^2}, \tag{3}$$

$$\frac{\partial T}{\partial t}+u\frac{\partial T}{\partial x}+v\frac{\partial T}{\partial y}=\alpha\frac{\partial^2 T}{\partial y^2}+\tau\left[D_B\frac{\partial C}{\partial y}\frac{\partial T}{\partial y}+\frac{D_T}{T_\infty}\left(\frac{\partial T}{\partial y}\right)^2\right]+\frac{D_B k_T}{c_s c_p}\frac{\partial^2 C}{\partial y^2}, \tag{4}$$

$$\frac{\partial C}{\partial t}+u\frac{\partial C}{\partial x}+v\frac{\partial C}{\partial y}=D_B\frac{\partial^2 C}{\partial y^2}+\frac{D_T}{T_\infty}\frac{\partial^2 T}{\partial y^2}+\frac{D_B k_T}{T_m}\frac{\partial^2 T}{\partial y^2}, \tag{5}$$

subject to boundary conditions (see Kuznetsov and Nield [16])

$$t<0,\ v=0,\ u=0,\ T=T_\infty,\ C=C_\infty \text{ for all x and y,}$$
$$t\ge 0,\ v=v_w,\ u=u_w(x)=\varepsilon\frac{Ax}{t},\ T=T_w,\ D_B\frac{\partial C}{\partial y}+\frac{D_T}{T_\infty}\frac{\partial T}{\partial y}=0 \quad \text{at } y=0, \tag{6}$$
$$u\to 0,\ T\to T_\infty,\ C\to C_\infty \text{ as } y\to\infty,$$

Where x and y are the Cartesian coordinate along and perpendicular to the plate with u and v are the velocity component in x and y directions, respectively, T is the temperature of the nanofluid, C is the nanoparticle fraction, v is the kinematic viscosity of the nanofluid, α is the thermal diffusivity of the nanofluid, ρ is the fluid pressure, D_B is the Brownian diffusion coefficient, D_T is the thermophoretic diffusion coefficient, D_m is the coefficient of mass diffusivity, C_p is the specific heat at constant pressure, T_m is the mean fluid temperature, k_T is the thermal diffusion ratio, C_s is the concentration susceptibility, $\tau=(\rho c_p)_f/(\rho c_p)_p$ where $(\rho c_p)_p$ is the heat capacity of the nanofluid and $(\rho c_p)_p$ is the heat capacity of the nanoparticle, respectively.

In order to find the similarity solution of Equations 1- 6, the similarity transformation parameters are introduced as follows

$$\psi=Ax(v/t)^{1/2}f(\eta),\ \theta(\eta)=\frac{T-T_\infty}{T_w-T_\infty},\ \varphi(\eta)=\frac{C-C_\infty}{C_\infty},\ \eta=\frac{y}{(vt)^{1/2}} \tag{7}$$

where η is the dimensionless similarity variable, the prime denotes differentiation with respect to η,ψ is the stream function which defines $\mu=\partial\psi/\partial y=(Ax/t)f(\eta)$ and $v=-\partial\psi/\partial_x=-A(v/t)^{1/2}f(\eta)$, $f(\eta),\theta(\eta)$and $\phi(\eta)$ are dimensionless stream, temperature and concentration functions of the fluid in the boundary layer, respectively and v_w is represented as $v_w=-A(v/t)^{1/2}s$, where s is the constant mass flux which s>0 for suction and s<0 for injection.

Using similarity transformation Equation 7, the partial differential Equation 1- 5 are transformed to ordinary differential Equation 8-10 subject to boundary conditions

$$f'''+A\left(ff''-f'^2\right)+f'+\frac{\eta}{2}f''=0, \tag{8}$$

$$\frac{1}{\Pr}\theta''+\left(Af+\frac{\eta}{2}\right)\theta'+Nb\phi'\theta'+Nt\theta'^2+Df\phi''=0, \tag{9}$$

$$\phi''+Le\left(Af+\frac{\eta}{2}\right)\phi'+\frac{Nt}{Nb}\theta''+LeSr\theta''=0, \tag{10}$$

$f(0)=s, f\eta(0)=\varepsilon, \theta(0)=1, Nb\phi(0)+Nt\eta(0)=0,$

$$f\eta(\eta)\to 0,\ \theta(\eta)\to 0,\ \phi(\eta)\to 0 \text{ as } \eta\to\infty \tag{11}$$

Where *Pr* is Prandtl number, is Brownian motion parameter, *Nt* is the thermophoresis parameter, *Le* is Lewis number, *Df* is Dufour number and *Sr* is the Soret number which are defined as

$$\Pr=\frac{v}{\alpha},\ Nb=\frac{\tau}{v}D_B C_\infty,\ Nt=\frac{\tau}{v}\frac{D_T(T_w-T_\infty)}{T_\infty},\ Le=\frac{v}{D_B},$$
$$Df=\frac{D_B k_T C_\infty}{vc_s c_p(T_w-T_\infty)},\ Sr=\frac{D_B k_T(T_w-T_\infty)}{vT_m C_\infty}. \tag{12}$$

The skin friction coefficient and the local Nusselt number are the quantities of physical interest in this problem and defined as

$$C_f=\frac{\tau_w}{\rho u_w^2} \text{ and } Nu_x=\frac{xq_w}{k(T_w-T_\infty)}, \tag{13}$$

Where τ_w *and* q_w are the shear stress and heat flux at the surface, respectively as given by

$$\tau_w=\mu\left(\frac{\partial u}{\partial y}\right)_{y=0} \text{ and } q_w=-k\left(\frac{\partial T}{\partial y}\right)_{y=0}, \tag{14}$$

Where μ is the dynamic viscosity of the fluid and k is the thermal conductivity of the nanofluid. Equations 7, 13, 14 we obtain

$$C_f Re_x^{1/2}=A^{-1/2}f\eta\eta(0) \tag{15}$$
$$Nu_x Re_x^{-1/2}=-A^{-1/2}\theta\eta(0),$$

Where $Re_x=u_w/v$ represents local Reynold number.

Stability Analysis

Weidman et al. [32], and Roşca and Pop [39,40] have shown that the lower branch solutions are unstable (not physically realizable), while the upper branch solutions are stable (physically realizable). We test these features by considering Equations 1-5. Thus, we introduce the new dimensionless time variable $\tau=\ln(t/t_o)$, where t_o is a characteristic time. The use of τ is associated with an initial value problem and is consistent with the question of which solution will be obtained in practice (physically realizable). Using the variables τ and Equation 7, we have

$$u=\frac{Ax}{t}\frac{\partial f}{\partial\eta}(\eta,\tau),\ v=-A\left(\frac{v}{t}\right)^{1/2}f(\eta,\tau),\ \theta(\eta,\tau)=\frac{T-T_\infty}{T_w-T_\infty},\ \varphi(\eta,\tau)=\frac{C-C_\infty}{C_\infty},$$
$$\eta=\frac{y}{(vt)^{1/2}},\ \tau=\ln(t/t_0). \tag{16}$$

As in Weidman et al. [32], we introduce the new variable $\tau=ln(t/t_o)$ as it is associated with initial value problem, where t_o is a characteristic time and we take $t_o=1$ so we have $\tau=ln(t)$ Thus, Equations 1-5 can be written as

$$\frac{\partial^3 f}{\partial\eta^3}+A\left[f\frac{\partial^2 f}{\partial\eta^2}-\left(\frac{\partial f}{\partial\eta}\right)^2\right]+\frac{\partial f}{\partial\eta}+\frac{\eta}{2}\frac{\partial^2 f}{\partial\eta^2}-\frac{\partial^2 f}{\partial\eta\,\partial\tau}=0, \tag{17}$$

$$\frac{1}{\Pr}\frac{\partial^2\theta}{\partial\eta^2}+\left(Af+\frac{\eta}{2}\right)\frac{\partial\theta}{\partial\eta}+Nb\frac{\partial\varphi}{\partial\eta}\frac{\partial\theta}{\partial\eta}+Nt\left(\frac{\partial\theta}{\partial\eta}\right)^2+Df\frac{\partial^2\varphi}{\partial\eta^2}-\frac{\partial\theta}{\partial\tau}=0, \tag{18}$$

$$\frac{\partial^2\varphi}{\partial\eta^2}+Le\left(Af+\frac{\eta}{2}\right)\frac{\partial\varphi}{\partial\eta}+\left(Le\,Sr+\frac{Nt}{Nb}\right)\frac{\partial^2\theta}{\partial\eta^2}-Le\frac{\partial\varphi}{\partial\tau}=0, \tag{19}$$

subject to the initial and boundary conditions

$$f(0,\tau)=s,\ \frac{\partial f}{\partial\eta}(0,\tau)=\varepsilon,\ \theta(0,\tau)=1,\ Nb\frac{\partial\varphi}{\partial\eta}(0,\tau)+Nt\frac{\partial\theta}{\partial\eta}(0,\tau)=0,$$
$$\frac{\partial f}{\partial\eta}(\eta,\tau)\to 0,\ \theta(\eta,\tau)\to 0,\ \varphi(\eta,\tau)\to 0 \text{ as } \eta\to\infty. \tag{20}$$

To determine the stability of the solution $f=f_0(\eta)$, $\theta=\theta_0(\eta)$ and $\phi=\phi_0(\eta)$ satisfying the boundary-value problem Equation 8-11, we write Weidman et al. [32], and Roşca and Pop [39,40])

$$F(\eta,\tau)=f_0(\eta)+e^{\gamma\tau}F_0(\eta),\ \theta(\eta,\tau)=\theta_0(\eta)+e^{\gamma\tau}G_0(\eta), \tag{21}$$

$$\Phi(\eta,\tau)=\phi 0(\eta)+e^{\gamma\tau}H_0(\eta),$$

Where γ is an unknown eigenvalue parameter, and $f_0(\eta)$, $G_0(\eta)$ and $H_0(\eta)$ are small relative to $f_0(\eta)$, $\theta_0(\eta)$ and $\phi_0(\eta)$ Substituting Equation 21 into Equation 17-19, and take $\tau=0$, we obtain the following linear eigenvalue problem

$$F_0{}''' + \left(Af_0+\frac{\eta}{2}\right)F_0{}'' + (1-2Af_0{}'+\gamma)F_0{}' + Af_0{}''F_0 = 0, \tag{22}$$

$$\frac{1}{\Pr}G_0{}'' + \left(Af_0 + Nb\phi_0{}' + 2Nt\theta_0{}' + \frac{\eta}{2}\right)G_0{}' + \theta_0{}'F_0 + Nb\ \theta_0{}'H_0{}' + Df\ H_0{}'' + \gamma G_0 = 0, \tag{23}$$

$$H_0{}'' + Le\left(Af_0+\frac{\eta}{2}\right)H_0{}' + \phi_0{}'F_0 + \left(LeSr + \frac{Nt}{Nb}\right)G_0{}'' + Le\,\gamma H_0 = 0, \tag{24}$$

subject to the boundary conditions

$F_0(0)$, $F_0\eta(0)=0$, $G_0(0)=0$, $Nb\,H_0\eta(0)+Nt\,G_0\eta(0)=0$,

$F_0\eta(\eta)\to 0$, $G_0(\eta)\to 0$, $H_0(\eta)\to 0$ as $\eta\to\infty$ (25)

Solving the eigenvalue problem Equations 22-25 we obtain an infinite number of eigenvalues $\gamma_1<\gamma_2<\gamma_3<\ldots$ If the smallest eigenvalue is positive the flow is stable and if the smallest eigenvalue is negative the flow is unstable.

According to Harris et al. [41], the range of possible eigenvalues can be determined by relaxing a boundary condition on $f_0(\eta)$, $G_0(\eta)$ or $H_0(\eta)$. For the present problem, we relax the boundary condition $F_0\eta(\eta)\to 0$ as $\eta\to\infty$ and for a fixed value of γ we solve the system of Equations 22-24 subject to [25] along with the new boundary condition $F_0\eta\eta(0)=1$.

Results and Discussion

In order to obtain numerical results of the reduced skin friction coefficient $F_0\eta\eta(0)$ and reduced Nusselt number $-\theta=(0)$ with the effect of various physical parameters, we fixed some parameter values such as Prandtl number Pr=0.71 (air), Lewis number $Le=1$, Brownian motion parameter $Nb=0.5$ and thermophoresis parameter $Nt=0.5$ and solve the system of ordinary differential Equation 8-10 subject to boundary conditions Equation 11 by applying bvp4c in Matlab. Further, the presence of Soret and Dufour parameters are now considered and the values of Soret, Sr and Dufour, D_f are chosen in such way so that the mean temperature T_m remains as a constant (see El-Kabeir [6]). A comparison with previous results (see Bachok et al. [38]) with the absence of Soret and Dufour effects is made in order to validate the numerical computation where the comparison shows a good agreement as can be seen in Table 1. Apart from that, the presence of Soret and Dufour effects doesn't affect the range of solution of ε since the value of ε_c are identical with the absence of both effects.

The effects of mass suction/ injection parameters on the reduced skin friction coefficient $F_0\eta\eta(0)$ and reduced Nusselt number $-\theta=\eta(0)$ with stretching/ shrinking parameter ε, when A=1 are shown in Figures 1 and 2. Moreover, Figures 3 and 4 present the effects of unsteadiness parameter, A on those two quantities physical of interest. Based on the Figures 3 and 4 the dual solutions are found when $\varepsilon\le\varepsilon_c$ and no solution can be obtained when $\varepsilon>\varepsilon_c$. From these figures, the range of solutions, $|\varepsilon_c|$ are expanded as the value of A and s increased.

Figures 1 and 4 show that the reduced skin friction coefficient increases when the ε decreases whereas the reduced Nusselt number slightly decreases for the same decreasing parameter. Besides, the range of solutions for higher value of mass flux parameter are wider compared to smaller value as can be noticed in Figures 1 and 2. Likewise Figures 3 and 4 demonstrates the ranges of solution for larger value of unsteadiness parameter are broader than lower value of A with ε.

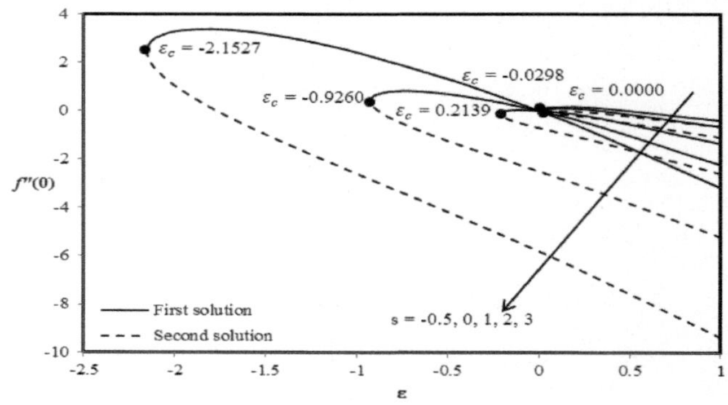

Figure 1: Variation of reduced skin friction coefficient with $\mathcal{E}$.

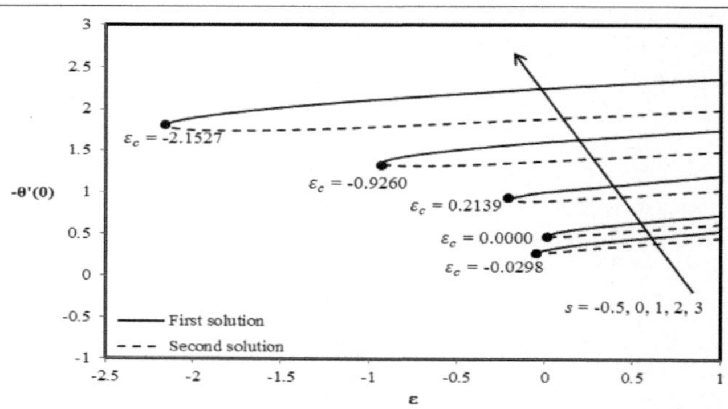

Figure 2: Variation of reduced Nusselt number with for different values of *s when Nt=Nb=0.5 Df=0.15, Le=1, Sr=0.4, A=1 and Pr=0.71.*

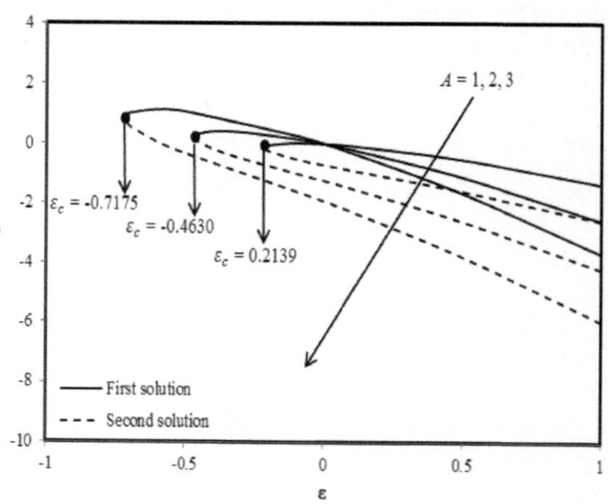

Figure 3: Variation of reduced skin friction coefficient with ε for different values of when *s=1*.

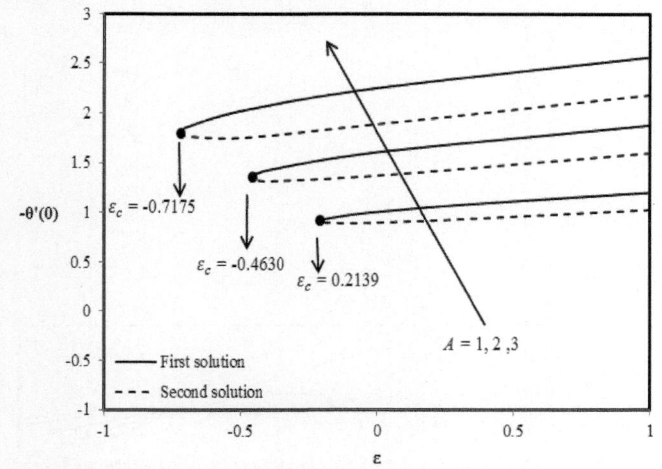

Figure 4: Variation of reduced Nusselt number with ε for different values of A when $Nt=Nb=0.5$, $Df=0.15$, $Le=1$, $Sr=0.4$, $A=1$, $s=1$ and $Pr=0.71$.

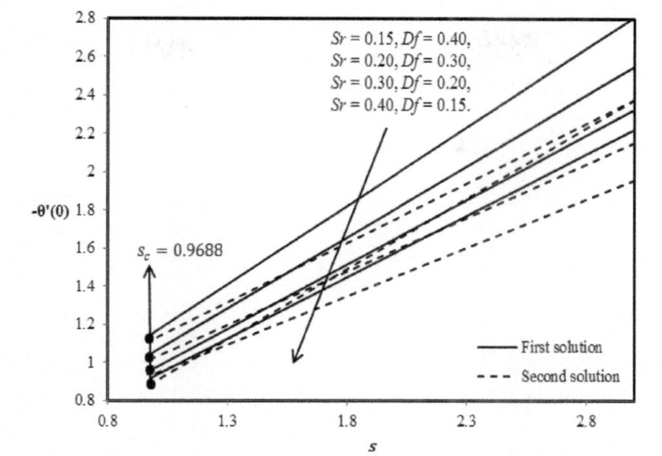

Figure 6: Variation of reduced Nusselt number with s for different values of Sr and Df when, $Nt=Nb=0.5$, $Le=1$, $A=1$, $Pr=0.71$ and $\varepsilon=-0.2$ (Shrinking).

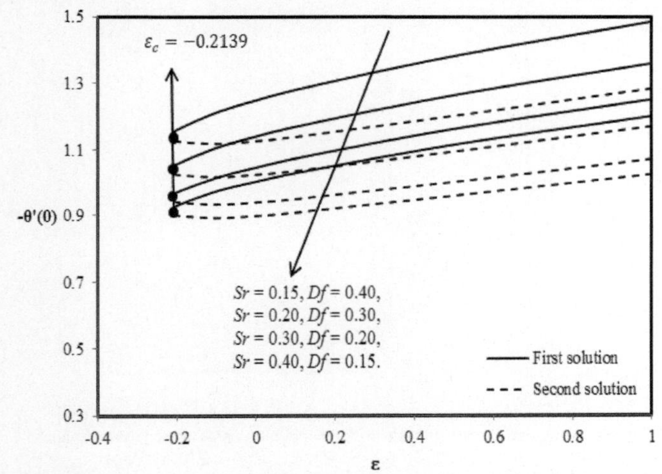

Figure 5: Variation of reduced Nusselt number with ε for different values of Sr and Df when $A=1$, $Nt=Nb=0.5$, $Le=1$, $s=1$ and $Pr=0.71$.

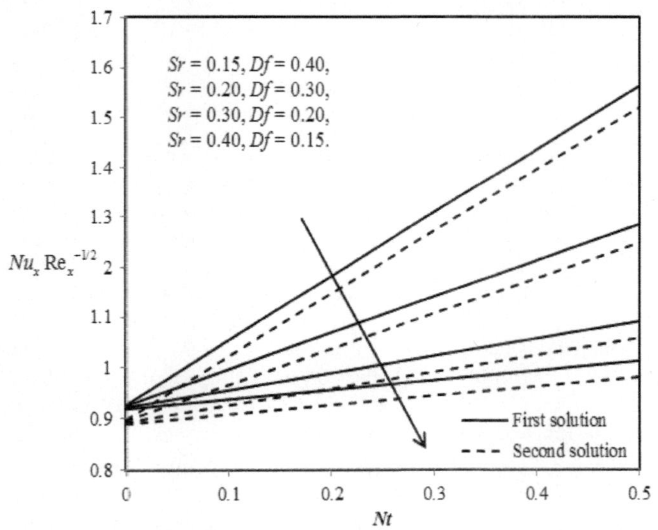

Figure 7: Variation of local Nusselt number with Nt for different values of Sr and Df when, $Nb=0.3$, $Le=1$, $A=1$, $Pr=0.71$ and $\varepsilon=-0.2$ (Shrinking).

Figure 1 illustrates that as mass flux increases, the values of $F_0\,\eta\eta(0)$ are found decelerated for $\varepsilon>0$ (stretching sheet) and opposite trend can be observed when ε (shrinking sheet). Moreover, the values $-\theta(0)$ are larger for higher values of mass flux parameter s as depicts in Figures 2 and 3 shows that a hike in unsteadiness parameter increases the skin friction at the surface for shrinking sheet but for stretching sheet $\varepsilon>0$, the skin friction declines. It is also can be mentioned from Figure 4, the $\varepsilon<0$ accelerating unsteadiness parameter increases the reduced Nusselt number which proportional to the heat transfer rate at the surface.

Figure 5 represents the effect of different values of Soret Sr and Dufour D_f on reduced Nusselt number with. From the figure, it is found that the increase in Sr (or decreasing D_f) doesn't expand the range of solution where the dual solutions are found when $\varepsilon>\varepsilon_c$=-0.2139 while no solution is reported when $\varepsilon>\varepsilon_c$=-0.2139. Apart from that, the increase in Sr (or decreasing D_f)) decreases the reduced Nusselt number and the decrease in $\mathcal{E}$ decreases the reduced Nusselt number. This shows that the heat transfer rate at the surface decreases for higher Soret (or lower Dufour) parameter.

The effects of Soret and Dufour with mass flux parameter s on reduced Nusselt number, $-\theta\ \eta\ (0)$ is shown in Figure 6. The increasing

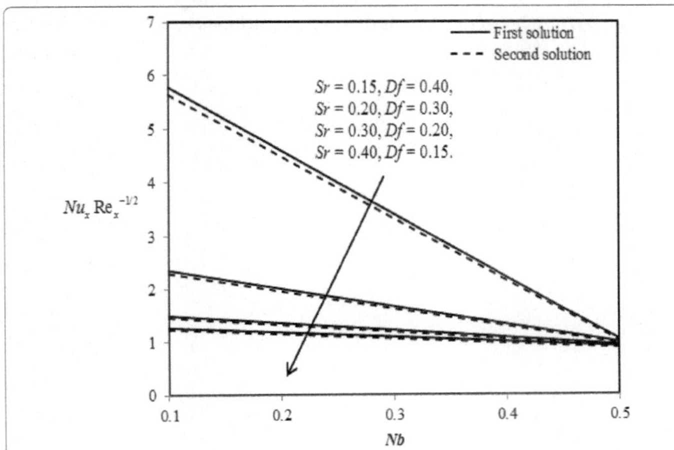

Figure 8: Variation of local Nusselt number with Nb for different values of Sr and Df when, $Le=1$, $A=1$, $Pr=0.71$ and $\varepsilon=-0.2$ (Shrinking).

Soret (or decreasing Dufour) decreases the reduced Nusselt number which means the rate of heat transfer decreases for higher Soret (or lower Dufour) parameter. While for increasing mass flux parameter *s* the reduced Nusselt number is increasing for both solutions. The presence of Soret and Dufour effects is found doesn't change the range of the solutions where the dual solutions exist when $s>sc=0.9688$ and no solution can be obtained when $s<sc=0.9688$. In addition, it is revealed that the solutions exist only for $s>0$ which represents mass suction parameter. Hence, the dual solutions can be obtained when the mass suction parameter is considered the study and no solution is existed as mass injection parameter is taken into account.

The variations of local Nusselt number with thermophoresis parameter *Nt* and Brownion motion parameter *Nb* for several values of Soret and Dufour are shown in Figures 7 and 8 respectively. Figure 7 shows that the local Nusselt number decreases with increasing Soret (or decreasing Dufour) which indicates the heat transfer rate at the surface decreases when the heat flux due to the concentration gradient is increased or mass flux generated by temperature gradient is decreased. In addition, the local Nusselt number is increased when the thermophoresis parameter increases. Physically, the rate of heat transfer is found increased as the migration of the particles increase. Figure 7 demonstrates the local Nusselt number which represents rate of heat transfer decreases when the Soret (Dufour) parameter increases (decreases) as well as the Brownion motion parameter *Nb* increases.

Figures 8-21 present the effect of unsteadiness, mass flux, stretching/shrinking velocity, and Sufor and Dufour parameters on velocity, temperature and concentration profiles. As can be seen in these profiles, the boundary conditions Equation 11 are satisfied asymptotically. Besides that, the dual solutions are obtained which support the variations of reduced skin friction coefficient as well as reduced Nusselt number as presented in Figures 1-8. These profiles also show that the boundary layer thicknesses for first solution are smaller than second solution in each profile. The effect of increasing *Sr* (or decreasing D_f) on temperature and concentration profiles can be observed in Figures 18 and 19 for shrinking parameter and Figures 20 and 21 for stretching parameter. It is found that increasing *Sr* (or decreasing D_f) decreases the temperature and the concentration at the surface for both stretching and shrinking cases.

The stability of dual solutions can be determined by performing the stability analysis and has been solved using bvp4c to obtain the smallest eigenvalue γ for some values of mass suction parameter *s* and shrinking parameter γ as can be seen in Table 2. According to the table,

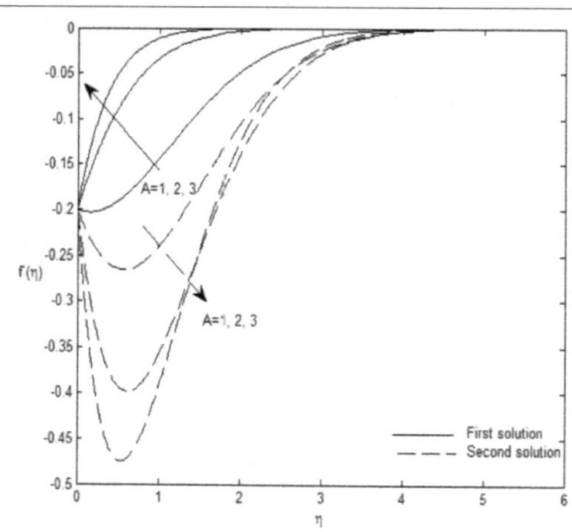

Figure 9: Velocity profile for different values of *A* when $s=1$ and $\varepsilon=-0.2$ (Shrinking).

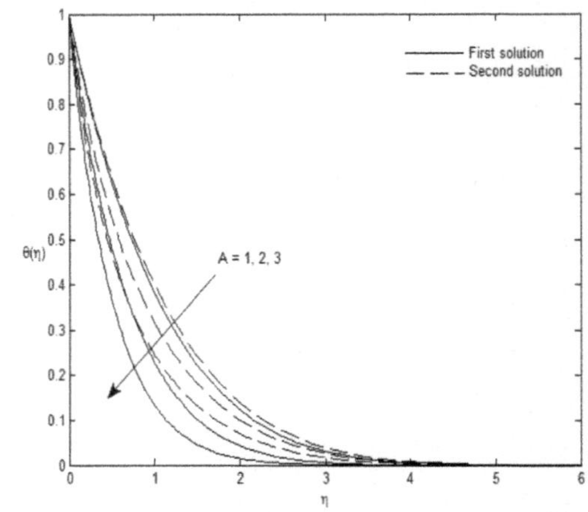

Figure 10: Temperature profile for different values of *A* when $Nt=Nb=0.5$, $Le=1$, $A=1$, $Df=0.15$, $Le=1$, $Sr=0.4$, $Pr=0.71$ and $\varepsilon=-0.2$ (Shrinking).

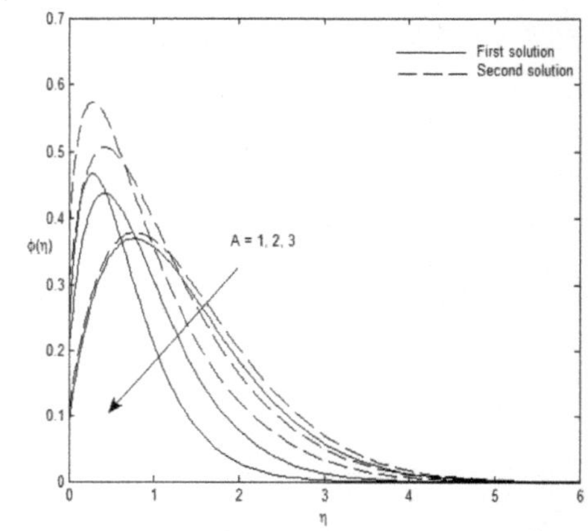

Figure 11: Concentration profile for different values of *A* when when $Nt=Nb=0.5$, $Le=1$, $A=1$, $Df=0.15$, $Le=1$, $Sr=0.4$, $Pr=0.71$ and $\varepsilon=-0.2$ (Shrinking).

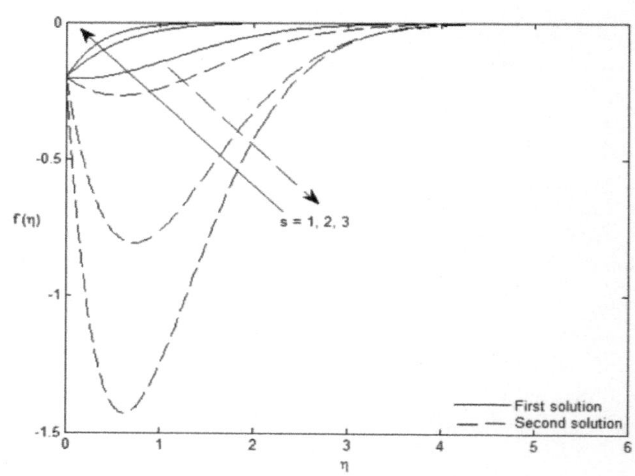

Figure 12: Velocity profile for different values of *s* when $A=1$ and $\varepsilon=-0.2$ (Shrinking).

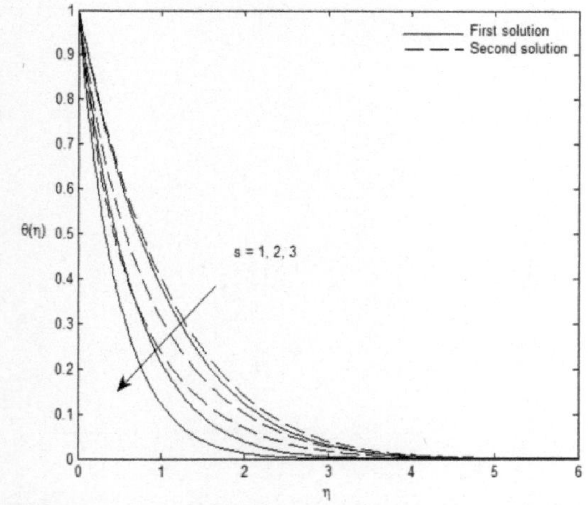

Figure 13: Temperature profile for different values of s when Concentration profile for different values of *A* when when *Nt=Nb=0.5, Df=0.15, Le=1, Sr=0.4, A=1, Pr=0.71* and *ε=-0.2* (Shrinking).

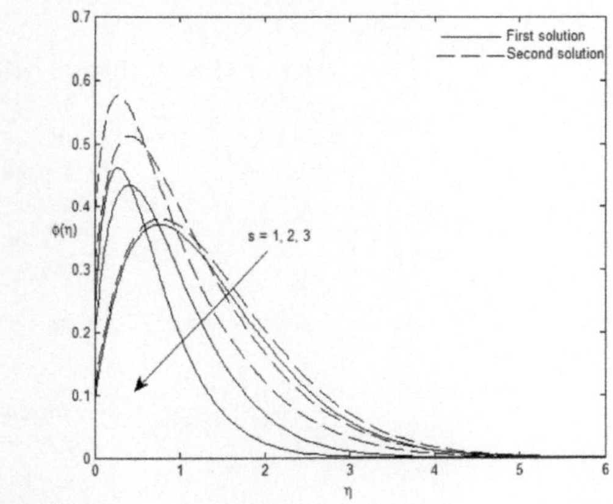

Figure 14: Concentration profile for different values of *s* when *Nt=Nb=0.5, Df=0.15, Le=1, Sr=0.4, A=1, Pr=0.71 and ε=-0.2* (Shrinking).

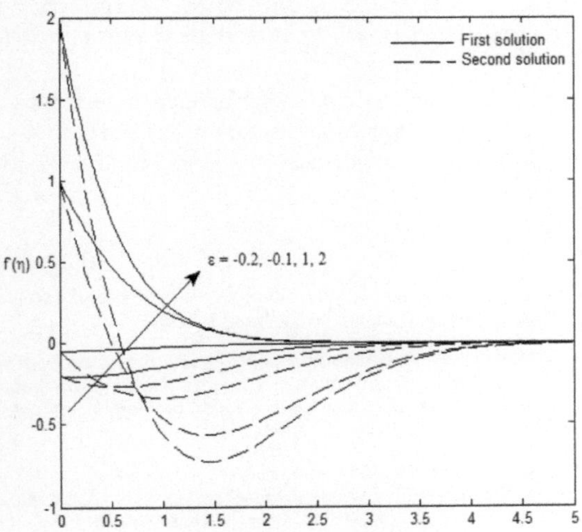

Figure 15: Velocity profile for different values of *ε* when *A=1 and s=1.*

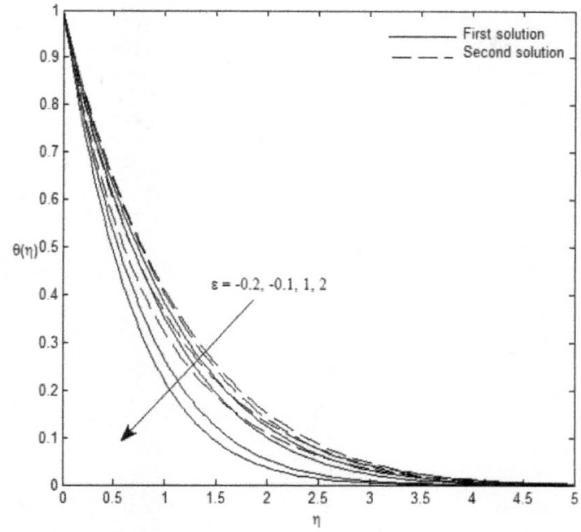

Figure 16: Temperature profile for different values of when *Nt=Nb=0.5, Df=0.15, Le=1, Sr=0.4, A=1, Pr=0.71* and *s=1*.

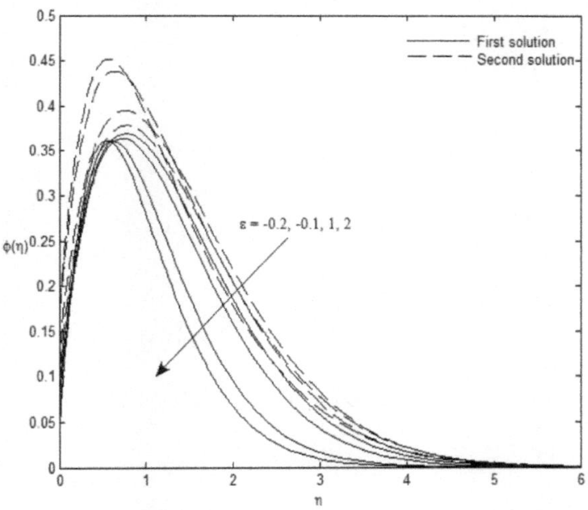

Figure 17: Concentration profile for different values of when *Nt=Nb=0.5, Df=0.15, Le=1, Sr=0.4, A=1, Pr=0.71* and *s=1*.

the smallest eigenvalues for upper branch solution is positive while for lower branch solution is negative. Thus, the first solutions with positive smallest eigenvalues indicate that the solution is stable and physically reliable while second solution is unstable and physically unreliable.

Conclusion

The effect of Soret and Dufour parameters as well as the unsteadiness, mass flux, thermophoresis, Brownian motion parameters on heat transfer characteristics for unsteady boundary layer flow over stretching/shrinking sheet in nanofluid is investigated numerically. It is found that the dual solutions are obtained in this study and presented graphically for reduced skin friction coefficient, reduced Nusselt number as well as the velocity, temperature and nanoparticle volume fraction profiles. The different values of Soret and Dufour parameters are found doesn't affect the range of ε and s parameters since the critical value of ε ($\varepsilon_{c=}$-0.2139) and s (s_c=0.9688) are unchanged even though the values of Soret and Dufour parameter are different. In contrast, the higher values of unsteadiness and mass flux parameters expand range of where the dual

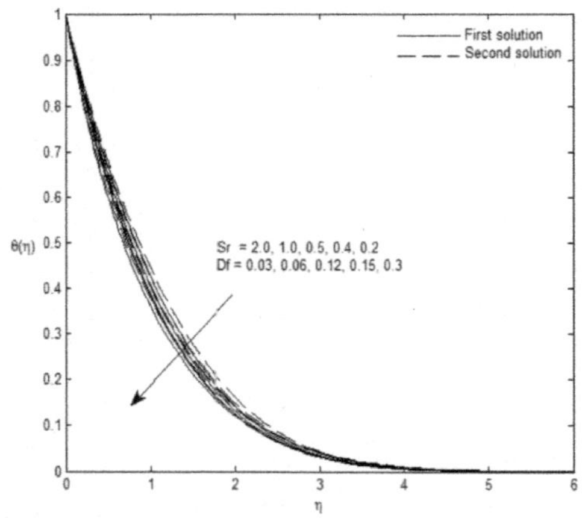

Figure 18: Temperature profile for different values of *Sr* and *Df* when *Nt=Nb=0.5, Le=1, s=1, A=1, Pr=0.71 and ε=-0.2* (Shrinking).

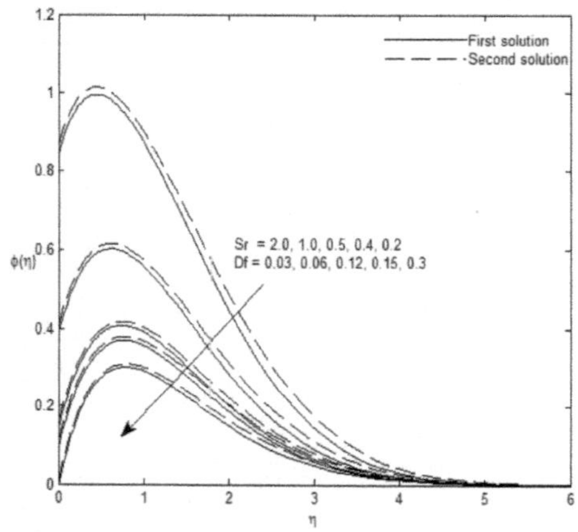

Figure 19: Concentration profile for different values of *Sr* and *Df* when *Nt=Nb=0.5, Df=0.15, Le=1, s=1, A=1, Pr=0.71 and ε=-0.2* (Shrinking).

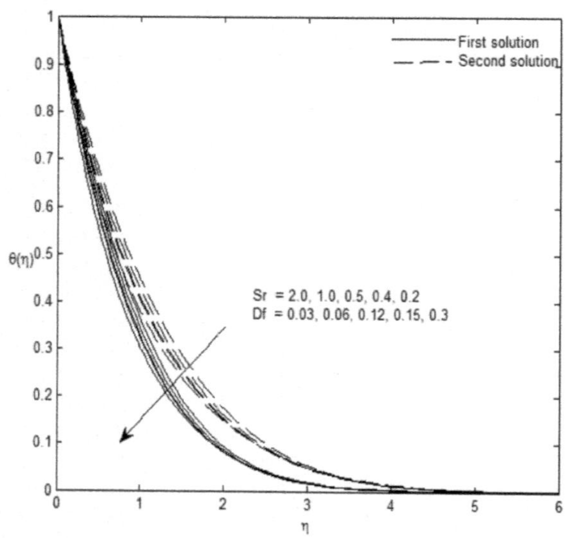

Figure 20: Temperature profile for different values of *Sr* and *Df* when *Nt=Nb=0.5, Le=1, s=1, A=1, Pr=0.71 and ε=-0.2* (stretching).

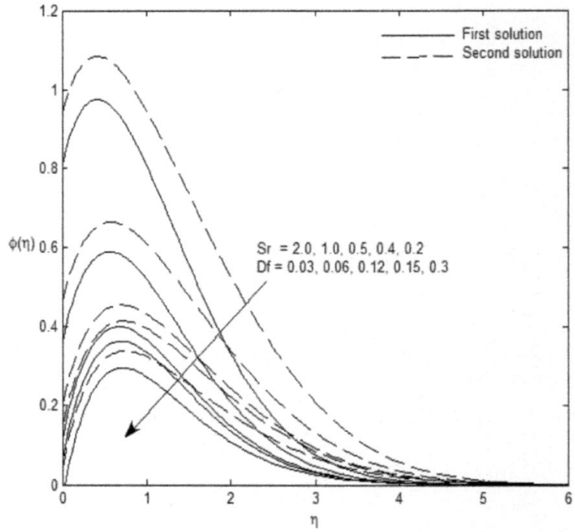

Figure 21: Concentration profile for different values of *Sr* and *Df* when *Nt=Nb=0.5, Le=1, s=1, A=1, Pr=0.71 and ε=-0.2* (stretching).

Df	Sr	s	Bachok et al. [38] ε_c	Present results ε_c
0	0	0	0.0000	0.0000
		1	-0.2138	-0.2139
		2	-0.9259	-0.9260
0.15	0.4	0	-	0.0000
		1	-	-0.2139
		2	-	-0.9260

Table 1: Comparison numerical values of critical epsilon, ε_c for different *s* when *Nt=Nb*=0.5, *Le*=2, *A*=1 and *Pr*=0.71.

s	ε_c	ε	γ (First solution)	γ (Second solution)
1	-0.2139	-0.2135	0.0314	-0.0311
		-0.213	0.0472	-0.0466
		-0.21	0.0991	-0.0964
2	-0.9260	-0.926	0.0024	-0.0007
		-0.92	0.1509	-0.1468
		-0.91	0.2482	-0.2373

Table 2: The smallest eigen values *γ* for some values of s and when *Nt=Nb*=0.5, *Df*=0.15, *Le*=1, *Sr*=0.4, *A*=1 and *Pr*=0.71.

solutions can be obtained when *ε>-22* for different mass flux parameter and *ε>- 88* for various unsteadiness parameter. The stability of the solutions is performed where the first solution is found stable while the second solution is unstable.

References

1. Joly F, Vasseur P, Labrosse G (2000) Soret-driven thermosolutal convection in a vertical enclosure. Int. Commun. Heat Mass Transf 27: 755-764.
2. Mansour A, Amahmid A, Hasnaoui M, Bourich M (2006) Multiplicity of solutions induced by thermosolutal convection in a square porous cavity heated from below and submitted to horizontal concentration gradient in the presence of Soret effect. Num. Heat Transfer 49: 69-94.
3. Chamkha AJ, Rashad AM (2014) Unsteady heat and mass transfer by MHD mixed convection flow from a rotating vertical cone with chemical reaction and Soret and Dufour effects. Canadian J. Chem. Engng 92: 758-767.
4. Kafoussias NG, Williams EW (1995) Thermal-diffusion and diffusion-thermo effects on mixed free-forced convective and mass transfer boundary layer flow with temperature dependent viscosity. Int. J. Engng. Sci 33: 1369-1384.

5. Alam MR, Samad M (2006) Dufour and Soret effects on unsteady MHD free convection and mass transfer flow past a vertical porous plate in a porous medium. Nonlinear Anal. Model. Control 119: 217-226.

6. Kabeir SME (2011) Soret and Dufour effects on heat and mass transfer due to a stretching cylinder saturated porous medium with chemically-reactive species. Latin Am. Appl. Res 41: 331-337.

7. Choi SUS (1995) Enhancing thermal conductivity of fluids with nanoparticles. In: Proceedings of the 1995 ASME International Mechanical Engineering Congress and Exposition, FED 231/MD, Vol. 66, ASME, San Francisco, CA, pp: 99-105.

8. Khanafer K, Vafai K, Lightstone M (2003) Buoyancy driven heat transfer enhancement in a two-dimensional enclosure utilizing nanofluids. Int. J. Heat Mass Transfer 46: 3639-3653.

9. Tiwari RK, Das MK(2007) Heat transfer augmentation in a two-sided lid-driven differentially heated square cavity utilizing nanofluids. Int. J. Heat Mass Transfer 50: 2002-2018.

10. Oztop HF, Abu NE (2008) Numerical study of natural convection in partially heated rectangular enclosures filled with nanofluids. Int. J. Heat Fluid Flow 29: 1326-1336.

11. Lai FH, Yang YT (2011) Lattice Boltzmann Simula tion of Natural Convection Heat Transfer of Al2O3/water Nanofluids in a Square Enclosure. Int. J. Therm. Sci 50: 1930-1941.

12. Aminossadati SM, Ghasemi B (2011) Natural convection of water-CuO nanofluid in a cavity with two pairs of heat source-sink. Int. Commun. Heat Mass Transf 38: 672-678.

13. Buongiorno J (2006) Convective transport in nanofluids. J. Heat Transfer 128: 240-250.

14. Kuznetsov AV, Nield DA (2010) Natural convective boundary-layer flow of a nanofluid past a vertical plate. Int. J. Thermal Sci. 49: 243-247.

15. Mansur S, Ishak A (2013) The flow and heat transfer of a nanofluid past a stretching/ shrinking sheet with a convective boundary condition. Abstract and Applied Analysis 2013: 1-9.

16. Kuznetsov AV, Nield DA (2013) The Cheng-Minkowycz problem for natural convective boundary layer flow in a porous medium saturated by a nanofluid: A revised model. Int. J. Heat Mass Transfer 65: 5792-5795.

17. Nemati H, Farhadi M, Sedighi K, Fattahi E, Darzi AAR (2010) Lattice Boltzmann simulation of nanofluid in lid-driven cavity. Int. Commun. Heat Mass 37: 1528-1534.

18. Cimpean D, Pop I (2012) Fully developed mixed convection flow of a nanofluid through an inclined channel filled with a porous medium. Int. J. Heat Mass Transfer 55: 907-914.

19. Jou RY, Tzeng SC (2006) Numerical research of nature convective heat transfer enhancement filled with nanofluids in rectangular enclosures. Int. Comm. Heat Mass Transfer 33: 727-736.

20. Das SK, Choi SUS, Yu W, Pradeep T (2008) Nanofluids: Science and Technology, Wiley, New Jersey, USA, p: 416.

21. Kleinstreuer C, Li J, Koo J (2008) Microfluidics of nano-drug delivery. Int. J. Heat Mass Transfer 51: 5590-5597.

22. Kakac S, Pramuanjaroenkij A (2009) Review of convective heat transfer enhancement with Nano fluids. Int. J. Heat Mass Transfer 52: 3187-3196.

23. Wong KV, Leon OD (2010) Applications of nanofluids: current and future. Adv. Mech. Engg. 519: 659.

24. Wen D, Corr M, Hu X, Lin G (2011) Boiling heat transfer of nanofluids: The effect of heating surface modification. Int. J. Therm. Sci. 50: 480-485.

25. Jaluria Y, Manca O, Poulikakos D, Vafai K, Wang L (2012) Heat transfer in nanofluids. Adv. Mech. Engng 972: 973.

26. Mahian O, Kianifar A, Kalogirou SA, Pop I, Wongwises S (2013) A review of the applications of nanofluids in solar energy. Int. J. Heat Mass Transfer 57: 582-594.

27. Nield DA, Bejan A (2013) Convection in Porous Media. 4th edn. Springer, New York, USA.

28. Sheikholeslami M, Ganji DD (2016) Nanofluid convective heat transfer using semianalytical and numerical approaches: a review. J. Taiwan Inst. Chem. Eng 65: 43-77.

29. Shenoy A, Sheremet MA, Pop I (2016) Convective Flow and Heat Transfer past Wavy Surfaces: Viscous Fluids, Porous Media and Nanofluids, CRC Press, Taylor & Francis Group, New York.

30. Merkin JH (1980) Mixed convection boundary layer flow on a vertical surface in a saturated porous medium. J. Engng. Math 14: 301-313.

31. Ishak A (2014) Dual solutions in mixed convection boundary layer flow : A stability analysis. Int. J. Math. Comput. Phys. Quantum Eng. 8: 1131-1134.

32. Weidman PD, Kubitschek DG, Davis AMHJ (2006) The effect of transpiration on self-similar boundary layer flow over moving surfaces. Int J Eng Sci. 44: 730-737.

33. Aleng NL, Bachok N, ArifinM N (2014) Boundary Layer Flow of a Nanofluid and Heat Transfer over an Exponentially Shrinking Sheet: Copper-Water. Math. Comput. Methods Sci. Eng 74: 137-142.

34. Nazar R, Noor A, Jafar K, Pop I (2014) Stability Analysis of Three-Dimensional Flow and Heat Transfer over a Permeable Shrinking Surface in a Cu-Water Nanofluid. World Acad Sci Eng Technol Int J Math Comput Stat Nat Phys Eng 1: 780-786.

35. Hafidzuddin E, Nazar R, Arifin NM, Pop I (2015) Stability Analysis of Unsteady Three-Dimensional Viscous Flow Over a Permeable Stretching / Shrinking Surface. J. Qual. Meas. Anal. 11: 19-31.

36. Bachok N, Najib N, Arifin N, Senu N (2016) Stability of Dual Solutions in Boundary Layer Flow and Heat Transfer on a Moving Plate in a Copper-Water Nanofluid with Slip Effect. WSEAS Trans. Fluid Mech. 11: 151-158.

37. Najib N, Bachok N, Arifin N, Senu N (2017) Boundary Layer Flow and Heat Transfer of Nanofluids over a Moving Plate with Partial Slip and Thermal Convective Boundary Condition : Stability Analysis. Int. J. Mech. 11: 18-24.

38. Bachok N, Ishak A, Pop I (2013) Unsteady boundary-layer flow and heat transfer of a nanofluid over a permeable stretching/shrinking sheet. Int. J. Heat Mass Transf. 55: 2102-2109.

39. Rosca AV, Pop I (2013) Flow and heat transfer over a vertical permeable stretching/ shrinking sheet with a second order slip. Int. J. Heat Mass Transfer 60: 355-364.

40. Rosca NC, Pop I (2013) Mixed convection stagnation point flow past a vertical flat plate with a second order slip: heat flux case. Int. J. Heat Mass Transfer 65: 102-109.

41. Harris SD, Ingham DB, Pop I (2009) Mixed convection boundary-layer flow near the stagnation point on a vertical surface in a porous medium: Brinkman model with slip. Transport Porous Med 77: 267-285.

Monitoring of Anti-Caking Solution Preparation and Loading during Loadout Treatment of Potash Using FT-NIR

Mohammad Hussein and Adam Donaldson*

Process Engineering and Applied Science, Dalhousie University, Halifax, Nova Scotia, B3H4R2, Canada

Abstract

Fourier transform near-infrared spectroscopy is presented as a novel method for monitoring anti-caking reagent preparation and loading ratio on granular potash samples during loadout. Using commercial anti-caking agents and granular samples obtained from an operational loadout facility, amine weight percent in the anti-caking solution was determined between 0 and 7 wt% to an accuracy of ± 0.1 wt%, while the loading ratio was determined between 0 and 4.3 lb/ton to within an accuracy of 0.1 to 0.3 lb/ton, depending on the sample heterogeneity. These errors are comparable to current analysis techniques, with minimal sample preparation and an analysis time of seconds. FT-NIR is thus presented as a novel monitoring method for these applications within the potash processing industry

Keywords: Amine content; FT-NIR; Anti-caking; Granular salt; Spectroscopy

Introduction

Potash is a potassium-rich fertilizer that facilitates plant growth, promotes pest and disease resistance, and improves the ability of plants to absorb and retain water and produce more crops [1,2]. Environmental conditions such as humidity and temperature impact the effectiveness of the stored fertilizers, where dust formation may occur over prolonged storage periods affecting the adsorption efficiency of plants [2]. Chemical additives such as long chain amine molecules, C_{15}-C_{20}, are introduced as anti-caking agents to prevent dust and cake formation during storage. The presence of liquid in fertilizers is one of the leading roles in cake formation, resulting vapour condensation due to temperature variations or liquid presence due to capillary forces [3-11]. Many studies have concluded that amine additives can be employed as a coating material to prevent dust and cake formation, such as the use of lauryl amine to reduce caking of ammonium perchlorate [12]. These additives function by inhibiting the growth of crystal bridges and the resulting caking that may occur [12-14].

In anticaking applications, the amine molecule is added to an light oil at weight percent ranging from 2 to 7%, after which the amine/oil mixture is spray-coated onto granular KCl within a loadout facility. The loading rate of the anti-caking agent applied may range by compound, but typically is near 1 to 2 kg of surfactant mixture per tonne of granular salt. This amount of added surfactant was enough to coat the entire surface area of the granular salt and prevent cake formation. Higher amounts of loading of the surfactant mixture could prove harmful since most amines used in this application have an LD_{50} of 2.4 g/kg, and are considered toxic to marine life. Increased loading of the surfactant mixture also inhibits plant uptake of the potassium due to the increased hydrophobicity. To low a dosing will result in dust or cake formation during transport and storage, which may be considered off-spec and require the prepared product to be reprocessed.

From an operational perspective, it is important to be able to monitor both the surfactant preparation and loading rate prior to and during loadout. Given the amine-base, the surfactant content can be determined by titration, Baume Scale, or total organic contents, (TOC). Titration measurements can be subjective due to color interpretation by a lab analyst, and may introduce errors from volumetric determination of the amount of acid and base added. The use of strong acids or bases makes this measurement technique less attractive. The Baume scale is a density meter that can provide the density and temperature of the ammonia solution. By utilizing these two properties, the weight percent of ammonia may be correlated from several factors that may introduce errors within the final reading. For determination of the average loading on a given particle sample, the TOC method involves the addition of an acid to oxidize the organic carbon and produce carbon dioxide. The formed carbon dioxide is then measured by a detector. This technique requires that the reactions take place in the aqueous phase, suggesting that the technique can be replaced with a faster method. TOC technique requires a minimum of 10-15 minutes to complete a single measurement, with samples diluted by a factor of 2-50 depending on the type of reaction chosen. Further, a catalytic reactor would also be required for the oxidation reaction of the carbon.

This work demonstrates the novel application of Fourier transform near infrared (FT-NIR) as a measurement technique with sufficient accuracy to determine both the amine weight percent as well as the oil loading onto KCl. FT-NIR has been successfully used to monitor and characterize a variety of chemical processes [15-17], having a unique advantage in its ability to measure granular samples and the strong response of amine groups within the wavenumber range of this spectrum (10,000-4,000 cm^{-1}). Of specific interest is the three NH_2 overtone regions as well as the carbon-carbon overtones for the oil. Of these regions, the first and second overtones are examined in detail to provide accurate amine weight percent and oil loading onto the samples.

***Corresponding authors:** Mohammad Hussein and Adam Donaldson, Process Engineering and Applied Science, Dalhousie University, 1360 Barrington St., Halifax, Nova Scotia, B3H4R2, Canada, E-mail: adam.donaldson@dal.ca

Experimental Setup

Anticaking mixtures were prepared through the combination of a Tallow alkyl amine and a commercial light oil. The dispersing agent selected was a product of AkzoNobel, having an average molecular weight of 267 and a neutralization equivalent ranging from 258 to 269.

Anti-caking solution preparation

Amine-oil calibration standards were prepared at concentrations ranging from 0 to 7 wt% and mixture temperatures of 40, 50, and 60°C, sufficient to melt and maintain the mixture its liquid state. At each condition, a sample was collected in a 1 cm path length quartz cuvette and placed in a heated transmission cell within the Antaris II FT-NIR spectrometer. Full spectra were recorded to measure the absorbance as a function of both temperature and concentrations. 50 samples were taken in which 40 were used for calibration model development and 10 were used for validation. After model development, additional batches were prepared and examined against the calibration model to determine its accuracy. 15 scans were collected for each sample with a resolution of 4 cm^{-1} in 6 seconds. The amine bond presented itself in the first (~5,000 cm^{-1}), second (~ 6,700 cm^{-1}) and third (~9,700 cm^{-1}) overtone regions.

Anti-caking loading on KCl

Granular product samples were provided by the Mosaic Company with variable surfactant loading ratios using the same surfactant/oil combination employed in this work. The loading rates tested were 0 (i.e., no treatment), 2.9 and 4.3 lb/ton. Additional heterogeneous samples were created through combinations of the received commercial samples (Table 1) to determine the impact of heterogeneity on the methods' ability to detect loading rates.

The samples were prepared in flat-bottom glass vials and placed on the integrated sphere module of the Antaris II FT-NIR system. 15 scans were collected for each sample with a resolution of 4 cm^{-1} in 6 seconds. A background reference was also collected before the beginning of the experiment to ensure that any environmental changes within the lab were accounted for in the sample spectra. On average, 10 spectral sets were collected for each sample with the sample shaken between each set to vary the powder fraction being analyzed. This provided 70 spectra in total, of which 55 were used within the calibration model and the remaining 15 were used as validation points.

Results and Discussion

Armeen T in oil dosing

The complete spectra of the prepared surfactant-oil solutions are presented in Figure 1 with the second and third overtone regions indicated. The second overtone region was selected for analysis of the amine content due to absorbance varying significantly between 0.1 and 1 (i.e., 10 to 90%). Using a partial least square model for the analysis of the amine/light oil mixture, with a Savitzky-Golay filter for smoothing, spectral analysis was performed on the second overtone region with a wavenumber range of 6,700 - 6,390 cm^{-1} (Figure 2).

Calibration models were subsequently developed using the TQ Analyst software for the first-derivative of the spectra with temperature corrections included, yielding an equivalency plot (Figure 3) and error residual plot for the range of temperatures and concentrations tested (Figure 4). Temperature was observed to impact the characteristic behaviour of the NH_2 bonds, decreasing light adsorption as temperature increased. A temperature-correction factor was applied within the TQ-Analyst software to counter the temperature's effect. The resulting cross-validation factor of the calibrated model was 0.0445, with an independent validation, RMSEP, of 0.0333. The correlation coefficients of both factors were near 1, confirming NIR was an excellent analytical tool for monitoring amine content within the anti-caking agent solution. The resulting method was capable of determining amine weight percent to within 0.1 wt%.

Anti-caking agent on KCl

Through similar analysis of the full spectra for powder samples, the second overtone region of the CH and the CH_2 bonds (5,500-5,800 cm^{-1}) was selected in the analysis of loading of anti-caking agent on granular samples. It should be noted that in the presence of significant water content, it may be necessary to shift the focus of this analysis to the second overtone region used for the anti-caking agent, or that a ratio of this region and the 6,700-6,390 cm^{-1} regions explored for amine content may enable single-point measurement of granular samples for both properties.

The CH/CH_2 region of interest exhibited two distinct peaks as loading ratios were increased (Figure 5), providing a viable reference from the raw spectra for calibration model development. Utilizing a partial least squares regression with four factors for the 5,500-5,800 cm^{-1} regions, a calibration model was obtained for homogeneous and heterogeneous samples with the equivalency and residual error plots shown (Figures 6 and 7). The calibration model yielded a cross-validation factor, RMSEC, of 0.169 and an independent validation, RMSEP, of 0.085. The correlation coefficients of both factors, 0.992, and 0.996 suggested a good fit of the validation points to the calibration model with margins of errors of ± 0.3 lb/ton of loading. Notice that the spread of the data was small for each homogenous sample mixture compared to the spread for the heterogenous mixtures, resulting in a larger overall error margin. If only the raw granular are used to develop the calibration model, the margin of error reduces to ± 0.1 lb/ton.

Conclusion

Fourier transform near infrared spectroscopy was successfully applied to quantify Armene T content from 0 to 7 wt% in light oil at temperatures from 40 to 60°C, and to determine the loading ratio of anti-caking agent on granular potash samples from 0 to 4.3 lb/ton. The prediction error obtained for both weight percent and loading ratio were ± 0.1 wt% and ± 0.3 lb/ton, respectively. Where the error for amine content is comparable to traditional methods, and a loading ratio error of 0.3 lb/ton is sufficient to detect variations within the acceptable loading ratios range from 2.2 to 4.4 lb/ton, the use of FT-NIR is presented as a novel monitoring method for these applications within the potash processing industry.

0 lb/ton (g added)	2.9 lb/ton (g added)	4.3 lb/ton (g added)	Resulting Loading
1	0	0	0 lb/ton
0	1	0	2.9 lb/ton
0	0	1	4.3 lb/ton
1	1	0	1.45 lb/ton
1	0	1	2.15 lb/ton
0	1	1	3.6 lb/ton
1	2	0	1.93 lb/ton
1	0.5	0	0.97 lb/ton

Table 1: Composition of heterogeneous powder samples.

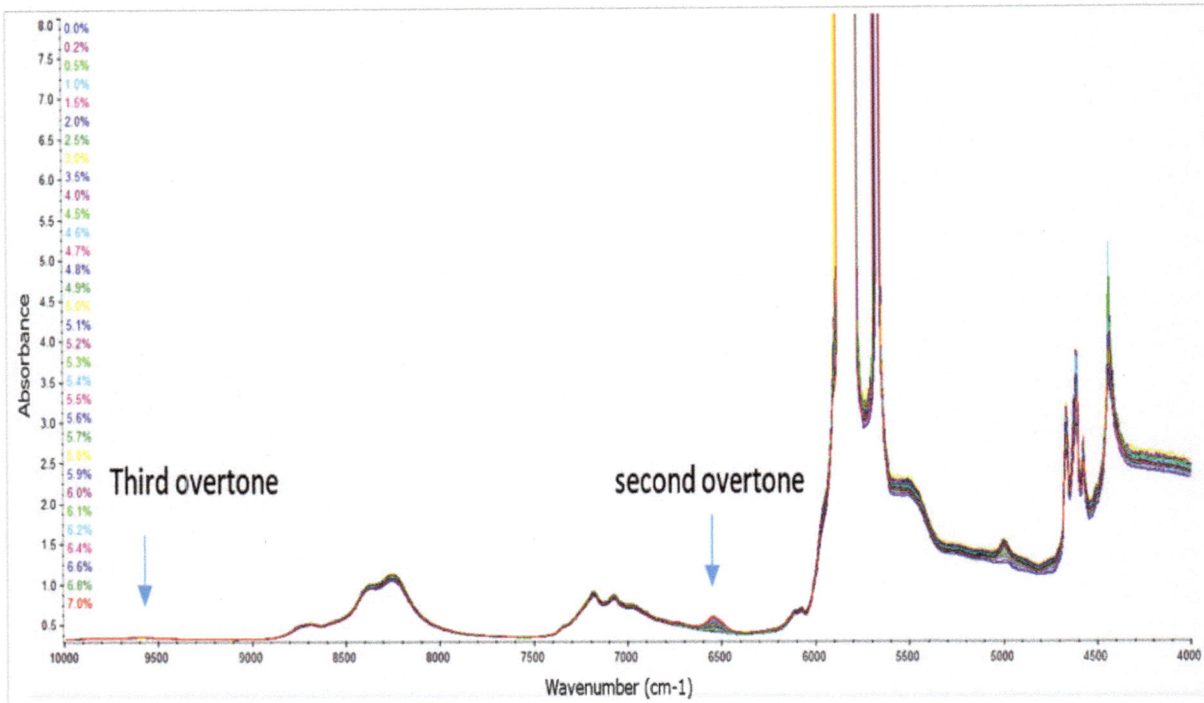

Figure 1: The entire spectra of the solution for various amine weight content.

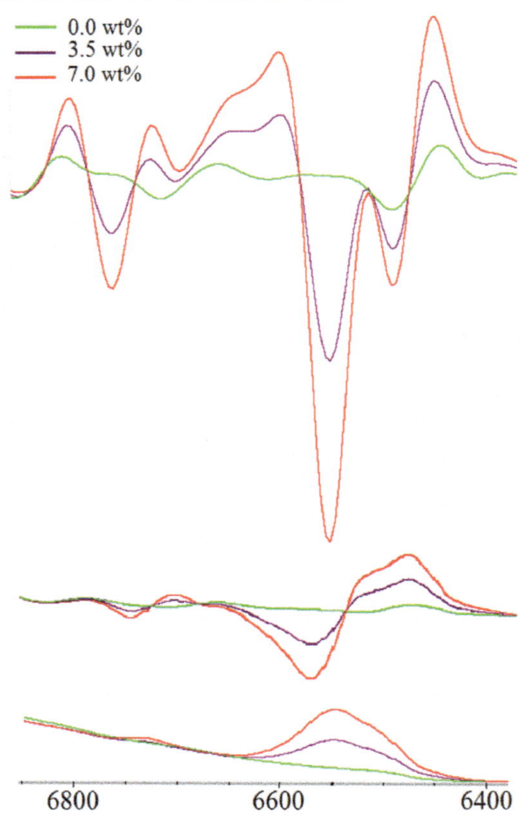

Figure 2: 2nd derivative (top), 1st derivative (middle) and raw spectra (bottom) observed in the second second overtone region of the anti-caking agent solution.

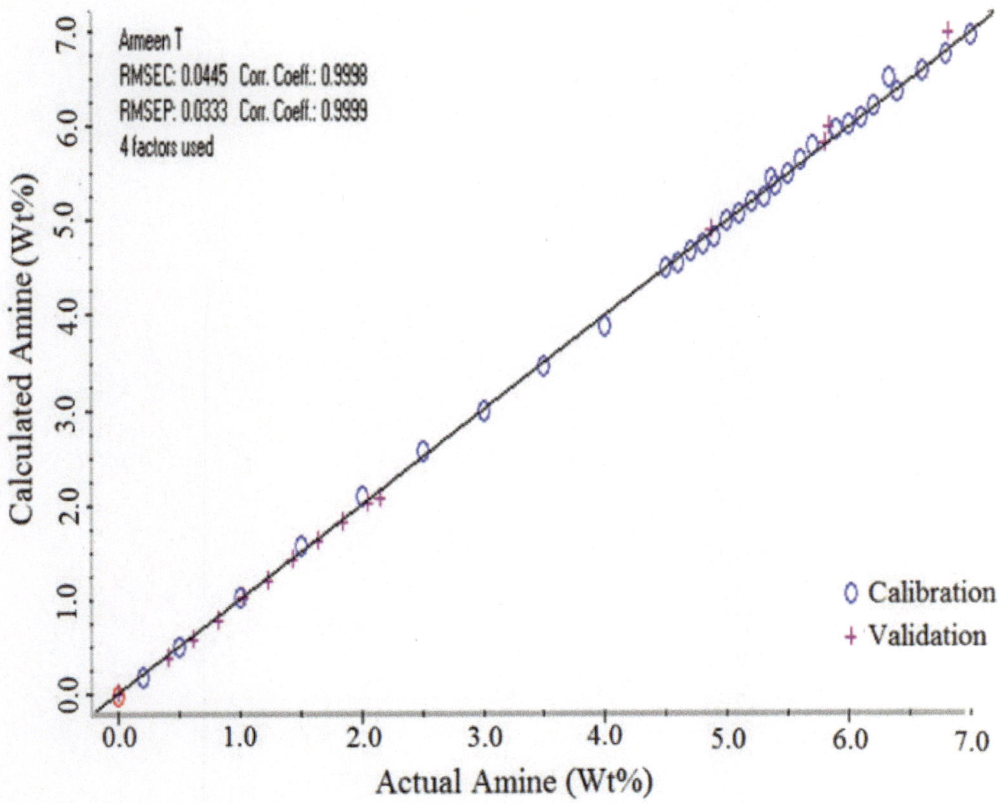

Figure 3: Calibration plot for Armeen T with temperature effect included.

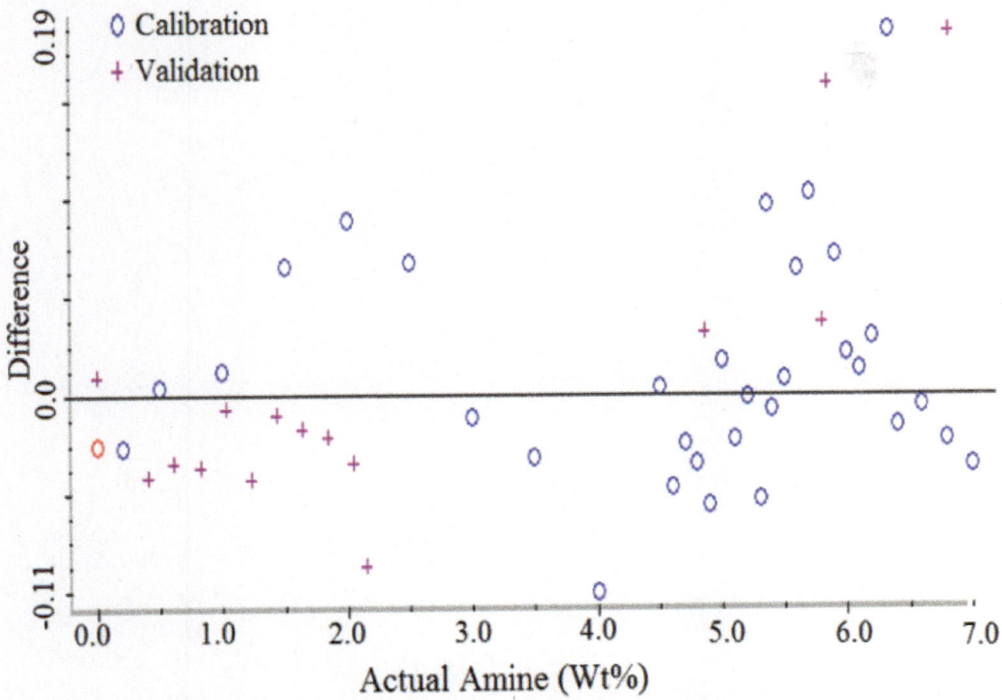

Figure 4: Residual error plot of actual vs. predicted for the 1st derivative analysis of the 2nd and 3rd overtones.

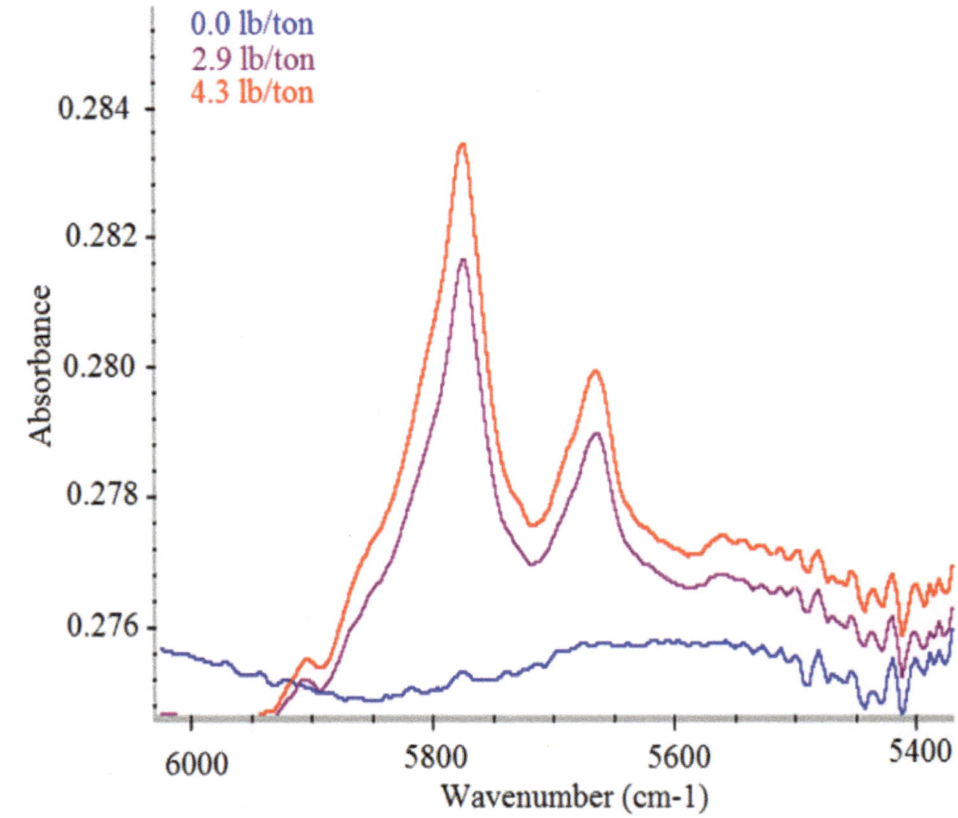

Figure 5: Absorbance spectra of the CH and CH_2 functional groups observed for the raw granular samples.

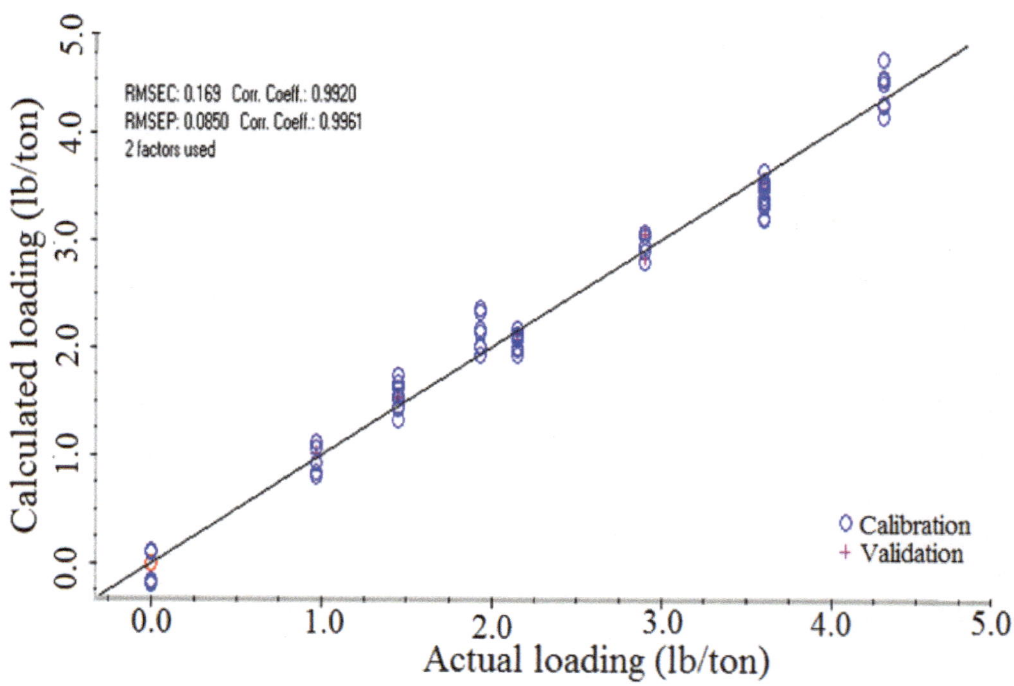

Figure 6: Equivalency plot of homogeneous and heterogeneous powder samples for variable loading ratios (lb/ton).

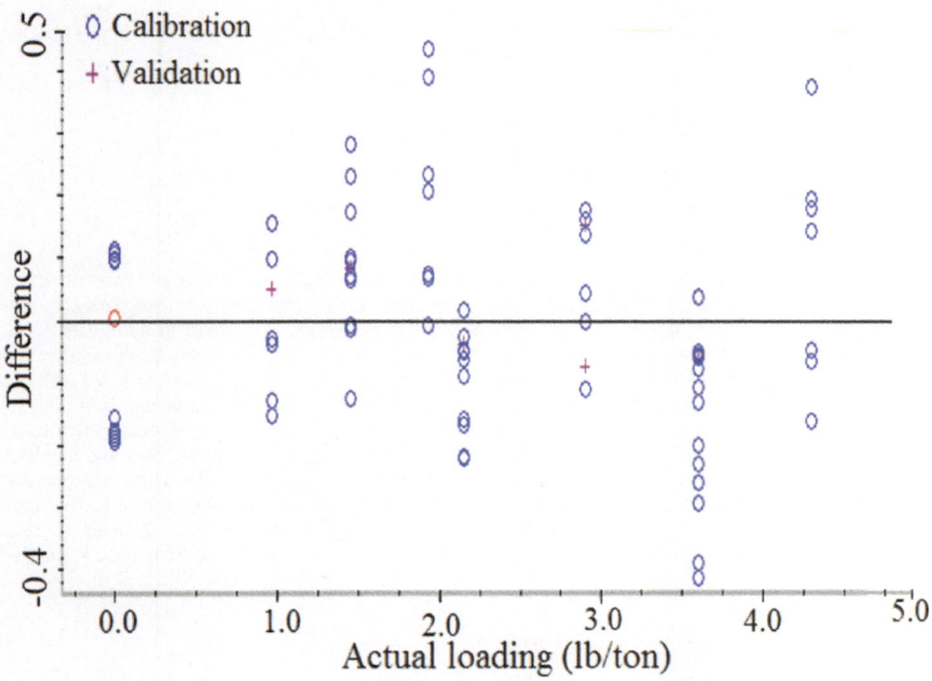

Figure 7: Residual error plot of actual vs. predicted loading per ton determined using FT-NIR.

References

1. Stefano M, Leonard L, Filippo R, Umberto Z (2013) Phosphorus and Potassium Fertilizer Effects on Alfalfa and Soil in a Non-Limited Soil. Agronomy J 105: 1613-618.
2. Zafar U, Calvert V, Cleaver G (2017) A Review of Bulk Powder Caking. Powder Technology 313: 389-401.
3. Wang C, Patel B, Leaper L (2006) Aggregation and Caking Processes of Granular Materials: Continuum Model and Numerical Simulation with Application to Sugar. J of Food Engineering 17: 543-565.
4. Pierrat P, Caram HS (1997) Tensile strength of wet granular materials. Powder Technol 9: 83-93.
5. Pietsch WB (1968) Tensile strength of granular materials. Nature 217: 736-737.
6. Seville JPK, Tüzün U, Clift R (1997) Processing of Particulate Solids. Blackie Academic and Professional, London, UK.
7. Pietsch WB, Hoffman B, Rumpf R (1969) Tensile strength of moist agglomerates. Ind Eng Chem Prod Res Dev 8: 58-62.
8. Rondeau X, Affolter C, Komunjer L, Clausse D, Guigon P (2003) Experimental determination of capillary forces by crushing strength measurements. Powder Technol 130: 124-131.
9. Rumpf H (1962) The strength of granules and agglomerates. International Symposium of Agglomeration, Interscience Publishers, Philadelphia, USA.
10. Schubert H (1984) Capillary forces - modelling and application in particulate technology. Powder Technol 37: 105-116.
11. Tanaker T (1978) Evaluating caking strength of powders. Ind Eng Chem Prod Res Dev 17: 241-246.
12. Chen YL, Chou JY (1993) Selection of anti-caking agents through crystallization. Powder Technol 77: 1- 6.
13. Hansen LD, Hoffmann F, Strathdee G (1998) Effects of anticaking agents on the thermodynamics and kinetics of water sorption by potash fertilizers. Powder Technol 98: 9779-9782.
14. Komunjer L, Affolter C (2005) Absorption-evaporation kinetics of water vapour on highly hygroscopic powder. Powder Technol 157: 67-71.
15. Kachko A, van der Ham LV, Bardow A, Vlugt TJH, Goetheer ELV (2016) Comparison of Raman, NIR, and ATR FTIR spectroscopy as analytical tools for in-line monitoring of CO2 concentration in an amine gas treating process. Int J Greenhouse Gas Control 47: 17-24.
16. Magalhães L, Machado S, Segundo MA, Lopes JA, Páscoa R (2016) Rapid assessment of bioactive phenolics and methylxanthines in spent coffee grounds by FT-NIR spectroscopy. Talanta 15: 460-467.
17. Wang B, Peng B (2017) A Feasibility Study on Monitoring Residual Sugar and Alcohol Strength in Kiwi Wine Fermentation Using a Fiber-Optic FT-NIR Spectrometry and PLS Regression. Journal of Food Science 82: 358-363.

Probing Thermal and Antibacterial Behaviors of Plant Oil-Based Alkyd Resins Loaded by Bio Reduced Ag Nanoparticles

Hossein Behniafar[1*] and Maryam Alipour Moghaddam[2]

[1]*Department of Chemistry, University of Damghan, Semnan Province, Damghan, Iran*
[2]*Department of Polymer Chemistry, Azad University of Damghan, Semnan Province, Damghan, Iran*

Abstract

In this research, extracts of *Elaeagnus angustifolia* leaf is used to reduce the lifetime of silver nanoparticles for the first time. The photosynthesized nanoparticles are homogeneously dispersed into Alkyd resin with 0.5 and 1.0 weight percentage. The two-stage Alcoholysis-poly esterification method was used to synthesis Alkyd resin in 210°C. First, mono glyceride is formed from triglyceride and glycerol, and then Alkyd resin is produced from mono glyceride and phthalic anhydride. The prepared samples including pure Alkyd resin and Alkyd resin with 0.5 and 1.0 weight percentage of silver are investigated by FT-IR, UV-vis, SEM, TGA and XRD analysis. In addition, to study on the antibacterial effects of the loaded nanoparticles, growth and reproduction of two well-known bacteria, i.e., *Staphylococus aureus* and *Klebsiella pneumoniae*, in a nutrient agar containing resultant disk-like nanocomposites are evaluated. The results showed that nano composites containing antibacterial silver has non-growth bacteria halo. While pure Alkyd resin don't have this capability. The more silver amount in polymer matrix, the less bacteria resistance against it and the non-growth halo diameter increases. Finally, the presence of silver nanoparticles in polymer matrix will lead to anti-bacterial property which can be used in the containers keeping food stuff, medical and dental equipment.

Keywords: Photosynthesis; Alkyd resin; Silver nano composites; Antibacterial behavior; Thermal analysis

Introduction

Alkid resins are products which are arisen by compaction or triglyceride oil. Alkid resins are derived by polyols and a dicarboxylic acid or anhydride acid carboxylic. The word "alkid" is the primary modified name of alcid and reflects this fact that they are derived by alcohol and organic acids. Fatty acid in these compounds forms a flexible cover [1].

There are two kinds of dried alkid resins (including semi-dried) and un-dried ones. Both of them are produced by acids or dicarboxylic anhydrides like phthalic anhydride or maleic anhydride and polyols such as trimethylolpropane, glycerin or pentaerythritol. Alkids are used in materials such as resins and colors too.

Alkid resins usually are classified by three methods including: long-oil, average oil and short-oil. These terms represent the percent of the ratio of existing dried oil compound in resins which are called short-oil if the used oil is less than 45% and average-oil in the range of 45%-55% and long-oil if greater than 55% [2].

Alkid resins are synthesized by a two-stage method alcoholosis and polyesterization. In this two-stage process, Oil reacts with glycerol and turns to mono glyceride cursor. Then it reacts with phthalic anhydride in order to synthesize alkid resins.

Due to the new properties and characteristics reflected by nanometer-dimension materials in industries, nowadays, there is much tendency towards to their process and application. Basically, characteristics relative to ratio of the surface and volume of the material in nano metric scale show significant changes. In other words, surface and volume characteristics of the material in nano metric scale are related and have relationship, surface molecules can cause so much hardness in metals and also can produce electronic machineries and pharmaceuticals with better performance and yield.

Using materials and nanostructures whose range is 1-100 nanometer are considered as a new scope of nanotechnology. Also these materials can bring about solutions for technological and environmental problems. Nanoparticles provide completely new or enhanced characteristics according to size, distribution and structure. Different physical and chemical methods are used to synthesize the silver and other metal nanoparticles.

The effect of Nanoparticles on nano-compounding properties can be summarized as follow. Of course, you must be noted that all these cases are not observed in a nano-compounding concurrently.

a) Hardness promotion, durability and dimension stability,

b) Toughness promotion and impact durability,

c) Thermal softness temperature promotion,

d) Mechanical loss promotion,

e) Electrical properties promotion,

f) Weight and value promotion.

Although polymers with anti-bacterial properties are known and produced, but anti-bacterial properties of alkid resin silver nano compounds are not studied. The main aim of the project is production of silver nano-particles in producing alkidresin-silver nano-compounds with studying their thermal and anti-bacterial behavior.

First of all, silver nano-particles are produced using the Extracts of *Elaeagnus angustifolia* leaf and $AgNO_3$ solution. Then produced nano-

***Corresponding author:** Hossein Behniafar, Department of Chemistry, University of Damghan, Semnan Province, Damghan, Iran
E-mail: h_behniafar@du.ac.ir

particle is added to alkid resin with different weights percent's, and then cooking agent i.e., Methyl-Ethyl Keton peroxide (MEKP(and Cobalt-Actoate is added to them and cooking reaction is done during a thermal stage. Finally, nano-compounding with different densities of silver nano particles will be put on the seeding Agar bacteria environment than the minimum preventing density for bacteria growth will be discussed.

Materials and Methods

Chemicals

The used materials in this research are big oil, glycerol, phtalic anhydride, lead oxide catalyzer (II), silver nitrate. Their specifications are shown in Table 1 completely. All above-mentioned materials are provided by Merck and Aldrich companies and are used as provided without any purification.

Deionized water and the Extract of *Elaeagnus angustifolia* leaf, also Whatman paper number 1 were used for filtering.

The necessary materials and organisms to make microbial tests

a) *Staphylococcus aureus* bacteria

b) *Klebsiella pneumoniae*

c) Disc Blanc 6 mm

d) Moller Hinton Agar planting environment

e) Ethanol 70% for disinfecting

The needed equipment for microbial test includes plate, bacteria insemination standard loop, sterile blade, sterile plastic dish and swap.

Devices and techniques

For distinguishing crystalline structure as well as calculating of nano-particles size, X-ray diffraction device model Brunker- D_8 ADVANCE with Nickel filter equipped with radiation source of Cu/K_α with wavelength of =1/54056λ in the angle range of 2θ=5-90° stationed in solid state research laboratory of Damghan university was used. To study morphology of polymer samples surface, the pictures of samples in powder form was provided after fixing on Carbon glue base by gold mark layer by Scanning Electron Microscope (SEM) model ZEISS-DSM-960A and Hitachi S4160 located in campus of science school and engineering-technical school of Tehran university. Thermogravimetric Analysis (TGA) on samples was made by TGA device model Bahar Termo Analyse Gmbh STA 503 under Argon atmosphere with thermal rate of 15°C/min in Damghan university. UV-Vis Spectroscopy of samples was done by model varian 50 spectrometer. Infrared spectrum of raw materials and target samples of this research was registered using FT-IR model Turmo 220 spectrometer.

Preparing the extracts of *Elaeagnus angustifolia* leaf

At First, the *Elaeagnus angustifolia* leaves were collected and washed with distilled water and dried for 5 days in the shadow and then were changed into powder by mill. 20 grams of plant powder was poured in 200 ml deionized water and boiled for 30 minutes in oil bath with 103°C. After cooling, the resulting extracts (Figures 2 and 3) filtered with Wahtmann Filter Paper No 1 and be kept in 4°C for the next tests.

Silver nanoparticles synthesis

First, a blend of 30 ml plant extract and 500 mL $AgNO_3$ 1 mM shall be ready and then was exposed to sunlight for 24 hours. The solution color got dark. The settlement was separated from the solution by centrifuge device, washed by deionized water and then was dried in a normal oven under 85°C temperature for two hours and hereby Nitrate nano-particles was prepared.

Alkid resin synthesis

Alcoholysis-Polyesterization reaction was done in a circle-bottom dish equipped with a mechanical mixer, a thermometer and an entrance door for Hydrogen gas. A mix of 33.6 gram (0.04 mol) big oil, 7.36 gram (0.08 mol) Glycerol and 0.05 wt% (regarding the oil) of PbO were exposed to heat in 230°C temperature and mixed with constant rate of 500 rpm under nitrogen atmosphere for 60 minutes to constitute mono glyceride. Monoglycerid constitution was confirmed by methanol solubility test.

One part of the resin was mixed with three part of Methanol in a sample glass and in the room temperature. Solving the resin completely in Methanol arises clear liquid indicative of mono glyceride constitution. Then the reaction mix will be cooled to 12°C and 0.12 mol phtalic anhydride will be added along with 1.98 gram extra glycerol (0.27%). Now, reaction temperature will increase to 210°C and the heating will continue until the acidity stay between 20-30 mg KOH/g.

Alkid resin nano-compounds synthesis containing 0.5 and 1 weight percentage of silver

Alcoholysis-Polyesterization reaction was done in a circle-bottom dish equipped with a mechanical mixer, a thermometer and a hydrogen gas input. A mix of 33.6 gram (0.04 mol) big oil, 7.36 gram (0.08 mol) glycerol and %wt 0.05 (with regard to oil) of PbO were exposed to heat in 230°C temperature and mixed with constant rate of 500 rpm under Nitrogen atmosphere for 60 minutes to constitute monoglycerid. Monoglycerid constitution was confirmed by Methanol solubility test.

One part of the resin was mixed with three part of Methanol in a sample glass and in the room temperature. Solving the resin completely in Methanol makes a clear liquid, indicative of monoglycerid constitution. Then the reaction mix will be cooled to 120°C and 0.12 mol phtalic anhydride will be added along with 1.98 gram extra glycerol (0.27%).

Item	Chemical Formula	Reagent Aspect	Mol Mass g/mol	Density g/cm³	Fusion Temperature °C	Boiling Temperature °C	Purity Percent
Silver nitrate	$AgNo_3$	White crystalline	169.87	4.35	212	444	≤ 99
Glycerol	$C_3H_8O_3$	Odorless & colorless liquid	92.09	1.261	17.8	290	≤ 99.5
Phthalic anhydride	$C_8H_40_3$	White flakes	148.1	1.53	131	295	≤ 98
Big Oil	-	Yellow and clear liquid	-	-	24	-	-
Lead (II)	PbO	Yellow or red	223.20	9.53	888	1477	-

Table 1: The specifications of raw materials.

Now, reaction temperature will increase to 210°C and the heating will continue until the acidity is between 20-30 mg KOH/g. This method was done for nano-compound containing 1% weight of silver.

Studying and measuring anti-bacteria properties of silver alkid-resin nano-compounds

Linear planting: In the linear planting method which is a qualitative separation method for getting separated colonies, plate is divided into four divisions. In this method, the aim is to dilute the number of bacteria. Hence, insemination loop will be heated after planting of each part. Heating the loop destroys the residual bacteria and reduces the number of them. Each part indicates decreasing in the number of thousands of bacteria cell existing in primitive insemination sample. The colonies are constituted after 24 hours incubation in 37°C.

Provision of microbial suspension

Klebsiella pneumonia bacteria (negative gram) and *Staphylococcus aureus* (positive gram) was provided by Pastor institute in lipophilized ampules. After bacteria survival, the mentioned bacteria were planted linearly on Moller Hinton Agar environment. For provision of microbial suspension from 24-hour linear planting of bacteria, 4 or 5 colonies were removed with laboratory loop and were completely mixed in a sterile tube containing 5 ml sterile physiology serum and the resulting mixture got vertex in order to get a uniform suspension from bacteria. Then the tube was put for 30 minutes in 37°C and the tube darkness was compared with standard darkness of McFarland Standard No. 0.5 (15×10^8 cfu/mL) and in case of difference, suspension darkness was reached to standard darkness of 0.5 McFarland adding sterile physiology serum or bacteria colony. For more certainty, the optic absorption of suspension and the standard solution in wavelength of 600 nm was read by Spectrophotometer. The Blanc of device is distilled water.

Determining the anti-bacteria properties of alkidresin-silver nano-compounds using the method of disc emission

In sterile conditions under Laminar Fluor with sampler 100 microliter from prepared microbial suspension was poured on plates containing planting environment of Moller Hinton Agar with 5 ml thickness and was distributed with L-shape and sterile on plate surface as the surface of plate was covered with a microbial layer completely. Then, the discs with 6 ml diameter (from Paadtan Company) were put on the plate surface with sterile pans with suitable distances and the discs were fixed in their situation using the tip of the pans. 50 micro liters from each one of the provided nano-compounds (with different densities) was poured in sterile distilled water on the discs.

After 24 hours incubation in 37°C, the diameter of non-growth halo was measured using ruler on the basis of ml. The test was done with three repetitions and the average of results was reported.

Results and Discussion

FT-IR analysis of samples of extracts of *Elaeagnus angustifolia* leaf and the solution of the extract and silver nitrate

The extracts of *Elaeagnus angustifolia* leaf analyzed by FT-IR. Figure 1a shows the FT-IR spectrum of the extract provided by *Elaeagnus angustifolia* leaf in the range of 500-4000 cm^{-1}. Photochemistry analysis of the extract showed that it consists of chemical compounds including: Flavonoids and Phenols. The Target spectrum shows the active groups presence in the extract. The peak of 3267 cm^{-1} show the absorption frequency of phenol OH which there is abundant in the extract and the peak of 1636 cm^{-1} show the absorption frequency of bending OH and the peak of 2107 cm^{-1} is related to C-H asymmetric pull vibration.

Figure 1b shows the FT-IR spectrum of nano-particle of silver suspended in the bed of the extracts of *Elaeagnus angustifolia* leaf in the range of 500-4000 cm^{-1}. Bio-reduction of silver ions is done by chemical compounds existing in plant extract. More amounts of Flavonoids play a significant role in silver ions reduction reaction. Also, Carbonyl group is oxidized to metal silver for silver ion reduction. Those peaks observed in the spectrum of extract FT-IR will be seen here. This spectrum shows that silver ions reduction might be due to the presence of polyols.

Analysis of alkidresin FT-IR spectrum and alkidresin-silver nano-composites

The resulting FT-IR spectrum in this research is studied in the range of 500-4000 cm^{-1} wavelength. The specified peaks were found in 3471 cm^{-1} which were due to pull vibration of O-H as well as in 2853-2924 cm^{-1} which were due to aliphatic pull vibration of C-H. The existing peaks are appeared in 1724 cm^{-1} for pull vibration of C=O of triglyceride, in 1599 cm^{-1} for C=C and in 1452 cm^{-1} for C-H bond. Also, the existing peaks are due to pull vibration of C-O-C in 1069 cm^{-1} and to kinetic vibration of Methylene in 741 cm^{-1} (Figure 2).

XRD spectrum analysis of silver nano-particles, alkidresin, 0.5 and 1 weight percentage of alkidresin-silver nano-composite

XRD spectrum observed in Figure 3 is indicative of presence of synthesized silver nano-particles crystalized from the extracts

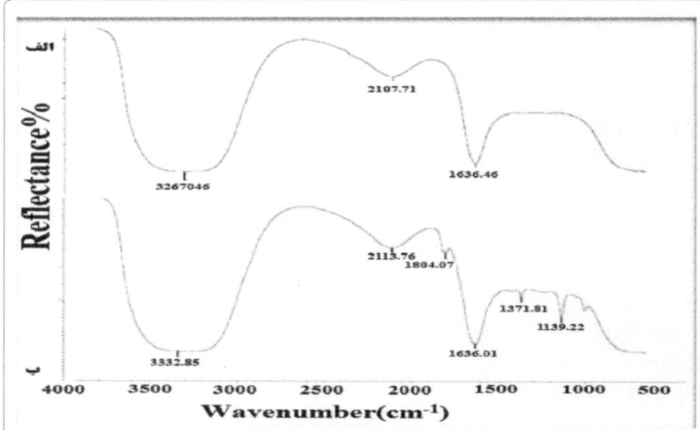

Figure 1: FT-IR Spectrum A) extract b) extract with silver.

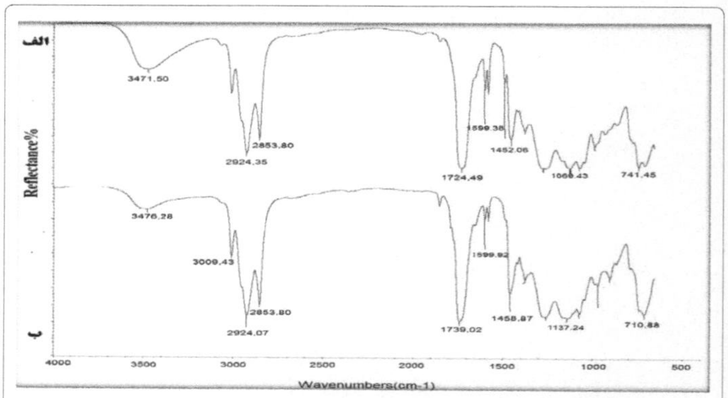

Figure 2: FT-IR spectrum A) alkidresin and b) alkidresin-silver.

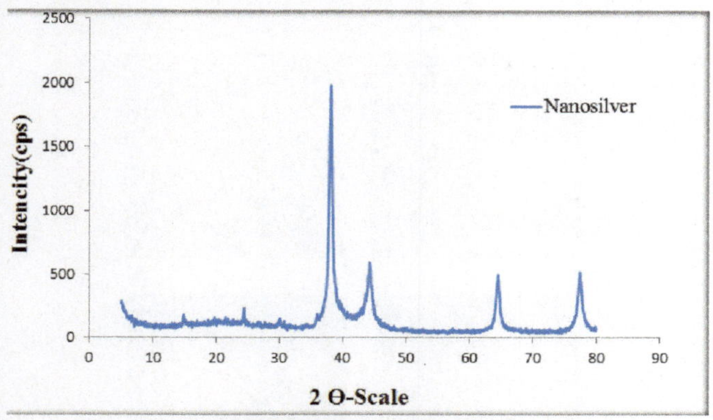

Figure 3: XRD spectrum of silver nanoparticles.

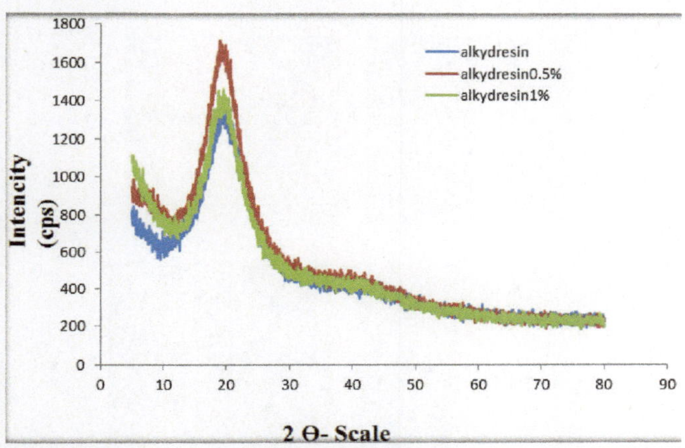

Figure 4: XRD spectrum, alkidresin, 0.5% silver alkidresin, 1% silver alkidresin.

of *Elaeagnus angustifolia* leaf. Generally, the Increase of the height of the peak and existence of sharp regions in spectrum is indicative of more crystallinity in the target composition. Figure 3 shows the XRD spectrum of silver nano-particles. As is shown in the figure, in this spectrum, a sharp peak is appeared in about $2\theta=38°$ which can be considered as a clear signal for nano-particle. As is clear from the figure, the existence of four index peaks in angles 38, 44, 64, 77 is a confirmation for the constitution of silver nano-particle.

Figure 4 shows the XRD alkidresin, alkidresin 0.5% and 1% weight percentage of silver. As is observed, a sharp peak is appeared about $2\theta=20°$ which can be considered as a clear signal for alkidresins and also the weak fluctuations in $2\theta=38°$ indicate the presence of silver nano-particle in polymer matrix.

TGA analysis of alkidresin and nano-composite alkidresin-silver 5%

This graph is considered as an important indicator of candescent for polymers. The resulting sample shows a three-stage weight loss from the room temperature to 600°C. Pure alkidresin meets weight loss in 78.9°C while exposing silver nano-particles in polymer matrix causes the increase of thermal resistance and the temperature which causes nano-composite weight loss is 205°C. Increasing the percentage of silver in matrix polymer associate with increased thermal stability of nano-composite and has more remaining coke than pure alkidresin (Figure 5).

According to Table 2, IDT is the temperature which only the 5% polymer chains of the samples is destroyed and changed into coke. With temperature increase in coke yield column (CY), the remained percentage of samples (95%), has turned into coke and in fact, the residue polymer for each one of samples in 600°C is mentioned in Table 2.

UV-Visible spectrum

According to spectrum reported, in the range of 405 nm is indicative of the presence of silver nanoparticles constituted in plant extract whereas it doesn't show any λ_{max} in this interval. This strong strip is related to surface Plasmon resonance of silver nano-particle. Silver has the most conductivity and in addition, its absorption peak is more strong and clear than other nobel metals. In visible spectrum of noble metals, when the particles size reaches to nanometer scale, a strong absorption is observed which its origin is surface Plasmon resonance (Figure 6).

SEM pictures

Figure 7 shows the pictures of Scanning Electronical Microscope (SEM) of silver nano-compounds in different magnifications. As the pictures show clearly, the resulting nano-particles are seen in the form of compact mass particles which have uniformity and lumpy surface and are sticked together (Figure 8).

Figures 9 and 10 shows morphology of alkidresin nano-composites containing 0.5 and 1% silver weight percentage. The picture (a) shows the morphology of alkidresin nano-composites containing

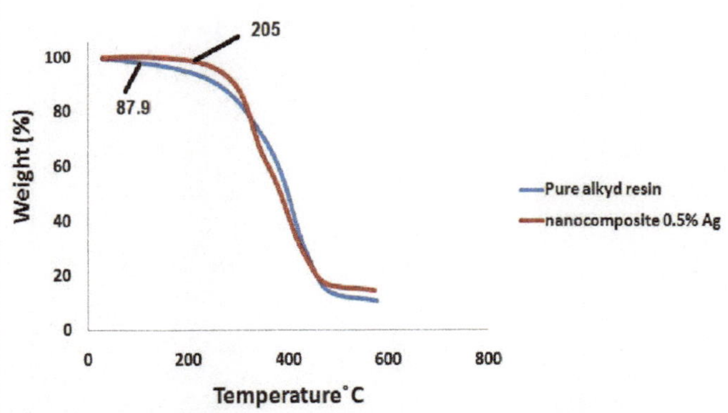

Figure 5: TGA analysis, alkidresin, alkidresin-silver nano composite.

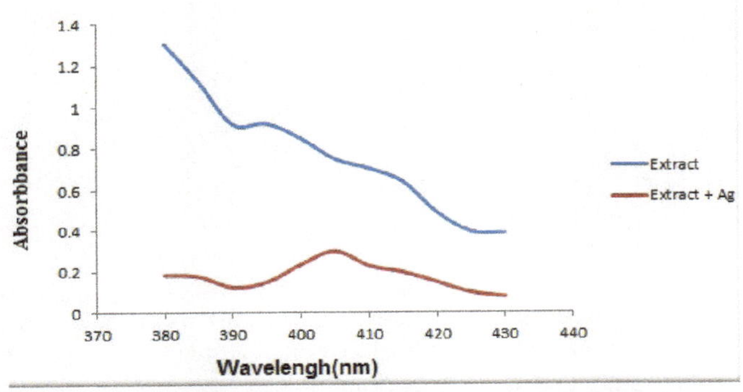

Figure 6: UV-visible spectrum, the extracts of *Elaegnus angustifolia* leaf and the extract with silver.

Figure 7: SEM micrograph of silver nanoparticle provided by the extracts of *Elaegnus angustifolia* leaf and $AgNO_3$ solution in different magnifications.

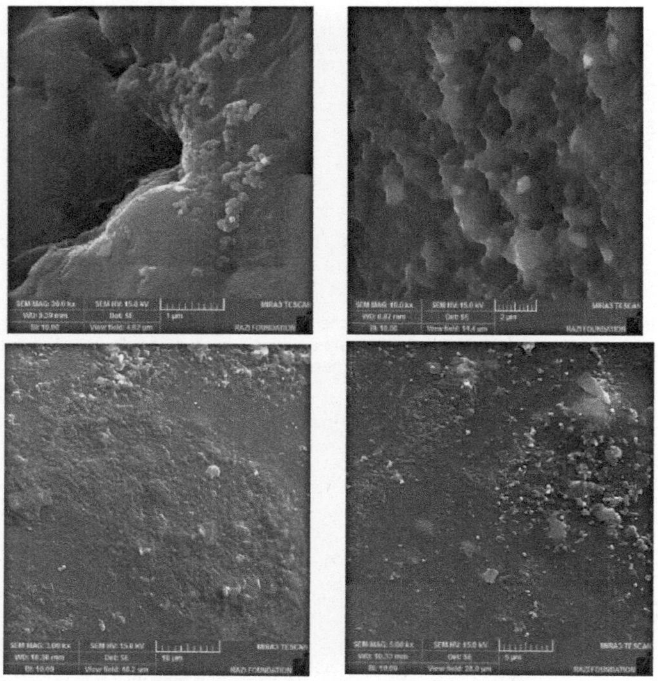

Figure 8: Alkid resin SEM micrographs.

0.5% silver weight percentage and the picture (b) shows Alkidresin nano-composites containing 1% silver weight percentage in different magnifications. In both pictures Nanoparticles are seen as spherical compact bullets in polymer bed and are dispersed pretty uniformly [3-26].

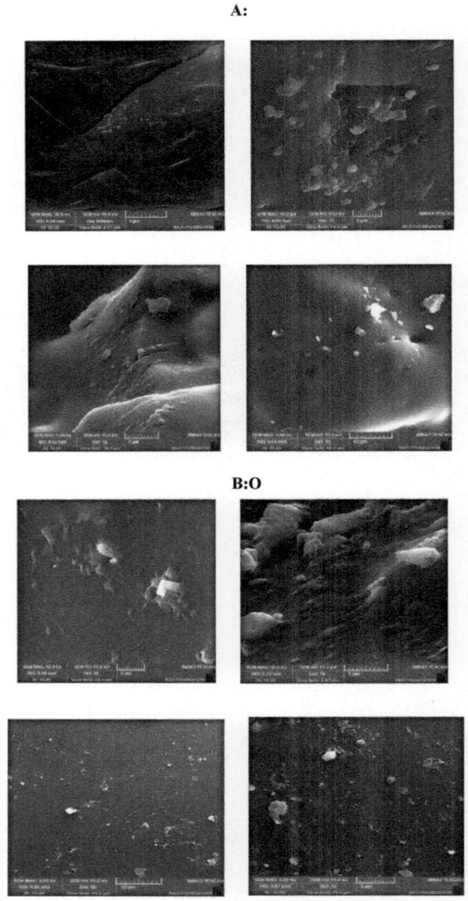

Figure 9: A) Alkidresin nano-composite of 0.5% silver. B) Alkidresin nano-composite of 1% silver.

Code	IDT°C	CY (%)
Alkid-resin	87.9	11
Silver alkid resin 0.5%	205	15

Table 2: An explanation of TGA analysis.

Biological studies

As it is seen in the picture, after 24 hours, we studied the amount of inhibition of bacteria growth. With constitution of non-growth halo and its measurement, we found that 1% nano composite has fewer diameters than 2.5% nano composite because. The reduction of resistance of bacteria versus silver (Table 3).

Conclusions

Silver nanoparticle was prepared by the *Elaeagnus angustifolia* leaf in size of 16-24 nm with the method of green chemistry and this method is important due to adaptability with environment. Also a two-stage method of alcoholysis-polyesterization is used for preparing nano-composites. This method evenly spreads the nanoparticles in polymer matrix and prevents the accumulation. Studying FT-IR Spectrums, the presence of OH group in the extract makes silver ion decrease into metal silver and because there is no definition of limitations in FT-IR for Ag ، its presence in polymer matrix hasn't induced change in IR peaks. Presence of silver in polymer matrix leads to polymer resistivity increase which was observed in TGA Diffract gram analysis clearly. The

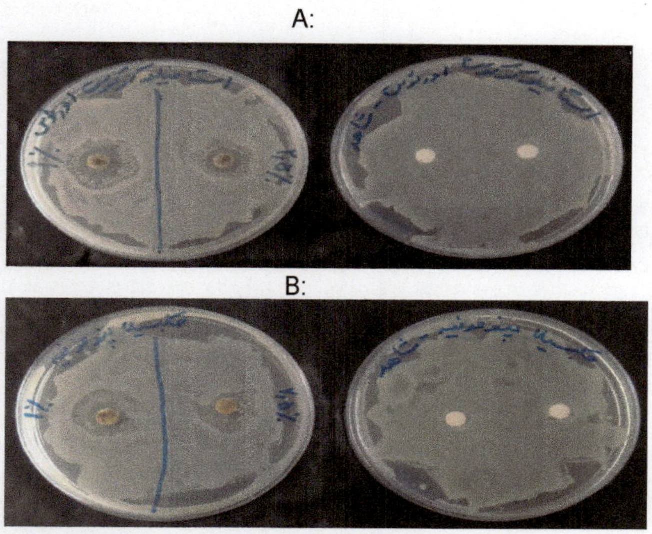

Figure 10: Anti-bacteria picture, a) *Staphylococcus aureus* and b) *Klebsiella pneumonia*.

Sample	Inhabitation Area (mm)	
	Klebsiella pneumoniae	***Staphylococcus aureus***
Pure polymer	-	-
Nano-compounding 1%	12	11
Nano-compounding 2.5%	13	12

Table 3: The size of non-growth halos.

remained coke in nano-composite containing 0.5% sliver was less than alkidresin.

Silver metal reflects absorption when it is in the path of UV beam due to surface vibrations and this process is the same with surface Plasmon resonance. As observed, UV-Visible spectrum in the range of 405 nm confirmed the presence of silver in the extract. When particle sizes reach to nanometer scale in visible spectrum of noble metals, a strong absorption with the origin of surface Plasmon resonance is observed.

Beam X diffraction was used to determine crystallinity amount of silver nanoparticles confirming silver presence in polymer matrix. Generally, the increase of peak and existence of sharp areas in diffract gram is indicative of more crystallinity in the target compound. 4 index peaks were appeared for silver nanoparticles, the most important of them was in the interval of $=\theta 238^\circ$. Alkidresin showed its respective peak in $2=\theta 20^\circ$. On the other hand, due to low concentration of nanoparticles in nano-composite samples, we can't expect the peaks related to nanoparticles appear in high resonance. Here, due to very low concentration, the peaks related to silver in polymer matrix have experience very weak vibrations. Using the Shearer equation, the size of nanoparticles was 16.04 nm.

$$t=0.9\lambda/B\cos\theta_B$$

SEM analysis of baked resins indicates a good surface property and as it is clear from the pictures, silver nanoparticles are dispersed in polymer surface evenly.

Silver showed the anti-bacteria properties, alkidresin didn't have any anti-bacteria property in the normal state, but with constitution of polymer matrix with silver and production of 1 and 2.5 weight percentages, nano-composites anti-bacteria property was appeared with constitution of non-growth halo. The more silver amount in polymer matrix, the less bacteria resistance against it and the non-growth halo diameter increases.

Finally, the presence of silver nanoparticles in polymer matrix will lead to anti-bacterial property which can be used in the containers keeping food stuff, medical and dental equipments. Green chemistry process is adaptable to environment, non-toxic and valuable.

References

1. Aigbodion AI, Okieimen FI (2001) An investigation of the utilization of African Locust bean seed oil in the preparation of alkyd resins. Industrial Crops and Products 13: 29-34.
2. Kurt I, Acar I, Guclo G (2014) Preparation and characterization of water reducible alkyd resin/colloidal silica nano composite coatings. Progress in Organic Coatings 77: 949-956.
3. Ahmed NM, Fatah HTMA, Youssef EA (2012) Corrosion studies on tailored Zn•Co aluminate/kaolin core-shell pigments in alkyd based paints. Progress in Organic Coatings 73: 76-87.
4. Ang DTC, Gan SN (2012) Novel approach to convert non-self-drying palm stearin alky into environmental friendly UV curable resins. Progress in Organic Coatings 73: 409-414.
5. Bora MM, Gogoi P, Deka DC, Kakati DK (2014) Synthesis and characterization of yellow oleander (Thevetia peruviana) seed oil-based alkyd resin. Industrial Crops and Products 52: 721-728.
6. Ezed IE, Umoren SA, Essien EE, Udoh AP (2012) Studies on the utilization of Hura crepitans L. seed oil in the preparation of alkyd resins. Crop Prod 36: 94-99.
7. Gade AK, Gaikwad SC, Tiwari V, Yadav A, Ingle AP, et al. (2010) Bio fabrication of silver nanoparticles by *Opuntia ficus-indica:* in vitro antibacterial activity and study of mechanism involved in the synthesis. Curr Nanosci 6: 370-375.
8. Guo L, Yuan W (2013) Polymer/ nano silver composite coatings for antibacterial applications. Journal Colloid Surface. A: Physico Chem Engineering Aspects 5: 1-15.
9. He W, Huang B (2011) A review of chemistry and bioactivities of a medicinal spice: *Foeniculum vulgare.* Journal Med Pl Res 16: 3595-3600.
10. Iser-caglar D, Basturk E, Oktay B, Kahraman V (2014) Preparation and evaluation of linseed oil based alkyd paints. Progress in Organic Coatings 77: 81-86.
11. Jacquelo E, Galy J, Gerard JF, Roche A, Chevet E, et al. (2009) Morphology and thermo-mechanical properties of new hybrid coatings based on polyester/ melamine resin and pyrogenic. Progress in Organic Coatings 66: 86-92.
12. Mallikarjun K, Narsimha G, Dillip GR, Praveen B, Shreedhar B, et al. (2011) Green synthesis of silver nanoparticles using *Ocimum* leaf extract and their characterization. Digest Journal Nanomat Biostruct 6: 181-186.
13. Mude N, Ingle A, Gade A, Rai M (2009) Synthesis of silver nanoparticles using callus extract of *Carica papya- A* first report. Journal Pl Biochem Bitechnol 18: 83-86.
14. Nagajyoti PC, Prasad TN, Shreekanth VM, Lee KD (2011) Bio fabrication of silver nanoparticles using leaf of *Saururus chinenis.* Digest Journal Nanomat Biostruct 6 1: 121-133.
15. Narayanan KB, Sakthivel N (2011) Green synthesis of biogenic metal nanoparticles by terrestrial and aquatic phototrophic and heterotrophic eukaryotes and biocompatible agents. Journal of advances in Colloid and Interface Science 169: 59-79.
16. Ogunniyi DS, Odetoye TE (2008) Preparation and evaluation of tobacco seed oil-modified alkyd resins. Bioresource Technology 99: 1300-1304.
17. Ploeger R, Scalarone D, Chiantore O (2009) Characterization and Stability Issues of Artists' Alkyd Paints Rebecca Ploeger and Oscar Chiantore. Polym Degrad Stabil 94: 2036-2041.

18. Raut RW, Lakkakula JR, Kolekar NS, Mendhulkar VD, Kashid SB (2009) Phyto synthesis of silver nanoparticle using *Gliricidia sepium* (Jacq). Curr Nanosci 5: 117-122.

19. Sadeghi B, Garmaroudi S, Hashemi M (2012) Comparison of the anti-bacterial activety on the nanosilver shapes: nanoparticles, nanorods and nanoplates. Journal Advance Powder Technology 23: 22-26.

20. Singh C, Sharma V, Naik PK, Khandelwal V, Singh H (2011) A green biogenic approach for synthesis of gold and silver nanoparticles using *Zingiber officinale*. Digest Journal Nanomat Biostruct 6: 535-542.

21. Song JY, Beom SK (2008) Rapid biological synthesis of silver nanoparticles using plant leaf extracts. Bioproc Biosyst Engineer 32: 79-84.

22. Tzeng SH, Tsai JL (2007) Size effect on fracture behaviors of epoxy composites with micron and nano alumina particles. Journal of Materials Science Letters 60: 869-873.

23. Yakutik IM, Shevchenko PG (2004) Self- organization of silver nanoparticles forming on chemical reduction to give mono disperses spheres. Journal Surface Science 566: 414-418.

24. Yin X, Duan H, Wang X, Sun L, Sun W, et al. (2014) An investigation on synthesis of alkyd resin with sorbitol. Progress in Organic Coatings 77: 674-678.

25. Yokoyama T, Huang CC (2007) Nano particle technology for the production of functional material. Journal of the American Chemical Society 23: 7-17.

26. Zhang X, Yan Y, Tyagi RD, Surampalli RY (2011) Synthesis of nanoparticles by microorganisms and their application in enhancing microbial teaction rates. Chemosphere 82: 48-49.

Recovery of Glutaric Acid Using Tri-N-Butyl Phosphate: Effect of Diluents and Temperature

Kanti Kumar Athankar*, Kailas L Wasewar, Mahesh N Varma and Diwakar Z Shende

Advance Separation and Analytical Laboratory (ASAL), Department of Chemical Engineering, Visvesvaraya National Institute of Technology (VNIT), Nagpur (Maharashtra), India

Abstract

Reactive separation of glutaric acid with tri-n-butyl phosphate in ethyl acetate and hexanol was studied to explain the effect of temperature on the extraction. Extraction was carried out at different temperature range of 308-338 K. The reactive extraction results were discussed in terms of overall distribution coefficients (K_D), degree of extraction (η%), loading ratio (φ), and equilibrium complexation constants (K_E). The results showed that an increase in temperature reduced distribution coefficients and extraction efficiency for both diluent used in the organic phase. Moreover, thermodynamic function viz enthalpy (ΔH), entropy (ΔS) and Gibbs free energy (ΔG) were calculated. Reactive extraction of glutaric acid by tri-n-butyl phosphate resulted in negative values of enthalpy. Thus, the reactive extraction process is an exothermic reaction.

Keywords: Reactive extraction; Glutaric acid; Tri-n-butyl phosphate; Ethyl acetate; Hexanol

Introduction

Glutaric acid is a linear di-carboxylic acid. It is produced naturally in the body during the metabolism of amino acids. It is white crystalline solid having the lowest freezing point (371 K) among carboxylic acids and very soluble in water and the solution is a medium strength, i.e., pK_{a1}=4.31 and pK_{a2}=5.41[1]. It can be used in production of polymer viz polyester polyols, polyamides, ester plasticizers and corrosion inhibitors, synthesis of pharmaceuticals, surfactants and metal finishing compounds. α-Ketoglutaric acid is a derivative of glutaric acid which can be used in dietary supplement. Current global market value of glutaric acid is approximately US$ 24000 /year [2]. Despite the immense commercial importance of glutaric acid, it is important to recover from aqueous solution. Glutaric acid can be extract from fermentation medium by calcium salt precipitation but it is very expensive and unfriendly to the environment. The fermented medium contains either the pure acid or its salt or a mixture of both. The search for an advantageous process allowing for the removal of carboxylic acid is based on an approach that removes acids from the fermentation broth or other mixture, while leaving the soluble salts behind in the fermentation broth. Industrial scale fermentors for production of carboxylic acids operate at different temperatures from 273-373 K according to product selectivity [3]. Reactive extraction is an efficient and clean technology for separation of carboxylic acid from fermentation broth and industrial stream. Phosphoric bonded oxygen containing extractants, dissolved in appropriate organic solvents are effective extractants for carboxylic acid [3].

In the present work, equilibrium and thermodynamic study for reactive extraction of glutaric acid has been explored and the result are presented in terms of distribution coefficient, equilibrium complexation constants, loading factor, degree of extraction, thermodynamic function viz. enthalpy, entropy, and Gibbs free energy were calculated.

Materials and Methods

Chemicals

The chemicals used in present work were procured from Merck, India with more than 98% purity whereas glutaric acid purchased from Acros Organics, India. Tri-n-butyl phosphate (TBP) was used as an extractant with ethyl acetate (EA) and hexanol (HX). Physiochemical properties of used chemicals are shown in Table 1. All the chemicals were used without any further purification.

Experimental procedure

Extraction experiments were conducted in 100 mL Erlenmeyer flasks at various temperatures 308-338 K. The initial aqueous solution of glutaric acid was prepared (0.19-1.10 mol/kg) in deionized water without pH adjustment. An organic solution was prepared by dissolving fix concentration of tri-n-butyl phosphate 1.46 (mol/kg) in ethyl acetate and hexanol. The volumetric ratio of the aqueous and organic phase was kept as 1:1 (15 mL of each phase). The flasks containing the mixture of aqueous and organic solution was shaken for 12 hours in a temperature-controlled water bath shaker RSB-12 (REMI Laboratory Instruments, Mumbai, India) and the solution was allowed to settle for 2-3 hours at room temperature (298 ± 0.5 K) and atmospheric pressure (760 mmgH). The bottom layer (aqueous phase) was used for pH measurement and acid concentration analysis. After equilibrium, the acid concentration of aqueous solution was determined by HPLC. The consistency was checked by repeating few experiments which were observed within the limit of ± 2%.

Analysis

Analysis was performed with Eclipse XDB-C18 (4.6 mm ID × 250 mm, 5 μm) column. Pure acetonitrile (25%) and 20 mM of acetic acid aqueous solution (75%) was used as a mobile phase in volume ratio 1:3 and the flow rate was set at 1 ml/min. Maximum absorption (λ_{max}) of glutaric acid in UV range was observed to be 275 nm thus the detector (DAD) wavelength was set at λ=275 nm. The column was operated at 308 K with the injection volume of 5 μL. Before injecting, the samples

***Corresponding author:** Kanti Kumar Athankar, Advance Separation and Analytical Laboratory (ASAL), Department of Chemical Engineering, Visvesvaraya National Institute of Technology (VNIT), Nagpur, Maharashtra, India
E-mail: kanti.kumar@rediffmail.com

chemicals	MW	MF	BP	MP	S_{water}	ρ	μ	RI	$logK_{ow}$	D_m
	g/mol		K	K		g /cm³	cP	η_D		Debye
glutaric acid (pK_{a1}=4.31, pK_{a2}=5.41)	132.12	$C_5H_8O_4$	200°C	98°C	430 g/L	1429	NA	1.4187	-0.47	NA
hexanol	102.17	$C_6H_{14}O$	482-432	220- 232	5.9 g/L	0.815	4.37	1.4178	1.858	1.55
ethyl acetate	88.11	$C_4H_8O_2$	350	189	8.3 g/100 ml	0.897	0.43	NA	NA	1.78
tri-n-butyl phosphate	266.31	$C_{12}H_{27}O_4P$	562	193	1ml/165 ml	0.975	3.8	1.424	2.5-4	NA

MW: molecular weight; MF: molecular formula; BP: boiling point; MP: melting point; S_{water}: solubility in water at 298 K;
ρ: density of pure liquid; μ: viscosity; RI: refractive index; $\log(K_{ow})$, water partition coefficient; D_m, dipole moment; NA, not available.

Table 1: Various Physicochemical Properties of Chemicals selected for recovery of Glutaric Acid.

were filtered through syringe filter with pore size 0.2 μm PVD filter media supplied by WHATMAN, U.S.A. Retention time of glutaric acid was 4.3 min. Each analysis was done in triplicates under identical conditions, with the average value of the considered parameters being used. The maximum experimental error was observed as below ± 2 %.

Uncertainty analysis

In this study, experimental uncertainty was evaluated as per the National Institute of Standard and Technology (NIST) protocols [4]. Few experiments were performed in triplicate and the reliability of replicated experiments was found within ± 2% having confidence interval 95%. Standard uncertainty was evaluated to be ± 0.001 by using following Eq.

$$u(x)=\sqrt{\frac{\sum_{i=1}^{N}\left(x_i-\bar{x}\right)^2}{N-1}} \quad (1)$$

Where, x_i=values of experimental observations, $\bar{X}$ =mean of three observations, and N=number of observations. For further calculations average of the replicated data was used.

Results and Discussion

Partition and dimerization coefficient

The physical extraction of glutaric acid with ethyl acetate and hexanol are shown in Figure 1. A linear relationship for equilibrium between organic and aqueous phase of glutaric acid were observed at low range of glutaric acid concentration. At higher concentration of glutaric acid, a parabolic relationship was obtained for all diluents. Henry's law type isotherm can be considered at low concentrations of glutaric acid, while deviation from Henry's law was observed at higher concentrations of glutaric acid where non-ideal behavior was overcome. The values of partition coefficient (P) and dimerization coefficient (D) were obtained by fitting experimental data in the following equation (3):

$$K_D=P+2DP^2[C_{GA}]_{aq} \quad (2)$$

Where, $[C_{GA}]_{aq}$ concentration of glutaric acid in aqueous solution at equilibrium. Partition and dimerization coefficient values for glutaric acid in ethyl acetate and hexanol was obtained as 0.275, 25.00, and 0.276, 61.92 respectively as shown in Figure 2. Thus, solute-solvent hydrogen bond in comparison to solute-solute bond leads to form dimmers.

Distribution coefficient and extraction efficiency

Equation 3 and 4 represents the distribution coefficient (K_D) and extraction efficiency (η%) for glutaric acid and it can be expressed as:

$$K_D=\frac{[C_{GA}]_{org}}{[C_{GA}]_{aq}} \quad (3)$$

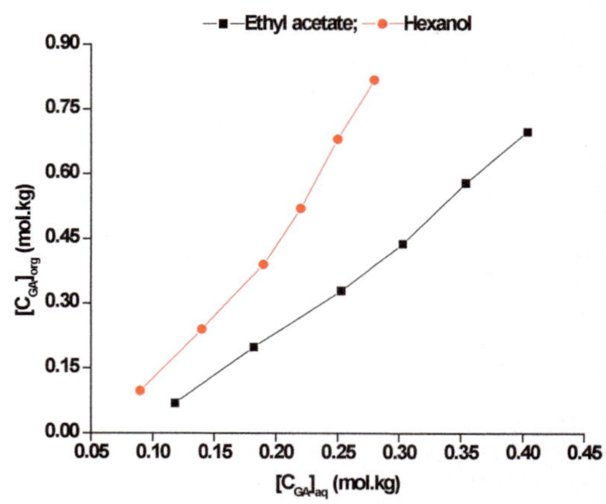

Figure 1: Physical Extraction of Glutaric Acid with Ethyl Acetate and Hexanol..

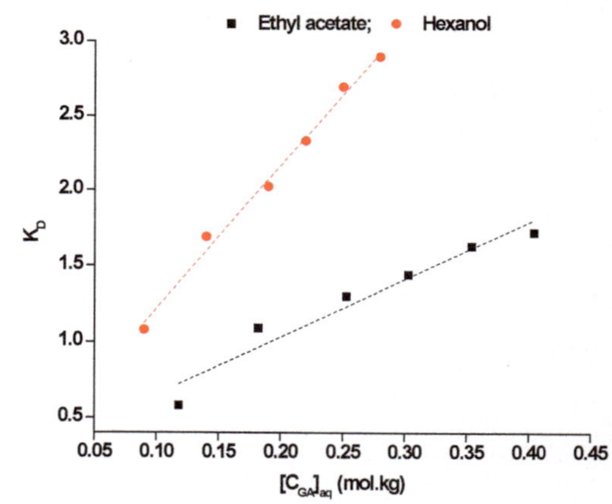

Figure 2: Partition and Dimerization coefficient for glutaric acid with ethyl acetate and hexanol.

$$\eta\%=\frac{K_D}{(1+K_D)}\times 100 \quad (4)$$

Where, $[C_{GA}]_{aq}$ and $[C_{GA}]_{org}$ are the total concentration of glutaric in aqueous and organic phase respectively. Physical extraction results were shown in Table 2. Distribution coefficients of glutaric acid with ethyl acetate and hexanol were obtained as 0.58 to 1.72 and 1.08 to 2.89 respectively. The average extraction efficiency was found to be ~ 55% in ethyl acetate and ~ 66.5% in hexanol. It can be due to hexanol is protic

diluent	$[C_{GA}]^o$	$[C_{GA}]_{aq}$	$[C_{GA}]_{org}$	K_D	η (%)	P	D	R^2
EA	1.1	0.404	0.696	1.72	63.3	0.275	25	0.93
	0.93	0.354	0.578	1.63	62			
	0.74	0.303	0.437	1.44	59.1			
	0.58	0.253	0.328	1.3	56.5			
	0.38	0.182	0.198	1.09	52.2			
	0.19	0.118	0.069	0.58	36.8			
HX	1.1	0.28	0.817	2.89	74.3	0.276	61.92	0.99
	0.93	0.25	0.679	2.69	72.9			
	0.74	0.22	0.518	2.33	70			
	0.58	0.19	0.388	2.02	66.9			
	0.38	0.14	0.239	1.69	62.8			
	0.19	0.09	0.097	1.08	51.9			

[a]Standard uncertainties u are u(T)= ± 1 K, $u(C_{GA})$= ± 0.001 mol/kg

Table 2: Determination of partition coefficient and dimerization constant for physical extraction of glutaric acid[a].

diluent contain hydrogen atoms bound to electronegative element O (-OH) is a hydrogen bond donor and large value of relative permittivity and transmission energy indicates its strong polar nature. Thus hexanol provide high distribution coefficient values over ethyl acetate since it is good anion solvators due to their hydrogen bonding ability.

Reactive extraction of glutaric acid with tri-n-butyl phosphate

Chemical (reactive) extraction of glutaric acid and organic solution (tri-n-butyl phosphate+ diluents) can be expressed as:

$$C_{GA,aq} + C_{TBP,org} \overset{K_E}{\longleftrightarrow} [C_{GA}:C_{TBP}]_{org} \quad (5)$$

The equilibrium complexation constant (K_E) can be calculated by Eq. 6.

$$K_E = \frac{[C_{GA}:C_{TBP}]_{org}}{[C_{TBP}]_{org}[C_{GA}]_{aq}} \quad (6)$$

From Figure 3a, it was observed that the distribution coefficient was inversely proportional to the temperature for both diluents due to the back extraction of the acid from the organic to fresh aqueous phase. It was decrease in the range of 2.72-2.28 for ethyl acetate and 2.28-1.84 for hexanol. Moreover, Figure 3b shows that the extraction efficiency was linearly decreases from 73.1-69.5 and 69.5-64.8 for ethyl acetate and hexanol respectively in 308-338 K temperature range. At higher temperature, the rapid enhancement of ionization of acid can be suppressed the unionization of acid so the distribution of glutaric acid in organic and aqueous solution was decreased. Table 3 shows the equilibrium complexation constant in studied temperature range average K_E value was 4.70 and 3.53 (kg/mol) for ethyl acetate and hexanol respectively.

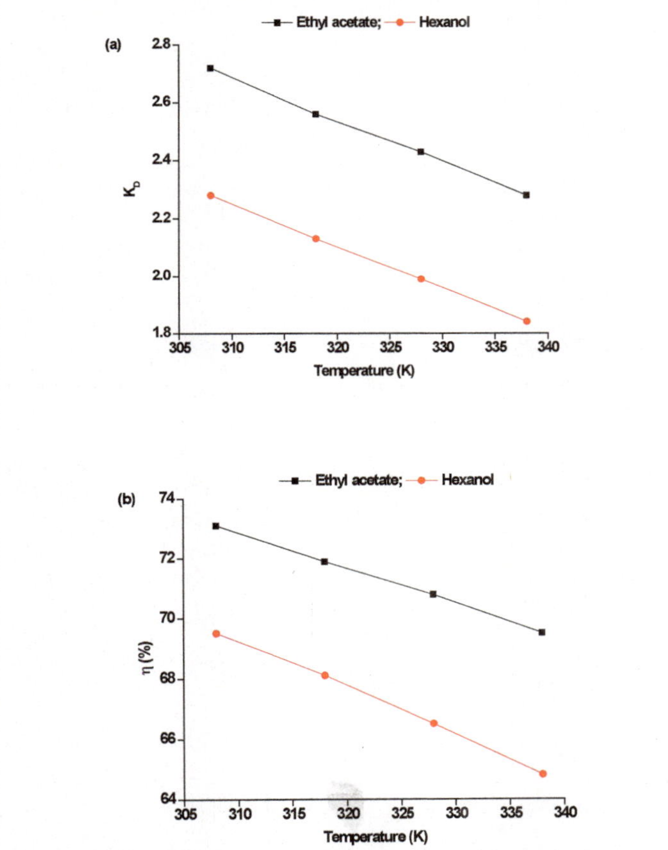

Figure 3: Effect of temperature: (a) on distribution coefficient; (b) on extraction efficiency.

Glutaric acid – tri-n-butyl phosphate complex in diluents

The solvation number for glutaric acid – tri-n-butyl phosphate system is same as the number of carboxylic group in acid [5]. Hence, it is assumed that only 2:1 glutaric acid – tri-n-butyl phosphate complex is present in both diluents. The formation of other complexes, 1:1, 3:1 depend on concentration of the species and loading ratio. The loading ratio (φ) is defined as the ratio of total concentration of acid in organic solution to the initial concentration of extractant it can be expressed as:

$$\phi = \frac{[C_{GA}]_{org}}{[C_{TBP}]_{org}} \quad (7)$$

1:1 glutaric acid – TBP complex:

$$\frac{\phi}{1-\phi} = K_E[C_{GA}] \quad \text{and } \varphi < 0.5 \quad (8)$$

2:1 glutaric acid – TBP complex:

$$\frac{\phi}{2-\phi} = K_E[C_{GA}]^2 \quad \text{and } \varphi > 0.5 \quad (9)$$

3:1 glutaric acid – TBP complex:

diluent	temperature	K_D	η	K_E	φ	ΔΓ	H	ΔΣ	R^2
	K		%	kg/mol		kJ	J/mol	J/mol.K	
EA+TBP	308	2.72	73.1	5.34	0.652	-518.34	-935.06	-1.35	0.996
	318	2.56	71.9	4.88	0.641	-504.81			
	328	2.43	70.8	4.49	0.631	-491.28			
	338	2.28	69.5	4.1	0.62	-477.75			
HX+TBP	308	2.28	69.5	4.1	0.62	-436.17	-1110.38	-2.19	0.991
	318	2.13	68.1	3.7	0.607	-414.28			
	328	1.99	66.5	3.33	0.593	-392.39			
	338	1.84	64.8	2.98	0.578	-370.5			

[a]Standard uncertainties u are u(T)= ± 1 K, $u(C_{GA})$= ± 0.001 mol/kg

Table 3: Effect of temperature and thermodynamic parameters of reactive extraction of glutaric acid[a].

$$\frac{\phi}{3-\phi}=K_E\left[C_{GA}\right]^3 \quad \text{and } \varphi >> 0.5 \qquad (10)$$

Only 2:1 glutaric acid–tri-n-butyl phosphate complex was formed in ethyl acetate and hexanol as the dilute concentrations of glutaric acid were used in present study. The nominal difference among K_E values in both diluents for glutaric acid shows that solvation is a crucial for reactive extraction of carboxylic acid. For higher concentrations of acid in aqueous phase, 3:1 complexes may be formed. The proportions of 2:1 to 3:1 complex depend on type of diluent and its properties [6]. The loading ratio value was found to be ~ 0.6 for all cases except some points as shown in Table 3.

Effect of temperature

Heat of mixing of organic solution using distribution coefficient can be calculated by Eq. 11 [6].

$$\frac{d(\ln K_D)}{d(1/T)}=-\Delta H_{transfer} \qquad (11)$$

Where, $\Delta H_{transfer}$ is the heat of transfer from aqueous to organic phase. In high temperature range, change in enthalpy (ΔH) and entropy (S) of the reaction are studied [7]. The equilibrium complexation constant is correlated with temperature, enthalpy and entropy and it can be expressed by the following equation.

$$\ln K_E=\frac{-\Delta H}{RT}+\frac{\Delta S}{R} \qquad (12)$$

Moreover Gibbs free energy (ΔG) can be calculated by the following correlation:

$$\Delta G = \Delta H - T\Delta S \qquad (13)$$

Eq. 12 revels that, the plot (Figure 4) between $\ln(K_E)$ and 1/T gives straight line. The slope and intercept of straight line gives enthalpy and entropy of reaction correspondingly. Overall equilibrium constants for both the diluents were found to be decrease as the temperature increase from 308-338 K this is probably due to the back extraction of the glutaric acid. Table 3 shows the negative magnitude of both thermodynamic functions viz. enthalpy and entropy for ethyl acetate and hexanol indicates that all the reactions are exothermic reaction. ΔH and was 935.06 and 1110.38 J/mol whereas, ΔS was -1.353 and -2.189 J/mol.K for ethyl acetate and hexanol respectively. This behaviour also confirms that at higher temperatures, the reactive system becomes more unstable and hence the possibility of back-extraction increases. Decrease in ΔS revels that, increase of the order of system due to complex formation. By the formation of complexes the number of degree freedom decreases as two molecules are combined to form one entity. The results of present study were compared with reactive extraction of other carboxylic acids which is tabulated in Table 3.

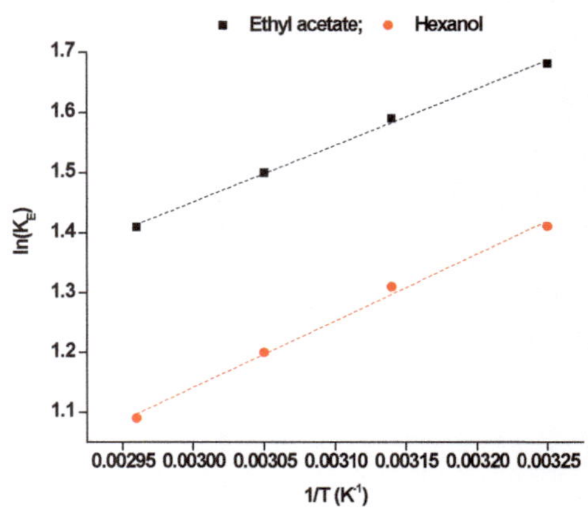

Figure 4: Plot of 1/T vs $\ln K_E$ for fixed concentration 1.464 (mol/kg) of acid tri-n-butyl phosphate.

The overall effect of temperature is attributed to the effects of different parameters such as pK_a, the acid-extractant interaction, solubility of the acid in both phases, the extractant basicity and water co-extraction [8]. The pK_a values of common carboxylic acids decrease slightly with an increase in temperature, leading to the dissociation of acid molecules in the aqueous phase and lower extraction efficiency. Similarly, the solubilities of acid in both aqueous and organic phases are affected by temperature. It is observed that the solubilities of carboxylic acids in water increase as the temperature increased from 278.15 to 338.15K [9].

Conclusion

Reactive extraction of glutaric acid with tri-n-butyl phosphate in ethyl acetate and hexanol was studied. P and D values for glutaric acid in ethyl acetate and hexanol were obtained as 0.275, 25.00, and 0.276, 61.92 respectively. The K_D values were found in the range of 0.58 to 1.72 for ethyl acetate and 1.08 to 2.89 for hexanol. The average extraction efficiency was found to be ~ 55% in ethyl acetate and ~ 66.5% in hexanol. The loading ratio value was found to be ~0.6 for all cases i.e., it is assumed that only 2:1 glutaric acid–tri-n-butyl phosphate complex is present in both diluents. Thermodynamic function enthalpy (ΔH), entropy (ΔS) and Gibbs free energy (ΔG) were calculated as negative magnitude of enthalpy confirms the glutaric acid extraction with both diluents is exothermic reaction. Increasing values of temperature are inversely proportional to the distribution coefficient and extraction efficiency. This behavior also confirms that at higher temperatures, the reactive system becomes more unstable and hence the possibility of back-extraction increases.

References

1. Phelivanoglu N, Uslu H, Kirbaslar IS (2009) Experimental and Modeling of Glutaric Acid by Trioctylamine. J Chem Eng Data 54: 3202-3207.
2. Market Research Report on Global and Chinese Glutaric acid Industry, December 2016.
3. Kertes AS, King CJ (1986) Extraction Chemistry of Fermentation Product Carboxylic Acids. Biotechnol Bioeng 28: 269-282.
4. Chirico RD, Frenkel M, Diky VV (2003) ThermoML – An XML-Based Approach for Storage and Exchange of Experimental and Critically Evaluated Thermophysical and Thermochemical Property Data. 2. Uncertainties. J Chem Eng Data 48: 1344-1359.
5. Niitsu M, Sekine T (1978) Solvent Extraction of Ionic Solutes in Aqueous Solutions. Bull Chem Soc Jpn 51: 705-717.
6. Tamada JA, Kertes AS, King CJ (1990) Extraction of Carboxylic Acids with Amine Extractants. 1. Equilibria and Law of Mass Action Modeling. Ind Eng Chem Res 29: 1319-1326.
7. Tamada JA, King CJ (1990) Extraction of Carboxylic Acids with Amine Extractants. 3. Effect of Temperature, Water Coextraction, and Process Considerations. Ind Eng Chem Res 29: 1333-1338.
8. Canari R, Eyal AM (2004) Temperature Effect on the Extraction of Carboxylic Acids by Amine-based Extractants. Ind Eng Chem Res 43: 7608-7617.
9. Apelblat A, Manzurola E (1987) Solubility of Oxalic, Malonic, Succinic, Adipic, Maleic, Malic, Citric, and Tartaric Acids in Water from 278.15 to 338.15 K. J Chem Thermodyn 19: 317-320.

Meta-Xylene Removal from Aqueous Solution Using Surfactant-Modified Bentonite: A Kinetic View

Chidi Obi* and Michelle Abonele Apemiye

Department of Pure and Industrial Chemistry, Faculty of Science, University of Port Harcourt, Rivers State, Nigeria

***Corresponding author:** Chidi Obi, Department of Pure and Industrial Chemistry, Faculty of Science, University of Port Harcourt, Rivers State, Nigeria

E-mail: zarasexcom@yahoo.com

Abstract

The bentonite used was a natural calcium bentonite from Enugu State, Nigeria. This was treated with Na_2CO_3 by ion exchange method to form a sodium bentonite described as unmodified bentonite (UB). Unmodified bentonite was treated with cetyltrimethylammonium bromide (CTAB) to produce a modified bentonite (MB). The performance of unmodified bentonite (UB) and modified bentonite (MB) were evaluated for the sorption of meta-xylene from aqueous solution. The resultant materials were characterized using kinetic model equations. The optimum contact time for the adsorption of meta-xylene was 120 minutes with a maximum percentage removal of 96.82 % for the modified adsorbent. Pseudo-second order equation gave the best correlation coefficient value (R_2) of 0.999 for the adsorption of meta-xylene on the modified bentonite (MB). These results obtained confirmed the efficiency and effectiveness of modified bentonite in the removal of petroleum wastes.

Keywords: Cetyltriammonium bromide (CTAB); Sodium bentonite; Meta-xylene; Kinetic model equations; Remediation

Introduction

Nigeria is the seventh largest oil producer of oil resources which has effect in her economy. Till date the petroleum industry in Nigeria relies on bentonite clay which produces approximately two million barrels of crude oil daily, thus, it's essential for drilling fluid application. Reports over the past years described the oil and gas industry in Nigeria as the country's live wire [1,2].

The term 'clay' is said to be a naturally endowed fine-grained minerals, that when moistened with water forms plastic, and tends to harden when fired or dried [3,4]. Clay has great potentials of enormous variety of uses which is broadly distributed for its richness of abundance of minerals it possesses. Attention has been drawn to bentonite in respect to its swelling capacity, high sorption, high ion exchange and large surface area [5,6].

Bentonites are classified under smectitic clays, an interesting group of clay minerals and used in various industrial applications. Therefore, there is need to subject them to chemical treatments for their potentials to be actualized. The petroleum industries are the major consumer of bentonite, making clay as thixotropic agent for drilling wells, binding agent and catalyst used in the refinery industries. Studies to remediate the polluted environment are done using several types of sorbent materials. Promising materials are obtained from these organophilic materials which are made up of smectite clay and long chain surfactants intercalated between the layers of clays [7].

Natural bentonite are increasingly used for waste water treatment in the production stage especially during extraction and refining to remove effluents (volatile organic compounds such as benzene, toluene, ethyl benzene and xylene (BTEX), phenols and heavy metallic ions such as copper, zinc and cadmium [8].

Structurally, bentonite consists of about sixty percent montmorillonite. The framework composes of two basic building of sheets of aluminium octahedral and silica tetrahedral which are stacked together in a 2:1 ratio (Figure 1). Montmorillonite clay unit cell is framed up by the assembling of an Al octahedral sheet inserted between two Si tetrahedral sheets [9]. Two outer tetrahedral layers comprises of silica and oxygen atoms stalked with an inner octahedral layer containing Al and Mg atoms which are bonded to oxygen or hydroxyl group.

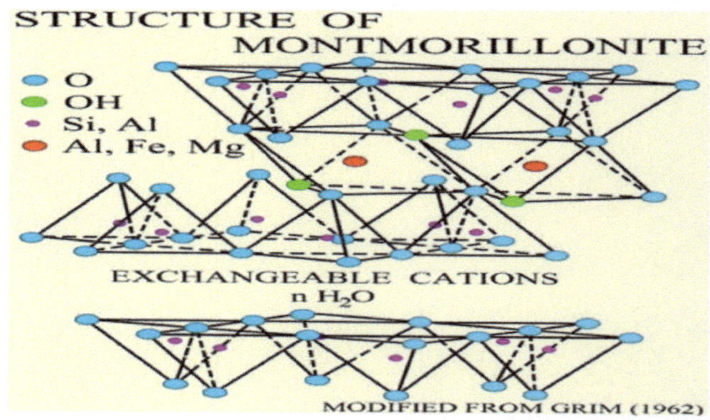

Figure 1: Structure of montmorillonite [10].

Organophilic bentonites are good absorbent for organic compounds or polymers and are formed when natural bentonite with exchangeable cations transforms to hydrophobic organic product by displacing the metal ion with organic surfactant like quaternary ammonium salts. The layers of these modified products tend to have low surface energy and large basal spacing. Thus, organophilic bentonites are utilized for contaminant retention and waste stabilization.

Environmental policy is increasing severely, due to the high pollution rates thereby industries are constantly on the move to establish standards to reduce the concentration of pollutants in petrochemical wastewater and are devising new techniques to eliminate or reduce toxic substances contained in their discharges [11]. In petroleum industry, the production stages (extraction and refining) generate a high level of effluents depending on the oil type, processes, complexity of refinery. Other processes include de-paraffinization, de-asphaltination, and processes involving chemical conversion by isomerisation, alkylation, catalytic reforming, etc.

Practice of the treatment of waste water is done to separate the derivatives from water being discharged to the surroundings. The hazard of environmental pollution and its harmful effect to the health of humans have motivated more research [12]. The smallest concentration of benzene, toluene, ethylbenzene and xylene (BTEX) into the human system leads to kidney disruption, cancer, liver, mucosal membrane irritation and impairment of central nervous system [13]. Consequently, bentonite as a noble adsorbent is very available at low costs; they are essential adsorbents for petroleum wastes treatment beforehand release into the water bodies.

The aim of this study is to investigate the effectiveness of Enugu bentonite when modified with surfactant for the removal of meta-xylene from aqueous solution and possibility evaluates the kinetic feasibility to enable its applicability as sorbent in the petroleum industry.

Experimental Materials

The natural bentonite sample used in this work was rich in calcium content and obtained from Enugu State, Nigeria

Figure 2: Sample of local Enugu bentonite clay.

Benefication of Unmodified Bentonite (UB) using ion exchange method

Natural bentonite of 40 g was measured in a one-neck bottom flask and soaked in 200 ml of distilled water for 24 hrs. A solution of 200 ml containing 0.2 M Na_2CO_3 was added and refluxed at 60°C-80°C with stirring for 2-3 hrs [14]. The mixture was left to settle completely, decanted and washed thrice with distilled water. The final decanted sediment was dried for 12 hrs at 105°C using oven. Finally, some portion of resultant product was ground with a pestle and mortar to achieve a homogeneous sodium bentonite fine powder and was labelled unmodified bentonite (UB) for beneficiation and characterization.

Figure 3: Set-up showing the beneficiation of sodium bentonite.

Surfactant modified bentonite (mb) from sodium bentonite slurry (ub)

The synthesis of CTAB-modified bentonite was carried following the steps in the flow chart below:

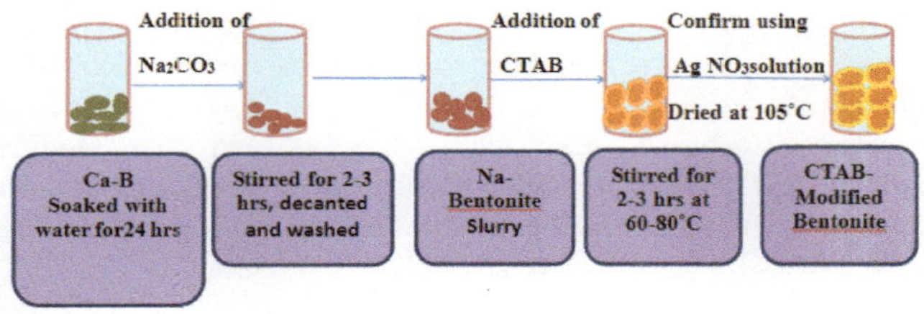

Figure 4: Schematic diagram of the preparation of CTAB-modified bentonite.

Kinetic analysis

This test was done in batch form with 0.1 g of unmodified bentonite added to 15 ml of meta-xylene aqueous solution of concentration 50 mg/L. This was repeated for modified bentonite. The mixture was stirred at 25°C using a mechanical shaker at a speed of 150 rpm at varying period of 30, 60, 90, 120 and 150 mins. The supernatant solution was analyzed using UV/VIS Spectrophotometer (Shimadzu, UV-160A) at 325 nm wavelength. The data was analyzed using kinetic model equations presented in Table 1. The amount of meta-xylene adsorbed q_e (mg/g) and removed are obtained from the expressions in equations 1 and 2.

$$q_e = \frac{C_o - C_t}{W} \times V \ (1) \%R = \frac{C_o - C_t}{C_o} \times 100 (2)$$

Where C_o represents the equilibrium concentration in mg/L, C_t represents the concentration at a particular time in mg/L, W represents the weight of the adsorbents in g and V represents the volume of meta-xylene in L.

Table 1: Showing adsorption kinetic model equations used in this study.

Models	Equation	References
Lagergren first-order	In (qe-qt)=In qe-K1t	[15]
Pseudo-second-order	$q^t = k^i dt^{\frac{1}{2}} + C$	[16]
Intra-particle diffusion	$\frac{t}{q_t} = \frac{1}{k_2 q_e^2} + \frac{1}{q_e}t$	[17]

Results and Discussion

Effect of contact time

At a given concentration, the adsorption of meta-xylene was experimented for five varying time intervals (30, 60, 90, 120 and 150 mins) as presented in Figures 2-5 and Table 2. The result showed that the percentage removal of meta-xylene increases steadily, rapidly and smoothly from the initial point to a contact time of 30 minutes as a result of more adsorption sites available [18]. It was observed further that there was gradual but continuous increase from 30 mins to 120 mins. The optimum contact time for the adsorption of m-xylene was 120 mins with a maximum percentage removal of 96.82% for the modified adsorbent. The range of percentage removal of xylene in the surfactant-modified adsorbent at contact time of 30-120 mins was between 54.00%-96.82% while unmodified adsorbent ranged from 29.84%-40.90%. This is possibly because of saturation of adsorbent surfaces with hydrocarbon follow by adsorption and desorption that occurred. Consequently, the adsorbents analyzed from the plot depicted that modified bentonite (MB) had greater inf uence due to its adsorption power in comparison with unmodified bentonite (UB). The order of percentage removal was MB (96.82%)>UB (29.84%).

The difference in percentage removal of meta-xylene between the unmodified bentonite (UB) and modified bentonite (MB) at 120 mins were far apart with about 56.62% while at 30 mins, it was about 24.16 % being that unmodified bentonite (UB) was far less superior to modified bentonite (MB). This is due to a change in the interlayers in MB which increased the effectiveness of contact between meta-xylene and surfactant-modified bentonite and enriches the affinity for the surface [19]. Hence, modified bentonite (MB) is an effective adsorbent for removal of non-polar and non-ionic organic compound. This observation is similar to the work according to Oladoja, et al. [20].

Table 2: Effect of contact time.

Time (min)	30	60	90	120	150
UB (% R)	29.8	34	39.7	40.2	40.2
MB (% R)	54	74	84.2	96.8	96.8

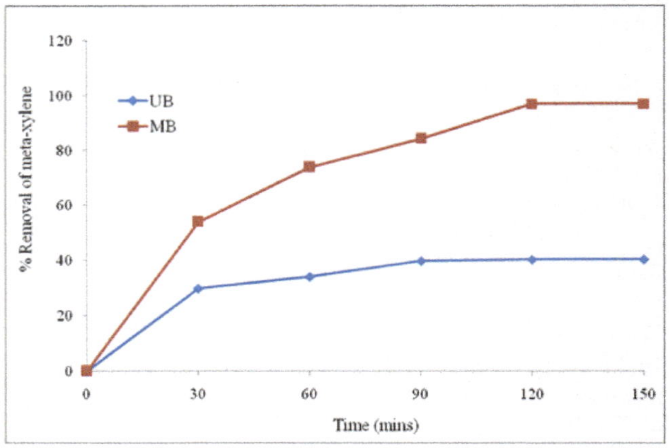

Figure 5: Percentage removal of meta-xylene by effect of contact time.

Kinetic models

The kinetic isotherm study was done by comparing the pseudo-first order, pseudo-second order and intra-particle diffusion plots which are used to depict adsorption, equilibrium rate constant and q_e as presented in Figures 6-8 and Table 3.

The experimental data when modelled followed pseudo-second order equation with high correlation coefficient (R^2) of 0.999. The other kinetic model equations showed relatively good correlation coefficient but pseudo-second order kinetic equation was adopted. This experimental result corresponds to previous studies according to Sidmara et al. [21] and Zhou et al. [22].

Additionally, the adsorption capacity is governed by k1 and k2, clearly shows that adsorption capacity depends on the nature of adsorbent and adsorbate [23] as shown in Table 2. Furthermore, adsorption capacity q_e, have the tendency to increase in this trend of UB (3.373 mg/g)<MB (9.328 mg/g), respectively. is obviously shows that modified bentonite (MB) has the highest adsorption capacity. e slope from the plot in Figure 6 is intra-particle diffusion constant K_{id}. e K_{id} increases as hydrocarbon adsorption increases. Previous studies [24] have reported that increase in the initial concentration of hydrocarbon in the bulk sample result to increase in $K_{i.d}$. e intercept from the Figure 6 identifies the thickness of boundary layer [25].

Consequently, the data for the unmodified and modified bentonite best fits pseudo-second order model [26]. is means that meta-xylene forms a chemical bond by sticking to the bentonite surfaces which tends to find sites that maximize their co-ordination number with the surface [27,28]. erefore, it is verified that surfactant modified bentonite affects the adsorption rate and adsorption capacity.

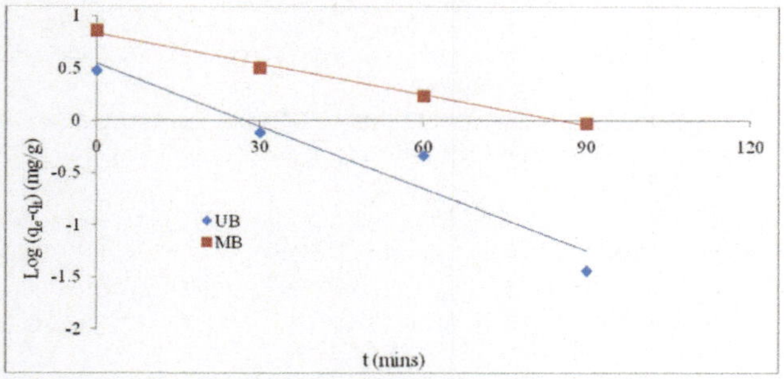

Figure 6: Pseudo-first order plot

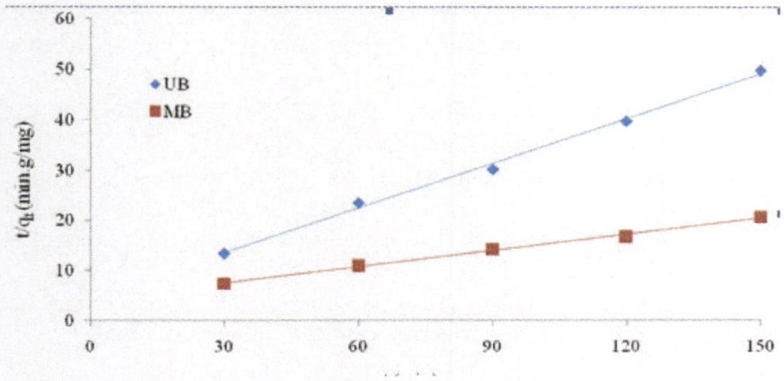

Figure 7: Pseudo second order plot.

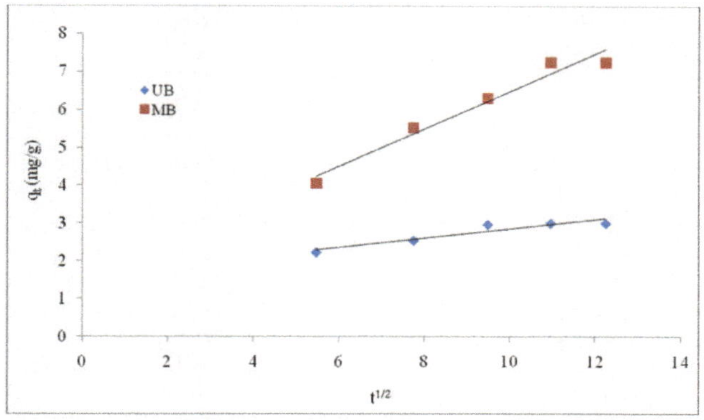

Figure 8: Intra-particle diffusion plot.

Acknowledgements

The authors are highly thankful to Merck KGAA Company for supplying cetyltrimethylammonium bromide (CTAB), JO CHEM Nigeria limited for meta-xylene and the Department of Pure and Industrial Chemistry, University of Port Harcourt for their laboratory equipment used during the course of this research work.

Table 3: Kinetic model parameter for unmodified and modified bentonite.

	Kinetic Model							
Pseudo first-order	Pseudo second order				Intra-Particle Diffusion			
Adsorbent	k_1(1/min)	q_e (mg/g)	R^2	k_2/mg.min)	q_e(mg/g)	R	k_{id}(mg/g.min$^{1/2}$)	R^2
MB	0.01	0.834	0.994	0.003	9.328	0.999	0.496	0.962

Conclusion

This study examined the capacity of unmodified and modified bentonite for treatment of meta-xylene from aqueous solution. The optimum contact time for the adsorption of meta-xylene was 120 mins with a maximum percentage removal of 96.82 % for the modified bentonite. The kinetic model result indicated pseudo-second order as the best kinetic model equation with correlation coefficient (R^2) of 0.999. Therefore, surfactant-modified bentonite is proficient in adsorption and elimination of organic pollutants from aqueous solution and subsequently waste waters.

References

1. Atakpu L (2007) Resource-based conflicts: Challenges of oil Extraction in Nigeria; paper presented at the European Conference hosted by German European Council Presidency, Berlin, Germany.
2. Odulari GO (2008) Crude oil and the Nigerian economic performance. Oil and Gas Business: World Trade Organization centre, Willian Rappard, Geneva.
3. Guggenheim S, Martin RT (1995) Definition of Clay and Clay Mineral: Joint Report of the AIPEA Nomenclature and CMS Nomenclature Committees. Clay Clay Miner 43: 255-259.
4. Souza SP (2002) Science and Technology of Clays. 2nd Edition, Sao Paulo, England Blucher.
5. Jacobs JD, Kaerner H, Heinz H, Farmer BL, Mirau P, et al. (2006) Alkyl ammonium intercalates within organically modified montmorillonite:

dielectric relaxation and ionic conductivity. J Phy Chem 110: 20143-20157.

6. Xue W, He H, Zhu J, Yuan P (2007) FTIR investigation of CTAB-AL-Montmorillonite complexes. Spectrochemica.Acta Pat. A. 67: 1030-1036.
7. Ghafari S, Aziz HA, Isa MH, Zinatizadeh AA (2009) Application of response surface methodology (RSM) to optimixe coagulation-flocculation treatment of leachate using polyaluminium chloride (PAC) and alum. J Hazard Mater 163: 650-656.
8. Al QMH, Mekhemer WK, Zaghloul AA (2005) The adsorption of Cu (II) ions on bentonite – A Kinetic Study. J Coll Interf Sci 283: 316-321.
9. Safa OA, Erdem O, Ozcan BA (2004) Adsorption of acid blue 193 from aqueous solution onto Na-bentonite and DTMA-bentonite. J Coll Inter Sci 280: 44-54.
10. Grim RE (1962) "Applied Clay Mineralogy". Second Edition. McGraw Hill Book Company, New York.
11. Zhu H, Du ML, Zhou ML, Xu CS, Fua YQ, et al. (2012) Dalton Transition. J Coll Interf Surf 41: 10465.
12. Aktar M, Hasary SN, Bhanger MI, Iqbal S (2007) Sorption potential of moringa oleifera pods for the removal of organic pollutants from aqueous solutions. J Hardard Mater 141: 546-556.
13. WHO (2004) "Guidelines for Drinking Water Quality". World Health Organization Distribution and Sales, Geneva, Switzerland.
14. Yadaz N, Calimi A (2002) Alteration of three Turkish bentonites by treatment with Na2CO3 and H2SO4. Turk J Chem 26: 393-401.
15. Lagergren S (1898) Zur theorie der sogenannten adsorption geloster stoffe, Kungliga Svenska Vetenskapsakademiens, Handlingar, 24: 1-3
16. Ho YS, Mckay G (1998) Kinetic models for the sorption of dye from aqueous solution by wood. Process Safety Environ. Protect 76: 183-191.
17. Weber Jr WJ, Morriss JC (1963) Kinetics of adsorption on carbon from solution. J Sanitary Eng Div Am Soc Civ Eng 89: 31-60.
18. Jiang MQ, Jin XY, Lu XQ, Chen ZL (2010) Adsorption of Pb (II), Cd (II) Ni (II) and Cu (II) onto natural kaolinite clay. Desalin 251: 33-39.
19. Chen R, Peng F, Su S (2008) Synthesis and characterization of novel swelling tunable oligomeric poly (styrene-co-acrylamide) modified clays. J Appl Polym Sci 108: 2712-2717.
20. Oladoja NA, Aboluwoye CO, Oladimeji YB (2008) Kinetics and isotherm studies on methylene blue adsorption onto ground palm kernel coat. Turk J Eng Sci 32: 303-312.
21. Sidmara B, Maria FO, Melissa GAV (2013) Adsorption of toluene in batch system in natural clay and organoclay. Ital Assoc Chem Eng, 32: 313-318.
22. Zhou L, Chen H, Jang X, Lu F, Zhou Y, et al. (2009) Modification on montmorillonite surfaces using a novel class of cationic Gemini surfactants. J Coll Interf Sci 332: 16-21.
23. Paiva LB, Morales AR, Diaz FRV (2008) Organoclays: Properties, preparation and applications. Appl Clay Sci 42: 24.
24. Wambu EW, Muthakia GK, Wa-Thiog'O JK, Shiundu PM (2011) Kinetics and thermodynamics of aqueous Cu (II) adsorption on heat regenerated spent bleaching earth. Bull Chem Soc Ethopia 25: 181-190.
25. Kannan N, Sundaram MM (2001) Kinetics and Mechanism of removal of methylene blue by adsorption on various carbons – A Comparative Study. Dyes Pigments 51: 25-40.
26. Sun S, Wang A (2006) Adsorption kinetics of Cu (II) ions using N, O-carboxyl methyl chitosan. J hazard mater 131: 103-111.
27. Aroke UO, El-Nafaty UA, Osha OA (2014) Removal of oxyanion contaminant from waste water by sorption onto HDTMA-Br modified organo-kaolinite clay, North-Eastern, Nigeria. Inter J Emerg Technol Adv Eng 4: 475-484.
28. Wang L, Wang A (2008) Adsorption properties of congo red from aqueous solution onto surfactant-modified montmorillonite. J Hazard Mater 160: 173-180.

23

Synthesis of Mass Exchanger Network Considering Piping and Pumping Costs Using Process Integration Principles

Deepshikha Singh and Shabina Khanam*

Department of Chemical Engineering, Indian Institute of Technology Roorkee, Roorkee, India

Abstract

This paper presents a procedure to optimize mass exchanger network considering piping and pumping costs. This optimization technique is based on the combination of Pinch technology with mathematical programming approach to obtain the optimum minimum allowable composition difference (ϵ) value. In the present work three examples are taken from open literature to study the contribution and effect of piping and pumping costs on optimum ϵ value and mass exchanger network design. The results obtained for Examples 1, 2 and 3 shows that ϵ changes from specified to optimized ϵ value with improved network configuration before considering piping and pumping costs. After accounting piping and pumping costs ϵ changes from optimized to new optimized ϵ value with the changes of network configuration. Further, based on results of Examples 1, 2 and 3 it is found that piping and pumping costs contributes around 12% and 3% towards TAC of a mass exchanger network.

Keywords: Synthesis of mass exchange network; Total annualized cost; Composite interval diagram; Piping cost; Pumping cost

Nomenclature:

G	Flow rate of rich stream, kg/s or kmol/s
H	Height of column, m
D	Diameter of column, m
L	Flow rate of lean stream, kg/s or kmol/s
Nr	Real number of stages or tray
X	Composition of lean stream, (mass fraction or mole fraction)
Y	Composition of rich stream, (mass fraction or mole fraction)
m	Coefficient in equilibrium relation, Dimensionless
u_v	Actual gas velocity, (m/s)
s	Tray spacing, (m)
ρ_l	Liquid density, (kg /m^3)
ρ_v	Gas density, (kg /m^3)
D_p	Pipe size or diameter, (m)
Q	Mass flow rate of the fluid flowing into the pipe, (kg/s)
ρ	Density of the fluid flowing into the pipe, (kg/m^3)
v	Velocity of fluid in pipe, (m/s)
R	Rich stream
S	Lean Stream
CID	Composite Interval Diagram
MSA	Mass Separating Agent
TOC	Total Operating Cost
TCC	Total Capital Cost
TAC	Total Annualized Cost
MEN	Mass Exchanger Network
N_{stages}	Number of actual stages or tray
b	Constant in equilibrium relation, dimensionless
C	Unit price of lean stream, ($/kg).
U_{min}	Minimum number of units, dimensionless

Greek letters	
ε	Minimum allowable composition difference
$\int o$	Cost annualized factor, (yr-1)
υ	Specific volume of fluid, (m^3/kg)
Superscripts	
in	inlet composition
out	outlet composition
supply	supply composition
target	target composition
Subscript	
I	Rich stream
j	Lean stream

Introduction

Mass exchanger networks (MENs) are mostly used in chemical, metallurgical and allied industries for the manufacturing of chemicals and food products, recovery of valuable materials, product finishing and hazardous waste and wastewater minimization. El-Halwagi and Manousiouthakis [1] introduced the concept of MEN, using the approach of pinch design method [2]. In the later work, El-Halwagi and Manousiouthakis [3] presented a two-stage procedure for automatic synthesis of the MEN. There are many approaches for the synthesis of MEN with their relative merits and limitations. These are Mixed Integer Linear Programming (MILP) / Mixed Integer Non Linear Programming Approach (MINLP), Process Graph Theory Approach [4], State- Space Approach [5], and Genetic Algorithm Approach [6].

***Corresponding author:** Shabina Khanam, Department of Chemical Engineering, Indian Institute of Technology Roorkee, Roorkee-247667, India
E-mail: shabinahai@gmail.com

These all approaches are based on mathematical programming (MP). For example initially, LP was used in the first stage to determine the pinch point as well as the minimum cost of MSAs, while a MILP was used to minimize the number of mass exchangers in the second stage. Later developed a procedure based on MINLP to overcome the gap of the sequential procedure in the LP. In this models optimization was done based on hyperstructure, which considered many network alternatives, in order to get a minimum total annual cost (TAC) with great amount of efforts for optimizing the network hyperstructure. Due to this fact hyperstructure did not take into account the thermodynamic bottleneck of the networks. Hallale and Fraser [7] proposed the concept of Supertargeting approach based on the Pinch Technology. These above approaches such as MILP, MINLP Approach, State –Space Approach etc., unlike the supertargeting a heuristic rule based on thermodynamic approach, do not use the concept of targeting, which is a less tedious step to monitor the feasibility of the solution, before taking up the rigorous design. Hallale and Fraser [8-11] presented a method in a series of papers for targeting the TAC along with designing of MENs to meet the targets. These papers also demonstrated one important fact that using the minimum number of units did not necessarily lead to a minimum TAC of a MEN design.

Till now many researchers developed different methods for mass integration and synthesis of MENs. However, they did not account piping network as well as pressure drop in the MENs. Moreover, Peters and Timmerhaus [12] pointed out that piping is a major item in the cost of all the type of chemical process plants. These costs in a process plant can run as high as 80% of the purchased equipment cost or 20% of the fixed capital investment". It is a usage amount and should be included in the synthesis of MEN. Akbarnia et al. [13] considered piping network for the synthesis of heat exchanger network (HEN) and also proposed a correlation for the estimation of piping costs for each stream passing through the heat exchanger. This correlation was formulated for accounting piping costs of HEN based on experimental data over a range of pipe diameter for piping associated to a single heat exchanger. To calculate the total piping cost for one stream, the calculated piping cost for one heat exchanger was multiplied by the number of heat exchanger units used for that stream. However, it can be analyzed for practical cases that piping length and pipe size both will affect the piping costs, so piping length should also be considered in piping cost along with pipe diameter. However, it appears from the literature that no study is available where piping cost as function of length and diameter of pipe is considered for the synthesis of MEN. Only piping cost based on length and diameter was accounted in the optimization of hydrogen network [14]. Further, it is found through the literature that no research work is available for the synthesis of MEN considering the pressure drop or pumping costs. But for the synthesis of heat exchanger network (HEN) Polley and Panjeh Shahi [15] were first addressed the issue of allowable stream pressure drops in the conceptual design phase. Many researchers proposed several methods to incorporate pressure drop effects into the optimum design of a HEN [16-19]. Serna [20] presented a mathematical method for the optimization of HEN, by considering the effects of pressure drop, which was account ted in terms of pumping cost. Thus, based on above backdrops the aim of the present paper is to synthesize the MEN considering piping as well as pumping costs in the total capital cost.

Example 1

This example is adapted from El-Halwagi and Manousiouthakis [1], which involves simultaneous removal of Hydrogen Sulfide from two gas streams. The removal of H2S is necessary because H2S is corrosive and produces gas pollutant SO2 while combustion. For this problem two MSAs are available: a process MSA and an external MSA. The initial minimum composition difference ϵ is specified as 0.0001 (Table 1).

Example 2

This Example 2 is adopted from El-Halwagi and Manousiouthakis [1], in which the removal of SO_2 from a set of four process gas streams is considered. Water is an external MSA which is used in a system of tray columns to absorb the SO_2. The initial minimum composition difference, ϵ is specified as 5×10^{-6}. The mass exchangers are carbon steel sieve tray columns (Table 2).

Example 3

This Example 3 is adapted from [21], which involves the absorption of phenol from two aqueous waste oil streams. For this purpose two process MSA and one external MSA is needed to extract phenol from the waste oil streams. The minimum composition approach (ε) is specified equal to 0.001 (Table 3).

Detailed Solution Methodology

The solution techniques used in the present work are the combination of Pinch technology with mathematical approach. Here real number of stages, number of units, column height, column diameters and tray

Stream	G (kg/s)	Y^{in} (kmol/kmol inert)	Y^{out} (kmol/kmol inert)	ρ (kg/m³)		
R1 (Sour coke oven gas)	0.9	0.07	0.0003	0.5		
R2(Tail Gas)	0.1	0.051	0.0001	0.48		
Stream	**L (kg/s)**	**X^{in} (kmol/kmol inert)**	**X^{out} (kmol/kmol inert)**	**ρ(kg/m³)**	**m**	**Cost ($/yr)**
S1(Aqueous Ammonia)	2.3					
S2(Chilled Methanol)		0.0002	0.0035	834	0.26	176040
Capital cost data						
Installed costs Shell+ Trays ($)	$20700D^{0.57}H$+ 250e0.66DFnNr		(F_n) tray number factor varies with the Nr			
Capital annualisation factor	0.2		**Nr**		**F_n**	
E_0 (stage efficiency)	50%		25		1	
			20		1.05	
			15		1.25	
			10		1.5	
			5		2.3	
			1		3	

Table 1: Data for the Streams of Example 1.

Rich streams	G (kmol/hr)	Yin (kmol/kmol	Yout (kmol/kmol inert)	ρ(kg/m3)		
R1	50	0.01	0.004	1.09		
R2	60	0.01	0.005	1.09		
R3	40	0.02	0.005	1.09		
R4	30	0.02	0.015	1.09		
Lean streams	**L (kmol/hr)**	**Xin (kmol/kmol inert)**	**Xout (kmol/kmol inert)**	**ρ(kg/m³)**	**m**	**b**
S1	∞	0		1000	26.1	-0.00326
Capital cost data						
Installed costs Shell+ Trays ($)			6400 H $^{0.95}$ D $^{0.6}$ + 304 e $^{0.8D}$ per tray			
Capital annualisation factor			0.2			
E0 (stage efficiency)			20%			
Water cost:			$ 0.64/ton			
Operating time:			8600 h/yr			

Table 2: Stream data of rich streams and lean stream for Example 2.

Rich Streams	G (kg/s)	Y^s (kmol/kmol inert)	Y^t (kmol/kmol inert)	ρ (kg/m³)			
R1	2	0.05	0.01	1000			
R2	1	0.03	0.006	1000			
Lean Streams	L (kg/s)	X^s (kmol/kmol inert)	X^t (kmol/kmol inert)	ρ (kg/m³)	m	b	Cost ($/kg)
S1(Gas oil)	5	0.005	0.015	880	2	0	0
S2 (Lube oil)	3	0.01	0.03	930	1.53	0	0
S3 (Light oil)	--	0.0013	0.015	830	0.71	0.001	0.01
Capital cost data							
Installed costs Shell+ Trays			$ 4552 per yr per equilibrium stage per tray				
E_o (stage efficiency)			100%				
100% Operating time:			8600 h/yr				

Table 3: Data of waste streams and MSAs for example 3.

spacing, piping cost, pumping cost and the distribution of units and trays will be considered for getting accurate capital costing result. To solve the mathematical model GAMS software is used. The different steps, encountered during targeting and optimum designing of a MEN is presented in a flowchart as shown in Figure 1.

Composite interval diagram (CID) formulation

Pinch technology used to develop a composite interval diagram (CID) for finding the minimum utilities demand and the location of the pinch point at a specified ε. The equilibrium relation for the transferable component is expressed as Eq.1, which is a linear relationship between the jth process MSA scales X_j, and the ith rich stream concentration scale Y_i.

$$Y_i = m_j X_j^* + b_j \quad (1)$$

Where Xj* is the theoretically attainable maximum equilibrium composition of jth lean stream. For avoiding the infinite size of mass exchangers, is necessary to employ ϵ. Therefore, the equilibrium relation can be expressed as in Eq.2.

$$Y_i = m_j(X_j + \epsilon_j) + b_j \quad (2)$$

In a MEN, the composition of rich stream decreases, whereas lean stream composition increases. Using Eq. (2), the corresponding composition scales of the component constraints for the jth lean stream and i^{th} rich stream Yi^{*out}, Yi^{*in}, Xj^{*out} and Xj^{*in} can be expressed by Xj^{in}, Xj^{out} and Yi^{in} , Yi^{out} with Eq.(2) respectively.

CID consisting of a series of "composition intervals", which corresponds to the supply or target composition of components for each stream. Generally, the number of composition intervals can be related to the total number of streams through the following expression.

$$n < 2X (N_R + N_S) - 1 \quad (3)$$

The CID for the ith rich stream and the jth lean stream is shown briefly in Figure 2.

From this CID Excess capacity of the external MSA can be calculated; To eliminate this excess capacity new shifted mass flow rate of external MSA calculated as;

$$New\ L_l = old\ L_l - \left(\left(Excess\ capacity / y_{sl}^{target} - y_{sl}^{supply}\right)\right) \quad (4)$$

Number of trays and units target

The ideal number of trays in each interval is computed analytically using Kremser equation, Eq. 5(a). Eq.5(b).

$$NTS = \frac{\ln\left[\left(1-\frac{1}{A}\right)\left(\frac{y_i - mx - b}{y_o - mx - b}\right) + A\right]}{\ln A} \quad when \quad A \neq 1 \quad (5a)$$

$$NTS = \left(\frac{y_i - y_o}{y_o - mx - b}\right) \quad when \quad A=1 \quad (5b)$$

$$A = \frac{L_j}{M_j \times G_i} \quad (6)$$

Where Y_i, Y_o, X denote the inlet and outlet compositions of the corresponding components of the ith rich stream and the jth lean stream passing through the mass exchanger.

$$N_{real,i} = \frac{N_i}{E_o} \quad (7)$$

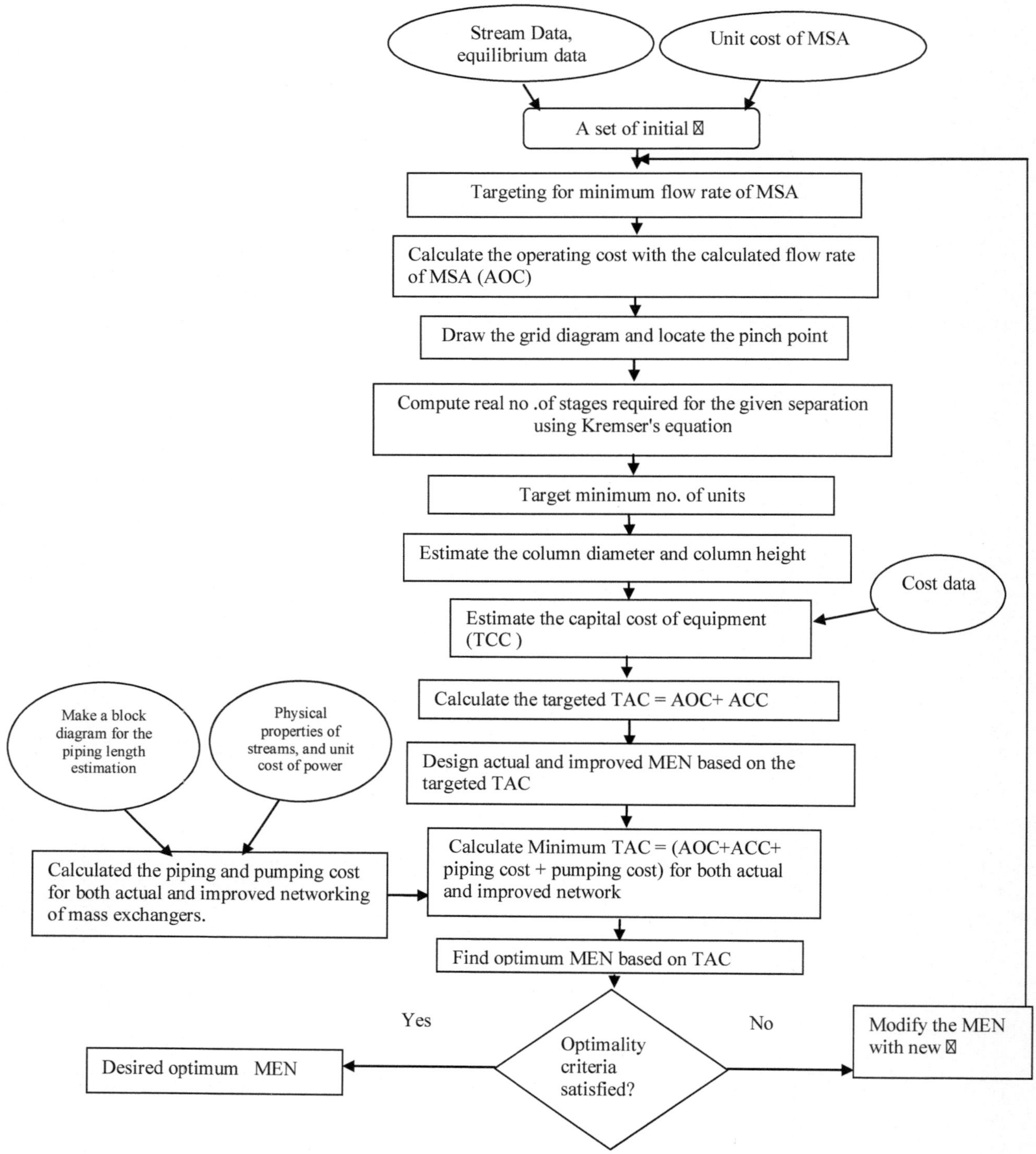

Figure 1: Flow chart of present work procedure.

The targeted minimum numbers of units are estimated by eq. 8a and 8b. These equations are applied above and below the pinch separately and then are summed up to get the total minimum number of units target required for the network.

$$U_{min,\ pinch} = U_{min,\ above\ pinch} + U_{min,\ below\ pinch} \quad (8a)$$

$$U_{min,below\ pinch=}\ U_{min,\ above\ pinch} = Ns-1 \quad (8b)$$

Column height: For each column, the height, *H*, is determined by

Figure 2: A sketch of composite interval diagram (CID).

multiplying the number of real trays, N_r, by the tray spacing (s) and adding an inactive height of 3 m to account for vapour disengagement space and a liquid sump [22,23]. For simplicity the tray spacing is assumed as 0.5 m.

Column diameter: Basically column diameter depends on the flow rates and properties of the streams passing through in to column. At the targeting stage, the MSA flow rates through each column are not yet known. However, for liquid–gas systems, the column diameter depends primarily on the gas stream [22]. In both example which are taken in this paper, the rich streams are gas streams and so each one can have an approximate column diameter assigned to it without knowing anything about the MSA flow rates. The column diameter, *D* (m), can then be calculated by the following Eq. 9.

$$D = \sqrt{\frac{4G}{\Pi \rho_v u_v}} \tag{9}$$

The maximum gas velocity to avoid excessive liquid entrapment or a high pressure drop, u_{max} (m/s), is given by Coulson [24] as:

$$u_{max} = \left(-0.711s^2 + 0.27s - 0.047\right)\sqrt{\frac{\rho_l - \rho_v}{\rho_v}} \tag{10}$$

Note that u_{max} is independent of liquid flow rate. The actual gas velocity, u_v, is taken as 80% of *u*max [22]. For simplicity s is assumed equal to 0.5 m.

$$s=0.5D^{0.3} \tag{11}$$

Thus, *s* and *D* are found by trial and error. An initial guess of 0.5 m is used for *s* and this is updated using Eq. (11) if *D* is greater than 1 m.

Design tool

By obeying the network design rules [1] we can design the actual MEN. If this actual MEN design does not give the optimum output then we have to improve this MEN design by adding one more unit in above or below the pinch point.

Piping cost estimation

Akbarnia et al. [13] presented a correlation for the piping costs estimation of a Heat Exchanger Network based on pipe diameter. However, in practical cases the piping length, piping material cost and physical properties of the fluid should also be considered in piping cost along with pipe diameter. Thus, piping cost per unit length of different pipe diameter is calculated using following expression [13], which is given by a correlation as a function of the pipe diameter and piping length;

$$Piping\ cost\ (per\ m\ length) = 3.2 + 11.42\ D_p^2 \tag{12}$$

where, D_p pipe diameter in inches. The length of piping for a mass exchanger depends on the distance between two streams, which are exchanging mass in that exchanger. For all streams pipe diameter can be calculated as:

$$D_p = \sqrt{\frac{4Q}{\Pi \rho p v}} \tag{13}$$

For MSAs streams pipe size is dependent on the variation of **ε** as the mass flow rate changes by increasing or decreasing concentrations. Therefore, we must calculate pipe size for the range of ε. The calculated pipe size shall be rounded to the nearest standard commercial pipe size such as 1/2, 3/4, 1, 1 1/2, 2, 3, 4, 6, 8, 10, 12, 14, 16, 18, 20, 22, 24, . . . inches. These pipe sizes are valid for all ranges of ε. For avoiding the hammering problem, liquid phase streams velocity is assumed low enough. So the maximum allowable velocity for lean and rich streams is assumed equal to 1.1 and 15 (m/s) respectively [13]. For annualizing piping cost, it is also assumed that interest rate *i*=10% and plant life *n*=5 years [13]. Material of construction of the pipelines is considered to be SS-304. Piping lengths are estimated with the help of flow diagram which is estimated based on MEN design configuration. For simplicity piping distance, in between the two units and storage tank to units are assumed 20 meters.

Pumping cost estimation

Once the external lean stream cost and piping cost of the modified network are computed, pumping cost for modified MEN is predicted based on the flow rates of lean streams in each connection and the required head. Pump head is equal to the height of the column in each unit. Pump and motor efficiency considered in the calculations are 65% and 90%; respectively [23]. Cost of electricity is considered as Rs. 6.67/ unit. The details of pumping cost calculation are given as below. Power consumption is calculated as;

Power cost=(flow rate × head ×density× g) / (efficiency of motor × 3.6 × 10^6) (14)

Fixed cost of pump, motors and valves

Fixed cost of pumps, motors and valves are calculated based on the flow rate of streams that has to be pumped. An approximate equation was formulated based on the values available in [23,25] as:

Cost of Pump ($)=(3.145 × f)+317.593 (15)

Cost of Motor ($)=(2.22 × f)+3.69 (16)

Cost of Valve ($)=(0.616 × f)+712.55 (17)

Where f=flow rate of liquid stream to be pumped (t/h)

Percent contribution of total pumping cost in TAC can be analyzed as:

$$\%Contribution\ of\ pumping\ cost = \frac{TAC\ with\ pumping\ cost - TAC\ without\ pumping\ cost}{TAC\ without\ pumping\ cost} \tag{18}$$

Mathematical model

Once the CID is constructed, it is modeled as a non-linear program (NLP) in order to optimize the TAC. In the CID, the entire composition

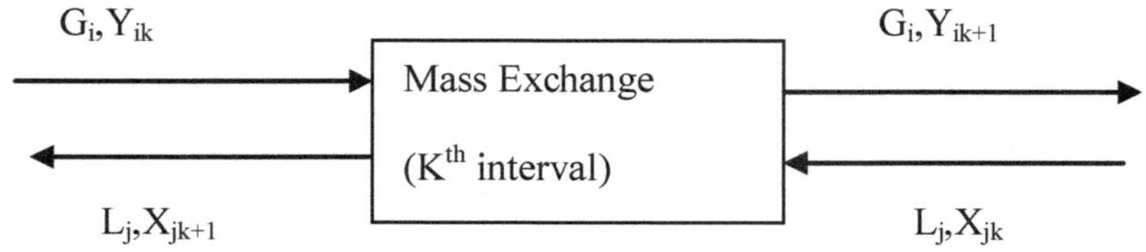

Figure 3: A mass exchanger in the kth interval of the CID.

range is supposed to be divided into *n* composition intervals, with the highest composition interval being denoted as *k*=1 and the lowest being denoted as *k*=*n*. The mass exchange sketch of the ith rich stream and the jth lean stream passing through the kth interval is denoted in Figure 3.

Objective function

In the present work, a non-linear programming (NLP) model based on CID and Nr is developed to minimize the total annual cost (TAC) of the MEN.

The model takes the TAC as the objective function, which is expressed as follows:

$$[TAC]_{min}=[cost]_{operating}+[cost]_{capital}+[cost]_{pipe}+[cost]_{pump} \quad (19)$$

The objective function contains operating cost, capital cost and piping cost for the minimum TAC of MEN.

Operating cost

The annualized operating cost (AOC) is targeted by multiplying the flow rate of MSAs, with corresponding unit costs given in Table 2. Total annualized operating cost is calculated as:

$$[cost]_{operating}(\$/\text{yr})=\sum\nolimits_j c_j S_j \quad (20)$$

Where c_j is the unit price of lean stream, generally C_j is known. L_j flow rate of MSA j.

Capital cost

For example (1) the total capital cost is computed by adding the cost of shell and trays which is the given in cost data Table 2. The following correlation is formulated for the calculation of total capital cost of the column as:

$$[cost]_{capital}(\$/\text{yr}) = \text{cost of shell}+\text{cost of trays} \quad (21)$$

Model equations

The mass load exchanged in the *k[th]* interval for the rich stream can be calculated using the following expression:

Mass exchange load for i[th] rich stream$=G_i \times (Y_{i,k}-Y_{i,k+1})=W_{i,k}$ (22)

Similarly, the mass load exchanged for the lean stream in the *k[th]* interval can be calculated as follows:

Mass exchange load for j[th] rich stream$=W_{i,k}=L_j \times (X_{j,k+1}-X_{j,k})$ (23)

Where Gi is assumed constant [1] for i[th] rich stream and Lj is the variable which we must determine using CID.

In the *k[th]* interval, the equality constraints representing successive material balance can be obtained by Eq. 24.

$$\Delta_{i,k}-\Delta_{i,k-1}=\sum\nolimits_i W_{i,k}-\sum\nolimits_j W_{j,k} \quad (24)$$

$$\Delta_{i,k}\ge 0(k=1,2,3.......n). \quad (25)$$

$$y_i^{out}\ge y_{i,n}^{out},\; x_j^{out}\ge y_{j,n}^{out},\; L_j^{max}\ge L_{i,n} \quad (26)$$

Results and Discussion

The results found for Example 1, 2 and 3 are analyzed here under:

Targeting

Minimum flow rates of MSAs

Minimum flow rates of MSAs for Example 1 are calculated using composite interval diagram (CID) [24]. From this CID the following data can be calculated;

Excess capacity of the aqueous ammonia to remove H_2S=0.00283 kg/s,

To eliminate this excess capacity new shifted mass flow rate of aqueous ammonia is calculated using Eq.4, which comes out as 1.522 kg/s.

After reducing the capacity flow rate of process MSA L_1 from 1.586 kg/s to 1.522 kg/s the improved CID is computed to obtain the minimum mass flow rate of external MSA (L_2) as:

Actual mass flow rate of aqueous ammonia L_1=1.522 × 1.45=2.207 kg/s

Minimum mass flow rate of external MSA (L_2)=0.000735 kg/s

Actual minimum mass flow rate of chilled methanol required,

L_2=0.000735/(0.0035-0.0002)=0.223 kg/s.

The pinch point at the composition of rich stream is 0.00102 and that of the lean stream is 0.0006.

Number of trays and units target

In Example 1 it is assumed that carbon steel is the construction material and sieve tray type absorption columns are used in MEN. The number of trays required for all columns in MEN, is targeted using grid diagram [9] which shows the stream population in each interval above and below the pinch point.

The ideal number of trays in each interval is computed using Kremser equation [8]. The ideal and real number of trays for accounting non-equilibrium trays in each interval are computed by the use of grid diagram. The tray contributions after rounding up and summing them for each rich stream above and below the pinch are given in column 4 of Table 4. It shows that total 70 real trays are required 48 above the pinch and 22 below it for Example 1. The targeted minimum numbers of units are estimated using method proposed by Hallale and Fraser [10] and found as four for Example 1. The total capital cost target for the network is \$532948.81. When annualised, it comes as \$106590 per year. The targeted TOC, TCC and TAC for Example 1 at different values of ε are shown in column no. 3, 4, and 5 respectively of Table 5. This cost profiles reveals the optimum ϵ value which is 0.00025 for Example 1. Corresponding the minimum ϵ (0.00025) optimum TAC is=\$398351/year.

Designing of MEN

Figure 4 shows a network designed to use targeted minimum number of units, which is four for Example 1 based on pinch design rules [9]. It shows that a poor driving force is used above the pinch which increases the number of trays by 41.4% to the targeted minimum value as shown in Figure 4. The capital cost of this design is \$647370.94, which is 21.47% above than that targeted. To improve the network design we can add one more unit in a network. Figure 5 shows a modified network design with a TAC of \$410323 per year, which is only 1.37% above the targeted TAC. This design is acceptable based on the TAC. In fact, for this design the total number of trays required is 63, which is 10% below the target. For actual as well as improved networks, number of trays and TAC at different values of ϵ is shown in Table 6 for Example 1. It shows that the minimum TAC for the desired output is achieved by preferring the improved network design at every specified ϵ values. The optimum TAC is obtained at ϵ (0.00025) when 51 numbers of plates are required. It is further noted from Table 6 that optimum ϵ value does not change when improved design is selected as final; however, it varies from 0.00025 to 0.0003 when actual network is considered.

Figure 6 represents the variation of TAC for Example 1 at different ϵ with and without piping and pumping costs consideration. This graph shows, that new optimum ϵ obtained after considering piping and pumping costs in TAC. New optimum value of ϵ for Example 1 is 0.0003 instead of 0.00025. Figure 7 shows the effect of piping and pumping costs consideration in TAC estimations on preferred network design to obtain synthesized MEN for Example 1. This shows that after considering piping and pumping costs the optimum network design obtained by preferring actual network design instead of improved network design. Because in improved network design one more mass exchanger unit is required from the targeted units. The TAC obtained after including piping and pumping costs for improved network is more due to this one more unit. By detailed cost estimation the piping and pumping costs required for improved network is approximately 31% more as compared to the piping and pumping costs required for actual network design for Example1.

Effect of piping cost on TAC

The effect of piping cost on TAC is observed after detailed capital costing for Example 1 show that piping cost affects the TAC of the MEN significantly and it alter the preferred network design as well as the optimum ϵ value of the MEN. Piping cost noticeably decreased with the increased value of ϵ. This is happened due to the fact that as ϵ value increases the lean stream phase concentration in the rich stream phase decreases. It shows that the low rich stream phase concentrations are preferable to assure maximum mass transfer rates. One way to keep low rich stream phase concentration is to use high rich stream mass/molar flow rate but here in both the Examples rich streams are gaseous streams, which are assumed to remain constant [1]. So to achieve maximum mass transfer rates mass/molar flow rate of lean streams can be reduced. As mass flow rate decreases piping size decreases and consequently the piping cost decreases. The % contribution of piping cost on TAC alone contributes around 8 to 12% in TAC for Example 1. The piping costs are a fraction of total costs, typically not of the same order of magnitude as the major equipment. These costs become more

MAC (ε)	Target				Actual network (A) U_{min}=4			Improved network (I) U_{min}=5			Preferred Network design
	Nr	TOC (\$/yr)	TCC	TAC	Nr	TCC (\$/yr)	TAC (\$/yr)	Nr	TCC (\$/yr)	TAC (\$/yr)	
0.0001	70	298192	106590	404782	99	129474.2	427667	63	112130	410323	I
0.00015	64	301780	99669	401449	89	118830	420610	57	104546	406326	I
0.0002	59	305368	93834	399201	83	112803.4	418171	55	100794	406161	I
0.00025	55	308955	89396	398351	78	107521.3	416477	51	97105	406060	I
0.0003	54	312543	88674	401217	73	102256	414799	48	94975	407518	I
0.00035	50	316131	84953	401084	70	99242.73	415373	48	93383	409513	I
0.0004	49	319718	82728	402446	70	99242.73	418961	46	91151	410870	I
0.00045	48	323306	82067	405372	65	93943.12	417249	43	87386	410692	I
0.0005	47	326894	80492	407385	65	93943.12	420837	43	87386	414280	I

Table 4: Preferred network design based on TAC of MEN

Example	Factors	Optimum (ϵ)			MEN Design		Contribution in TAC%
		Specified	Target	With Factor	Target	With Factor	
1	Pumping Cost	0.0001	0.00025	0.00025	Improved	Improved	0.5-1
	Piping Cost	0.0001	0.00025	0.0003	Improved	Actual	8-12
	With piping and pumping	0.0001	0.00025	0.0003	Improved	Actual	
2	Pumping Cost	0.000005	0.000005	0.0000455	Improved	Actual	2-3
	Piping Cost	0.000005	0.000005	0.0000455	Improved	Actual	7-9
	With piping and pumping	0.000005	0.000005	0.0000455	Improved	Actual	
3	Pumping Cost	0.001	0.00075	0.00035	Improved	Actual	1-2
	Piping Cost	0.001	0.00075	0.00035	Improved	Actual	1-2
	With piping and pumping	0.001	0.00075	0.00035	Improved	Actual	

Table 5: Consolidated result of 3 taken example.

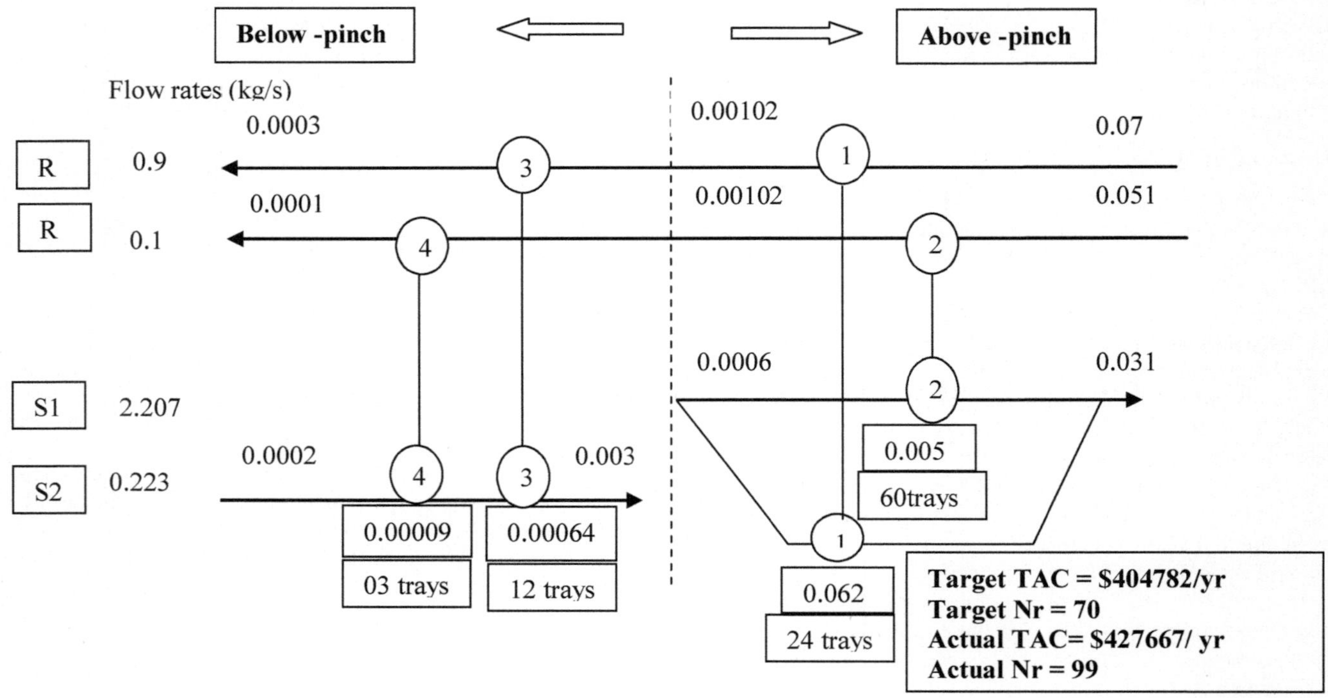

Figure 4: Actual network design at specified ϵ=0.0001.

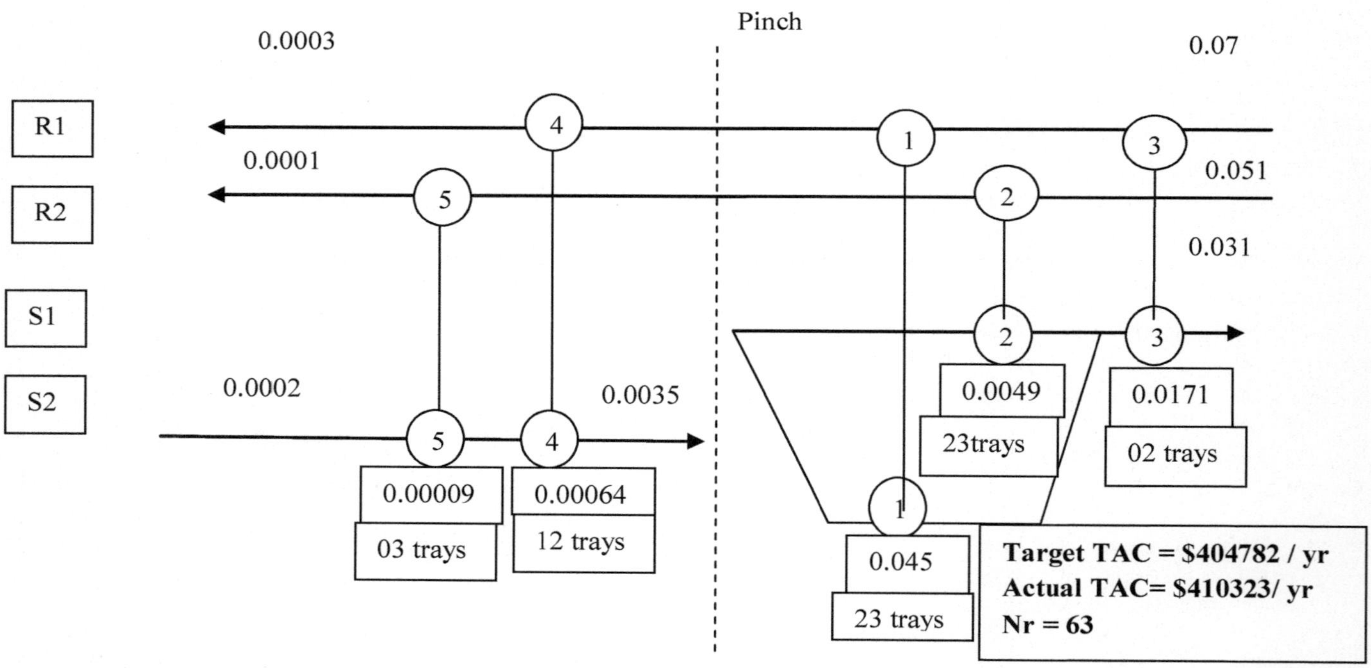

Figure 5: Improved network design at initially specified value of ϵ=0.0001.

Author	Process used	(ε)	Nr	TOC *10⁴	TCC ($/yr)	TAC *10⁴ ($/yr)	Piping cost ($/yr)	Pumping cost ($/yr)	TAC ($/yr)
El-Halwagi et al. [1]	Pinch Analysis specified, ϵ	0.0001	50	29.844	227600	52.604	--	--	--
Hallale et al. [10]	Pinch analysis	0.0001	50	29.844	227600	52.604	--	--	--
Hallale et al. [7]	Super-target method	0.00031	25	29.844	113800	42.706	--	--	--
Cheng-Liang et al. [23]	MINLP	0.0001	25	31.59	113800	42.97	--	--	--
Hallale and Fraser part 2 [11]	Detailed capital costing models, specified, ϵ	0.0001	63	29.8 44	112415	41.085	--	--	--
Present work	CID with MP, specified, ϵ	0.0001	63	29.8192	112130	41.032	47186	2211.2	463602
Present work CID	CID with MP, optimized, ϵ	0.00025	51	30.896	107521	40.606	32552	1525.4	450554
Present work	CID with MP, after including piping cost	0.0003	73	31.254	102256	40.752	32544	----	447343
Present work	CID with MP, after including pumping cost	0.00025	51	30.896	97105	40.606	---	1907.82	407968
Present work	CID with MP, after including piping and pumping costs	0.0003	73	31.254	102256	40.752	32544	1509.1	448852

Table 6: Comparison of results of present work with that of published work (Example 1)

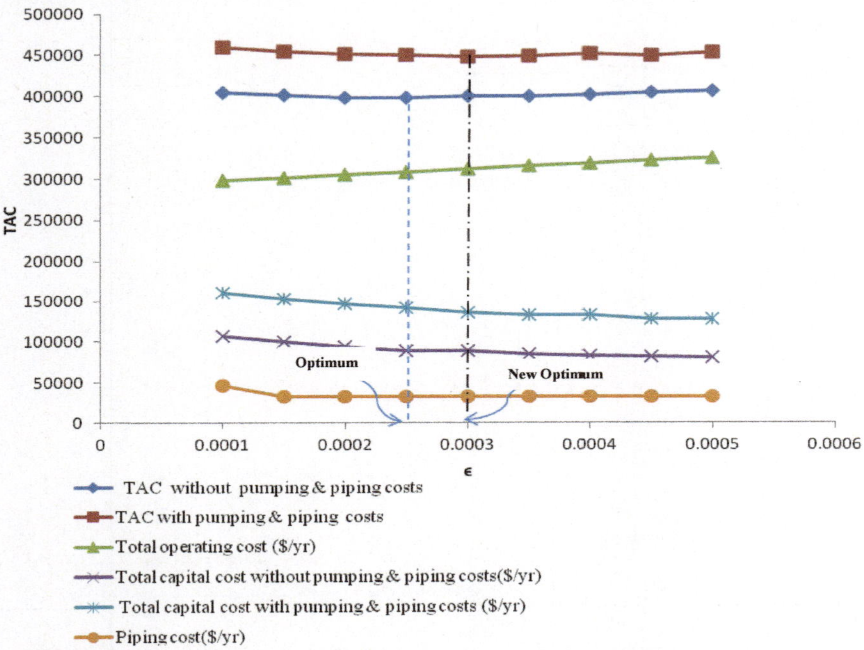

Figure 6: Results of supertargeting with and without piping and pumping costs for example 1.

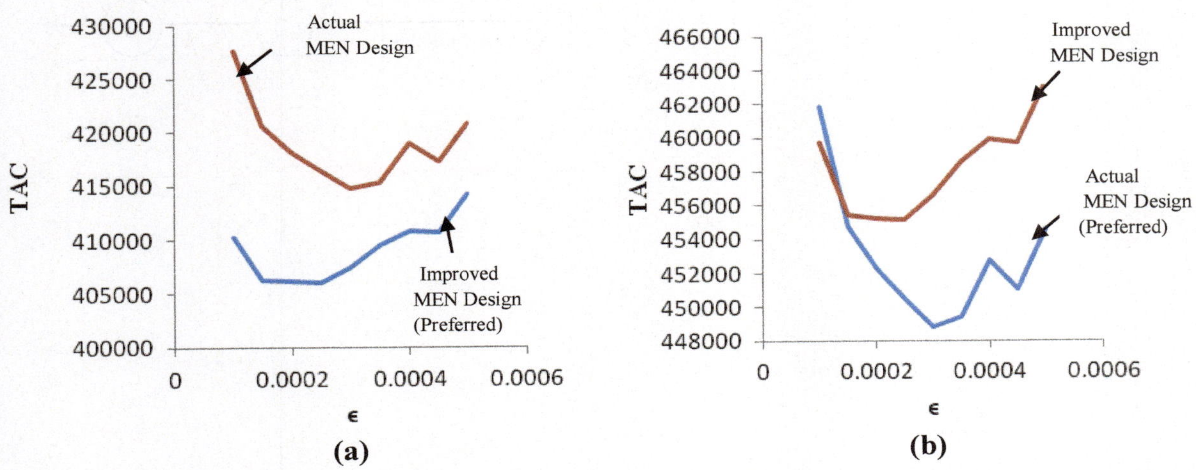

Figure 7: Comparison of the results found (a) without and (b) with considering piping and pumping costs in TAC estimation on preferred MEN design (Example 1).

important when piping dominates most of the equipment, such as water distribution networks. Hence piping cost is an important factor and must be considered in the design of MEN.

Effect of pumping cost on TAC

The effect of pumping cost on TAC is analyzed after detailed costing that pumping cost consideration in TAC is not affecting the preferred network design and the optimum ϵ values for Example 1. The % contributions of pumping cost in TAC shows that pumping cost alone contributes very less its around 0.5% in TAC for Example 1. Figure 8 shows the variation of pumping cost with a range of ϵ values for the Example 1. The pumping cost is noticeably decreased with the increased values of ϵ. This is happened due to the fact that as ϵ increases lean stream flow rate decreases and by the detailed cost estimation pumping cost decreases because it is mainly depend on the stream flow rate.

Similarly following the same targeting and designing tools to solve the Example 2 and 3 and the final results obtained are shown in Figures 9 and 10 for Example 2 and in Figures 11 and 12 for Example 3.

For comparison the final result obtained in present work are compared with the published work as shown in Tables 6-8 for Example 1, 2 and 3 respectively. From these tables it can be seen that no one considered piping and pumping costs for the optimization of MEN. The values of ϵ specified and optimized by other researchers are presented in the third column of the table. The different approaches used to solve these examples are presented in the second column. The method presented in the present work gives more convenient and more precise results by the other methods. The optimum TAC obtained in the present work before considering pumping and piping costs is $ 406060 per year which is 1.2% below the TAC obtained by Hallale

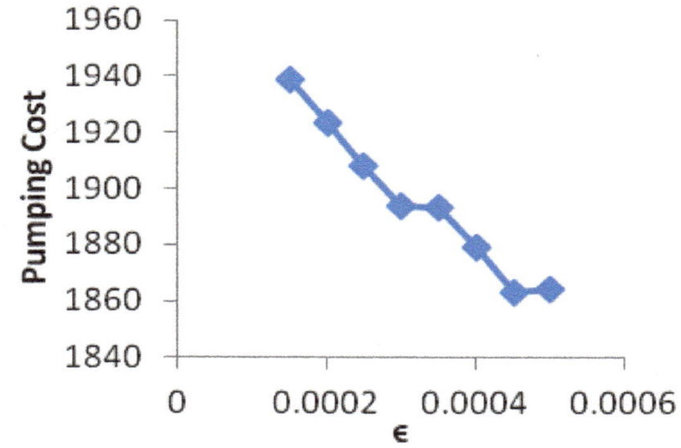

Figure 8: Pumping cost variation with the range of ϵ for Example 1.

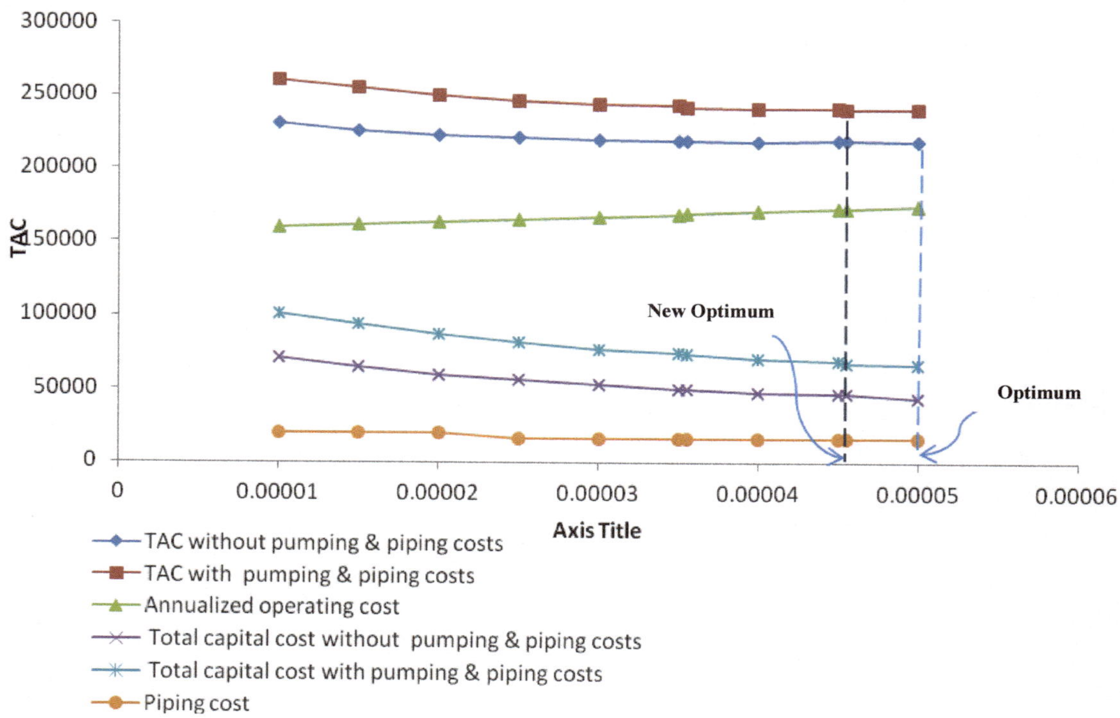

Figure 9: Results of supertargeting with and without piping and pumping costs (Example 2).

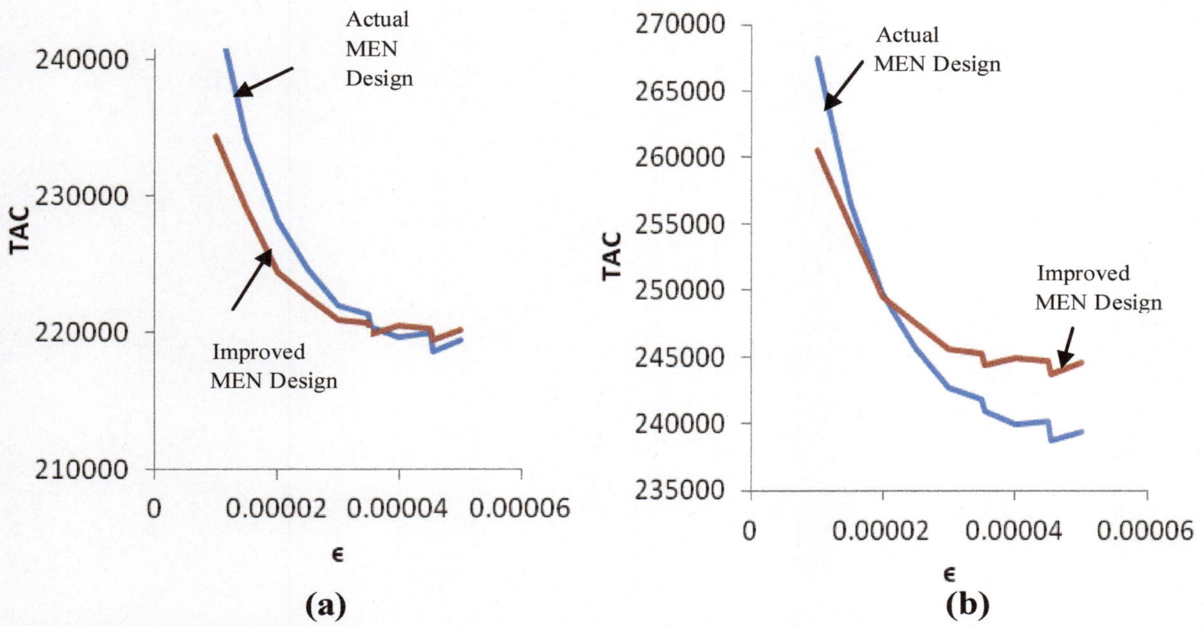

Figure 10: Comparison of the results found (a) without and (b) with considering piping and pumping costs in TAC estimation on preferred MEN design (Example 2).

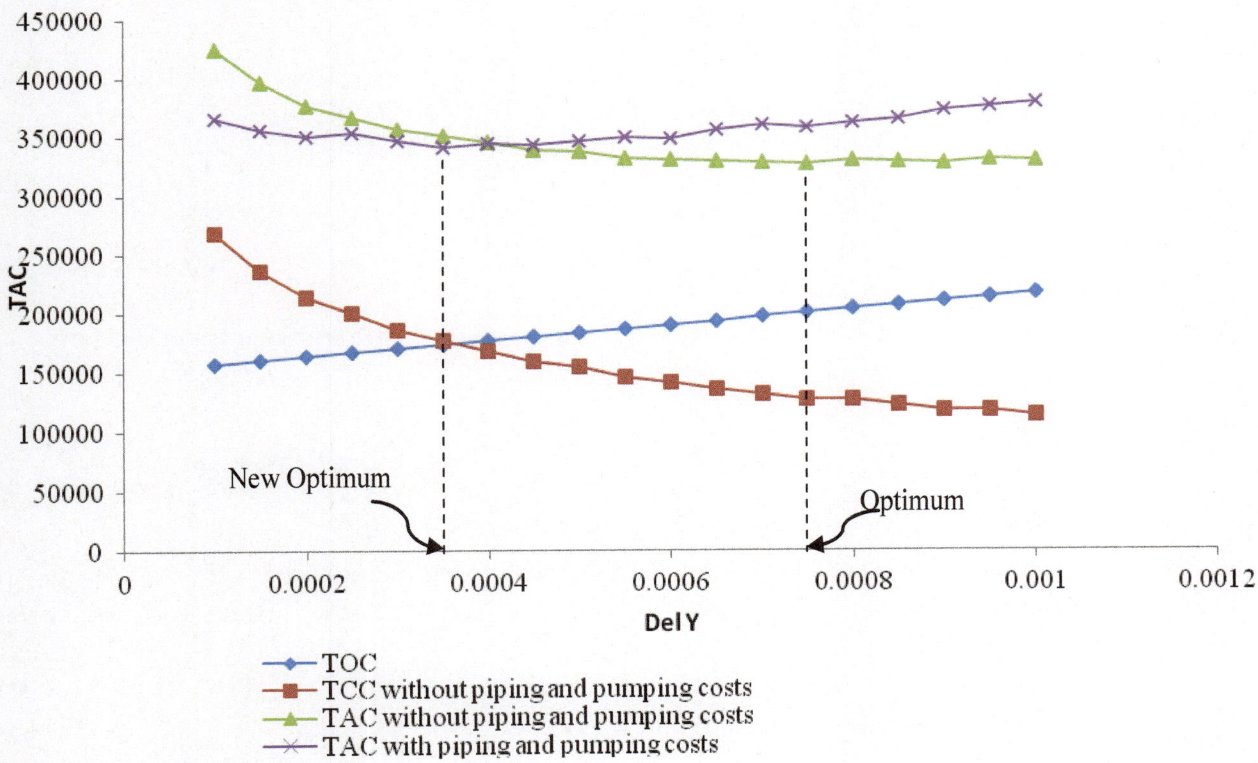

Figure 11: Results of supertargeting with and without piping and pumping costs (Example 3).

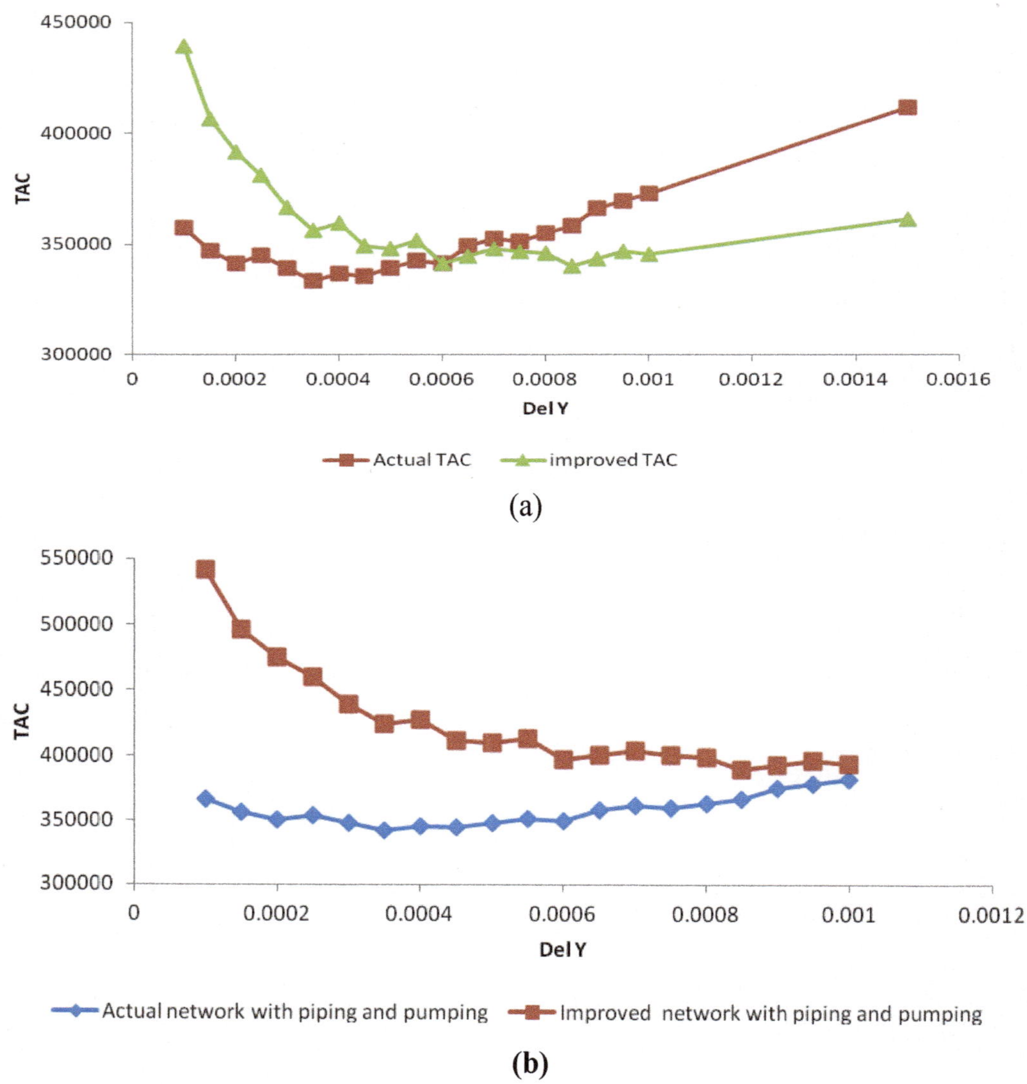

Figure 12: Comparison of the results found (a) without and (b) with considering piping and pumping costs in TAC estimation on preferred MEN design (Example 3).

Author	Process used	MAC (ε)	L_{tot} kmol/h	Nr	TOC ($/yr)	TCC ($/yr)	TAC ($/yr)	Piping cost ($/yr)	Pumping cost ($/yr)	TAC ($/yr)
N Hallale and DM Fraser [9]	Pinch Analysis specified, ϵ	0.000005	1593	140	157962	86000	243962	--	--	--
N Hallale and DM Fraser [9]	Pinch Analysis optimized, ϵ	0.00005	1749	65	178311	40689	219000	--	--	--
Present work	CID with MP, specified, ϵ	0.00005	1590	140	157555	86020	243576	177560	7175	428310
Present work	CID with MP, optimized, ϵ	0.00005	1747	65	173034	46352	219386	158315	4154	381855
Present work	CID with MP, after including piping cost	0.0000455	1730	67	171351	47313	218608	158312	---	376920
Present work	CID with MP, after including pumping cost	0.0000455	1730	67	171351	47313	218608	---	8438	227046
Present work	CID with MP, after including piping and pumping cost	0.0000455	1730	67	171295	47313	218608	158312	8438	385358

Table 7: Summary and comparison of final result with the published work for example 2.

Author	Process used	Minimum composition approach	U_{min}	Nr	TOC ($/yr)	TCC ($/yr)	TAC ($/yr)	Piping cost ($/yr)	Pumping cost ($/yr)	TAC (Rs/yr)
Hallale [10]	Detailed capital costing models, specified, ϵ	0.001	7	28	217960	159320	345416	--	--	--
Present work	CID with MP, specified, ϵ	0.001	7	28	218225	127456	345681	3985	5116	354782
Present work	CID with MP, after including piping cost	0.00035	6	35	174158	159320	333478	3985	--	354782
Present work	CID with MP, after including pumping cost	0.00035	6	35	174158	159320	333478	--	5146	338624
Present work	CID with MP, after including piping and pumping cost	0.00035	6	35	159320	333478	3918	5146	342542	159320

Table 8: Summary and comparison of example 3 final result with the published work.

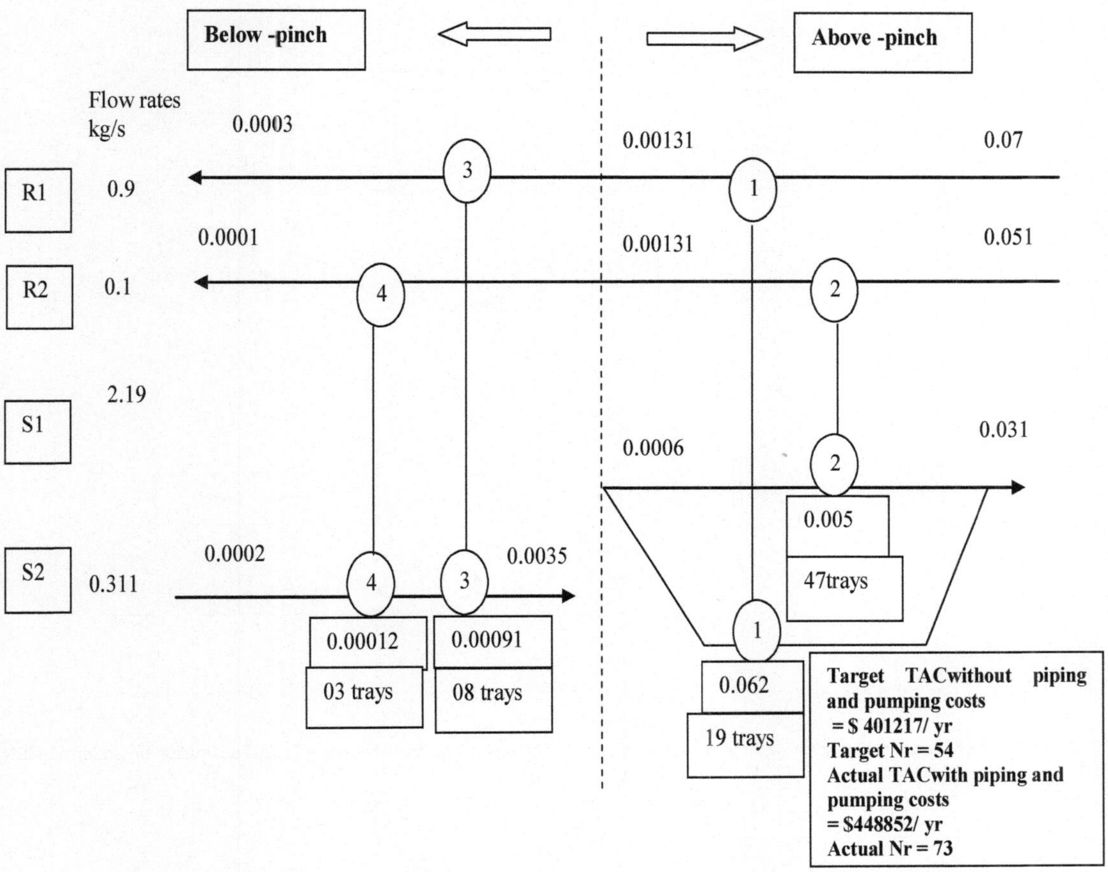

Figure 13: Final network design for example 1 at optimal value of ϵ=0.0003.

[10] at specified ϵ value. After including piping and pumping costs in TAC the optimum MEN is obtained at (0.0003) ϵ which is 17% above the optimized ϵ value given by Hallale [10]. At (0.0003) ϵ the TAC obtained by present method is smallest than the other results which are presented in column 7 of Table 4. Similarly from Table 5 it can be seen that the optimum TAC obtained in the present work is $219386 per year for Example 2 before considering piping and pumping costs, which is 0.2% above the TAC given by Hallale [9]. After considering piping and pumping costs the optimum result is obtained at (0.0000455) ϵ which is 9% below the ϵ value given by Hallale [9]. At optimized ϵ the TAC of the MEN is $407520 per year which is also the smallest cost than the other methods. From Table 8 it can be seen that the optimum TAC obtained by the present method before considering pumping and piping costs is $ 345681 per year which is obtained at 0.001 (ϵ) it is only 0.1% above the result obtained in Hallale and Fraser [10]. After considering pumping and piping costs the optimum TAC is achieved at ϵ equal to 0.00035 which is 65% below the specified value and optimum TAC is $ 342542 /yr. According to the calculating results, the optimal

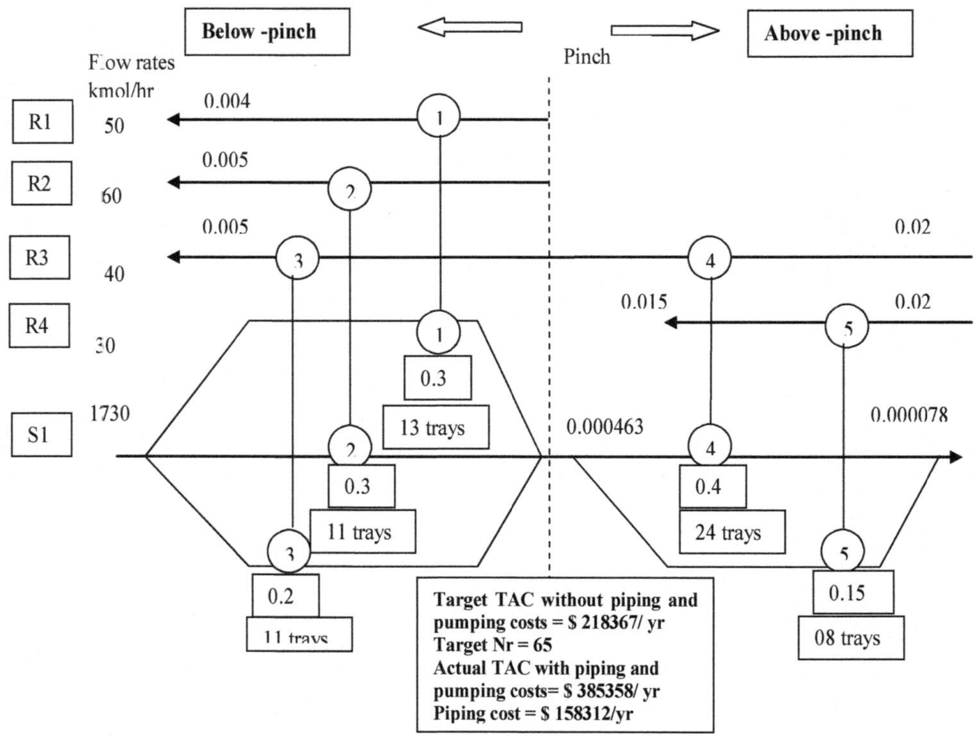

Figure 14: Final network design for example 2 at optimal value of ϵ=0.0000455.

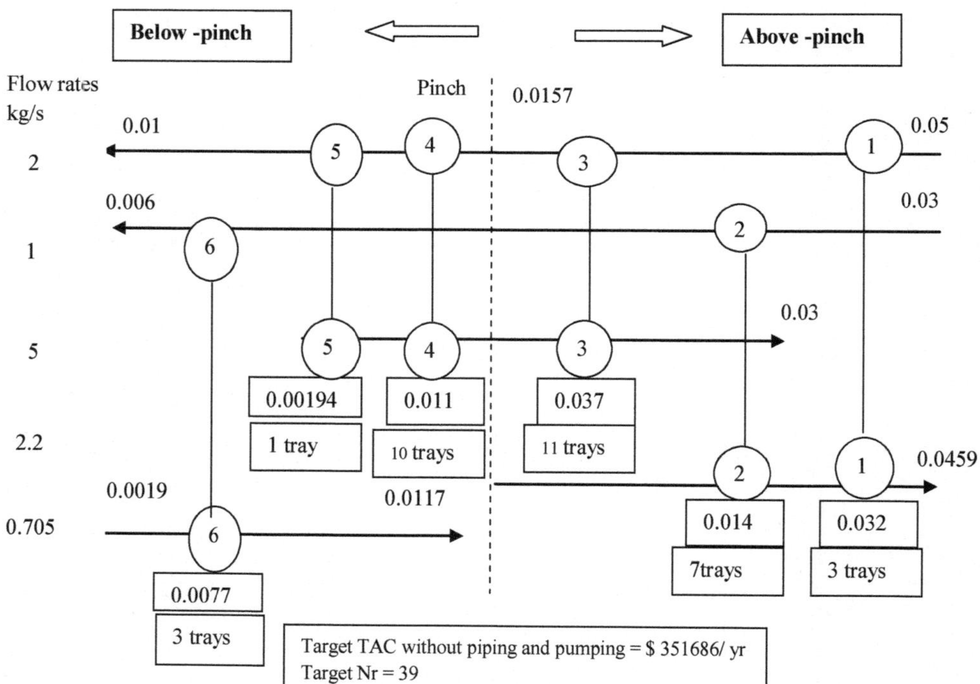

Figure 15: Final network design for example 3 at optimal value of ϵ=0.00035.

MEN can be drawn as presented in Figures 13-15 for Example 1, 2 and 3 respectively.

Conclusion

The optimization procedure presented in this study is cost effective. It is obvious that the optimization of ε values is highly important to synthesize a MEN. Less than 1% contributions of piping and pumping cost in TAC of a MEN does not affect the network design and optimum value of minimum allowable composition difference for any type of MEN problems. But more than 1% contribution of piping and pumping cost will alter the preferred network design and affects the optimum ϵ value. The inclusion of piping as well as pumping cost in TAC gives more realistic results.

Acknowledgements

The facility provided by the Indian Institute of Technology Roorkee is gratefully acknowledged.

References

1. Halwagi EMM, Manousiouthakis V (1989) Synthesis of mass exchange networks. AIChE Journal 35: 1233-1244.
2. Linnhoff B, Hindmarsh E (1983) The pinch design method for heat exchanger networks. Chemical Engineering Science 38: 745-763.
3. Halwagi EMM, Manousiouthakis V (1990) Automatic synthesis of mass-exchange networks with single-component targets. Chemical Engineering Science 45: 281-2831.
4. Lee S, Park S (1996) Synthesis of mass exchange network using process graph theory. Computers & chemical engineering 20: S201-S205.
5. Bagajewicz MJ, Robert P, Manousiouthakis V (1998) On the state space approach to mass/heat exchanger network design. Chemical Engineering Science 53: 2595-2621.
6. Garrard A, Fraga ES (1998) Mass exchange network synthesis using genetic algorithms. Computers & chemical engineering 22: 1837-1850.
7. Hallale N, Fraser D (2000) Supertargeting for mass exchange networks: Part I: Targeting and design techniques. Chemical Engineering Research and Design 78: 202-207.
8. Hallale N, Fraser D (2000) Supertargeting for mass exchange networks: Part II: applications. Chemical Engineering Research and Design 78: 208-216.
9. Hallale N, Fraser D (1998) Capital cost targets for mass exchange networks A special case: Water minimisation. Chemical Engineering Science 5: 293-313.
10. Hallale N, Fraser D (2000) Capital and total cost targets for mass exchange networks: Part 1: Simple capital cost models. Computers & Chemical Engineering 23: 1661-1679.
11. Hallale N, Fraser D (2000) Capital and total cost targets for mass exchange networks: Part 2: Detailed capital cost models. Computers & Chemical Engineering 23: 1681-1699.
12. Peters MS, Timmerhaus KD, West RE, Timmerhaus K, West R (1968) Plant design and economics for chemical engineers. McGraw-Hill, New York.
13. Akbarnia M, Amidpour M, Shadaram A (2009) A new approach in pinch technology considering piping costs in total cost targeting for heat exchanger network. Chemical engineering research and design 87: 357-365.
14. Hallale N, Liu F (2001) Refinery hydrogen management for clean fuels production. Advances in Environmental Research 6: 81-98.
15. Polley GT, Panjeshahi MH (1991) Interfacing heat exchanger network synthesis and detailed heat exchanger design. Trans IcemE.
16. Frausto HS, Rico RV, Jiménez GA, Hernández CS (2003) MINLP synthesis of heat exchanger networks considering pressure drop effects. Computers & chemical engineering 27: 1143-1152.
17. Panjeh SM, Khoshgard A (2006) Heat exchanger networks targeting and design with unequal heat transfer coefficient regarding allowable pressure drop of streams. Heat transfer engineering 27: 36-43.
18. Serna GM, Ponce OJM, Jiménez GA (2004) Two-level optimization algorithm for heat exchanger networks including pressure drop considerations. Industrial & engineering chemistry research 43: 6766-6773.
19. Zhu X, Nie X (2002) Pressure drop considerations for heat exchanger network grassroots design. Computers & chemical engineering 26: 1661-1676.
20. Serna GM, Ponce OJM, Burgara MO (2010) Total cost trageting for heat exchanger networks including pumping costs. Computer Aided Chemical Engineering 28: 1135-1140.
21. Halwagi MM (1997) Pollution prevention through process integration: systematic design tools.
22. Coulson J, Richardson J (1991) Chemical engineering–particle technology and separation process. Pergamon Press, Oxford.
23. Dakwala M, Mohanty B, Bhargava R (2009) A process integration approach to industrial water conservation: a case study for an Indian starch industry. Journal of Cleaner Production 17: 1654-1662.
24. Younes N, Abdallah SM, Alim MA (2014) Fuzzy approach for the synthesis of mass exchange network. International Journal of Computer Applications 104: 5-11.
25. Ulrich GD (1984) A guide to chemical engineering process design and economics. Wiley, New York.

Production of Hydrogen and Aluminum Hydroxide by Electrochemical Dispersion of Aluminum Alloys D16 in Sodium Hydroxide Aqueous Solution

Reznichenko AV*, Rybalchenko VV, Badaev FZ, Ponomarev SG, Vasin AA and Smirnov AD

Federal State Educational Institution of Higher Education, Moscow Polytechnic University (Moscow Polytech), Moscow, Russia

Abstract

Kinetics experimental results of the interaction of an aluminum alloy D16 with a sodium hydroxide aqueous solution under the alternating electric current are presented. It is stated that the reaction rate under the AC is much larger than the reaction rate without applying an electric current. The use of an alternating electric current makes it possible to obtain smaller particles of $Al(OH)_3$ with a narrower particle size distribution range.

Keywords: Aluminum; Aluminum alloys; Aluminum hydroxide; Hydrogen production; Sodium hydroxide solution; Ceramics

Introduction

One of the most promising ways to solve the problems of depletion of natural energy carriers and pollution of the environment as a result of their use may be the transition to alumina hydrogen power [1]. To organize the wide use of hydrogen as fuel, not only hydrogen production is needed, but also the accumulation, storage and transportation to places of consumption, which causes significant technical problems. These problems can be solved by using metallic aluminum as an energy carrier at the point of consumption, that is, on autonomous power plants and mobile generators of hydrogen production, for example, on a car board [2].

At normal conditions aluminum is coated with a thin oxide film thus is inert. The oxide film on the aluminum surface is easily removed by reaction with an aqueous solution of sodium hydroxide:

$$Al_2O_3+2NaOH+3H_2O=2Na\,[Al(OH)_4] \qquad (1)$$

Without the oxide film, the interaction of aluminum with the alkali solution is described by the reaction:

$$2Al+2NaOH+6H_2O=2Na\,[Al(OH)_4]+3H_2 \qquad (2)$$

At the point of consumption, hydrogen generated from this reaction can be used to produce energy without storing, storing and transporting it. The easily soluble complex compound $Na[Al(OH)_4]$ decomposes reversibly according to the reaction:

$$Na[Al(OH)_4] \rightleftarrows Al(OH)_3 \rightleftarrows +NaOH \qquad (3)$$

Thus, sodium hydroxide is not eliminated from the reaction sphere. According to these reactions, the products of the interaction of aluminum and its alloys with an aqueous solution of sodium hydroxide are hydrogen and aluminum hydroxide $Al(OH)_3$. On the one hand, the obtained aluminum hydroxide can be used for the production of metallic aluminum, on the other hand, it can produce an ultradisperse aluminum oxide powder widely used as a raw material for the production range of materialssuch as ceramics, composites, electrical insulating, abrasives, adsorptives, catalytics, medical and others [3,4].

The kinetics of the interaction of aluminum and its alloys with aqueous hydroxides solutions of alkaline elements has been studied in detail in a number of papers [5,6]. In these studies, the influence on the speed of the process of alloying alloy components has been determinated [7-9].

Using hydrogen as an energy carrier in mobile systems, the issue of controlling the hydrogen evolution rate arises. In the author's opinion, one of the ways to control the rate of the hydrogen evolution reaction can be the effect of electric current on the process of aluminum dispersion.

The aim of this work is to study the influence of alternating electric current on the process of chemical dispersion of aluminum alloy D16 in a sodium hydroxide solution and on the properties of the reaction products.

Experimental Method

The reaction rate of an aluminum-based alloy in a sodium hydroxide solution was determined by a volumetric method by the volume of hydrogen evoluted at 25°C.

Disk ssamples with the same surface area, were made of alloy D16 and connected to the electrode-holder. Samples were put in pairs in a thermostated reactor, which was a conical flask containing 0.5 L sodium hydroxide solution at a concentration of 1 mole/litre. The reactor was hermetically connected to a volumetric system.

Beforehand, without passing an alternating electric current, 2 samples connected to the electrodes and contained in a solution of sodium hydroxide were chemically dispersed, and the volume of hydrogen gas evaluated was measured. Then, the same samples in the same solution were dispersed by an alternating current of 50 Hz, a voltage of 6.4 V and a current strength of 0.3 A and 1.5 A, and the volume of hydrogen evaluated was also measured.

The reaction rate r was calculated from the equation:

$$r=\frac{\Delta V}{S\cdot\Delta t}$$

where ΔV is the volume of hydrogen evoluted in time Δt, S is the area of the electrode.

***Corresponding author:** Reznichenko AV, Federal State Educational Institution of Higher Education, Moscow Polytechnic University (Moscow Polytech), Moscow, Russia, E-mail: reznichenko@sde.ru

Results and Discussion

Kinetic curves of hydrogen evolution under different experimental conditions are shown in Figure 1.

In the process of electrolysis of an aqueous sodium hydroxidesolution NaOH with electrodes made of an aluminum-based alloy, the following reactions occur:

Cathode (-):

$2H_2O+2\bar{e}=2OH^-+H_2$;

Anode (+):

$Al+4OH^- -3\bar{e}=[Al(OH)_4]^-$

In the process of electrolysis (comparing with the process without the flow of electric current), the rate of aluminum dispersion increases, as it is subjected to disperssion according to the reactions as electrochemical:

Anode (+): $Al+4OH^- -3\bar{e}=[Al(OH)_4]^-$, and chemical:

$2Al+2NaOH+6H_2O=2Na\ [Al\ (OH)_4]+3H_2$

1 - Chemical dispersion; 2 - passing an alternating electric current I=0.3 A; 3 - passing an alternating electric current I=1.5 A.

The consumption of OH^- ions during electrochemical dispersion is completely compensated by their formation by the cathodic reaction proceeding at the same electrode when the electric pole is changed:

Cathode (-): $2H_2O+2\bar{e}=2OH^-+H_2$

The flowing cathodic process of water recovery on the aluminum electrode causes additional hydrogen evolution and an increase in the total rate of its release. On the cathode, simultaneously with the evolution of gaseous hydrogen, OH^- ions are formed, which causes a suddenly increase in their concentration in the near-electrode layer. In the same layer, the concentration of Al^{3+} ions suddenly increases due to the anodic reaction occurring in the next phase of the alternating current. The Al^{3+} and OH^- ions react with the formation of a complex ion $[Al(OH)_4]^-$, which decays to form $Al(OH)_3$ aluminum hydroxide. Consequently, when electric current is passed through the solutionincreasesnot only the rate of hydrogen gas evolution, but also the dispersion rate of aluminum and, accordingly, the rate of synthesis of $Al(OH)_3$.

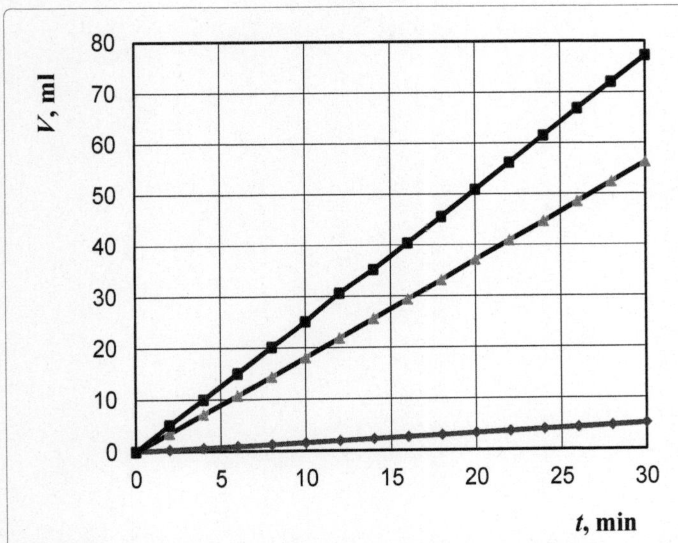

Figure 1: Dependence of the volume of evoluted hydrogen V per 1 cm^2 surface area versus time t.

With electrochemical dispersion at 0.3 Acurrent strength, the rate of hydrogen evolution compared to chemical dispersion increases approximately 5 times and is 9 l/(m^2·min). With 1.5 A current strength, the hydrogen evolution rate is approximately 12 l/ (m^2 min). In the same way, the speed of dispersion of the alloy increases.

It is known that when precipitating from solutions of high concentration and under conditions of sufficiently low temperatures, ultradispersed and slowly settling particles are formed [10]. In the process of electrochemical dispersion of aluminum and its alloys, the concentration of both Al^{3+} ions and OH^- ions in the near-electrode layer increases significantly in comparison with chemical dispersion. For this reason, in our opinion, in the electrochemical dispersion of an aluminum alloy, products in the form of $Al(OH)_3$ with more dispersed particles should be formed.

The particle size was studied on a FRITSCH Analysette 22 Nanotec plus laser particle analyzer. The results of the studies are presented in Table 1.

It has been stated that in the $Al(OH)_3$ products obtained by chemical dispersion, 90% are particles with a size of less than 29 μm. In this case, a four-modal particle size distribution is observed (Figure 2).

In products of electrochemical dispersion with 0.3 A alternating electric current, 90% particles are smaller than 18 μm. In this case, a narrower three-modal particle size distribution is observed (Figure 3). When the AC power is increased to 1.5 A, 90% of the dispersion products of the D16 alloy are

Impact type	d_{10}, μm	d_{50}, μm	d_{90}, μm
Chemical dispersion	1.8	18	29
Electrochemical dispersion 0.3 A	3.7	8.3	18
Electrochemical dispersion 1.5 A	1.8	5	11

*Here the value of dn is such that the number of particles with a characteristic size less than dn is n% of all particles

Table 1: Granulometric characteristics of powders.

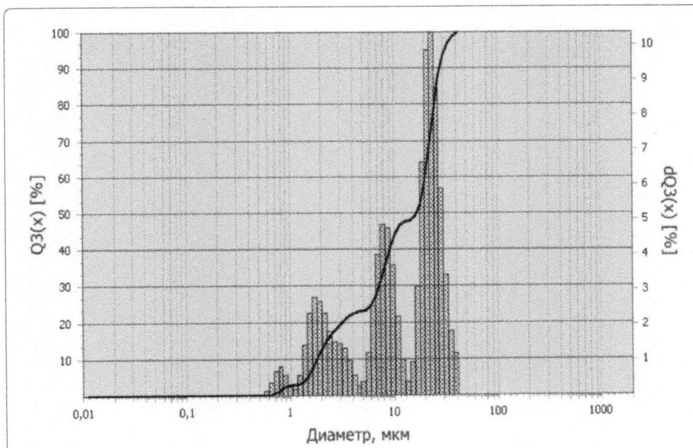

Figure 2: Volumetric distribution of particles in products of chemical dispersion of alloy D16.

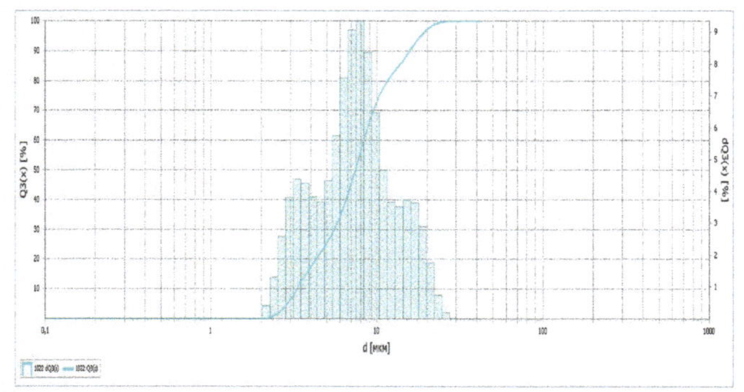

Figure 3: Volumetric distribution of particles in products of electrochemical dispersion of alloy D16 (current strength 0.3 A).

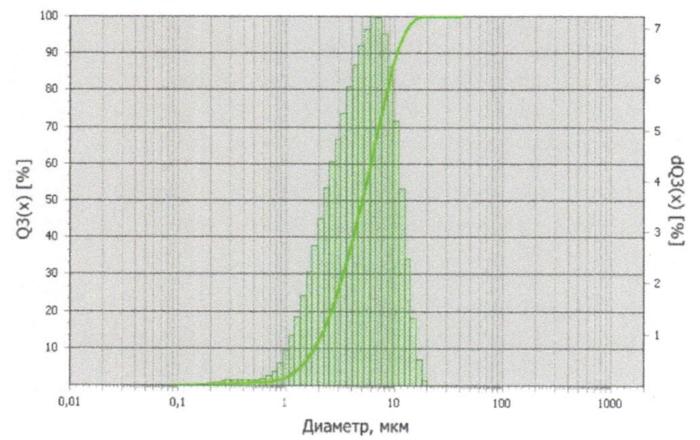

Figure 4: Volumetric distribution of particles in products of electrochemical dispersion of alloy D16 (current 1.5 A).

particles smaller than 11 µm. An even narrower single-modal particle size distribution is observed in this case (Figure 4).

Conclusions

1. The possibility of accelerating the process of dispersion of aluminum and its alloys in an aqueous solution of sodium hydroxide at a concentration of 1 mole/litre and controlling the rate of generation of hydrogen gas by an alternating electric current has been experimentally studied and established.

2. Electrochemical dispersion of aluminum alloy D16 produces aluminum hydroxide with particles of smaller size than aluminum hydroxide during chemical dispersion of this alloy.

3. An increase in the AC power from 0.3 to 1.5 A leads to a decrease in the average particle size. The distribution by particle size becomes narrower.

References

1. Zhuk AZ, Kleimenov BV (2007) Aluminous hydrogen power engineering. Sheindlin AE (ed.), Moscow: OIVT RAS, p: 278.
2. Kozlyakov VV, Stukalova NP, Omarov AY (2009) Methods of obtaining hydrogen as fuel for cars Izvestiya Moskovskogo State Industrial University 4: 35-42.
3. Romanova RG, Dresvyannikov AF, Abdullina AR (2012) Development of compositions of ceramic materials using nano-sized precursors. Bulletin of Kazan Technological University 16: 276-280.
4. Badaev FZ, Tarasovskii VP, Novoselov RA (2015) Preparation of Aluminum Oxide Ceramic Precursors by Dispersion of Aluminum Alloy D16. Refractories and Industrial Ceramics 56: 11-13.
5. Lurie BA, Chernyshov AE, Perova NN, Svetlov BS (1976) Kinetics of interaction of aluminum with water and aqueous solutions of alkalies. Kinetics and Catalysis 6: 1453-1458.
6. Sysoeva VV, Artjugina ED, Gorodilova VG, Berkman EA (1985) On the problem of the corrosion of aluminum in alkalies. Journal of Applied Chemistry 4: 921-924.
7. Khairi AK, Badaev FZ, Omarov AY, Ayrich AI (2012) Investigation of the kinetics of the interaction of aluminum-magnesium alloys with an aqueous solution of sodium hydroxide. Proceedings of the Moscow State Industrial University 25: 42-45.
8. Badaev FZ, Khayri AK, Kasatova NA, Ayrikh AI (2013) Kinetics of interaction of aluminum-silicon alloys with aqueous sodium hydroxide solution. Mechanical Engineering and Engineering Education 35: 44-48.
9. Badaev FZ, Khayri AK, Novoselov RA, Tarasovsky VP (2015) Kinetic features of the interaction of aluminum-copper alloys with aqueous solutions of sodium hydroxide. Mechanical Engineering and Engineering Education 43: 34-39.
10. Kreshkov A, Yaroslavtsev AA (1982) Course of Analytical Chemistry. Quantitative analysis. Moscow: Chemistry, p: 312.

Technology of Obtaining Composite Sorbents for Water Purification

Politaeva Natalia Anatolievna, Taranovskaya Elena Aleksandrovna*, Slugin Vladimir Valentinovich and Prokhorov Vladislav Vladimirovich

Peter The Great St. Petersburg Polytechnic University, 194021, Polytechnic Street-29, Saint Petersburg, Russia

Abstract

The article describes the main applications of biopolymer chitosan, the most important of which are medicine and food industry. In recent times, many works devoted to the application of chitosan for wastewater treatment because it has a flocculation and sorption properties. The market price of chitosan is high, so it is proposed to create a granular composite sorption materials based on chitosan and waste of agricultural processing, which will reduce the cost and improve the sorption properties. As waste agricultural processing is proposed to use heat-treated threshing of millet which has high sorption properties. The composites, where the binder is chitosan and the filler - heat-treated threshing millet with different content (10%; 20%; 30%; 40% of the total weight) were obtained. The adsorption isotherms of zinc ions on the composite adsorbent materials with different content of filler were constructed and the values of maximum sorption capacity were calculated. The mechanical properties (abrasion and grindability) of the obtained composite sorption materials were determined and it was shown that the best sorption characteristics of composite material with the addition of the filler for 30%. The technological scheme of production of composite materials of the heat-treated millet threshing and chitosan for wastewater treatment was developed. The microstructural study of the obtained materials showed that heat-treated additive threshing of millet increases. The economic indicators of production of composite sorption materials were calculated and the methods of disposal were examined.

Keywords: Sorption materials; Wastewater; Chitosan; Zinc ions; Millet husk

Introduction

As early as XIX century, French scientists separated chitin from acyl and got chitosan in substance. Since then, a lot of chitosan fundamental research has been conducted and tree Nobel Prize winners' academic works appeared on the subject 1903, 1929 and 1939. This natural polymer is still of interest nowadays and its unique properties attract attention of different experts with various specialties [1-5]. At present, we have many ways to use chitosan in different branches of industry where the most important ones are medicine, food industry and water purification. The main advantage of chitosan is its safety for a man and environment. In nature, chitosan fully decays. As it is ecological, its usage has good prospects for achieving ecological goals. Chitosan is also known for its great sorption and coagulation properties might be used as a flocculant, antibacterial properties make it possible to use chitosan for water purification in production process [6-9]. It is not justified to use chitosan in substance for waste waters purification because of its high cost price starting from 2000 rubles/kg. Obtaining composites on basis of chitosan as binding and cheaper raw materials as filling will reduce the cost and help use these materials more extensively. It makes the research aimed at production long-range and economically sound sorbents using secondary raw materials and chitosan relevant and scientifically and practically important.

Purpose of the research: development of technologies to obtain composite sorption materials on basis of chitosan and cellulose agricultural processing waste millet threshing for wastewaters purification of zinc nuclei.

Materials and Methods

To increase sorption capacity and reduce chitosan cost production it is offered to produce composite sorption materials where chitosan is used as binding and heat-treated millet threshing as filling. In Russian Federation most regions are agricultural ones where a great quantity of lignine and cellulose agricultural processing wastes is generated (sunflower husk, wheat husk, millet husk etc.,) which when modified acquire high sorption properties [10,11]. Vast areas are needed for the waste storage. Using them as secondary materials will solve two problems at a time water purification and waste utilization. It makes combining of chitosan and heat-treated millet threshing (HTMT) rational and cost-effective. The authors [10] describe in details sorption properties of HTMT at 300°C within 20 minutes. The thermal treatment and burning result in formation of a porous structure with the pores size ranged between ~0.8 and 4-5 nm. The obtained sorbate possesses rather high sorption properties: A≈17 mg/g, specific surface S=188 M^2/g, the total volume of pores by water, V=0.3 CM^3/g.

To get composite sorption materials was chosen 6% chitosan solution in 3% acetic acid solution [12] with addition of HTMT powder (composite material chitosan-millet (CMCM)) with quantity 10%; 20%; 30%; 40% of a total mass and without HTMT addition and (composite material chitosan (CMC)). The resultant mixture is stirred homogeneously for one hour. The resultant mixtures are poured with syringes in 5% solution of sodium hydroxide (NaOH) with the following water washing to the value pH 7.0 -7.5 and drying at room temperature within 24 hours.

The Granules obtained were analyzed for ability to recover zinc ions (Zn^{2+}) from simulated wastewaters with initial concentration from 5 to

***Corresponding author:** Taranovskaya Elena Aleksandrovna, Peter The Great St. Petersburg Polytechnic University, 194021, Polytechnic Street-29, Saint Petersburg, Russia, E-mail: conata07@list.ru

100 mg/l with a step 5 mg/l. The granules were added to simulated solutions with quantity 20 g per 1 liter and sorption process was conducted under static conditions within 20 minutes (time for sorption equilibrium) [12] with stirring and thermo stating in the temperature range 293+2 K. For comparison, the same sorption process was conducted with granulated chitosan without HTMT. After wastewater purification process, the sorbate was filtered and the final concentration of heavy metals ions was determined by means of voltamperometric analysis.

Results and Discussion

By initial and final concentrations sorption capacity Ap was determined and adsorption isotherms built (Figure 1).

Adsorption isotherms Zn^{2+} for CMCM with different proportions of HTMT were used to determine values of maximum volume capacity (Table 1). Comparing sorption capacity of modified materials with different content of heat-treated threshed millet showed that maximum sorption capacity (50 g/g) is for CMCM with HTMT40 and 20%. With HTMT additive 40% granules mechanical strength is much lower than with additive 20%. Granules with additives 40% and 30% are not of accurate form and decay during purification process (Figure 2).

Physical-mechanical properties (abradability and grindability) of the obtained composite materials (CMCM) were studied. The properties were determined using method [13] (Table 1). As shown in Table 1, CMCM with HTMT 10 and 20% and CMC without HTMT comply with requirements of GOST P 51641-2000 (abradability no more than 0.5%, grindability no more than 4%).

Analyzing the data, it might be concluded that HTMT additive increases sorption capacity. CMCM with HTMT additive 20% shows the best parameters: high sorption capacity (50 g/g), abradability 3%, grindability 0.3%. That is why this composed material is supposed to be used for drains purification from HMI. Granules without HTMT additive are mechanically strong but have low sorption capacity, A=36 mg/g.

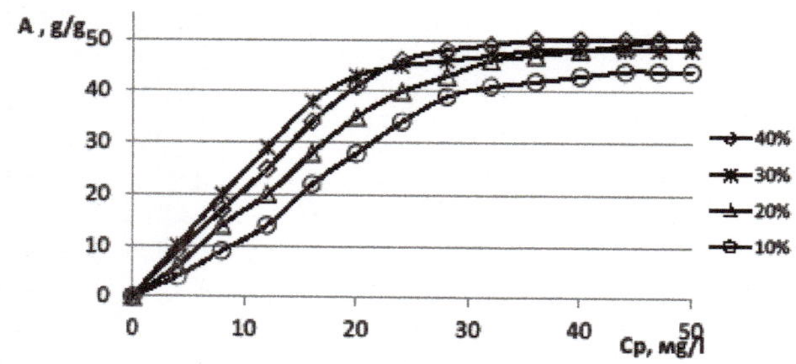

Figure 1: Adsorption isotherms Zn^{2+} for CMCM with different proportions of HTMT additive.

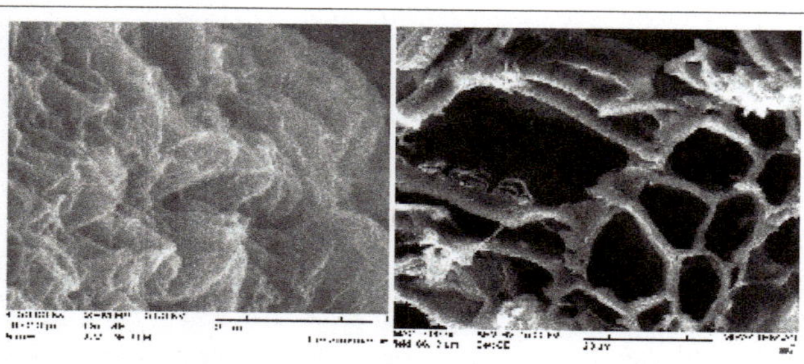

Figure 2: Microphotography of the granules surface: a-CMC; b-CMCM (20%).

Surface morphology analysis of CMCM containing HTMT 20% and without it revealed that with HTMT additive appears porous surface area. Porous structure of the surface results from the increase of CMCM sorption capacity with HTMT additive.

To obtain CMCM a process flowsheet has been elaborated (Figure 3). The chitosan flakes from the chitosan container (3) go through the doser (4) to the mixer (5). At the same time 3% solution of the acetic acid goes through the doser (6) to the mixer (5). The acid goes from the container (7). A composition is formed as a result of 24 hours chitosan and acetic acid mixing in the mixer (5). Millet threshing previously goes to the stove (2) from the container (1) where it goes through the heat treatment (T=300°C, t=20 мин.). By the use of the doser (4) HTMT goes to the mixer (5) where the solution of chitosan and acetic acid is ready and being mixed for 1 hour. Prepared mixture by the use of the doser (8) drop by drop goes to the granules settling chamber, which is already filled with 5% solution of alkali. The alkali goes from the container (9) by the use of the doser (10). Granules of black colour are formed in the granules settling chamber. Size of the granules might be varied depending on the volume of the solution dropped through the doser. Water from the tank (18) is used for washing the granules. The water goes to the alkali tank (9) until the value pH reaches 7-7.5. Measuring of the pH is done with the pH tester (19). The alkali is supplied to the tank (9) until the concentration of the solution NaOH equal 5%. To separate the obtained granules from the solution a mesh filter made of inert material (stainless steel, chemically persistent polymer materials) is used (12) with mesh size less than granules size (1-2 mm). The filter with the granules is moved to the drying box (14) and is used for drying within 4 hours. To clean air of alkali vapor an absorber (15) is used. To make the process for granules obtaining continuing two replacing mesh filters (12), (13) are used. After drying, the granules are poured into the container for granules storage. An empty filter replaces the filter with newly obtained granules. The obtained granules from the container (17) by use of the doser are packaged (5 to 50 kg sacks on terms of the customer).

In the article [14] it is offered to utilize waste sorption materials on basis of chitosan as soil fertilizers if they were used to purify heavy

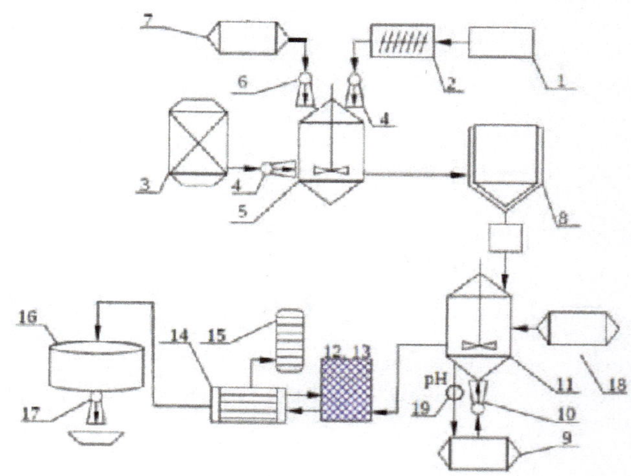

Figure 3: A scheme of CMCM obtaining: 1- a container for millet husk storage; 2- a stove for millet husk heat treatment; 3 - a container for chitosan storage; 4- a chitosan doser; 5- a mixer; 6- an acetic acid doser; 7- a container for acetic acid storage; 8- a doser; 9- a container for alkali storage; 10- an alkali doser; 11- granules settling chamber; 12, 13 - mesh filters; 14- a drying chamber; 15- an absorber; 16- a container for granules storage; 17- a doser of the obtained granules; 18- a water tank; 19- pH tester.

Millet husk quantity for CMCM %	Abradability	Grindability	Sorption Capacity A Zn^{2+}, mg/
0	0.3	3	36
10	0.3	3	44
20	0.3	3	48
30	0.9	6	50
40	1.2	10	50

Table 1: Physical-mechanical and sorption properties of CMCM depending on the composition.

metals ions- microelements (Cu^{2+}; Zn^{2+}). It is proved that CMCM used as soil fertilizer results in significant increase of germinating capacity (12%) and plants growing.

In the article [14] it is offered to utilize waste sorption materials on basis of chitosan as soil fertilizers if they were used to purify heavy metals ions- microelements (Cu^{2+}; Zn^{2+}). It is proved that CMCM used as soil fertilizer results in significant increase of germinating capacity (12%) and plants growing.

Calculation of economic indexes was made for CMCM production with quantity 150 tons of sorbate a year. The selling price was 117.3 rub/kg. Initial chitosan price is 2 thousand rub/kg. Investments to produce CMCM were 867027 rubles with the payback period more than two years.

Conclusions

It is proved that high sorption composite materials might be obtained on basis of chitosan and HTMT. It is established that sorption capacity, abradability, grindability CMCM depends on the percentage of HTMT. It is established that CMCM with HTMT 20% is of high sorption capacity (50 mg/g) and optimal mechanical properties. Microstructure study revealed that HTMT additive causes the formation of a porous CMCM surface structure with possible physical adsorption during HMI recovering process. A technology for obtaining and using CMCM for the wastewaters purification process was elaborated. Calculation of economic indexes for CMCM production showed that HTMT additive significantly reduces the cost of sorption materials on basis of chitosan (CMCM- 117.3 rub/kg, chitosan as substance- 2 thousand rub/kg).

References

1. Skryabin KG, Vihoreva GA, Varlamov VP, Nauka M (2002) Obtaining, properties and use /under the editorship. Chitin and Chitosan, p: 368.
2. Galbraih LC (2001) Chitin and chitosan: Structure, properties. Sorovsky Educational Magazine 7: 51-56.
3. Bikova VM, Krivosheina LI, Glazunov OI, Ezhova EA (2003) Chitosan use in medical cosmetics. New achievements in chitin and chitosan study: Materials of the 7th International Conference, pp: 231-233.
4. Safronova TM (2002) Chitosan use in food Obtaining, properties and use. pp: 346-359.
5. Dubinskaya AM (1989) Chitin and its derivatives use in pharmacy (review). A chemical and pharmaceutical magazine 23: 623-628.
6. Seliverstov AF (1997) Composite materials on the basis of chitin for metals ions sorption from water solutions. Applied chemistry magazine 70: 148.
7. Gorovoi LF (2002) Chitin and its derivatives sorption properties M. Nauka, pp: 217-246.
8. Bikova VM (2004) Some aspects of chitin and chitosan use as flocculants. Agricultural Russia 5: 30-31.
9. Gil G (2004) Selective antimicrobial activity of chitosan on beer spoilage bacteria and brewing yeasts. Biotechnology Letters 26: 576-574.
10. Sobgaida NA (2010) Influence of the modified wheat husk on its sorption properties to ions Pb^{2+}, Cd^{2+}, Zn^{2+} and Cu^{2+}. Higher Institutions bulletin. Sec Chemistry and Chemical Technology 53: 36-40.
11. Chikina C (2008) Use of Sorbent on the basis of polyurethane foam and buckwheat husk to reduce environmental load for the water area. Environment protection in the oil and gas complex 10: 53-57.
12. Taranovskaya EA (2016) Technology of obtaining and use of granulate sorbets on the chitosan basis. Chemical and Oil and Gas Machine Industry 5: 42-44.
13. Gost P (2000) Filter granular materials. General specifications. M. Standartinform, p: 14.
14. Politaeva NA (2016) Utilization of waste sorption materials on basis of chitosan. Technological University Bulletin 19: 25-28.

NIR-Based Simultaneous Measurement of Amine Loading and Degree of Neutralization during Flotation Reagent Preparation for Potash Processing

Mohammad Hussein and Adam Donaldson*

Dalhousie University, Process Engineering and Applied Science, 1360 Barrington St. Halifax, Nova Scotia, B3H4R2, Canada

Abstract

A continuous spectroscopy-based monitoring approach of the amine reagent preparation prior to flotation processing was investigated. The process involved the neutralization of a long carbon chain amine, C16-C20, with hydrochloric acid. Current methods of monitoring the degree of neutralization, controlling the HCl addition, and monitoring the amine content were based on pH measurements, with out-of-line sample lab-validation carried out using titration-based methods. Industry feedback indicated this method was prone to error and had resulted in challenges in process control. This work demonstrates a novel method of measuring both amine content and degree of neutralization using a mini-fluidic reactor and FT-NIR system equipped with a flow-through heated transmission cell. Analysis of the spectral response for wavenumbers in the range of 4,258-4,400 cm^{-1} yielded amine content and degree of neutralization measurements accurately to within ± 0.065 wt%, and ± 6.0% margin of errors, respectively. This led to monitoring intensification to increase KCl recoveries in flotation processing.

Keywords: Sorption materials; Wastewater; Chitosan; Zinc ions; Millet husk

Introduction

Aqueous amine solutions have been neutralized with HCl as part of the reagent conditioning process to separate KCl from NaCl in flotation vessels. The amine reagent, C-16 to C-20 hydrocarbon chain, is neutralized by the addition of HCl at an equimolar ratio to enable the protonated amine molecule to bind to the crystal defects in KCl, increasing its ability to attach to air as part of the flotation process. Batch conditioning of the amine reagent has demonstrated foaming tendencies and a lack of homogeneity of the final solution with respect to the degree of neutralization, (DON) of the solution. Due to mixing irregularities, complete protonation of the amine solution ($R\text{-}NH_2$ to $R\text{-}NH_3^+$) could not be achieved due to the formation of micelle globules that served as a protective layer against the protonation process [1]. The protective layer can be broken down by increasing the dispersion of the amine throughout the solution and maintaining constant agitation [2]. Batch to continuous conversion of this process is an on-going trend [3], but requires accurate monitoring of the neutralized amine solution to maintain suitable flotation reagent activity. Determination of the amine content and the amount neutralized has been done using titration-based measurements or the Baume Scale method. These two methods require added factors that affect the final measurements of the weight percent and the percent neutralized. There is also a need for dilution, drying, addition of additives, and pre-heating of the sample all in which will increase the margin of error (MOE) of the amine content and the DON. These techniques are also time consuming, labor intensive, and hazardous. Titration tests are inaccurate when the amine content must be within ± 0.001 g/g error margin. Baume scales are also as accurate as the introduced factors and cannot reach less than ± 0.01 g/g error margin.

Over 50% of potash production is achieved through flotation separation [4]. Many optimization studies have been done on the attachment of amine/KCl globule to air bubbles to enhance the separation [4]. It is crucial to note however that the amine must be homogenously neutralized for the attachment to take place. Amine molecules that are not neutralized will not bind to KCl crystals, and the amine recovery of KCl will significantly decrease across the flotation processes. Therefore, this paper focuses on the assurance of complete neutralization of the amine solution, along with a continuous monitoring method to regulate the flow of the reagent for recovery enhancement purposes. The near infrared spectral region has been used to monitor primary, secondary, and tertiary amines, with absorbance peaks for the first and second amine overtones occurring across the spectra for primary amines [5]. These absorbance peaks could be used to develop calibration models for the amine weight percent and it's DON. A primary amine loading onto silica as part of a molecular weight determination study has demonstrated the absorbance peaks and its variation with different loadings across the first overtone region ~4,950 cm^{-1} [6]. It has also illustrated the use of the partial least of square statistical method to develop a multivariate calibration model of the amine-containing loading in its liquid state onto silica [6]. This has demonstrated the applicability of using the Near Infrared region to measure the amine absorbance in the liquid form, and develop a calibration model to correlate between the absorbance and the amine content. The inline monitoring of amine has also been achieved through utilizing fiber optic cables to determine the reaction kinetics of amine in epoxy [7]. This is achieved through examining the characteristics of the absorbance peaks, and their respective appearance/disappearance as the reaction progressed [6]. The use of multivariate calibration models as monitoring methods has demonstrated durability and accuracy when compared to reference methods of monitoring [8-9]. High degrees of precision have been achieved in terms of monitoring pH, amine solutions, and hydrocarbon concentrations [6-11]. The use of in-line monitoring throughout continuous mixing of solution has been established in several studies [12]. Given that the FT-NIR can

***Corresponding author:** Adam Donaldson, Associate Professor, Dalhousie University, Process Engineering and Applied Science, 1360 Barrington St Halifax, Nova Scotia B3H4R2, Canada, E-mail: adam.donaldson@dal.ca monitor the homogeneity of a solution, [13-16] it provides the means for testing out the homogeneity of solutions during mixing by in-line monitoring [12,16]. The Antaris II FT-NIR analyzer will be utilized to monitor both amine weight percent and DON in a mini-fluidic system specifically configured for continuous production partially and fully neutralized solutions. The measured absorbance will be used to develop a calibration model that will focus on the amine bond overtones as well as stretches in the carbon chain, C-C. The desired outcome is to intensify the monitoring process to calculate the degree of neutralization to increase the recovery of KCl in flotation separation.

Experimental Setup

Mini-fluidic reagent preparation

Flotation reagents of varying amine weight percent and DON were generated in a minifluidic system comprised of a water reservoir, dosing pumps for both amine and HCl, a shear-based mixer and 1/8" PFA tubing flowing into a transmission cell for spectral analysis (Figure 1). The chemicals involved were a C-16 to C-20 amine hydrocarbon and a hydrochloric acid solution with CAS# of 7647-01-0. For each experiment, tap water was heated to a set temperature between 50 and 70°C and stored in a reservoir sufficient to service each experiment. The amine weight percent and the DON parameters were varied through adjusting the dosing rate of the two syringe pumps. The amine weight percent was adjusted from 1.0 to 3.0 wt% with a dosing rate of 0.5 wt%, while the DON was varied for each chosen weight percent from 0% to 110% DON with a dosing rate of 10%. A DON of 100% represented a neutralization equivalence of ~260. The molecular weight of the reagent is obtained by multiplying the neutralization equivalence by the number of basic functional groups that require neutralization within the reagent. Since there is only one basic functional group in the reagent, R-NH_2, and that the molecular weight is equal to the neutralization equivalence, then a DON of 100% represented a molar ratio of 1:1 between amine and HCl. The flotation reagent was initially heated to a matching temperature and drawn into an electrically-traced steel syringe. The molten reagent was then injected into the water stream prior to entering a shear-based mixer (Micropump P23 head installed on a magnetically coupled drive operating at 300 mL/min). A diluted 4.5 M HCl solution was injected at variable rates depending on the desired DON. Hot water was initially run through the PFA tubing to pre-heat the system and prevent any solidification of amine throughout the lines. Following pre-heat, HCl was injected at an equimolar flow rate to that of the amine, where it was adjusted then to reach the desired degree of neutralization. The solution then passed through a gear pump and 3 m length of 1/16" ID PFA tubing with a residence time of ~25 seconds for neutralization to occur. Note that initially a fiber optic probe was utilized at multiple points along the PFA tube to determine if the neutralization reaction could be tracked, but it was quickly determined that the spectra did not vary along the 3.0 m length, suggesting that neutralization between the amine reagent and the HCl solution was instantaneous. The reagent solution was thus directed through a flow-through heated transmission cell with a path length of 1 mm for subsequent analysis in the Antaris II FT-NIR unit.

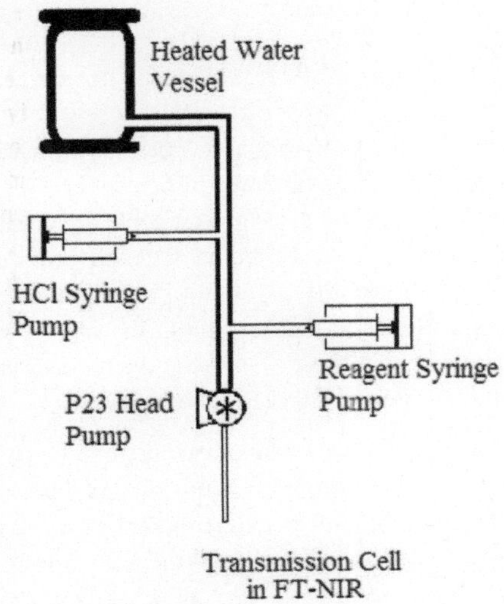

Figure 1: Schematic of the mini-fluidic setup of the reagent preparation and its monitoring.

Spectral analysis

Since the solution was comprised of an amine molecule, the focus then was on the three overtones of the R-NH_2 bond. The examined wavenumber range was 10,000-4,000 cm^{-1}. Water was included in the background reference to negate its effect on the amine absorbance. The focus of the spectra was on the first overtone region (~4,900 cm^{-1}), the second overtone region (~6,600 cm^{-1}), and the third overtone region (~10,000 cm^{-1}). Furthermore, the absorbance region in the range of 4,000 cm^{-1}-4,500 cm^{-1} was thoroughly examined because of the presence of a distinguishable trend in the spectra. For each experimental set, samples were collected and analyzed using the accepted quality-assurance methods of the industry partner to determine the amine weight percent as well as the neutralization percentage using both diluted HCl and KOH solutions. These measurements were used and referred to in this paper as the actual amine weight percent and the actual degree of neutralization. The water flow rate was held constant throughout the experiment, but the amine and the HCl flow rates were adjusted to provide a wide range of data. The use of the C-C bond chain wave number region was also explored to examine its capability in developing a model. A background spectra of heated water was initially taken to ensure that any environmental changes within the lab were accounted for in the sample spectra. 50 samples were taken in for each weight percent, with 5-10 validations points utilized for each. 10 scans were collected then averaged for each measurement with a resolution of 4 cm^{-1}. A partial least square analysis (PLS) was utilized for the basis of absorbance calibration on the wavenumber region chosen for the analysis, 4,400-4,250 cm^{-1}.

Results and Discussion

The amine overtones were examined to determine the amine weight percent in the neutralized solution (Figures 2-4) represent the three overtones of the amine bond for 1, 2, and 3 wt% at varying DON. The data was analyzed using the Antaris II software to provide a calibration model that predicts the weight percent and DON, but none of the models provided the minimum required accuracies with respect to the two parameters. These three overtones were utilized separately and together to obtain a best-fit model. A smoothing Savitzky-Golay polynomial filter was used to decrease the amount of noise. In the first overtone spectra, a shift in the absorbance minimum peak was observed from wavenumbers 5,250 cm^{-1} to 5,350 cm^{-1}. This shift took place as more of the amine was neutralized that also resulted in a decrease in the absorbance magnitude as seen in the 1st overtone across Figures 2-4. Across the 2nd overtone for the three figures, an increase in the absorbance magnitude around ~6,750 cm^{-1} was observed with higher degrees of neutralization, but the opposite trend was observed across an amine content of 1 wt%.

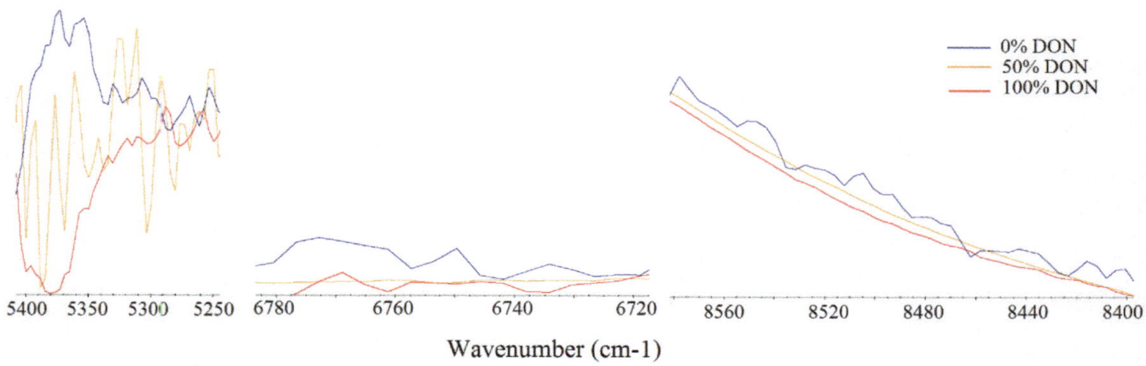

Figure 2: The three overtones for 1 wt% of amine at three different DON%. 1st overtone (left), 2nd overtone (middle), and 3rd overtone (right).

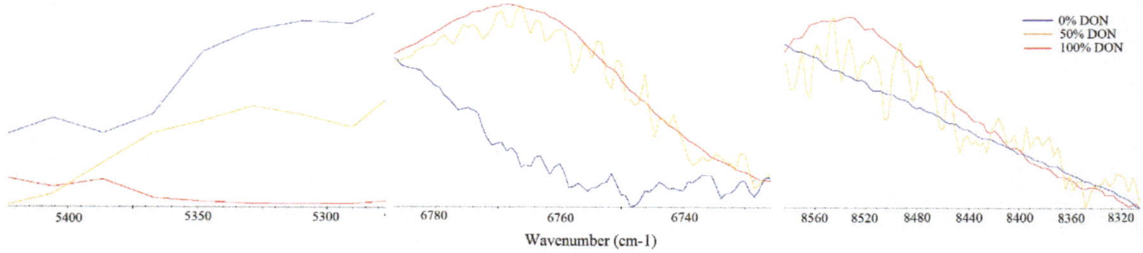

Figure 3: The three overtones for 2 wt% of amine at three different DON%. 1st overtone (left), 2nd overtone (middle), and 3rd overtone (right).

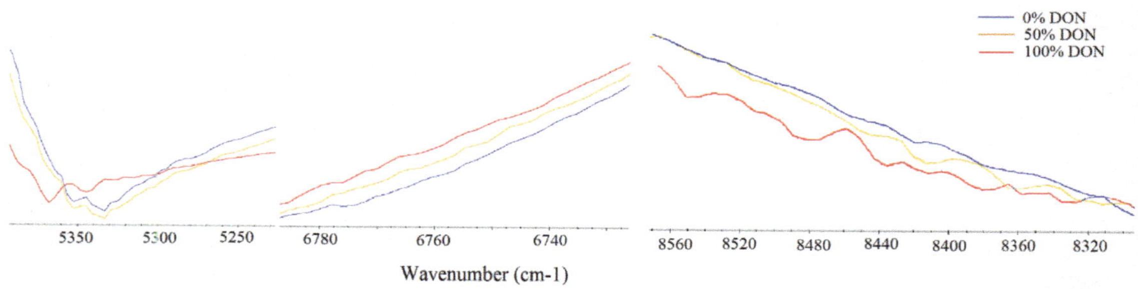

Figure 4: The three overtones for 3 wt% of amine at three different DON%. 1st overtone (left), 2nd overtone (middle), and 3rd overtone (right).

Lastly, the most noticeable change of absorbance was observed at the 3rd overtone region. In Figure 4, 0% DON had the highest peaks at 8,650 cm^{-1}, whereas higher degrees of neutralizations had lower magnitudes.

A calibration model using these three overtone regions was developed to assess how useful these overtones could be in determining the degree of neutralization of the solution. Figures 5 and 6 represent the calibration model developed along with its residual error plot. The calibration model was unable to predict the degree of neutralization within the required minimum margin of error of ± 5% DON obtained through titration-based measurements. Although there were trends observed as the degree of neutralization changed, these trends were not consistent throughout the absorbance spectra. Given that a constant flow rate of water was taken within the background reference, minimal changes to that flow rate would have a visible effect on the absorbance. Due to these minimal changes, the absorbance of amine and water overlapped across the 1st, 2nd and 3rd overtones. This overlap resulted in the observed noise in Figures 2-4. A different wavenumber region was explored to calculate the amine weight percent and it's DON. It was noted that the region of 4,400-4,250 cm^{-1} was not affected water overtones, improving signal to noise ratios within that region. Increases in the DON increased the absorbance of the C-C bond present in that region. Over additions of hydrochloric acid did not impact the spectra which made it more appealing to use as the basis to develop the multivariate calibration model rather than the use of the three amine overtones.

In theory, neutralization of the amine to an ammonium ion increases the attraction forces between the ammonium ion and the carbon electrons due to increases in electronegativity differences [17-18]. This pulling of the electrons towards to the ammonium ion shrinks the carbon chain and is thought to decreases its absorbance, as seen at wavenumbers 4,292 cm^{-1} and 4,400 cm^{-1}. The decrease of the absorbance in this region however, and not across other regions, is not clearly understood yet.

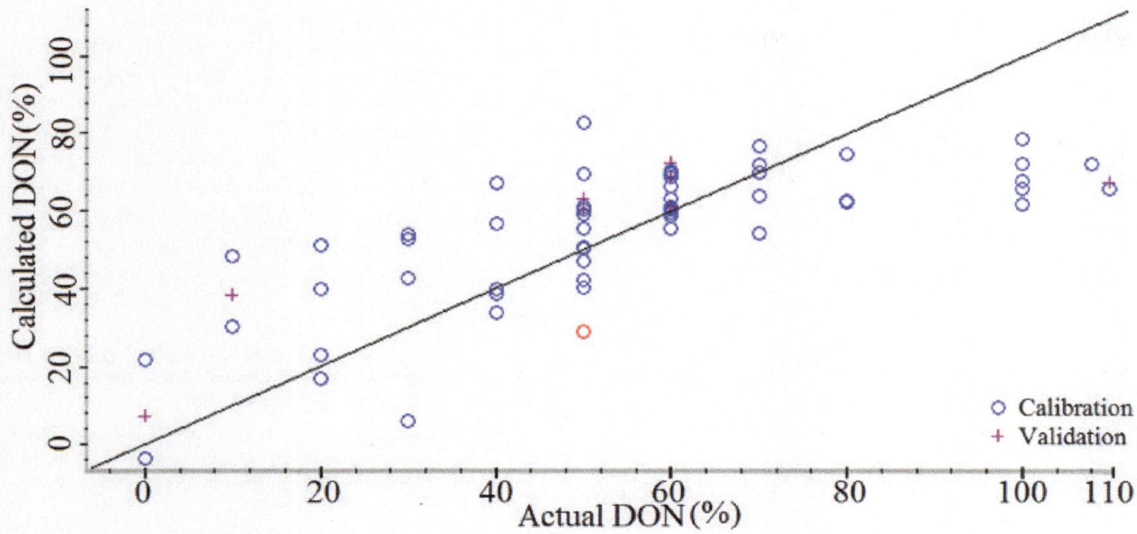

Figure 5: A calibration model using the three amine overtone regions after smoothing filters were applied.

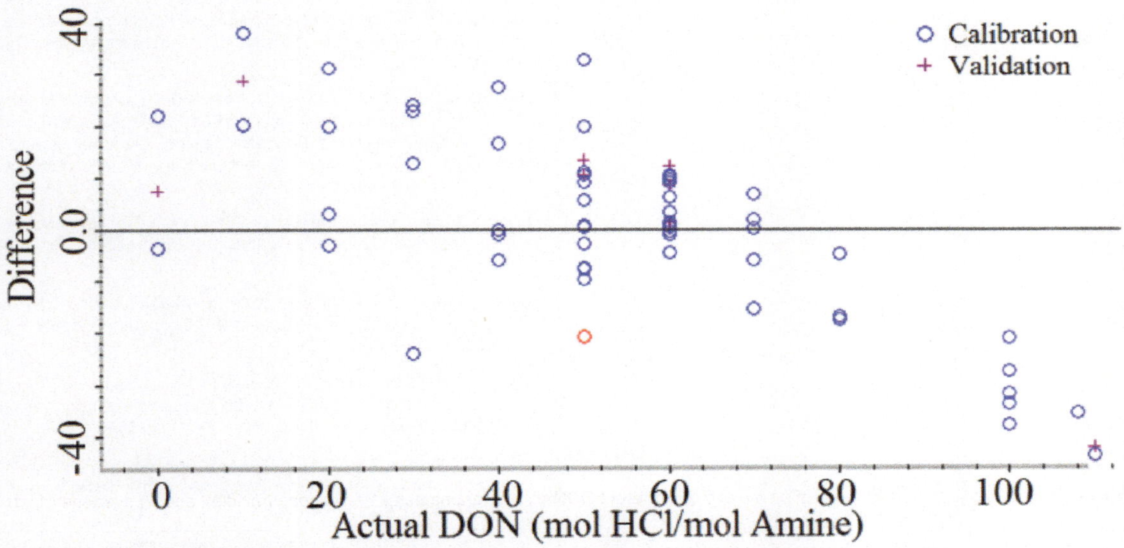

Figure 6: Residual error plot using the three overtone regions of amine after smoothing filters were applied.

The model for the three amine weight percent managed to predict the DON with a root mean square error of calibration of 7.17% and a root mean square error of prediction of 6.33. This was achieved by utilizing the Savitzky-Golay polynomial filter to reduce the noise in the spectra. The correlation coefficients for each were 96.6% and 99.0% respectively. The TQ Analyst software model predicted the DON with a MOE of ± 6%. Figures 7 and 8 represented the calibrated Antaris model for the DON and its residual error difference. The weight percent calibrated model through the software provided resulted that were within the required accuracy. As seen in Figures 9 and 10 a margin of error of $\pm 6.5 \times 10^{-4}$ g/g was achieved.

The amine weight percent and the DON multivariate calibration models provided in-line monitoring methods of the neutralization reaction with precision results that are better than that of the reference methods, but with no sample manipulation and minimal labor work. The modification of the monitoring method from a titration-based one to an in-line continuous spectroscopic-based one intensifies the reagent preparation process, and in return increases the recovery of KCl in the flotation process.

Conclusion

The Antaris II FT-NIR analyzer was utilized to develop multiple models to predict both the degree of neutralization and the amine weight percent as part of the reagent conditioning in flotation processing. The absorbance was measured using a built-in heated transmission module equipped with a flow-through cell to measure the absorbance of the solution. The results demonstrated a calibration model that relied on

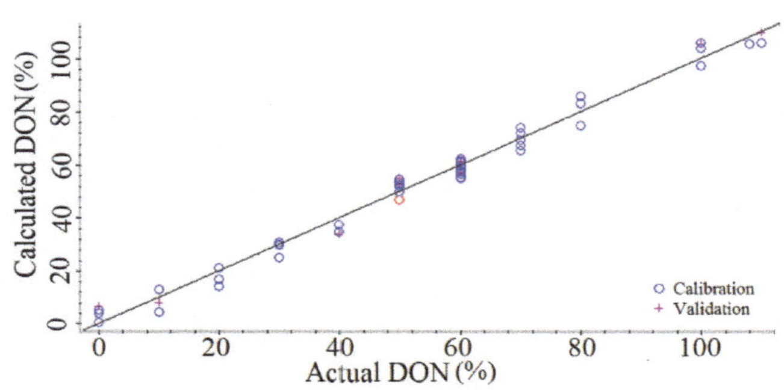

Figure 7: Calculated *vs.* actual DON for amine weight percent 1, 2, and 3 wt% using Antaris software.

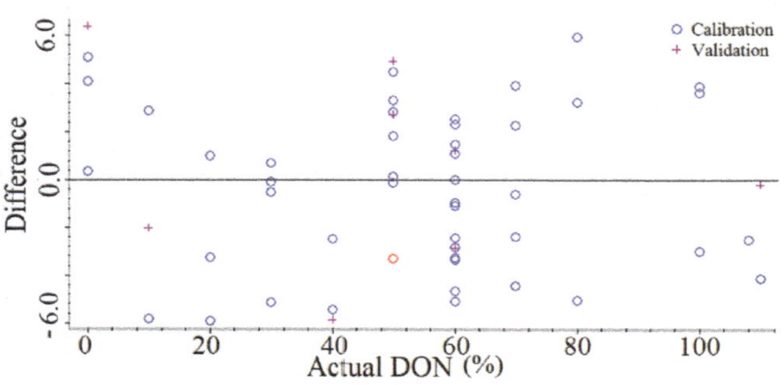

Figure 8: Residual error difference of the DON data for mass fractions 1, 2, and 3 wt%.

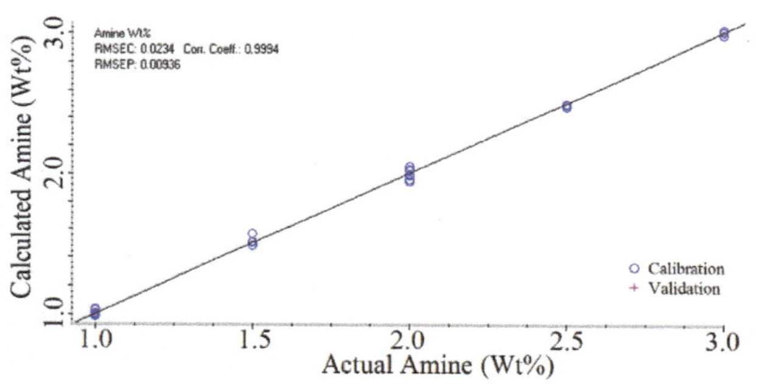

Figure 9: Calculated *vs.* actual weight percent using the Antaris software.

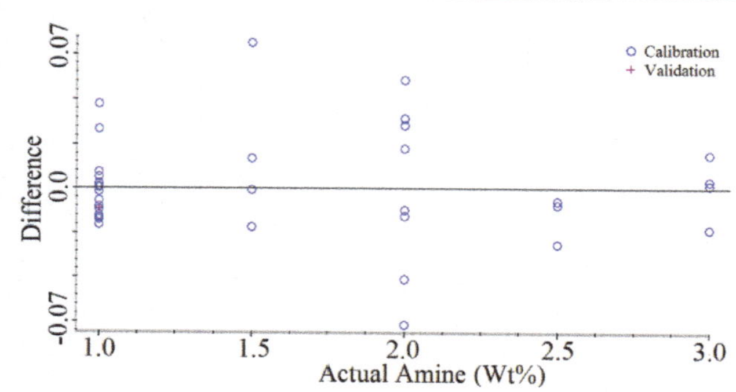

Figure 10: Residual error plot for the amine weight percent calibrated model using the Antaris software.

the absorbance in the region of 4400-4258 cm^{-1} to calculate the amine weight percent and its degree of neutralization to within 0.065 wt% and 6% DON, respectively. The continuous in-line monitoring could lead to process intensification where deviations of the DON away from the set point could be instantly corrected without out-of-line sampling taking place. Currently, deviations of DON away from the set point translate into low recoveries of KCl within the flotation process, where the average recovery is ~70%. The potential to increase these recoveries lies within increasing the DON of the amine solution to 100%. Continuous monitoring of the amine reagent preparation, and across the flotation process, would increase the recovery of KCl beyond an average of 70%.

References

1. Yoshioka T, Ikeuchi K, Hashida M, Muranishi S, Sezaki H (1982) Prolonged release of bleomycin from parenteral gelatin sphere-in-oil-in-water multiple emulsion. Chem Pharm Bull 30: 1408-1415.
2. Silva-Cunha A, Grossiord JL, Puisieux F, Seiller M (1997) W/O/W multiple emulsions of insulin containing a protease inhibitor and an absorption enhancer: Preparation, characterization and determination of stability towards proteases in vitro. Int J Pharm 158: 79-89.
3. Hussein M, Donaldson A (2017) Mixing strategy effect on dispersion of amine during reagent production for floatation applications. Can J Chem Eng.
4. Cao, Wang, Miller, Cheng, Jiao (2011) Bubble Attachment Time and FTIR Analysis of Water Structure in the Flotation of Sylvite, Bischofite and Carnallite. Miner Eng 24: 108-114.
5. Sauvageau P, Sandorfy C (1960) Hydrogen Bonding in the Amine Hydrohalides: Iii. Near-Infrared Spectra of Aliphatic Amine Hydrohalides. Can J Chem 38: 1901-910.
6. Heig N, Bachmann S, Petter CH, Marchetti-Deschmann M, Allmaier G, et al. (2009) Near-infrared Spectroscopic Study on Guest-host Interactions among G0-G7 Amine-terminated Poly(amidoamine) Dendrimers and Porous Silica Materials for Simultaneously Determining the Molecular Weight and Particle Diameter by Multivariate Calibration Techniques. Anal Chem 81: 5655-5662.
7. Mijović, Jovan, Saša A, José MK (1996) In Situ Real-time Monitoring of Epoxy/amine Kinetics by Remote near Infrared Spectroscopy. Polym Advan Technol 7: 1-16.
8. Sileoni, Marconi, Perretti, Fantozzi (2013) Evaluation of Different Validation Strategies and Long Term Effects in NIR Calibration Models. Food Chem 141: 2639-2648.
9. Sileoni, Valeria, Frans VDB, Ombretta M, Giuseppe P (2011) Internal and External Validation Strategies for the Evaluation of Long-term Effects in NIR Calibration Models. J Agr Food Chem 59: 1541-1547.
10. Ramírez-Morales, Rivero, Fernández-Blanco, Pazos (2016) Optimization of NIR Calibration Models for Multiple Processes in the Sugar Industry. Chemometr Intell Lab Syst 159: 45-57.
11. Tabasi, Simin H, Raafat F, Dennis B, Charles O, et al. (2008) Quality by Design, Part II: Application of NIR Spectroscopy to Monitor the Coating Process for a Pharmaceutical Sustained Release Product. J Pharm Sci 97: 4052-4066.
12. Quiñones, Leonel, Carlos V, Luis O (2014) A Novel Multiple Linear Multivariate NIR Calibration Model-based Strategy for In-line Monitoring of Continuous Mixing. AIChE J 60: 3123-3132.
13. Allan, Pamela, Luke JB, Alison N, David L (2010) Non-invasive Monitoring of the Mixing of Pharmaceutical Powders by Broadband Acoustic Emission. Analyst 135: 518-524.
14. Ilan, Pamela, Bellamy, Luke J, Nordon, et al. (2013) In Situ Monitoring of Powder Blending by Non-invasive Raman Spectrometry with Wide Area Illumination. J Pharmaceut Biomed Anal 76: 28-35.
15. Allan, Bellamy, Nordon, Littlejohn, Andrews, et al. (2013) In Situ Monitoring of Powder Blending by Non-invasive Raman Spectrometry with Wide Area Illumination. J Pharmaceut Biomed Anal 76: 28-35.
16. Vanarase, Alcalà, Jerez R, Muzzio, Romañach (2010) Real-time Monitoring of Drug Concentration in a Continuous Powder Mixing Process Using NIR Spectroscopy. Chem Eng Sci 6: 5728-5733.

Improved Fault Detection and Process Safety Using Multiscale Shewhart Charts

M Ziyan Sheriff[1,2] and Mohamed N Nounou[1*]

[1]*Chemical Engineering Program, Texas A&M University at Qatar, Doha, Qatar*

[2]*Artie McFerrin Department of Chemical Engineering, Texas A&M University, College Station, TX 77843, USA*

Abstract

Process safety is a critical component in various process industries. Statistical process monitoring techniques were initially developed to maximize efficiency and productivity, but over the past few decades with catastrophic industrial disasters, process safety has become a top priority. Sensors play a crucial role in recording process measurements, and according to the number of monitored variables, process monitoring techniques can be classified into univariate or multivariate techniques. Most univariate process monitoring techniques rely on three fundamental assumptions: that process residuals contain a moderate level of noise, are independent, and are normally distributed. Practically, however, due to a variety of reasons such as modeling errors and malfunctioning sensors, these assumptions are violated, which can lead to catastrophic incidents. Fortunately, multiscale wavelet-based representation of data inherently possesses characteristics that are able to deal with these violations of assumptions. Therefore, in this work, multiscale representation is utilized to enhance the performance of the Shewhart chart (which is a well-known univariate fault detection method) to help improve its performance. The performance of the developed multiscale Shewhart chart was assessed and compared to the conventional chart through two examples, one using synthetic data, and the other using simulated distillation column data. The results of both examples clearly show that the developed multiscale Shewhart chart provides lower missed detection and false alarm rates, as well as lower ARL_1 values (i.e., quicker detection) for most cases where the fundamental assumptions of the Shewhart chart are violated. Additionally, the relative simplicity of the proposed algorithm encourages its implementation in practice to help improve process safety.

Keywords: Process monitoring; Fault detection and diagnosis; Shewhart chart; Multiscale representation; Wavelets

Introduction

Statistical process monitoring (SPM) plays a critical role in most process industries. Primary functions of SPM are to ensure that plants run safely within minimal down-time for maintenance, and to ensure that the end product is of a desired quality. SPM is generally carried out in two phases: fault detection, during which the faults in a particular process system are first identified, and fault diagnosis, during which the root cause of the fault is isolated and determined, after which action is taken to bring the process back to normal operating conditions [1]. Sensors play a crucial role as they need to ensure that the process operates as designed, and this is accomplished by monitoring different process variables to ensure the process are running smoothly. Quick, efficient, and accurate fault detection is required because it can help prevent major safety incidents from occurring, thus ensuring safety of life, property and economy. This paper currently only looks to examine and improve the fault detection aspect of process monitoring.

There are a number of methodologies that can be used to classify fault detection methods. Fault detection methods can be classified according to their dependency on models used. These can be quantitative model-based methods, such as observers or parity space [2], qualitative model-based methods, such as fault trees, digraphs, or even process engineering experts [3-6], or process-history (data-based) methods, such as Principal Component Analysis (PCA) and neural networks [7,8]. Accurate qualitative and quantitative model-based methods may not always be available, especially for complex processes with multiple process variables, and therefore data-based methods are often employed. Data-based methods rely on collecting fault-free data (under normal operating conditions), that are then used to design fault detectors, i.e., define the control limits. These limits are then applied on new process measurements (testing data), to detect potential faults (deviations or abnormalities) in processes. Data-based monitoring methods have been applied to detect anomalies in a wide variety of applications, from medical sensors [9], to wind turbines [10], to the petroleum industry [11]. Deviations in process variables can be due to a number of reasons, from faulty sensors to malfunctioning alarms, and therefore it is essential to address these concerns in order to prevent catastrophic incidents.

Anomalies in processes can lead to catastrophic incidents. One example of a catastrophic incident is the Texas City refinery explosion, at the British Petroleum's (BP) plant in Texas City, Texas, where an explosion killed 15 people and caused nearly 180 injuries. Although there were a number of factors that led to the incident, one of the main contributing factors was the fact that multiple level indicators and alarms in the raffinate tower malfunctioned, and provided incorrect readings [12]. This led the tower to over-pressurize and eventually explode once it came into contact with an ignition source [13]. Data-based techniques are an efficient way of observing a process to monitor deviations in multiple key process measurements, and this paper aims to improve the fault detection capability of a very simple monitoring chart, the Shewhart chart,in order to widen its applicability in practice, and hopefully prevent major industrial disasters such as the incident mentioned.

***Corresponding author:** Mohamed N Nounou, Department of Chemical Engineering, Texas A&M University at Qatar, Doha, Qatar
E-mail: Mohamed.nounou@qatar.tamu.edu

Data-based techniques can be used to study trends of single process variables (univariate), or to analyze trends across multiple variables (multivariate). The Shewhart chart was one of the earliest (univariate) charts used for process monitoring purposes [14]. The chart is applied on process residuals, without the utilization of any filters. It is computationally simple to implement [15]. Other univariate charts, such as EWMA (Exponentially Weighted Moving Average) and CUSUM (Cumulative Sum) charts utilize the application of linear filters, before carrying out fault detection and can be more computationally expensive [16-18]. These charts may also require one or more tuning parameters to be modified, thus making the design of these charts increasingly computationally expensive and tedious. Process noise is often a concern in most process industries and can adversely affect the accuracy of fault detection techniques [4,19,20]. Charts with process memory such as CUSUM and EWMA may be able to address concerns related to process noise as they utilize filters. However, these filters are linear and may not be very effective at removing different types of noise or errors as they define a frequency threshold and anything above this threshold may be considered noise. The presence of process noise is not the only concern of these control charts. They also assume that process residuals being analyzed are independent (uncorrelated), and follow a normal (Gaussian) distribution [21]. However, in practice, real measurements might not necessarily satisfy these assumptions. In the past, many authors have looked into improving the fault detection abilities of particular techniques when process data violate one of these assumptions [22,23]. However, in practice, it may be possible that actual process data violate one or more of these assumptions. Therefore, more effective fault detection techniques that are able to simultaneously deal with process data that violate multiple assumptions are required. In addition, it is desirable that the developed technique is computationally simple, and relatively easy to implement in order to broaden its applicability.

One way to deal with these limitations of the conventional univariate fault detection methods is using wavelet-based multiscale representation. Multiscale representation has been shown to effectively deal with real process data as it allows efficient separation of important features from stochastic noise and provides wavelet coefficients that are approximately decorrelated and more Gaussian at multiple scales. Thus, it can help address most of the assumptions of the conventional univariate fault detection or control charts [24,25]. These advantages of multiscale representation will be utilized in this work to develop a multiscale Shewhart chart algorithm that will provide improved performance. Even though the ideas presented in this paper can be directly extended to other univariate control charts, this work focuses on the Shewhart chart due to its popularity and its computational simplicity [26]. These advantages of multiscale representation have been exploited by [27] who developed wavelet-based multiscale CUSUM and EWMA charts with improved fault detection abilities using autocorrelated process data. These techniques have been shown to improve the out-of-control average run length (ARL_1), which measures how long a particular technique takes to identify the presence of a fault. This is particularly important in terms of safety in process industries because faster detection allows more time to take corrective measures to bring a process back to its normal operating conditions. Other important metrics that can be used to assess the effectiveness of various fault detection techniques include the missed detection rate or the false alarm rate, which for certain applications can be as critical as the ARL_1.

Therefore, a proper evaluation of any fault detection technique requires assessing its performance with respect to all of these metrics. Unfortunately, the effect of multiscale representation on the performance of univariate control charts (especially under violation of their basic assumptions) and using all evaluation metrics (i.e., missed detection rate, false alarm rate, and ARL_1) has not been thoroughly studied. Therefore, the main objective of this work (which focuses on the Shewhart chart as an example of a univariate fault detection method) is to develop a multiscale form of the Shewhart chart that is easy to implement, and then assess its performance in comparison to its conventional counterpart under the violation of its three fundamental assumptions using various evaluation metrics. Such analysis should pave the road for similar ones using different univariate as well as multivariate fault detection methods.

This paper is structured as follows. First, an introduction to conventional univariate process monitoring is provided, along with a brief description of the Shewhart chart. Then multiscale wavelet-based data representation is explained, along with its advantages in process monitoring followed by a description of the developed multiscale Shewhart chart algorithm. The performance of the developed multiscale Shewhart chart is then analyzed through two illustrative examples, using simulated synthetic data, as well as process data from a simulated distillation column to validate its applicability and effectiveness in addressing safety concerns in the process industry. Finally, concluding remarks and future directions for research are presented.

Univariate Process Monitoring

Since this work focuses on extending the computationally simple Shewhart using multiscale representation, a brief overview of statistical and univariate process monitoring is needed. Most industries collect readings of process measurements through the utilization of multiple sensors. These sensors can monitor a number of different variables, such as temperature and pressure, to ensure that they are within acceptable limits for the purposes of plant safety and product quality. Univariate process monitoring methods can be implemented by following the algorithm illustrated in Figure 1.

As illustrated through Figure 1, process measurements collected from the sensors are initially compared to their model predictions (or desired values), creating process residuals. The process residuals are then used to compute a detection statistic. For the case of the Shewhart chart, the detection statistic is the process residuals themselves. For other univariate charts, such as the CUSUM or EWMA charts, the appropriate filter is applied in order to compute their respective detection statistics [28]. Using the distribution of the detection statistic, control limits (that can be used to determine the presence of absence of faults) can be computed. To test if new process measurements are faulty, the detection statistic is computed once again using the process residuals from testing data, which is then compared to these control limits. If the detection statistics computed from the testing data violate the control limits, a fault is declared. If the control limits are not violated by the detection statistic, the process is assumed to be operating under normal operating conditions, and no action is required.

For the Shewhart chart, the upper and lower fault detection control limits may be computed as follows [28]:

$$UCL, LCL=\mu_0 \pm 3\sigma \quad (1)$$

Where, μ_0 and σ are the mean and standard deviation of the detection statistic (which is the residuals for the Shewhart chart) obtained under normal operating conditions, respectively. The 3σ limits are commonly used for the Shewhart chart as they account for nearly 99.74% of all deviations in the data [29]. As stated previously,

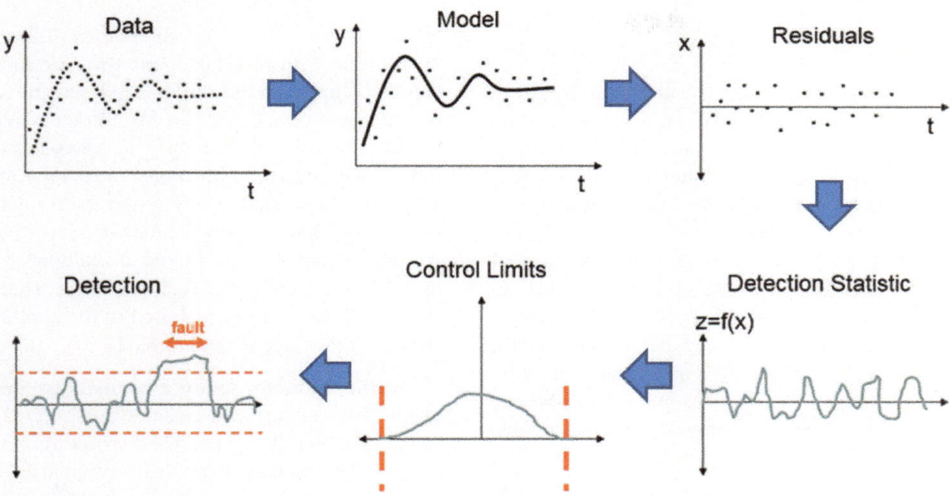

Figure 1: Algorithm for univariate process monitoring.

multiscale representation can help deal with violation in assumptions of most univariate monitoring methods. In the next section, multiscale wavelet-based representation of data is introduced, its advantages in fault detection are discussed, and a method to utilize these advantages to improve the effectiveness of the Shewhart chart is presented.

Multiscale Wavelet-Based Representation of Data

Practical process data, especially those collected from industrial processes are known to possess multiscale characteristics, i.e., they may contain noise or features that span wide ranges in both the frequency and time domains. For example a sharp change in the process data may span a narrow range in the time domain while spanning a much larger range in the time domain. Similarly, a slower change in the process data may span a wide range in the time domain, and a narrow range in the frequency domain [30]. Unfortunately, most commonly employed fault detection techniques in the process industry only operate on a single scale since they are applied on the time domain data, and therefore are unable to efficiently account for process data that may be multiscale in nature.

Multiscale wavelet-based representation of data is a powerful tool that has been used in data analysis. It is able to provide efficient separation of deterministic and stochastic features [31], and it has been successfully used to improve the accuracy of various process modeling, filtering, and state estimation techniques [32-34]. In multiscale representation, wavelet and scaling basis functions, which have the following form are used to represent time-domain functions and data at multiple resolutions and scales [35,36]:

$$\theta(t)=\frac{1}{\sqrt{s}}\theta\left(\frac{t-u}{s}\right) \tag{2}$$

where, the dilation and translation parameters are represented by s and u, respectively. A variety of families of basis functions have been used to carry out wavelet decomposition, e.g., Daubechies and Haar [37-39]. Several researchers utilized the Haar wavelet in various applications, and it will be utilized in the examples presented in this work [35,39,40]. Given a time domain data set (or a signal), a coarser approximation of the signal (often called a scaled signal) can be obtained by convoluting the original signal with a low pass filter h, which is derived from a scaling basis function of the following form [41]:

$$\phi_{ij}(t)=\sqrt{2^{-j}}\phi\left(2^{-j}t-k\right) \tag{3}$$

where, the dilation and translation parameters are represented by j and k, respectively. The difference between the original signal and the scaled signal is the detail signal, and can be obtained by convoluting the original and subsequent scaled signal with a high pass filter g, which is derived from a wavelet basis function of the following form [41]:

$$\psi_{ij}(t)=\sqrt{2^{-j}}\psi\left(2^{-j}t-k\right) \tag{4}$$

After repeating these approximations, the original signal can be reconstructed by taking the sum of the final scaled signal and all detail signals, as follows [41]:

$$x(t)=\sum_{k=1}^{n2^{-J}}a_{Jk}\phi_{Jk}+\sum_{j=1}^{J}\sum_{k=1}^{n2^{-J}}d_{jk}\psi_{jk}(t) \tag{5}$$

A number of researchers have utilized multiscale wavelet-based representation of data to improve fault detection. For example, [40] used multiscale representation to pre-filter the data, and then applied Principal Component Analysis (PCA) using the pre-filtered data. Pre-filtering of the raw data, before employing a fault detection technique, improves the effectiveness of the approach. However, pre-filtering may remove features in the data that are important for fault detection. Multiscale representation of data has also been used to develop a multiscale PCA (MSPCA) algorithm with improved performance over the conventional PCA method [35]. This is primarily due to multiscale being able to efficiently separate feature from noise making the MSPCA algorithm more sensitive to anomalies in process measurements than the conventional one. On the other hand, the performances of the many conventional univariate monitoring techniques have been limited by their inability to efficiently handle violations in their fundamental assumptions, like the presence of measurement noise, autocorrelation, as well as non-normality of evaluated residuals. Dealing with these limitations will help satisfy the assumptions of the various univariate fault detection techniques, which will reflect on their performances. Multiscale representation is an effective tool that is able to help deal with these challenges and its advantages are described in the next section.

Advantages of multiscale wavelet-based data representation

Ability to separate features from noise: A main assumption of the Shewhart chart is that the process data contains a moderate level of noise. One main benefit of multiscale representation is its ability to naturally separate noise from important features that may be present in the process data, which can be helpful from a fault detection perspective. This characteristic can be attributed to the successive application of high and low pass filters on the data in multiscale decomposition as illustrated in Figure 2. The noise-feature separation characteristic of multiscale representation and its advantages over other conventional linear filtering methods have been utilized in various applications, such as filtering time-series genomic data [41]. In this work, this separation ability of multiscale representation will also be employed in order to improve the performance of the conventional Shewhart chart, to widen its applicability in practice.

Ability to decorrelate autocorrelated data at multiple scales: Another fundamental assumption of the Shewhart chart is that the evaluated residuals are independent or uncorrelated. An important advantage of multiscale representation is that the wavelet coefficients obtained in multiscale decomposition are approximately decorrelated at multiple scales [42]. This can be illustrated by comparing the autocorrelation function (ACF) of an autocorrelated signal to the ACF of its details signals at multiple scales. The ACF, which plots the correlation between any two samples in a data set as a function of their separation, is usually used to quantify the autocorrelation in the data [43]. ACF measures the stochastic process memory in the data. Therefore, a white noise signal (which has independent samples) will have an autocorrelation function of zero for positions of lags other than zero, and a value of unity at lag zero, which indicates that the data set consists of completely uncorrelated samples. Conversely, correlated signals, such as those represented by autoregressive (AR) or autoregressive moving average (ARMA) models have non-zero ACF values at lags other than zero, which indicates that there is autocorrelation in the data [42]. To illustrate this characteristic, autocorrelated data are generated using the following AR (1) model [43]:

$$x_t = ax_{t-1} + \varepsilon \tag{6}$$

where, $a = 0.7$ *and* ε is white noise following a standard normal distribution with zero mean and unit variance. The obtained data are then decomposed at multiple scales and the ACF is computed for the time domain data as well as for all detail signals as shown in Figure 3. The first column of Figure 3 shows the time domain and detail signals, while the second column shows the ACF for all signals. Figure 3a clearly shows that, even though the time domain data are strongly correlated as indicated by the ACF shown in Figure 3b, the detail signals are approximately decorrelated at multiple scales as indicated by their ACFs shown in Figures 3c-3h. The decorrelation of detail signals at multiple scales can be attributed to the application of high-pass filters during wavelet decomposition [42].

Ability to better follow a normal distribution at multiple scales: The Shewhart chart also assumes that process residuals obtained from fault-free data follow a normal distribution. In practice, measured data may not necessarily follow a normal distribution. Modeling errors that leave non-modeled process variations in the model residuals can also be a source of non-Gaussian errors. Fortunately, multiscale decomposition is able to help address the issue of non-Gaussian errors as it provides detail signals that are closer to normal (Gaussian) at different scales.

To illustrate the effect of multiscale representation of data on the distribution of the detail signals, histograms of a chi square distributed data set and the detail signals obtained from its multiscale decomposition are compared as shown in Figure 4. Figure 4a shows that, even though the distribution of the time domain data is far from normal (which is expected since they follow a chi- square distribution) as indicated by Figure 4b, the histograms of the detail signals are closer to normality as indicated by Figures 4c-4h. Distributions other than chi-square were also used to validate this observation [30,44].

Furthermore, it should be noted that the multiscale representation provides wavelet coefficients that are also closer to being stationary for non-stationary data [24,35]. In this section, it has been shown that multiscale representation possesses advantages that can help

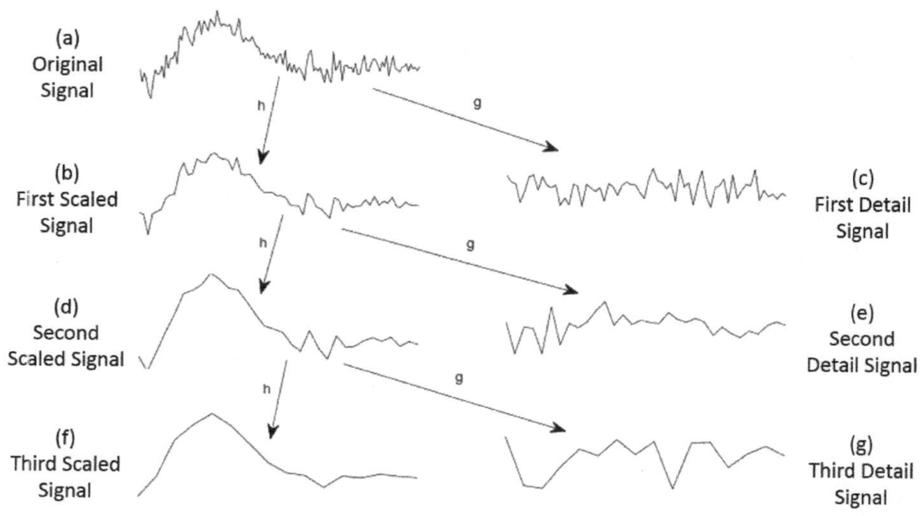

Figure 2: Schematic illustration of multiscale wavelet-based representation of data.

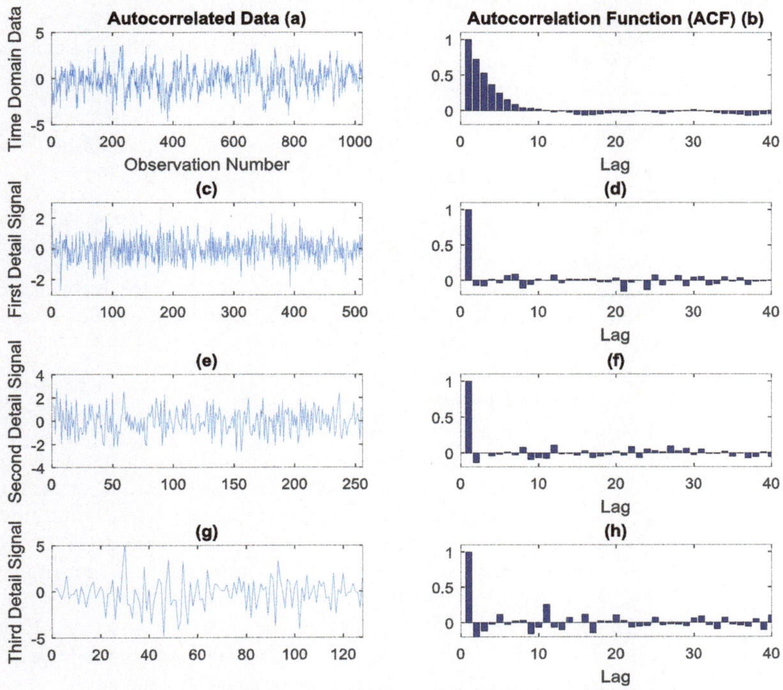

Figure 3: Decorrelating auto correlated data at multiple scales.

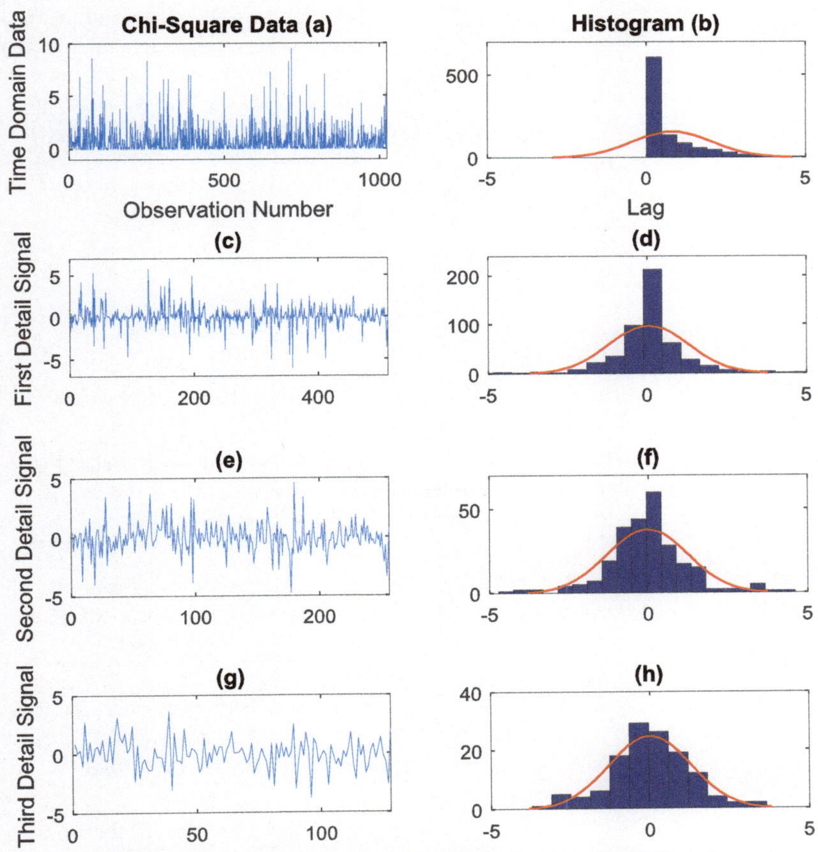

Figure 4: Distribution of chi-square data becomes Gaussian at multiple scales.

address many of the limitations that hinder the wide applicability of the Shewhart chart in practice. These advantages brought forward by multiscale representation are employed in this work to develop a multiscale Shewhart chart algorithm, which is described next.

Multiscale Shewhart chart fault detection algorithm

This section presents a multiscale Shewhart chart algorithm that can be used to provide improved performance over the conventional Shewhart chart. The idea behind this algorithm is to apply the conventional Shewhart using the detail signals as well as the final scaled signal obtained from the multiscale wavelet-based decomposition [45,46]. The algorithm consists of two phases, training and testing, as illustrated in Figure 5. In the first phase (training), fault-free training data are decomposed at multiple scales. The Shewhart chart is then applied using the detail signals at different scales and the final scaled signal, during which the control limits at each scale are computed (these limits will be used later in the testing phase). These control limits are used to threshold the wavelet coefficients (the detail signals) at each scale and the final scaled signal, and for any scale where any violation of limits is detected, all wavelet coefficients from that scale are retained. All retained signals are then reconstructed, and the Shewhart chart is applied once again on the reconstructed data set, where the control limits are computed once again. In the second phase (testing), the data are again decomposed at multiple scales (using the same wavelet filter utilized during the training phase). The control limits computed for each scale from the training phase are applied to the detail signals (and final scaled signal) for the testing data, and only the retained coefficients (after thresholding) are used to reconstruct the data set back in the time domain. Finally, the control limits obtained from the reconstructed training data are applied to the reconstructed testing data in order to detect any possible faults. This multiscale Shewhart chart algorithm is schematically represented in Figure 5.

It is important to note that for the wavelet reconstruction phase of the multiscale Shewhart fault detection algorithm, there are theoretically four alternative methodologies that can be used when deciding how to retain wavelet coefficients that violate the threshold limits at a particular scale [47]:

1. Retaining the entire scale of wavelet coefficients for both the training and testing data sets.

2. Retaining the entire scale of wavelet coefficients for the training data set, but only retaining the wavelet coefficients from each scale for the testing data set.

3. Retaining only the wavelet coefficients from each scale for both the training and testing data sets.

4. Retaining only the wavelet coefficients for the training data set, but retaining the entire scale of wavelet coefficients for the testing data set.

Methodology 2 was selected for the fault detection algorithm for the following reasons:

- Retaining the entire scale of wavelet coefficients during the training phase of the algorithm increases the number of observations available in the reconstructed training data set. This is particularly important from a fault detection perspective for the multiscale Shewhart chart, as a larger number of observations in the reconstructed training data set will give a more accurate estimation of the true standard deviation of the data. Retaining the entire scale of wavelet coefficients for scales that violate the threshold limits has the added advantage of only retaining scales that have important features and disregarding scales that do not contribute important features to the data set, thus acting similar to a filter.
- Since the standard deviation of the reconstructed data set is not an issue for the testing data, and we are only concerned with capturing faulty observations, retaining an entire scale when a threshold limit violation occurs is not necessary. If a fault is present in the testing data set, the final scaled signal will be able to detect the fault accurately, and reconstructing only coefficients that violate the threshold limits will show up as a clear pulse in the reconstructed testing data set, thereby violating the fault detection limits of the multiscale Shewhart chart.

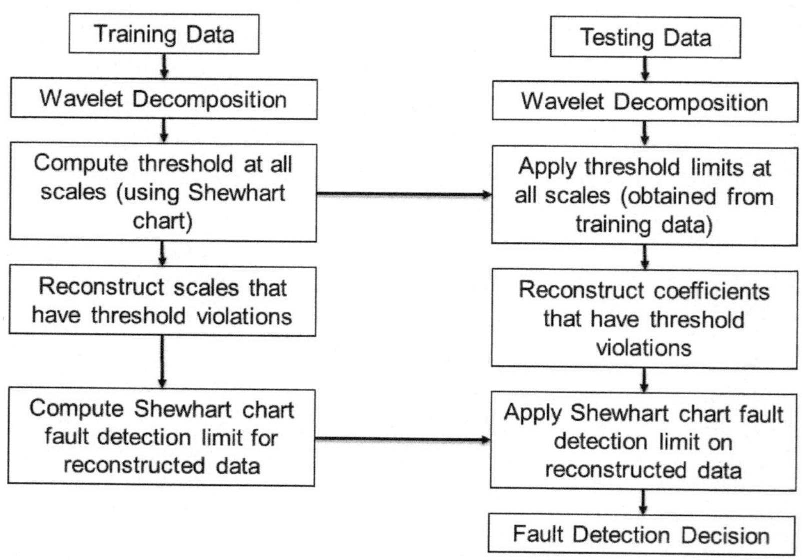

Figure 5: Schematic illustration of the developed multiscale Shewhart chart fault detection algorithm.

Illustrative Examples

This section compares the performance of the developed multiscale Shewhart chart to that of the conventional one using two illustrative examples: using synthetic data and using process data generated from a simulated distillation column. The simulated synthetic example examines the performance of both charts using data generated under a variety of conditions, in order to examine the robustness of the developed Shewhart chart, especially when the fundamental assumption of the conventional chart are violated.

Performance assessment under different levels of noise

Process data are usually contaminated with errors or noise. An essential feature of any monitoring technique used in the process industry is that it needs to be able be able to handle process data that are contaminated with noise, since it can severely deteriorate the performance of the technique. The higher the noise level, the more challenging it becomes to distinguish the difference between random variations in the process and faults. Therefore, this section examines the performance of both Shewhart charts at different noise levels.

In order to carry out this assessment, training data (consisting of 512 observations) that follow a standard normal distribution (of zero mean and unit variance), is used to compute the fault detection limits of both charts. Similarly, testing data (consisting of 512 observations) is generated, with additive steps faults of ± 3 added at two locations between observations 101-150 and 401-450, respectively. The simulation is then repeated using different noise standard deviations (σ=0.01- 5). This wide range of noise allows assessing the robustness of both charts to the presence of noise. A Monte-Carlo simulation of 5000 realizations is carried for each level of noise to ensure that accurate results are obtained and meaningful conclusions can be drawn. At each noise level, the performance of both charts are assessed by computing three fault detection metrics: missed detection rate (%), false alarm rate (%), and out-of-control average run length (ARL_1). Examining the performance of both charts, utilizing all three metrics allows us to assess their advantages and limitations.

The results of missed detection and false alarm rates for both charts are illustrated in Figure 6. These results show that even though the noise level does not change the false alarm rate, the missed detection rate increases at higher noise levels and can reach nearly 100% for very high levels of noise. Furthermore, it can also be seen that the false alarm rates are comparable for both techniques, and that the missed detection rate is consistently lower for the multiscale method than the conventional method. This can be attributed to the noise-feature separation characteristic of the multiscale method highlighted previously. with improve the detection ability of the Shewhart chart providing a reduction in missed detection rate of approximately 40% for certain noise levels.

The ARL_1 results on the other hand, are illustrated in Figure 7. These results show that for low to moderate levels of noise (σ=0.01-2), the conventional method provides comparable or slightly lower ARL_1 values than the multiscale method. However, for higher levels of noise (σ=2-5), it is evident that the multiscale method provides much lower ARL_1 values, which are due to the noise-feature separation advantage of multiscale representation, especially at high noise levels. The advantage of using the multiscale Shewhart chart over the conventional one when dealing data that are contaminated with noise can be further demonstrated by comparing the time series evolutions of the two charts at different levels of noise as illustrated in Figure 8. The results shown in Figure 8 demonstrate that at low noise levels, e.g., 0.5, the performances of both charts are comparable, with both charts achieving nearly 100% detection see Figures 8a and 8b. However, at moderate levels of noise, e.g., σ=2.5, the multiscale Shewhart outperforms the conventional one with much better detection (Figures 8c and 8d). At even higher noise levels, e.g., σ=4.5, the multiscale Shewhart chart shows even better fault detection abilities and even quicker detection (i.e., smaller ARL_1) as shown in Figures 7, 8e and 8f. These results clearly show that when process measurements are contaminated with noise, utilizing the multiscale Shewhart chart results in faster fault detection (lower ARL_1), which would allow quicker response thereby increasing the chance of averting any catastrophic disasters.

Performance assessment under different levels of autocorrelation

The second main assumption of the Shewhart chart is that the

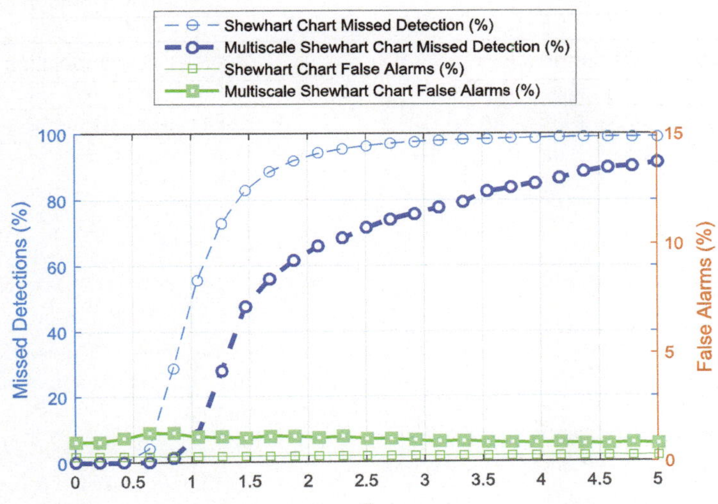

Figure 6: Performance comparison for the conventional and multiscale Shewhart charts for different noise levels (Missed Detection and False Alarm rates).

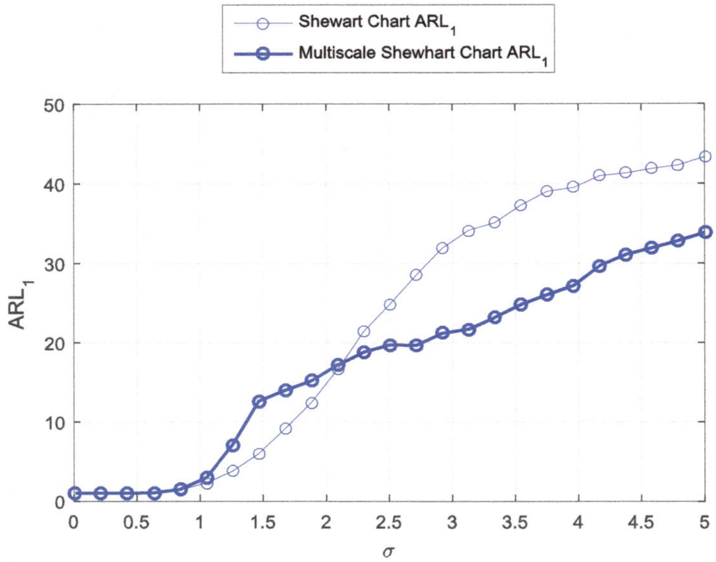

Figure 7: Performance comparison of the conventional and multiscale Shewhart charts for different noise levels (ARL$_1$).

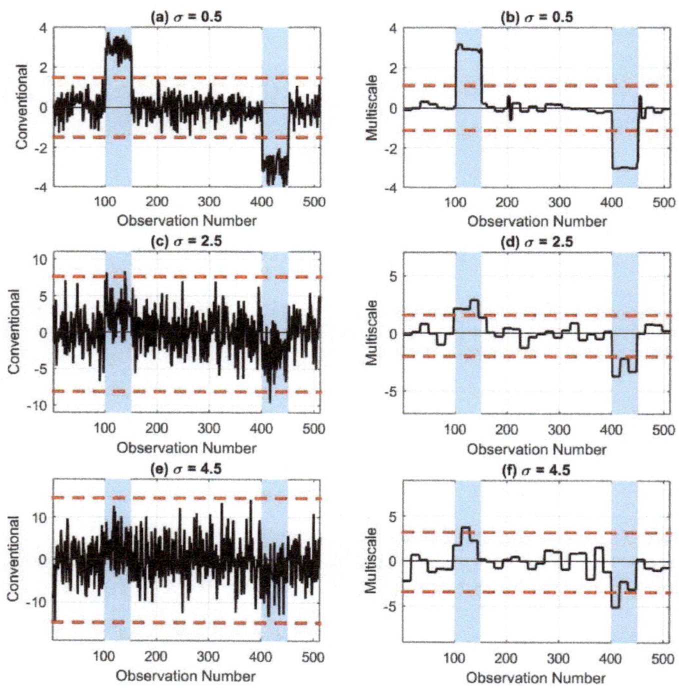

Figure 8: Time evolution comparison of the conventional and multiscale Shewhart charts for different noise levels.

process residuals being evaluated are uncorrelated (consecutive observations are independent). In practice, however, this assumption may not hold due to possible modeling errors or malfunctioning sensors that introduce correlation between consecutive observations, which can greatly affect the performance of the Shewhart chart. As described previously, multiscale, multiscale representation helps decorrelate autocorrelated data, which should provide the multiscale Shewhart chart an advantage when dealing with autocorrelated process measurements. This section presents a comparison and an assessment of the performances of both charts at different degrees of autocorrelation.

In order to carry out this assessment, an autoregressive AR (1) model (Equation 6) is used to generate training data (consisting of 512 observations). Similarly, testing data (consisting of 512 observations) are generated, but with additive step faults of magnitude 3 added at two locations between observations 101-150 and 401-450, respectively.

Then, the conventional and multiscale Shewhart charts are used to detect these faults. To provide a thorough assessment of the effect of autocorrelation and the robustness of both charts to autocorrelated data, the simulation is repeated for a wide range of autocorrelation (a = 0.01-0.99). Once again, a Monte-Carlo simulation of 5000 realizations is carried out for each level of autocorrelation, to ensure that accurate results are obtained and meaningful conclusions can be drawn.

The results of the simulation for both charts with regards to the missed detection and false alarm rates are illustrated in Figure 9. As Figure 9 demonstrates, although both charts show that the false alarm rate remains relatively constant (except at extremely high values of autocorrelation), the missed detection rate increases (to nearly 100%) as the degree of autocorrelation increases for both charts. Even though the false alarm rates are comparable for both charts, the missed detection rates obtained by the multiscale Shewhart chart are consistently lower than those obtained by the conventional chart (providing nearly 40% lower values of missed detection for a wide range of autoregressive coefficient). This advantage of the multiscale method is due to the ability of multiscale representation to decorrelate data at multiple scales as described in Section 3.1.3. Furthermore, Figure 10 that illustrates the ARL_1 values for both charts, shows that they provide comparable values over the entire range of autoregressive coefficient. This means that even though the multiscale Shewhart chart provides a better missed detection rate, it does not provide clear advantages in terms of the speed of detection.

The advantage of using the multiscale Shewhart chart over the conventional one when dealing when data that are autocorrelated, can be further demonstrated by comparing the time series evolutions of the charts at different levels of autoregressive coefficient as illustrated in Figure 11. The results illustrated in Figure 11 show that the multiscale Shewhart chart outperforms the conventional one at low and moderate autocorrelation levels (i.e., *a*=0.1 and *a*=0.5) as illustrated in Figures

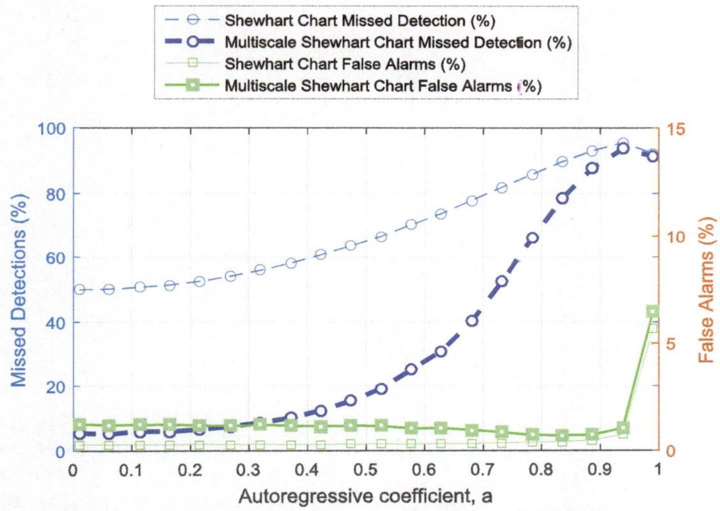

Figure 9: Performance comparison for the conventional and multiscale Shewhart charts for different autocorrelation levels (Missed Detection and False Alarm rates).

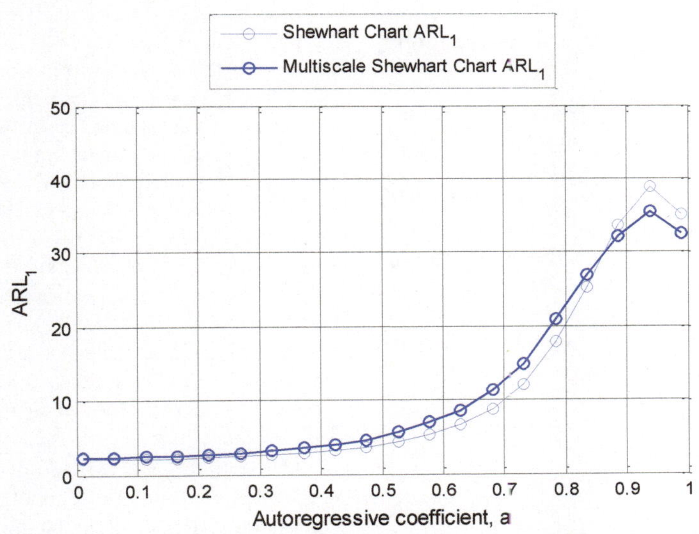

Figure 10: Performance comparison for the conventional and multiscale Shewhart charts for different autocorrelation levels (ARL_1).

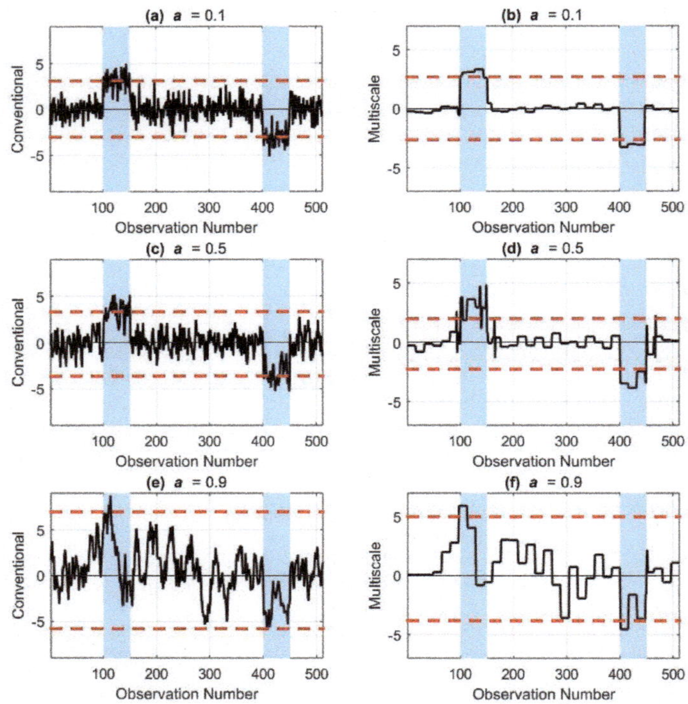

Figure 11: Time evolution comparison of the conventional and multiscale Shewhart charts for different autocorrelation levels.

11a and 11b for the case of *a* 0.1 and in Figures 11c and 11d for the case of *a*=0.5. At very high level of autocorrelation, however, as in the case where *a*=0.9, both charts fail to properly detect both faults in the data as illustrated in Figures 11e and 11f. This very high of autocorrelation is not very commonly encountered in practice, which makes the multiscale technique applicable in most practical situations. Therefore, utilizing the developed multiscale Shewhart when dealing with autocorrelated data is advantageous. Even though the response time is comparable (similar ARL_1 values) for both techniques, the multiscale chart is able to provide significantly lower missed detection rates over a wide range of autocorrelation coefficient, proving to be more robust that the conventional chart.

Performance assessment with distributions of varying degrees of non- normality

The third main assumption of the Shewhart chart is that it assumes that the fault-free process residuals being evaluated follow a Gaussian (normal) distribution. Practical data, however, might not always satisfy this assumption, thereby degrading the performance of the Shewhart chart. This section assesses and compares the performances of both charts under violation of this assumption. In practice, non-Gaussian noise can be a result of malfunctioning sensors, which introduce a certain bias or skewed randomness to the process observations being monitored. Non-Gaussian noise may also be introduced to the data by modeling errors that leave non-modeled process variations in the model residuals. Therefore, this assumption also needs to be dealt with in order to carry out efficient and safe process monitoring. To assess the effect of the distribution on the quality of fault detection, distributions of different degrees of non-normality are needed, along with a metric to quantify the level or degree of normality. A chi-square distribution can be utilized for this purpose in order to obtain data sets with different degrees of normality. The chi square distribution has the following probability density function [29]:

$$f(x)=\frac{1}{2^{\frac{k}{2}}\Gamma(\frac{k}{2})}x^{\frac{k}{2}-1}e^{-\frac{x}{2}} \qquad (7)$$

where, *k* represents the number of degrees of freedom. Adjusting the parameter *k* produces distributions with different degrees of normality. There are a number of metrics that can be utilized to test the normality of a given distribution. The Shapiro-Wilk metric is often considered to be one of the most powerful univariate normality tests [44], and hence it is used in this work to quantify the deviation from normality. The Shapiro-Wilk metric varies from zero to one, where a value closer to one implies that the distribution is closer to normality, and vice versa. Further detail on the Shapiro-Wilk and other metrics commonly used to quantify the deviation from normality can be found in [30,44]. In order to perform this assessment, the chi-square distribution is used to generate training data (consisting of 512 observations), which are used to compute the control limits for both charts. Testing data are also generated in a similar manner, and additive step faults of magnitude $\pm3\sigma$ (where σ is the standard deviation of the generated training data) are added at two locations between observations 101-150 and 401-450, respectively. For this simulation it is important to generate the faults with opposite signs (one positive and one negative) in order to ensure that effect of the direction (bias) of skewness in the non-Gaussian distribution is eliminated when assessing both techniques. To examine the performance of both charts at different levels of non-normality, this simulation is repeated for a wide range of Shapiro-Wilk values (between approximately 0.75-0.99). Once again, a Monte-Carlo simulation of 5000 realizations is carried out for each value of Shapiro-Wilk, to ensure that accurate results are obtained and meaningful conclusions can be drawn. The results of the simulation for both charts with regards to the missed detection and false alarm rates are illustrated in Figure 12.

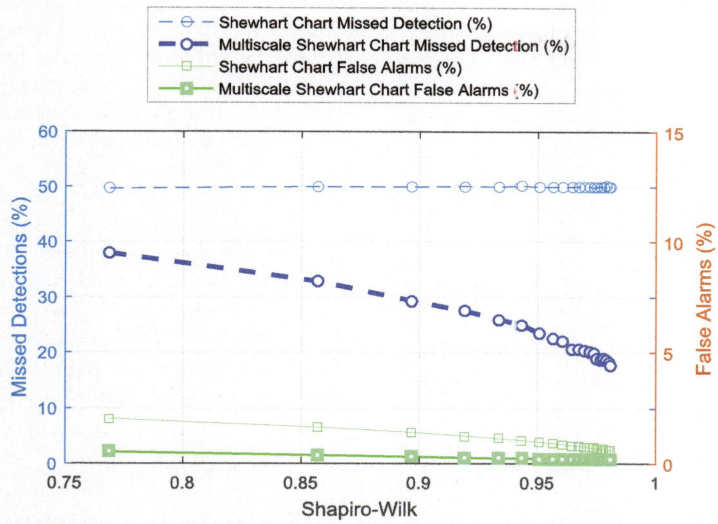

Figure 12: Performance comparison for the conventional and multiscale Shewhart charts for different cases of non-normality (Missed Detection and False Alarm rates).

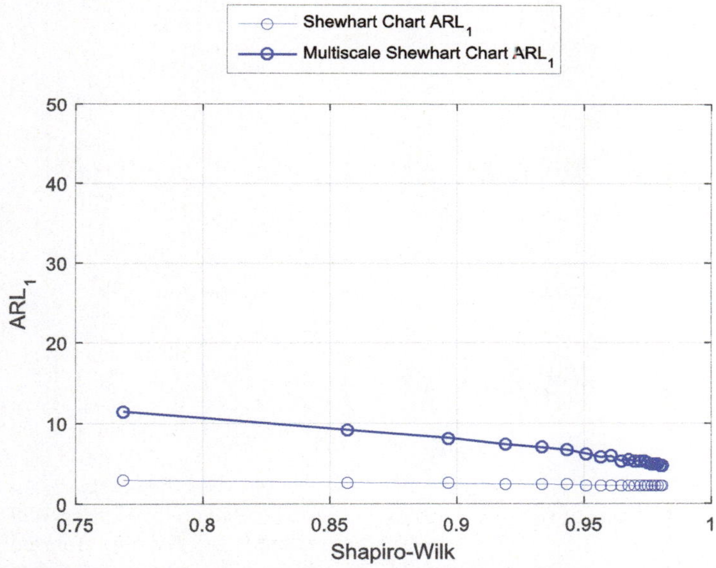

Figure 13: Performance comparison for the conventional and multiscale Shewhart charts for different cases of non-normality (ARL_1).

Figure 12 demonstrates that, for both charts, the missed detection and false alarm rates remain relatively constant at different degrees of non-normality. This will be explained and visually illustrated later in Figure 14. Figure 12 also shows that even though the false alarm rates are somewhat comparable for both charts, the missed detection rates for the multiscale methods are consistently better than the conventional method (with a reduction of nearly 30%) in the missed detection rate at some values of Shapiro-Wilk). These results can be attributed to the fact that multiscale representation provides detail signals that are approximately Gaussian at multiple scales even for non-Gaussian data [48]. On the other hand, Figure 13, that compares the ARL_1 for both charts, shows that even though the multiscale Shewhart chart provides a better detection rate, it is unable to provide lower ARL_1 values than the conventional chart, especially at lower values of Shapiro-Wilk. However, it should be noted that this increase in ARL_1 for the multiscale Shewhart chart is for data which deviate considerably from normality, which is not very common in practice. This issue may be remedied by fine-tuning the decomposition depth of the multiscale technique to ensure that both techniques provide similar ARL_1 values. An analysis of the choice of decomposition is presented in the upcoming section.

The advantage of using the multiscale Shewhart chart over the conventional one when dealing with data that deviate from normality, can be further demonstrated by comparing the time series evolutions of the charts at different degrees of deviation from normality (values of Shapiro-Wilk) as illustrated in Figure 14. These results indicate that even when the data are approximately normal, the multiscale Shewhart chart is able to outperform the conventional one, providing a reduction in the missed detection rate of nearly 30%. This advantage can also be seen for higher deviations from normality (i.e., for Shapiro-Wilk=0.87

and 0.75), but with narrow margins as illustrated in Figure 14c and 14d for the case of a moderate deviation from normality (Shapiro-Wilk=0.85) and in Figures 14e and 14f for the case of a high deviation from normality (Shapiro-Wilk=0.75). Additionally, it is important to note that the performance of the Shewhart chart is not highly affected by the deviation from normality of the data. This can be explained as follows: as the distribution of the data starts deviating from normality, the data becomes increasingly one-sided, and the control limits adjust to the direction of the distribution, which causes the performance of the conventional Shewhart chart to remain relatively constant over a wide range of non-normality as illustrated in Figure 14.

Effect of decomposition depth

An important decision when it comes to implementing the multiscale Shewhart chart in practice is the choice of decomposition depth, i.e., the number of scales to be used in the wavelet decomposition of the data. This section studies the effect decomposition depth has on the performance of the developed multiscale Shewhart chart. This analysis is performed by first generating training data (consisting of 512 observations) using a standard normal distribution (having zero mean and unit variance), which are then used to compute the control limits of the chart. The testing data are generated in a similar manner, and additive step faults of magnitude $\pm 3\sigma$ are introduced at two locations between observations 101-150 and 401-450, respectively, where σ represents the standard deviation of the training data set. This simulation is repeated for different decomposition depths, and a Monte-Carlo simulation of 5000 realizations is run for each decomposition in order to obtain accurate results and to draw meaningful conclusions. As with the previous examples, the performance of the chart is evaluated

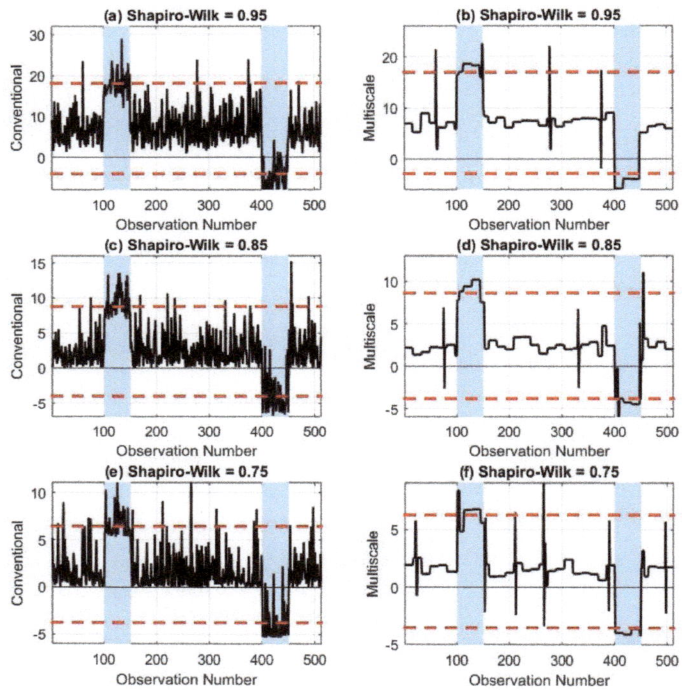

Figure 14: Time evolution comparison of the conventional and multiscale Shewhart charts for cases of non-normality.

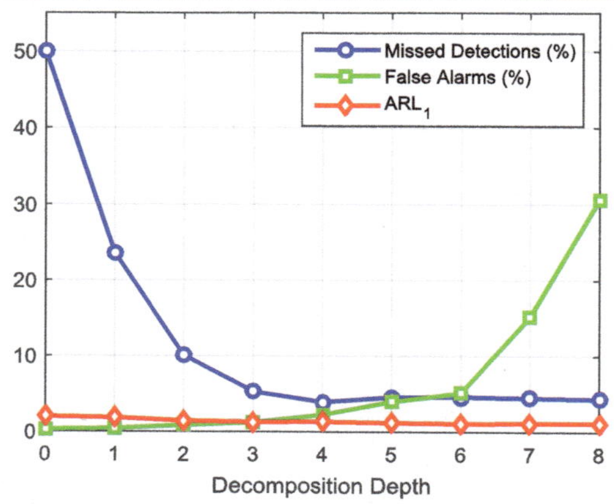

Figure 15: Analysis of the effect of choice of decomposition on the performance of the multiscale Shewhart chart.

using all three fault detection metrics. The mean value of all three metrics is computed from all 5000 realizations at each decomposition depth and the results are shown in Figure 15. The results show that the missed detection rate decreases as the number of scales increases, but up to a certain scale after which it settles to an almost constant value. Conversely, the false alarm rate increases as the number of scales increases. These observations are due to the fact that at coarser scales, more noise is removed and features are better detected, which decreases the missed detection rate. Also, at coarser scales, the control limits get tighter and tighter, which results in all the detail coefficients being retained. This would therefore not add any further detection advantage, but results in a slight increase in the false alarm rate. Figure 15 also shows that the ARL_1 is not greatly affected by the choice of decomposition depth. Therefore, in this example, an optimum depth of four can be selected. From practical experience, a depth of around half of the maximum possible decomposition depth is recommended.

Monitoring of a simulated distillation column using the Multiscale Shewhart chart

Although the illustrative examples presented thus far have made a strong case for using the developed multiscale Shewhart chart especially when dealing with process data that violate the assumptions of the conventional method, it is essential to examine how the charts perform in a practical situation. This section provides such an example where the performance of both Shewhart charts are assessed by applying them using process data from a simulated distillation column. Separation processes are a crucial part in many chemical plants, and distillation columns are often among the most commonly used separation units. The operation of distillation columns needs to be monitored to ensure that quality of products in the different streams meet the sought standards, and to ensure that the process is running safely.

It is possible for explosions to occur at distillation columns, if improper monitoring measures are in place. One such example is an explosion that took place on July 30, 2000, at a manufacturing company that produces specialty gases located in Dayton, NV. Although, no staff were injured at the facility, the building structure that housed the equipment was severely damaged [49]. The incident occurred in the distillation column that was used to separate residual fluorine and nitrogen gas from condensed nitrogen trifluoride. The incident occurred because the facility received liquid nitrogen that was much cooler than the nitrogen stored at the facility. The liquid nitrogen was used as a coolant in the heat exchanger and the colder temperature of the coolant forced the fluorine to liquefy. The incident investigation report concluded that condensed fluorine had reactor with the packing material made of stainless steel, and had caused an exothermic reaction to occur inside the reboiler/condenser of the distillation column. Combustion between the steel material and liquid nitrogen trifluoride was initiated, and the energy that was released from the exothermic reactions caused pressure and temperature to build up in distillation column, leading to an eventual explosion [49]. Therefore, it essential for streams in the distillation column, and other equipment in industrial facilities to be continuously monitored efficiently, to ensure incidents such as these do not occur or are minimized and handled before leading to more catastrophic ones. In this example, Aspen Tech 7.2 was utilized in order to simulate distillation column data. The goal of the simulation was to monitor the distillate stream of the column. The distillation column consisted of 32 theoretical layers, including both a total condenser and a reboiler. The feed stream (which has a composition of 60 mole% isobutene and 40 mole% propane), enters the column at a flow rate of 1 kmol/s as a saturated liquid at at stage 16. The nominal operating conditions for this distillation column are provided in [50]. In order to generate dynamic data, the reflux and feed flow rates are perturbed from the nominal values at which they operate, and step changes (with magnitude ± 2%) are introduced to the feed flow rate. Once the process has settled to a new steady state, similar perturbations are introduced in the reflux flow rate, and it is allowed to settle to a new steady state. A data set of 1024 observations is generated, which consists of the temperatures at different trays and the compositions in the product streams. These data are assumed to be noise-free and therefore are contaminated with Gaussian noise of zero mean. From a fault detection perspective, the goal of the simulation is to monitor the propane composition in the distillate stream (x_D). It is essential that all streams in a chemical process meet the required

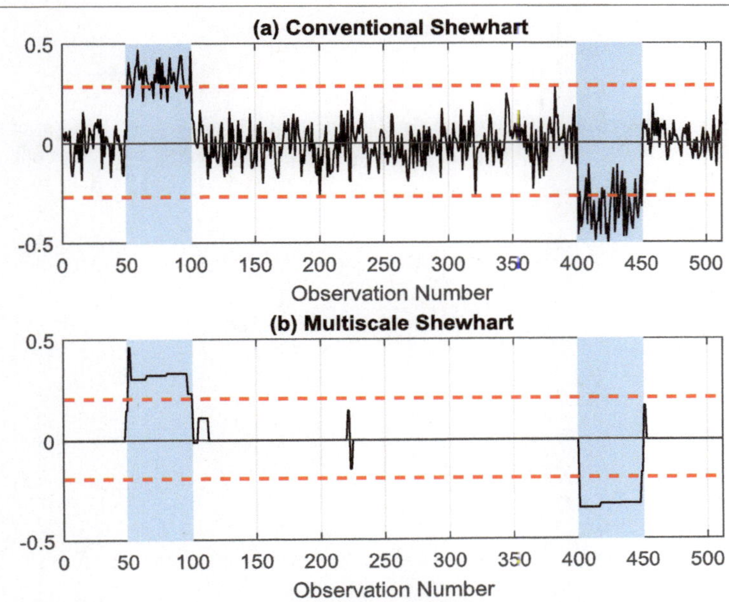

Figure 16: Comparison in performance of the conventional and multiscale Shewhart charts for a step fault of magnitude σ=3.5 in the process residuals.

specifications to avoid disastrous consequences.

The available process observations need to be compared to their reference values in order to compute the process residuals so that monitoring charts can be applied. A Partial Lest Squares (PLS) regression model was constructed and utilized in order to obtain the required process residuals. The process residuals are then split into two sets, 512 observations each, to be used as training and testing data sets. The training data are used to compute the fault detection (control) limits, which would then be utilized to detect faults in the testing data. Two step faults of magnitude ± σ 3.5 are added to the testing data at two locations between observations 51-100 and 401-450, respectively. Figure 16 illustrate the fault detection results for both the conventional Shewhart chart Figure 16a and the multiscale Shewhart chart Figure 16b. As demonstrated in Figure 16, the multiscale Shewhart chart is more able to detect the fault, with a significantly lower missed detection rate (~ 0%) when compared to the conventional Shewhart chart, which also detects most of the faults. Figure 16 also shows that the multiscale Shewhart chart does provide a lower false alarm rate when compared to the conventional chart.

The same simulation was also repeated for a step fault of magnitude ± 2.5 σ, again added between observations 51-100, and 401-450, respectively, and the results are illustrated in Figure 17. These results show that the conventional Shewhart chart is unable to detect most of the fault Figure 17a while the multiscale Shewhart provided nearly 100% detection. Figure 17 also shows that the multiscale Shewhart chart provides lower false alarm rates than the conventional one.

As illustrated through this simulated distillation column example, the developed multiscale Shewhart chart does outperform the conventional chart for two different magnitudes of step faults, as it is able to better detect the fault in both scenarios. Therefore, implementing the multiscale Shewhart chart for monitoring real chemical industries is advantageous.

Conclusion

Most univariate process monitoring techniques are known to rely on three main assumptions: that the process residuals are normally distributed, independent, and only contain moderate levels of noise. Data collected from sensors may violate one or more of these assumptions, which affects the performance of most univariate monitoring techniques. Multiscale representation has been shown to improve noise-feature separation in data, approximately decorrelate autocorrelated data, and transform data so that they better follow a normal distribution and multiple scales. In this work, these advantages of multiscale representation were utilized to enhance the performance of the Shewhart chart, through developing a multiscale Shewhart chart algorithm that can deal with these assumption violations. The multiscale algorithm relies on decomposing the data at multiple scales and applying the Shewhart chart using all detail signals and the last scaled signal, before being reconstructed back to the time domain. The performance of both charts, conventional and multiscale was then compared through simulated examples, in which the fundamental assumptions of univariate charts were violated at different levels using Monte-Carlo simulations. The performance was evaluated using three different criteria: the missed detection and false alarm rates, and as well as the out-of-control average run length, ARL_1. The results showed that the impacts of noise level and autocorrelation were most severe, as they led to missed detection rates of approximately 100% (at high levels of noise and autocorrelation) for the conventional chart. The impact of process residuals deviation away from normality was less critical. The results also showed that the multiscale technique was able to provide a reduction of approximately 40% in the missed detection rate over a wide range of noise levels, with a comparable false alarm rate, and lower ARL_1 than the conventional technique. The multiscale technique also provides a reduction of over 40% in the missed detection rate for a wide range of autocorrelation (except at high levels of autocorrelation) and lower ARL_1 values, while maintaining comparable false alarm rates, when compared to the conventional technique. Under violation of the normality assumption, however, the multiscale technique results in reductions in both the missed detection and false alarm rates for a wide range of non-normality, but in slightly higher ARL_1 values than those obtained by the conventional technique. The choice of decomposition

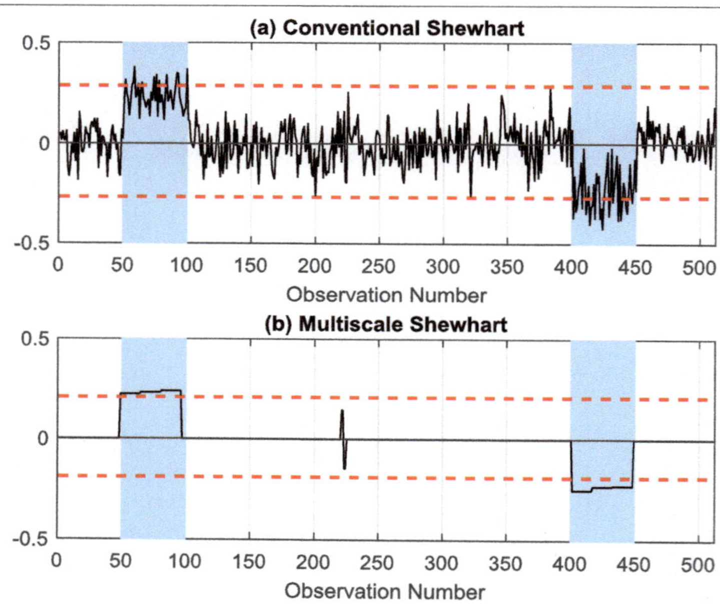

Figure 17: Comparison in performance of the conventional and multiscale Shewhart charts for a step fault of magnitude σ=2.5 in the process residuals.

depth is also an important factor in the implementation of the developed multiscale technique, and after evaluation of both charts using the three performance criteria for a range of decomposition depths, it was concluded that the optimal decomposition depth would generally be approximately half of the maximum possible decomposition depth for a given data set. The developed technique was additionally applied to a real world application, in which it is used to monitor the operation of a distillation column using simulated data. When used to monitor the distillate stream composition at two different fault sizes (2.5σ and 3.5σ), the multiscale approach was able to detect both faults with fewer missed detections and false alarms than the conventional technique. The relative computational simplicity of the proposed multiscale algorithm definitely encourages its implementation to improve process safety and product quality in a wide range of industrial processes. Although the developed algorithm does demonstrate significantly improved performance over the conventional technique, especially when the assumptions are violated, there are a few directions for future developments. The advantages obtained by the multiscale Shewhart chart algorithm can be extended to develop other univariate and multivariate multiscale monitoring techniques as well.

Acknowledgements

This work was made possible by NPRP grant NPRP7-1172-2-439 from the Qatar National Research Fund (a member of Qatar Foundation). The statements made herein are solely the responsibility of the authors.

References

1. Sarrate R, Aguilar J, Nejjari F (2007) Event-based process monitoring. Eng Appl Artif Intell 20: 1152-1162.
2. Venkatasubramanian V, Rengaswamy R, Yin K, Kavuri SN (2003) A review of process fault detection and diagnosis: Part I: Quantitative model-based methods. Comput Chem Eng 27: 293-311.
3. Venkatasubramanian V, Rengaswamy R, Yin K, Kavuri SN (2003) A review of process fault detection and diagnosis: Part II: Qualitative models and search strategies. Comput Chem Eng 27:313-326.
4. Nan C, Khan F, Iqbal MT (2008) Real-time fault diagnosis using knowledge-based expert system. Process Saf Environ Prot 86: 55-71.
5. Chetouani Y (2014) A sequential probability ratio test (SPRT) to detect changes and process safety monitoring. Process Saf Environ Prot 92: 206-214.
6. Rengaswamy R, Venkatasubramanian V (1995) A syntactic pattern-recognition approach for process monitoring and fault diagnosis. Eng Appl Artif Intell 8: 35-51.
7. Venkatasubramanian V, Rengaswamy R, Yin K, Kavuri SN (2003) A review of process fault detection and diagnosis: Part III: Process history based methods. Comput Chem Eng 27: 293-311.
8. Harrou F, Kadri F, Khadraoui S, Sun Y (2016) Ozone measurements monitoring using data-based approach. Process Saf Environ Prot 100: 220-231.
9. Yang Y, Liu Q, Gao Z (2015) Data fault detection in medical sensor networks. Sensors 15: 6066-6090.
10. Santos P, Villa LF, Reñones A (2015) An SVM-Based Solution for Fault Detection in Wind Turbines. Sensors 15: 5627-5648.
11. Martí L, Sanchez-pi N, Molina JM (2015) Anomaly detection based on sensors data in the petroleum industry applications. Sensors 15: 2774-2797.
12. Chakrabarty A, Mannan S, Cagin T (2015) Multiscale modeling for process safety applications. 1st edn. Butterworth-Heinemann.
13. US Chemical Safety Board (2007) Investigation Report Refinery Explosion and Fire. Chemical Safety and Hazard Investigation 341.
14. Shewhart WA (1930) Economic quality control of manufactured product. Bell Syst Tech J 9: 364-389.
15. Wang J, Luo C (2012) Long-period fiber grating sensors for the measurement of liquid level and fluid-flow velocity. Sensors 12: 4578-4593.
16. Page ES (1961) Cumulative Sum Charts. Technometrics 3: 1-9.
17. Roberts SW (1959) Control chart tests based on geometric moving averages. Technometrics 1: 239-250.
18. Hunter JS (1986) The exponentially weighted moving average. J Qual Technol 18: 203-210.
19. Cinar A, Undey C (1999) Statistical process and controller performance monitoring: A tutorial on current methods and future directions. Proc Am Control Conf, pp: 2625-2639.
20. Cinar A, Palazoglu A, Kayihan F (2007) Chemical Process Performance Evaluation. 1st edn. CRC Press, Boca Raton, FL, USA.
21. Cinar A, Parulekar SJ, Under C, Birol G (2003) Batch Fermentation. 1st edn. Marcel Dekker, Inc., New York, USA.
22. Cai L, Tian X (2013) A new fault detection method for non-Gaussian process based on robust independent component analysis. Process Saf Environ Prot 92: 645-658.
23. Lu CW, Reynolds MRJ (2001) CUSUM Charts for Monitoring an Autocorrelated Process. J Qual Technol 33: 316-334.
24. Ganesan R, Das TK, Venkataraman V (2004) Wavelet-based multiscale statistical process monitoring: A literature review. IIE Trans 36: 787-806.
25. Ganesan R (2007) Real-time monitoring of complex sensor data using wavelet-based multiresolution analysis. Int J Adv Manuf Technol 39: 543-558.
26. Reynolds MR, Stoumbos ZG (2004) Should observations be grouped for effective process monitoring? J Qual Technol 36: 343-366.
27. Guo H, Paynabar K, Jin J (2011) Multiscale monitoring of autocorrelated processes using wavelets analysis. IIE Trans 44: 312-326.
28. Montgomery DC (2013) Introduction to Statistical Quality Control. 7th edn. John Wiley & Sons, Hoboken, NJ, USA.
29. Montgomery DC, Runger GC (2011) Applied Statistics and Probability for Engineers. 5th edn. John Wiley & Sons Inc., Hoboken, NJ, USA.
30. Sheriff MZ (2015) Improved Shewhart chart using multiscale representation. Texas A&M University.
31. Mallat S (1989) A theory for multiresolution signal decomposition: the wavelet representation. Pattern Anal Mach Intell IEEE Trans II: 674-693.
32. Nounou MN (2006) Multiscale finite impulse response modeling. Eng Appl Artif Intell 19: 289-304.
33. Nounou HN, Nounou MN (2006) Multiscale fuzzy Kalman filtering. Eng Appl Artif Intell 19: 439-450.
34. Nounou MN, Nounou HN (2007) Improving the prediction and parsimony of ARX models using multiscale estimation. Appl Soft Comput 7: 711-721.
35. Bakshi B (1998) Multiscale PCA with application to multivariate statistical process monitoring. AIChE J 44: 1596-1610.
36. Davis J, Piovoso M, Hoo KA, Bakshi BR (1999) Process data analysis and interpretation. Adv Chem Eng 25: 1-103.
37. Daubechies I (1988) Orthonormal bases of compactly supported wavelets. Commun pure Appl Math XLI: 909-996.
38. Daubechies I (1992) Ten Lectures on Wavelets. Library of Congress Cataloging-in-Publication Data, Philadelphia, PA, USA.
39. Zhou S, Sun B, Shi J (2006) An SPC monitoring system for cycle-based waveform signals using Haar transform. Autom Sci Eng IEEE Trans 3: 60-72.
40. Kosanovich KA, Piovoso MJ (1997) PCA of wavelet transformed process data for monitoring. Intell Data Anal 1: 85-99.
41. Nounou MN, Nounou HN, Meskin N (2012) Multiscale denoising of biological data: a comparative analysis. IEEE/ACM Trans Comput Biol Bioinform 9: 1539-1544.
42. Nounou MN, Bakshi BR (2000) Multiscale methods for denoising and compression. In: Walczak B (ed), Wavelets Chem. Elsevier Science BV, pp: 119-150.
43. Box GEP, Jenkins GM, Reinsel GC (2008) Time Series Analysis: Forecasting and Control. 4th edn. John Wiley & Sons, Inc., Hoboken, NJ, USA.

44. Thode HCJ (2002) Testing for Normality. 1st edn. CRC Press, New York, NY, USA.

45. Sheriff MZ, Harrou F, Nounou M (2014) Univariate process monitoring using multiscale Shewhart charts. (2014) Int Conf Control Decis Inf Technol (IEEE), pp: 435-440.

46. Sheriff MZ, Nounou MN (2016) Enhanced performance of shewhart charts using multiscale representation. In: 2016 Am. Control Conf. IEEE, Boston, MA, USA, pp: 6923-6928.

47. Sheriff MZ, Nounou M (2016) Fault detection using multiscale Shewhart charts: Algorithms and applications. 2016 AIChE Spring Meet. Fuels Petrochemicals Div.

48. Aradhye H, Bakshi B, Strauss R, Davis J (2003) Multiscale SPC using wavelets: theoretical analysis and properties. AIChE J 49: 939-958.

49. Dekermenjian M, Reza A, Koonce M, Poblotzki J (2003) Exothermic reactions between cryogenic fluorine, nitrogen trifluoride and stainless steel. 37th Annu. Loss Prev. Symp. AIChE Spring Natl. Meet.

50. Madakyaru M, Nounou M, Nounou H (2012) Linear inferential modeling: theoretical perspectives, extensions, and comparative analysis. Intell Control Autom 3: 376-389.

The Impact of Stripping Gas Flow Rate on Triethylene Glycol Losses from Glycol Regeneration Unit: Simulation Study

Ibrahim TK[1], Abdulrahman RK[2], Khalaf FH[1] and Kamal IM[3]*

[1]Petroleum Engineering Department, Faculty of Engineering, Knowledge University, Kurdistan, Iraq
[2]Chemical Engineering Department, Faculty of Engineering, Koya University Ribwar, Kurdistan, Iraq
[3]Chemical Engineering Department, Faculty of Engineering, Soran University, Kurdistan, Iraq

Abstract

Natural gas can be considered as one of the most important fossil fuels that is increasingly consumed around the world. It is an environmental friendly fuel of high heating value. However, the unrefined natural gas produced from deep reservoirs may contain some impurities including sulfur and nitrogen compounds as well as water vapour. The presence of these undesirable compounds may induce corrosion and environmental pollution, because of that it is necessary to minimize or remove those impurities from natural gas streams. Triethylene glycol (TEG) with 99 wt.% is typically utilized to reduce the water vapor contact to less than 1 ppm. Indeed, glycol purities up to 99.9 wt% can be achieved by using stripping gases include Nitrogen. However, glycol losses is behind some technical problems in the dehydration process. The research is aimed at simulation the prospective gas dehydration process using Aspen HYSYS simulator. The effect of Nitrogen stripping rate on TEG losses from glycol regenerator tower, TEG lean wt.%, and water content in natural dry gas were investigated and correlated. The design of the dehydration system was simulated adequately to achieve a reduction in natural gas water content to less than 0.1 ppm.

Keywords: Natural gas dehydration; Aspen hysys; Process optimisation; Stripping rate; Triethylene glycol losses

Introduction

Natural gas is a major industrial and domestic fuel worldwide. Water vapour is probably the most common undesirable impurity in natural gas streams. Usually, it is not the water vapour itself, but rather the liquid or solid phase that may precipitate from the gas when it is compressed or cooled. Water condensation and solid gas hydrates formation in natural gas pipelines may cause blocking of the pipeline flow. Thus, natural gas should be dehydrated to a controlled water content to avoid hydrates, as well to minimize the corrosion problems due to the acids formed by carbon dioxide or hydrogen sulphide regularly found in natural gas streams [1].

The industrial dehydration process is carried out using several methods including direct cooling (condensation), adsorption and absorption [2]. Recently, membrane dehydrators are used. However, It is no doubt that absorption is the most important dehydration processes [3-5].

In direct cooling or (condensation), natural gas can be efficiently cooled using the Joule-Thompson effect [2]. Compression of the hot gases saturated with water at the lowest temperature value will resulted in prevention further condensation of water. Thus, well removing of water could be carried out [5]. In adsorption dehydration, removal of water vapour from a gas stream is conducted by using solid desiccant [6,7] The solid desiccants commonly used are molecular sieves [8]. In membrane dehydration, thin film polymer composite membranes are used in order to overcome the problem of emissions of hazardous volatile organic compounds (VOCs) may result using glycol dehydrators [9,10]. Liquid desiccants are used in absorption dehydration to remove water vapour from the gas. The desiccants used most frequently in absorption processes are solvents which have physical or chemical constituents. The physical solvents are glycols, methanol, and (a mixture of dimethyl ethers of polyethylene glycol (DEPG). The chemical solvents include amines, potassium carbonate and caustic. Hybrid solvents include both physical and chemical constituents are also used [11]. However, Glycols are used primarily to dehydrate gas streams [12]. Glycols are the favourite solvents because they have low solubility in natural gas, high hygroscopic characteristics, low vapor pressure and low volatility [13]. The glycols mostly used include ethylene glycol (EG), diethylene glycol (DEG) and triethylene glycol (TEG). However, triethylene glycol (TEG) is the most effective common sorbents for natural gas dehydration [14]. The advantage of using (TEG) is that, it can greatly reduce the emission of benzene, toluene, ethyl benzene and xylene (BTEX).

On the other hand, designing and optimization of TEG dehydration processes is widely carried out using simulation. In general, the more effective simulation software should enable to evaluate the operating conditions and optimize the design configurations and to allow solving many problems within a reduced time and a minimum investment [15]. The available commercial process simulators can offer different facilities to examine the entire process easily and with short time. Aspen HYSYS and ProMax are commonly used process modelling and simulation software in gas dehydration [16-18] as well as other gas processing operations [19]. Using [20,21] the component based framework could be easily customized, updated and maintained to meet the requirements of users. Moreover, implantation of other models of separation processes could be run along with its solution procedure by using computer programs such Visual Basic (VB), C^{++} subroutine, etc. The present work focuses on the study of the effect of stripping gas flow rate on TEG losses. The main platform used for this study was the simulation software: Aspen HYSYS. An adopted case study was conducted with a proposed stripping gas circulation rates. The TEG system operates at

***Corresponding author:** Kamal IM, Chemical Engineering Department, Faculty of Engineering, Soran University, Kurdistan, Iraq
E-mail: Ibtisam.kamal@soran.edu.iq

atmospheric pressure, and it is of 98% purity. The aim of the work is to study the effect of the flow rate of stripping gas on glycol loss takes place in glycol still of regenerator.

Materials and Methods

The absorption dehydration unit

In TEG absorption process, the wet gas is brought into contact with dry glycol in an absorber (a tray column or packed bed) [22]. Water vapour is absorbed in the glycol and consequently, its dew point reduces. The wet rich glycol then flows from the absorber to a regeneration system in which the entrained gas is separated and fractionated in a column and reboiler. The heating allows boiling off the absorbed water vapour and the water dry lean glycol is cooled by a heat exchange and pumped back to the absorber [22]. The entire process is illustrated in Figure 1.

Nevertheless, stripping gas is used when very pure glycol (up to 99.9% TEG) is required and cannot be achieved by the standard regeneration system. Stripping gas is typically dry fuel gas. It strips the remaining water from the glycol when bubbled through the hot glycol. Stripping gas can be introduced directly into the reboiler vessel or it can be admitted at a base of a packed column installed between the reboiler and the surge drum [23]. This method improves the stripping efficiency. It prevents the oxidation of glycol by prevention air from coming into contact with the dry glycol. The most important prerequisite for the cost effective operation of a glycol dehydration unit is the avoidance of glycol losses. In fact, there are several causes for glycol losses or carryover including the excessive foaming in absorber, glycol ageing, as well as due to technical problems relevant to equipment corrosion and, inaccurate absorber temperature and pressure. The use and rates of stripping gas should be adequately evaluated because the stripping gas that is added to the glycol in the regeneration unit is typically emitted from the still vent with the released water vapour into the atmosphere, it may significantly affect greenhouse gas emissions due to its high methane concentration [24]. The aim of the work is to study the impact of stripping gas flow rate on TEG loses, the concentration of the regenerated lean TEG, and dry gas water content using Aspen HYSYS. The importance of the study is the application of the simulation software Aspen HYSYS in modeling and optimization the parameters affecting the efficiency of one of the units in natural gas processing plants (regeneration unit).

Natural gas composition and operating conditions

Table 1 shows the sweet natural gas composition and operating conditions as well. It seems from Table 1 that the given natural gas stream is loaded with considerable amounts of water vapor that should be removed from the gas stream.

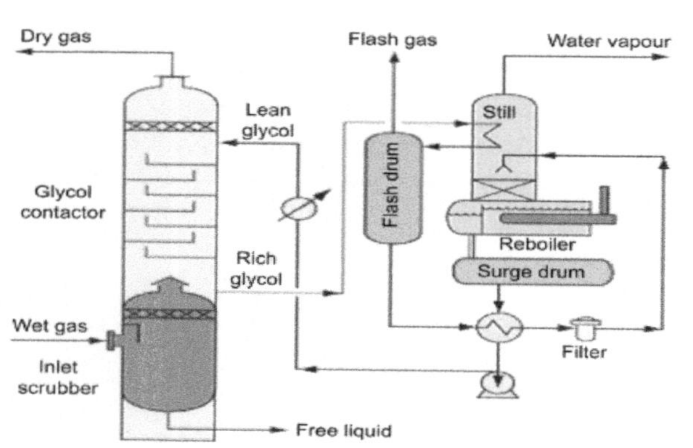

Figure 1: Scheme of absorption dehydration process (Bahadori).

Components	Mole %
Methane	0.7164
Ethane	0.1565
Propane	0.0606
i-Butane	0.0153
n-Butane	0.0276
i-Pentane	0.0114
n-Pentane	0.0083
n-Hexane	0.0013
Water	0.0023
Carbon dioxide	0.00004
Nitrogen	Trace
Hydrogen sulphide	Trace
	Operation Condition
Pressure	3555 kPa
Temperature	40°C
Flow rate	250 Million standard cubic feet of gas per day (MMSCFD)

Table 1: Natural gas composition and operation conditions.

Natural gas dehydration process simulation

The expected gas dehydration process is simulated by adopting Aspen HYSY software. The TEG is adopted as an absorbent to absorb water vapor that exists in the gas stream. Firstly, the first simulation step can be achieved by adding the gas stream compositions and operation conditions as given in Table 1. Figure 2 shows Hysys components list menu. Secondly, the Hysys fluid package can be chosen which should be (Glycol Pkg). Figure 3 shows Hysys fluid package menu.

The simulation environment is considered the main simulation area, that deals with the plant and shows the FPD for the process. The gas inlet separator should be added to remove any undesirable impurities for instance, solid particulars and liquids. Glycol absorber tower is also important part from the plant that it also require some specifications for instance, streams pressure and the TEG concentration (99% is used). Figure 4 shows glycol absorber menu. Rich glycol needs to be regenerate and that could be achieved by adding a regenerator. Figure 5 shows regenerator menu (Figure 6).

Finally, the simulation process has been run successfully. Figure 7 shows process flow diagram of natural gas dehydration process. The fresh TEG from the regenerator is cooled up and sends to the absorber tower. That can be achieved by installing a logical recycle operator that can be inserted between the two streams. Indeed, some of the process TEG could be lost from several process units for example separator and absorber. The lost TEG need be replaced. In the component splitter the TEG is separated from the dried gas and creating a stream of pure TEG that can be transferred back to the TEG stream. A mixer is also required to mix the pure TEG that recovered by the splitter with the fresh TEG from the regenerator. Figure 6 shows the splitter element menu.

The process regenerator possesses five trays and provided with a condenser and a boiler as well. Glycol purities up to 99.9 wt. % can be achieved by using stripping gas from the top of the stripping column. The stripping gas is usually nitrogen [25]. Water can be removed from the stripping gas by cooling it well below waters dew-point [26].

Results and Discussion

Researchers have envisioned the reasons of glycol carryover from natural gas dehydration units focusing on the various operation parameters

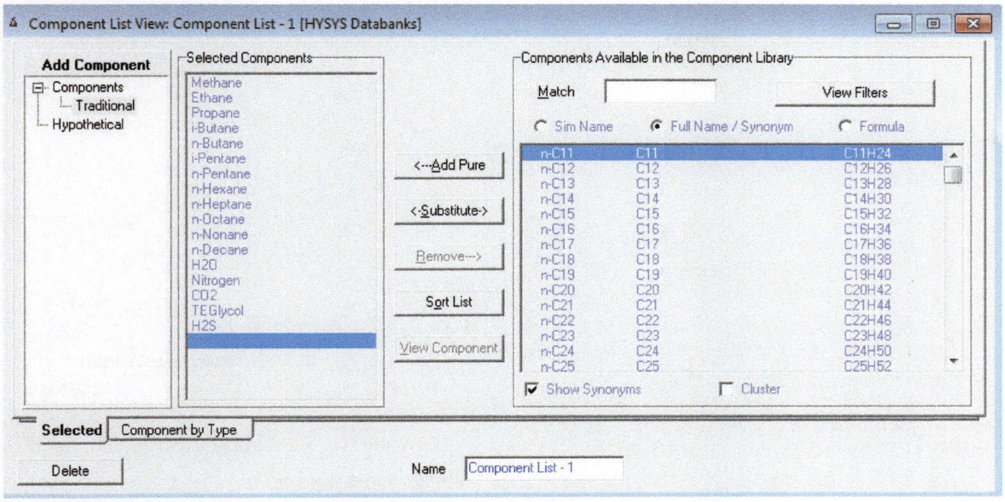

Figure 2: Hysys components list menu.

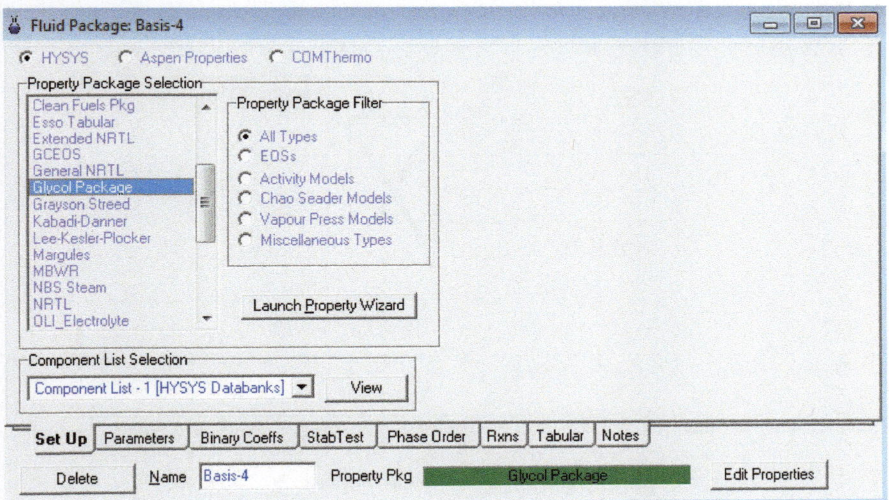

Figure 3: Hysys fluid package menu.

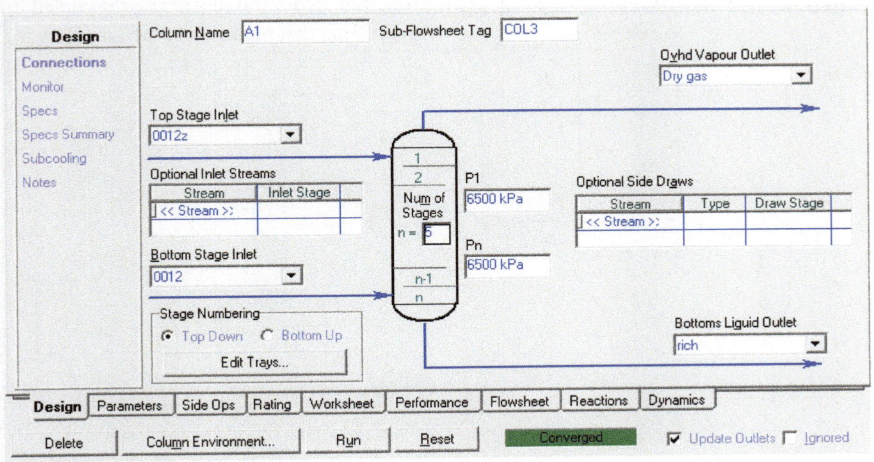

Figure 4: Hysys column design menu.

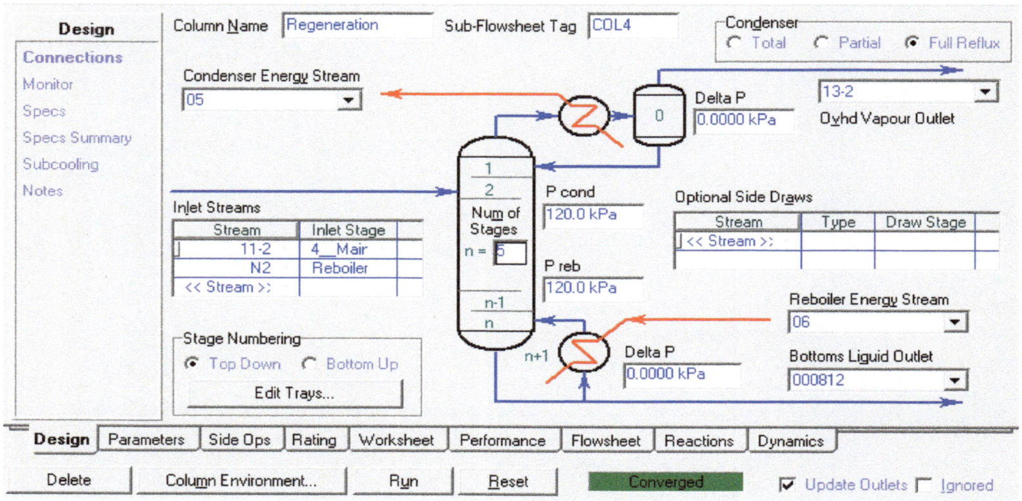

Figure 5: Hysys amine regenerator menu.

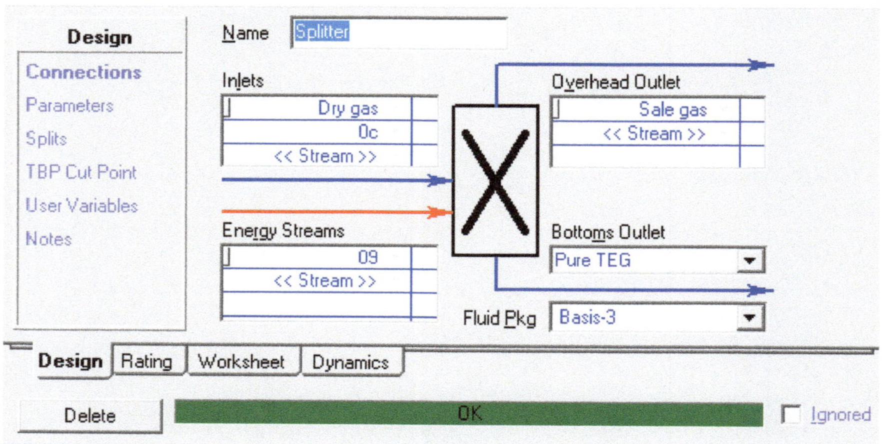

Figure 6: The splitter element menu.

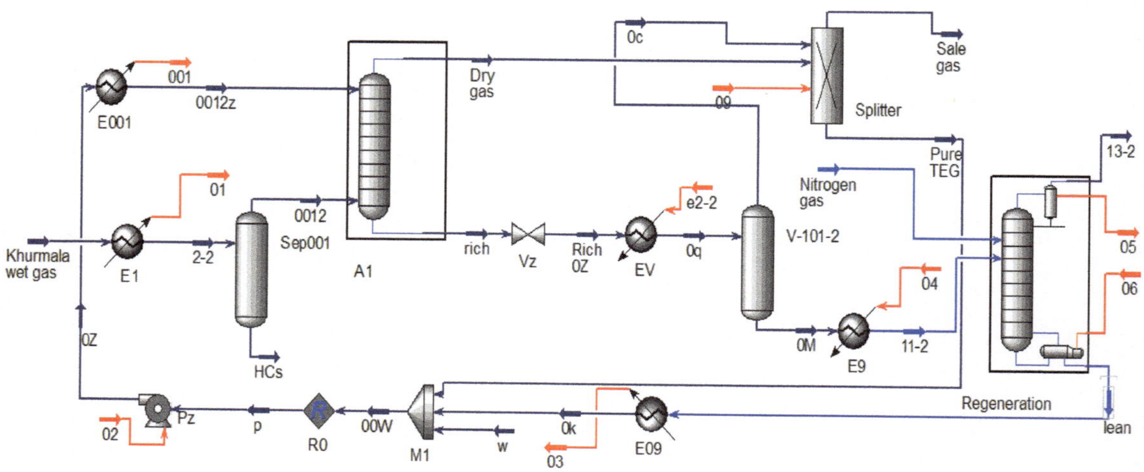

Figure 7: Process flow diagram of gas dehydration plant.

affecting the dehydration process including TEG circulation rate, lean TEG temperature and the dehydration operation conditions. However, studies about the effect of stripping gas flow rate on TEG losses from gas dehydration unit are still rare. In order to study the effect of the flow rate of the stripping gas on TEG losses from the top head stream of glycol regenerator, different flow rate values for Nitrogen as stripping gas have been applied in HYSYS software. They are: 28, 84, 160 and 420 Kg/hr.

The results of the plant simulation are illustrated in plots that correlate the design variables. Figure 8 shows the relationship between stripping gas flow rate and TEG losses from glycol regenerator top stream. It is well noticed that increasing the stripping gas flow rate will result in increasing the TEG losses from glycol regenerator outlet top stream. Moreover, the TEG loses could be excessive and could reach 2 Kg/hr at a stripping gas flow rate of about 400 kg/hr. A TEG flow rate of about 450 kg/hr could be applied to achieve a water content value in the dry gas of less than 1 ppm.

Variation of stripping gas flow rate and lean amine wt.% produced from the regnartor tower down stream is shown in Figure 9. It is illustrated that incresing stripping gas flowrate resulted in ncreasing the lean amine wt.%. However, the lean amine weight percent reaches equilibrum at a stripping gas flowrate of about 300 kg/hr The correlation between stripping gas flow rate (kg/hr) and water vapor amount in dry gas stream that produced from glycol absorber tower top stream is shown in Figure 10.

It is well noticed that when stripping gas flow rate increases, the water vapor amount in dry gas stream decreases, However, the glycol loss from glycol regenetor top stream increases (Figure 10).

TEG concentration required at the absorber to meet the dry gas dew point specification, and at the regenerator are crucial parameters for efficient dehydration. TEG vaporization loss may take place both in glycol contactor as well as in glycol still of regenerator. The vaporization loss from contactor is primarily due to higher gas inlet temperature, while the vaporization loss in glycol regenerator still is due to higher still temperature and high stripping gas rate [27,28].

Optimization the regenerator temperature as well as the stripping gas flow rate are key factor to minimize the TEG loss. The parameters will be optimized in a future study in order to realize what flow rate of stripping gas to use in order to minimize the TEG loss, and not to exceed the water content limit in natural gas stream.

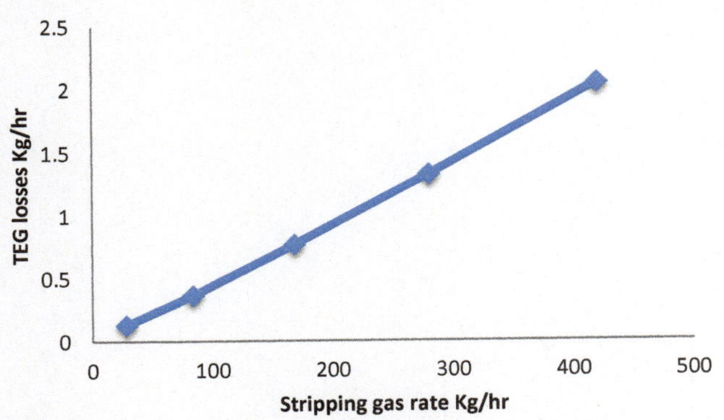

Figure 8: Variation of stripping gas flow rate and TEG losses from glycol regenerator top stream.

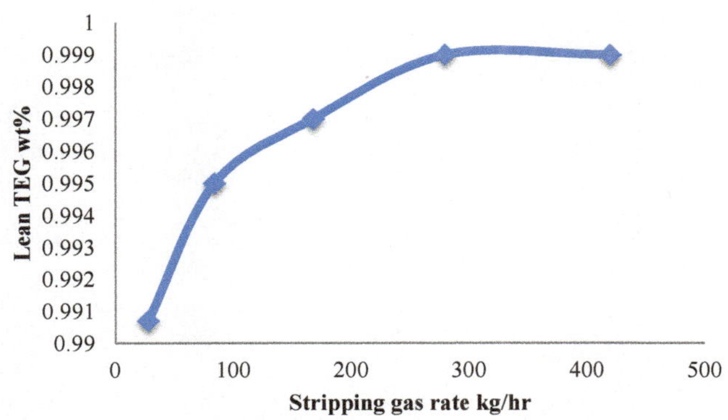

Figure 9: Stripping gas flow rate (kg/hr) versus the lean TEG wt. %.

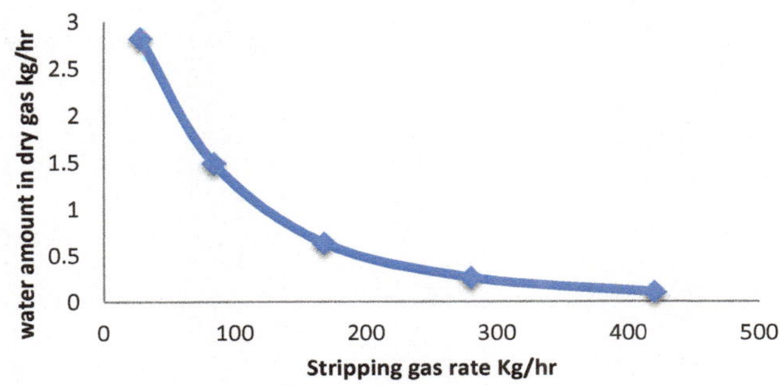

Figure 10: Effect of the stripping gas flow rate (kg/hr) on amount of water in dry gas stream.

Conclusions

In the current work, the impact of stripping gas flow rate on TEG loses, the concentration of the regenerated lean TEG, and dry gas water content was studied and correlated using Aspen HYSYS modeling and simulation software. To achieve the desired level of the parameters under investigation, series of plots and graphical correlations are prepared and demonstrated. The charts presented are based on the intensive calculations carried out by computer simulations which can be used for pilot plant TEG dehydration units to optimization the operating conditions as well as trouble shooting may face the operating units. The results obtained confirmed that increasing of stripping gas flow rate to 400 kg/h resulted in increasing the concentration of lean glycol up to 0.999 wt.%, and decreasing the water content value in the dry gas of less than 1 ppm, but increasing the TGE loss to about 2.2 kg/hr.

References

1. Partho SR, Ruhul AM (2011) Aspen-HYSYS Simulation of Natural Gas Processing Plant. Journal of Chemical Engineering IEB 26: 62-65.
2. Netusil M, Ditl P (2011) Comparison of three methods for natural gas dehydration. Journal of Natural Gas Chemistry 20: 471-476.
3. Bahadori A, Vuthaluru HB (2009) Simple methodology for sizing of absorbers for TEG (triethylene glycol) gas dehydration systems. Energy 34: 1910-1916.
4. Yildirim O, Kiss AA, Hüser N, Lemann K, Kenig EY (2012) Reactive absorption in chemical process industry: A review on current activities. Chemical Engineering Journal 213: 371-391.
5. Kolmetz K (2012) Gas Dehydration. Engineering Design Guideline. KLM Technology Group 1: 1-70.
6. Rohani SS (2009) Natural Gas Dehydration Using Silica Gel: Fabrication of Dehydration Unit. University Malaysia, Pahang.

7. Wang L, Yang RT (2014) New nanostructured sorbents for desulfurization of natural gas. Frontiers of Chemical Science and Engineering 8: 8-19.

8. Gholami M, Talaie MR, Roodpeyma S (2010) Mathematical modeling of gas dehydration using adsorption process. Chemical Engineering Science 65: 5942-594.

9. Runhon DJ, Liu L, Chakma A, Feng X (2010) Using poly (N,N-dimethylaminoethyl methacrylate)/polyacrylonitrile composite membranes for gas dehydration and humidification. Chemical Engineering Science 65: 4672-4681.

10. Lin H, Thompson SM, Serbanescu MA, Wijmans JG, Amo KD, et al. (2013) Dehydration of natural gas using membranes. Part II: Sweep/countercurrent design and field test. Journal of Membrane Science 432: 106-114.

11. Lunsford K, Mcintyre G (1999) Decreasing Contactor Temperature Could Increase Performance. Proceedings of the Seventy-Eigth GPA Annual Convention 78: 365-377.

12. Anyadiegwu CIC, Kerunwa A, Oviawele P (2014) Natural gas dehydration using triethylene glycol (TEG). Petroleum & Coal 56: 407-417.

13. Gandhidasan P, Farayedhi AA, Mubarak AA (2001) Dehydration of natural gas using solid desiccants. Energy 26: 855-868.

14. Isa MA, Eldemerdash U, Nasrifar K (2013) Evaluation of potassium formate as a potential modifier of TEG for high performance natural gas dehydration process. Chemical Engineering Research and Design 91: 1731-1738.

15. Carnelli L, Lazzari C, Caretta A, Pandolfi G, Valli F, et al. (2013) Experimental activity of distillation, thermodynamic model and simulation for performance analysis of a glycol reclaiming unit. Journal of Natural Gas Science and Engineering 10: 89-98.

16. Darwish NA, Al-Mehaideb R, Braek AM, Hughes R (2004) Computer simulation of BTEX emission in natural gas dehydration using PR and RKS equations of state with different predictive mixing rules. Environmental Modeling & Software 19: 957-965.

17. Darwish NA, Hilal A (2008) Sensitivity analysis and faults diagnosis using artificial neural networks in natural gas TEG-dehydration plants. Chemical Engineering Journal 137: 189-197.

18. Jacob NCG (2014) Optimization of triethylene glycol (TEG) dehydration in a natural gas processing plant. International Journal of Research in Engineering and Technology 3: 346-350.

19. Tang J, Chen J, Zhang C, Guo Q, Chu J (2013) Exploration on process design, optimization and reliability verification for natural gas deacidizing column applied to offshore field. Chemical Engineering Research and Design 91: 542-551.

20. Aspen HYSYS (2010) Aspen HYSYS customization guide. Aspen Technology Inc., Burlington, MA, USA.

21. Pro Max 3.2 (2012) Bryan Research and Engineering Inc., Bryan, Texas, USA.

22. Abed AKH (2015) Petroleum and gas field processing. Chemical Industries. 2nd edn. CRC Press.

23. Stewart M (2011) Gas Dehydration Field Manual. 1st edn. Gulf Professional Publishing.

24. Hall SM (2011) Rules of Thumb for Chemical Engineers. 4th edn. Gulf Professional Publishing.

25. Khan MA, Maruf A (2012) Optimizing Effective Absorption during Wet Natural Gas Dehydration by Tri Ethylene Glycol. IOSR Journal of Applied Chemistry 2: 1-6.

26. Jokar S, Rahimpour H, Abbasfard H (2014) Simulation and feasibility analysis of structured packing replacement in absorption column of natural gas dehydration process: A case study for Farashband gas processing plant, Iran. Journal of Natural Gas Science and Engineering 8: 336-350.

27. Gupta A, Ansari NA, Rai R, Sah AK (1996) Reduction of Glycol Loss from Gas Dehydration Unit at Offshore Platform in Bombay Offshore - A Case Study. Proceedings of Abu Dhabi International Petroleum Exhibition and Conference, 13-16 October, Abu Dhabi, United Arab Emirates.

28. Bahadori A (2014) Natural gas processing: technology and engineering design. 1st edn. Gulf Professional Publishing, Elsevier, USA.

Photocatalytical Treatment of Synthetic Wastewater Containing Chlorophenols by TiO_2 Nanoparticles Sensitized with Cobalt Phthalocyanine Under Visible Light

Azadeh Ebrahimain Pirbazari*

Fouman Faculty of Engineering, College of Engineering, University of Tehran, Iran

Abstract

In this paper, sensitization TiO_2 nanoparticles with sulphonated cobalt phthalocyanine(CoPcS-TNP) has been reported. The prepared samples characterized by different analysis such as: XRD, FT-IR, EDX, nitrogen physisorption, TG and TEM techniques. The XRD, TEM and nitrogen physisorption results showed the prepared samples containing only the pure anatase phase with crystal sizes in the range of 7-10 nm and high surface areas of 133-280 m^2/g. The presence of CoPcS in photocatalysts confirmed by FT-IR and EDX analysis. The CoPcS-TNP samples showed efficient activity for photocatalytic degradation of 4-chlorophenol (4-CP) and 2,4-dichlorophenol (2,4-DCP) under visible light. The sample CoPcS-TNP (2.24) was the best photocatalyst for degradation of chlorophenols under visible light. The degradation reactions follow first-order kinetics. Reusing experiments of the photocatalyst showed a reduction activity from 95% to 60% for chlorophenols degradation after four cycles. The major intermediates of degradation were oxalic acid, pyruvic acid and malonic acid which identified by gas chromatography-mass spectrometry (GC/MS).

Keywords: TiO_2; Nanoparticles; Phthalocyanine, Degradation; Chlorophenols

Introduction

The production and usage of chemicals in industry has led to the entry of many xenobiotics into the environment. Chlorophenols (CPs) are used as flea repellents, fungicides, wood preservatives, fungus inhibitors, antiseptics, disinfectants and antigumming agents for gasoline during the industrial production of pesticides, herbicides, dyes, pigments, phenolic resins and paper. Also, some. Wastewater of paper and pesticides industries contains chlorophenols [1-3]. The widespread occurrence of phenols in wastewater and associated environmental hazards has heightened concern over public health [4].

Activated carbon adsorption, membrane filtration, chemical coagulation, ion exchange on synthetic adsorbent resins are the conventional wastewater purification systems that generate secondary wastes during the treatment of contaminated water, which requires additional steps and cost. Heterogeneous photocatalytic oxidation (HPO) process as a promising new route for the degradation of persistent organic pollutants, employing TiO_2 and UV light has emerged, and produced more biologically degradable and less toxic substances [5,6]. This process is largely dependent on the *in-situ* generation of hydroxyl radicals under ambient conditions which are capable of converting a wide spectrum of toxic organic compounds including the non-biodegradable ones into relatively innocuous end products such as CO_2 and H_2O. Destruction of recalcitrant organics in the HPO process, is governed by the combined actions of an energetic radiation source, an oxidizing agent and a semiconductor photocatalyst [7]. The sun produces 0.2-0.3 mol photons m^{-2} h^{-1} in the range of 300-400 nm with a typical UV flux of 20-30 Wm^2 near the earth's surface, this suggests using sunlight as an economically and ecologically sensible light source [8]. Many attempts have been made to develop of an efficient wastewater treatment process for large scale applications have received considerable attention. Recently photocatalytic degradation of phenol and substituted phenols in wastewater has been widely investigated. Several reviews [9,10] focusing briefly on the available technologies for the abatement of phenolic compounds in wastewater have recently been published. TiO_2-type semiconductor photocatalysts have attracted considerable attention because of their good physicochemical stability and their ability to catalyze the effective elimination of organic refractory contaminants from the aquatic environment [7,11-14]. The main focus of the recent TiO_2 photocatalysis research is How to improve the photocatalytic efficiency of TiO_2 in the visible region? [15,16]. Researches have been made to achieve the utilization of visible light for TiO_2 material, such as transitional metal ion doping [17-19], nonmetal element doping [20,21] and dye sensitization [22,23]. In the dye-sensitized photocatalysis process, the dye absorbs the light and a subsequent electron transfer from the excited state of the dye to the conduction band of TiO_2 semiconductor. The electron transferring from excited state of the dye to the conduction band of TiO_2 depends strongly on the adsorption property of dye molecules [24]; and by the competitive adsorption of other coexisting species in the solution.. Because of low cost and efficiency relative to the conventional solid-state conversion of visible light into chemical energy, the use of phthalocyanines and porphyrins has attracted significant attention. The main steps in a photocatalytic process based on TiO_2-phthalocyanine nanoparticles can be described as below [25]:

(1) A macrocycle compound attached to the TiO_2 surface absorbs visible light, thereby

(2) Promoting electron injection into the TiO_2 conduction band.

(3) direct oxidation of the organic compound

(4) Internal cross systems processes at the excited dye

***Corresponding author:** Azadeh Ebrahimain Pirbazari, Fouman Faculty of Engineering, College of Engineering, University of Tehran, Iran
E-mail: aebrahimian@ut.ac.ir

(5) Singlet oxygen produced by the interaction of dye in an excited triplet state with molecular oxygen

(6) Reduction of dissolved oxygen produced hydroxyl radical, hydroxyl anion and singlet oxygen

(7) Superoxide radical reacts with the sensitizer cation and produced hydroxyl radical

(8) Singlet oxygen reduced at the sensitizer and produced hydroxyl radical

(9) Environmental global reaction: organic compound oxidation by the two main oxidants: hydroxyl radicals and singlet oxygen.

To the best of knowledge, there is some works have been devoted to the characteristics and application of TiO_2 nanoparticles containing phthalocyanines and in these reports the photocatalysts prepared by the impregnation method [26-30]. A common disadvantage of this method is a low degree of molecular distribution of the dye molecules in the TiO_2 structure. Incorporation of phthalocyanines into TiO_2 gel has also been reported [31-33]. Due to the incomplete characterization and uncertain validation of the photocatalytic activity mechanism, complementary study is still needed. The enhanced photocatalytic activity of TiO_2 sentisized by iron (III) phthalocyanine has been ascribed to the effective formation of hydroxyl radicals generated from both components [31]. The photocatalytic activity of TiO_2 containing copper phthalocyanine has been assigned to a dye-sensitized TiO_2 photocatalytic mechanism [33]. Although this photocatalyst functions under visible light but it cannot achieve complete degradation of methyl orange as a model pollutant. The aim of this work is, preparation of TiO_2 nanoparticles containing different amounts of sulphonated cobalt phthalocyanine (CoPcS) by the hydrothermal method. The prepared samples were characterized by different analysis and used for the heterogeneous photocatalytic degradation of 4-chlorophenol (4-CP) and 2,4-dichlorophenol (2,4-DCP) under visible light, which revealed enhanced photocatalytic activity as compared to the pure TiO_2. The kinetic of chlorophenols degradation were investigated and the reaction intermediates during degradation were identified by GC-MS analysis.

Experimental

Materials and reagents

Cobalt phthalocyanine complex (CoPc) was purchased from (Alfa Aesar No. A16658). The sulphonated form of complex was obtained by a conventional procedure [34]. Sulphonation of CoPc carried out by chlorosulfuric acid, $HClSO_4$ (Fluka No. 26388). Tetraisopropylorthotitanat (Merck No. 8.21895), ethanol, acetylacetone and deionized water were used for photocatalyst synthesis. High-purity 4- chlorophenol(4-CP) (Merck No. 802254) and 2,4-dichlorophenol (2,4-DCP) (Merck No. 803774) were used as a model of organic pollutant for photocatalytic tests. H_2O_2 (Merck No. 1.08597) was used as accelerating oxidant in the photocatalytic tests.

Cobalt phthalocyanine sulphonation

According to reported procedure in Ref. [34], 0.5 g of CoPc and 25 mL $HClSO_4$ were treated and the mixture was heated to 150°C for 2 h. The resultant slurry was quenched over crushed ice. A deep blue solid was isolated by suction filtration and hydrolyzed to yield CoPcS. This compound was characterized by UV-Vis spectroscopy.

Photocatalyst synthesis

TiO_2 nanoparticles containing different amounts of CoPcS and pure TiO_2 were synthesized by the described procedure in Ref. [33]. 20 mL $Ti(OC_3H_7)_4$, 20 mL ethanol and 1.62 mL acetylacetone was prepared and stirred for 30 min at room temperature (solution A). Then, a new mixture (solution B), containing X mg CoPcS (X: 30, 60, 120 and 180 mg), 80 mL ethanol and 2 mL H_2O was added into the solution A, which led to a coloured solution. This solution was transferred into an autoclave, and then heated to 240°C at a heating rate of about 2°C min^{-1}. Finally, the temperature was kept at 240°C for 6 h. After cooling, the obtained solid washed with ethanol and water and dried at 150°C for 2 h in air. From now on, these photocatalysts are labelled as CoPcS-TNP(x), where (x) is the nominal weight percentage of CoPcS in the final solid. A solution without CoPcS was also preparedto obtain a control sample. This photocatalyst contains only TiO_2 prepared by an identical procedure.

Characterization

A Siemens D500 automatic diffractometer with a Cu K_α monochromatized radiation source was used for recording of the X-ray diffraction (XRD) patterns of various solid phases. The peak intensities were fitted by a pseudo-Voigt function assuming a linear background. The standard error was ~ ± 5%. Transmission electron microscopy (TEM) studies were carried out with a Hitachi-H- 8100 instrument (accelerating voltage up to 200 kV, LaB6 filament). The samples were prepared by dispersing in an ultrasonic bath for 20 min in ethanol. The nitrogen physisorption measurements were carried out with a Quantachrome Autosorb-1-MP. The BET areas were determined by static nitrogen physisorption at ~196°C subsequent to out-gassing at 180°C, until the pressure was lower than 5 mbar. Diffuse reflectance spectra were recorded with a Perkin Elmer Spectrophotometer equipped with an integrating sphere assembly. The spectra were recorded at room temperature against barium sulphate. Fourier-transform infrared (FT-IR) spectra were recorded with a Shimadzu FT-IR spectrometer (Model 8400). Elemental chemical analysis (Vario EL, Elementar) was performed to determine the carbon and sulphur contents of the samples. The thermogravimetric analyses of the samples were carried out with a Cahn TG 2131 thermobalance in the temperature range of 25-900°C at a heating rate of 5°C/min in 100% pure helium. The reaction intermediates were identified by GC-MS in an Agilent 190915-433 instrument equipped with a HP-5MS capillary column (30 m × 0.25 mm). The column temperature was programmed at 50°C for 2 min, and from 50 to 250°C at a rate of 10°C min^{-1}. The samples used for GC-MS analysis were prepared according to the following procedure: The obtained degradation product was acidified to pH 1 and subsequently extracted with dichloro-methane. After dichloromethane was evaporated to dryness under vacuum, 10 mL methanol was added to dissolve the residue. Then, 1 mL concentrated sulfuric acid was added and the combined solution was refluxed for about 3 h. The solution was further extracted with dichloromethane followed by concentrating to about 1 mL under reduced pressure. The released chloride ions originating from the degradation of chlorophenols were identified and determined by the $AgNO_3$ method.

Photocatalytic degradation of chlorophenols

The photcatalytic activity of prepared photocatalysts was investigated for degradation of chlorophenols. The suspension containing optimum dose of photocatalyst and 50 mL aqueous solution of each chlorophenols (Conc: 40 mg/L) was stirred first in the dark for 10 min to establish adsorption/desorption equilibrium. Irradiation experiments were carried out in a self-built reactor. In some experiments, H_2O_2 (as an oxidant) was also added to the solution before light irradiation. A visible (Halogen, ECO OSRAM, 500W) lamp was used as irradiation source. At certain intervals, small aliquots (2 mL) were withdrawn and filtered to remove the photocatalyst particles. These aliquots were used for monitoring the degradation progress, with Biochrom WPA, Biowave II spectrophotometerat λ_{max}=224 nm.

Results and Discussion

Characterization of synthesized photocatalysts

X-ray diffraction analysis: The XRD patterns of synthesized samples were shown in Figure 1. The patterns indicated that TiO_2 nanoparticles and the CoPcS-TNP samples consist of the pure anatase phase (PDF 24-1272). The XRD patterns showed that presence of CoPcS complex into TiO_2 structure didn't change the diffraction positions and lattice parameters of TiO_2 (Figure 1). The diffraction patterns of pure TiO_2 and CoPcS-TNP samples shown considerable line width, indicating small crystal. Based on the Scherrer equation, the crystal size of the prepared photocatalysts can be calculated from the width of the anatase (101) diffraction width. The crystallite dimensions were determined using the equation:

$D_{(hkl)}$=0.9 λ/ β Cos θ (1)

Where $D_{(hkl)}$ is the crystal thickness in the h k l direction, λ is the wavelength of the X-ray radiation, β is the peak width at half-maximum corrected for instrumental broadening, and θ is the Bragg angle of the diffraction peak. The crystal size of the pure TiO_2 and CoPcS-TNP photocatalysts are in the range from 7.5 to 9.5 nm (Table 1).

Diffuse reflectance spectroscopy: The diffuse reflectance spectra of the synthesized samples are shown in Figure 2. Pure TiO_2 has no significant absorbance in the visible region because of the large band gap energy (3.1 eV), and exhibits only a fundamental absorption band in the UV region (Figure 2a). The diffuse reflectance spectra of CoPcS-TNP samples exhibit absorption bands in the 600-700 nm regions (Figure 2b-2e). The two maxima in the visible region correspond to the Q absorption bands originating from $\pi \rightarrow \pi^*$ transitions of CoPcS, which demonstrate that the absorption spectra of CoPcS-TNP samples extend to the visible region. The monomeric and aggregated form of CoPcS complex exhibited the absorption bands at about 680 and 620 nm, respectively [35]. The band gap energy was calculated from the DR spectra according to below equation [36] for the pure TiO_2 and CoPcS-TNP samples.

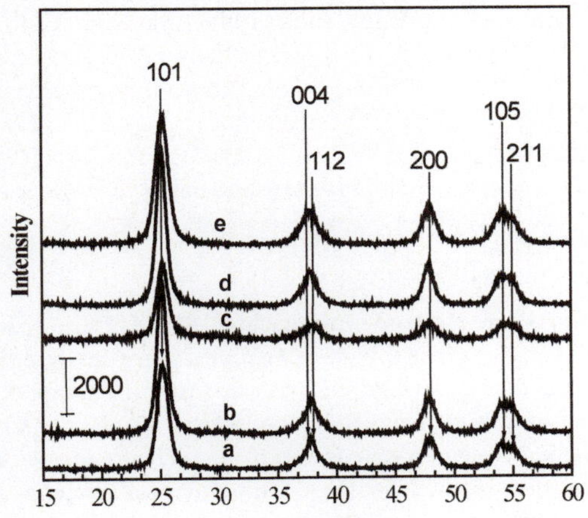

Figure 1: Powder XRD patterns of (a) TiO_2, (b) CoPcS-TNP (0.56), (c) CoPcS-TNP (1.12), (d) CoPcS-TNP (2.24) and (e) CoPcS-TNP (3.36).

Sample	Phase	Crystal Size (nm)
TiO_2	Anatase	9.0
CoPcS-TNP (0.56)	Anatase	9.0
CoPcS-TNP (1.12)	Anatase	8.0
CoPcS-TNP (2.24)	Anatase	9.5
CoPcS-TNP (3.36)	Anatase	7.5

Table 1: Crystalline phase and crystal size identified by XRD.

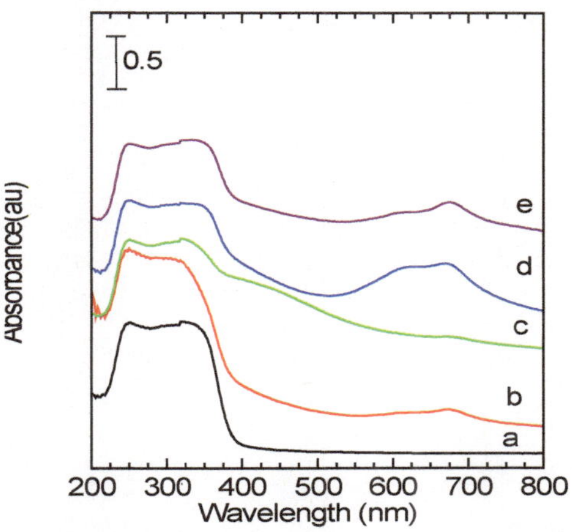

Figure 2: Diffuse reflectance spectra of (a) TiO_2, (b) CoPcS-TNP (0.56), (c) CoPcS-TNP (1.12), (d) CoPcS-TNP (2.24) and (e) CoPcS-TNP (3.36).

E_{bg}=1240/λ (2)

Where E_{bg} is the band gap energy (eV) and λ is wavelength (nm) obtained from the DR spectra. The E_{bg} data and the colour of samplesare summarized in Table 2. The band gap energy for all the samples except CoPcS-TNP (1.12) are within 3.10-3.11 eV. The lower band gap energy of 2.97 eV observed for CoPcS-TNP (1.12) may be due to the very low degree of incorporation of CoPcS into TiO_2. It may be concluded part of CoPcS molecules have been decomposed during the hydrothermal synthesis, and some of the cobalt species were inserted into the TiO_2 bulk decreasing the band gap energy.

FT-IR spectroscopy: The FT-IR spectra of the CoPcS, pure TiO_2 and CoPcS-TNP (2.24) sample are shown in Figure 3a and Figure 3b Some of the characteristic bands of CoPcS (Figure 3c) appeared in the 1700-1000 cm^{-1} regions [37]. The spectrum b in Figure 3 originates from the CoPcS-TNP (2.24) sample. The vibration about 1400 cm^{-1} can be attributed to cobalt phthalocyanine species in the synthesized photocatalysts [38], which is labelled with an asterix in Figure 3. The intensity of the 1400 cm^{-1} vibration is weak, because the amount of CoPcS incorporated into TiO_2 is low. The vibrations at about 670 cm^{-1} (Figure 3c) due to the sulfonic groups of CoPcS [39] disappeared in the CoPcS-TNP (2.24) sample, which is most likely due to the formation of $TiO\text{-}SO_3^-$ bond (Figure 4). Covalent grafting of sulphonated iron (III) phthalocyanine onto TiO_2 nanoparticles was carried out by Sorokin et al. [40] through sol-gel procedure. Their obtained characterization results showed that sulfonic groups are chemisorbed onto the TiO_2 nanoparticles and sulphonated iron (III) phthalocyanine complex acts as a bridging ligand and controls the molecular self-assembly hydrolytic process. Recently Núñez and etal reported modification of TiO_2 surface with copper(II) phthalocyanine-tetrasulfonic acid tetrasodium (CuPc-ulfunic acid) and their FT-IR data showed, the S-O-Ti characteristic stretching signals at 1150 and 1210 cm^{-1} suggest chemisorptions of CuPc complex through acid bond formation ($TiO\text{-}SO_3^-$) between Ti-OH and

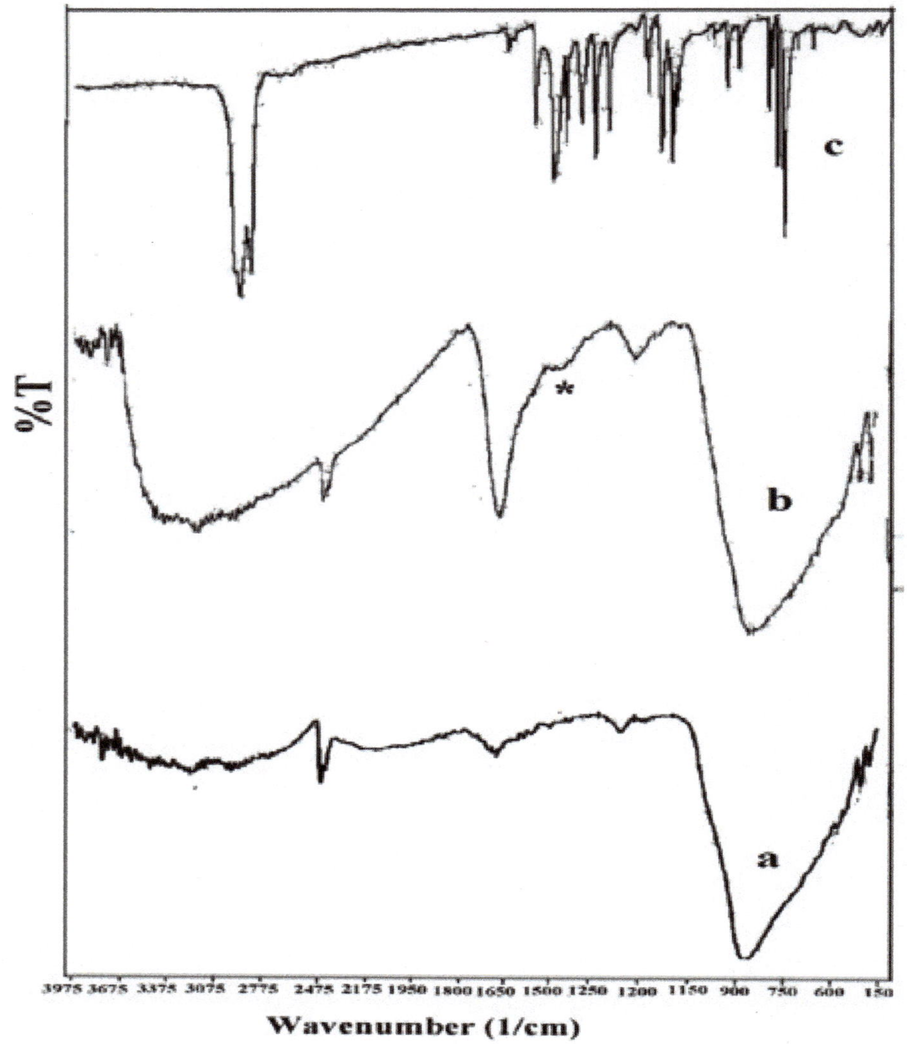

Figure 3: FT-IR pactra of a) TiO_2, b) CoPcS-TNP (2.24) and c) CoPcS.

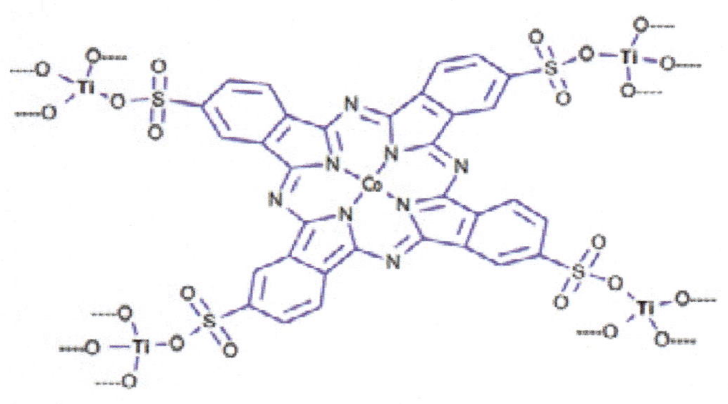

Figure 4: Chemical CoPcS adsorption on the TNP surface.

CuPc-sulfonic [25]. Their adsorption isotherm investigation showed that each CuPc molecules adsorbed onto TiO_2 surface nanoparticles using two or more sulfonic acid moieties. Therefore, the CuPc molecules may have more planar distribution on the TiO_2 surface and prevents CuPc molecule multi anchoring at the TiO_2 surface [25].

Sample	Colour	λ (nm)	Band gap energy (eV)
TiO_2	White	399	3.11
CoPcS-TNP (0.56)	Green	399	3.11
CoPcS-TNP(1.12)	Yellow	418	2.97
CoPcS-TNP(2.24)	Dark Green	400	3.10
CoPcS-TNP (3.36)	Pale blue	400	3.10

Table 2: Colour, wavelength and band gap energy of the prepared samples.

Transmission electron microscopy (TEM): The particle size and morphology of the prepared samples were analyzed by TEM. The TEM micrographs (Figure 5) revealed that the samples consist of nanoparticles with diameters in the range from 7 to 10 nm. The samples are not agglomerated, which is useful to increase the photocatalytic activity of the prepared samples. The chemical analysis results are summarized in Table 3. The obtained data from sulfur content suggest an average of two sulphonate groups per CoPc complex. Also the carbon amount is relatively high for CoPcS-TNP (1.12) that may be indicated

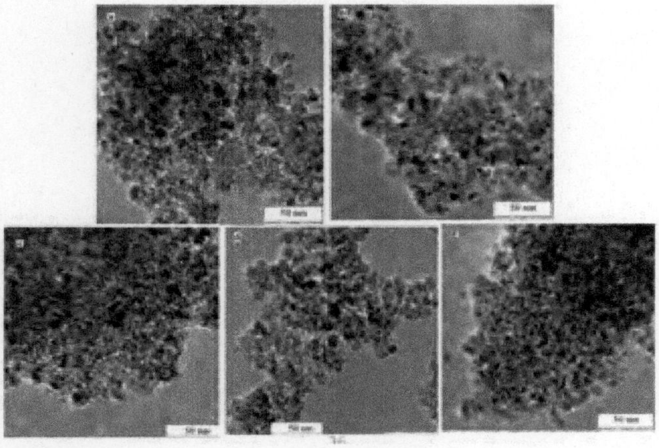

Figure 5: TEM micrograph of (a) TiO_2, (b) CoPcS-TNP (0.56), (c) CoPcS-TNP (1.12), (d) CoPcS-TNP (2.24) and (e) CoPcS-TNP (3.36).

Sample	%S	%C
TiO_2	0.00	1.14
CoPcS-TNP (0.56)	<0.10	1.87
CoPcS-TNP (1.12)	0.10	6.45
CoPcS-TNP (2.24)	0.20	2.04

Table 3: Elemental chemical analysis of the prepared samples.

decomposition of CoPcS complex during the hydrothermal synthesis and fractions of other organic compounds (such as acetylacetone or ethanol) remained as separated phase.

N_2 adsorption-desorption: Figure 6 shows N_2 adsorption-desorption isotherms of the pure and CoPcS-TNP samples. The sorption isotherms for all of the samples except CoPcS-TNP (1.12) correspond to the type IV isotherm according to the IUPAC classification [41]. Textural and structural parameters of the CoPcS-TNP samples are summarized in Table 4. Specific surface areas were calculated according to the BET method, pore volumes, and radii were derived from the desorption branch according to the BJH model. The specific surface areas of the CoPcS-TNP samples are lower than that of pure TiO_2 suggesting that incorporation of the CoPcS complex reduces the surface area of TiO_2 nanoparticles. Reduction of the surface area for CoPcS-TNP samples does not involve distortion of the TiO_2 crystal structure as shown by XRD analysis.

Thermogravimetric analysis: Thermogravimetric analysis curves of the prepared samples in helium atmosphere are shown in Figure 7. In the thermogram for pure TiO_2, three weight loss stages can be observed. The first weight loss is at about 90°C, which corresponds to desorption of physically adsorbed species such as H_2O and CO_2. The second weight loss at about 250°C is ascribed to the removal of hydroxyl groups from the surface of TiO_2 [42]. The third weight loss at about 370°C is for organic species decomposition, which used for preparation of TiO_2. We also observed three weight loss stages in the thermograms of the CoPcS-TNP samples (Figure 7). The first and the second weight loss at about at 90 and 370°C can again be attributed to desorption of water and decomposition of organic species, respectively. However, a new weight loss starting at about 500°C for CoPcS-TNP (2.24) and CoPcS-TNP (3.36) can be attributed to the decomposition of CoPcS complex. The phthalocyanines and their metallo derivatives are stable up to 400°C [43]. These results may indicate that CoPcS molecules immobilized onto TiO_2 nanoparticles through covalent bonding are more stable and decomposed at higher temperature (Figure 4). We did not observe dehydroxylation weight loss for CoPcS-containing TiO_2 samples at 250°C. This also is an indication of the strong interaction between TiO_2 and CoPcS through hydroxyl groups.

Photocatalytic degradation of chlorophenols

The photocatalytic efficiency of the CoPcS-TNP samples was investigated for degradation of chlorophenols. Elimination time of chlorophenols in the presence of the prepared photocatalysts is shown in Table 5. All of the CoPcS-TNP samples are more efficient in the degradation of chlorophenols compared to pure TiO_2, and the degradation rate of chlorophenols is the highest in presence of CoPcS-TNP (2.24). Obtained results can be confirmed the role of CoPcS complex as sensitizer on the surface of TNP: By the photon absorption, the electrons in the ground state of CoPcS complex are excited and then injected into the conduction band of TiO_2. Highly active species such as $OH^{\bullet}$ and $O_2^{\bullet -}$ are generated at the TiO_2 surface, where the photoelectrons and holes become effectively separated. The aggregated form of the CoPcS complex shortens the excited triplet state lifetimes by enhancement of internal conversion and therefore reducing their photosensitizing efficiency [44]. Consequently, a monomolecular distribution of the dye on the TiO_2 surfaces favors improved photocatalytic activity.

Figure 8 show the photocatalytic degradation of 4-CP (Figure 8a) and 2,4-DCP (Figure 8b) under various conditions and in the presence of the best photocatalyst (CoPcS-TNP (2.24)) respectively. Control experiments showed that the optimal dosage of photocatalyst is 0.3 g L^{-1}, and t a H_2O_2 concentration of 0.01 mol L^{-1} is adequate for photocatalytic degradation of chlorophenols. Complete elimination of 4-CP and 2,4-DCP are possible within 90 and 180 min in the presence of the CoPcS-TNP (2.24), H_2O_2, and visible light. Thus, the enhancement of the catalytic activity observed for the CoPcS-TNP samples compared to pure TiO_2 can be demonstrated synergetic effect of the sensitizer (CoPcS complex). CoPcS complex has a strong absorption around 600-650 nm (Figure 2), it has strong Q-band absorption and it can be excited under visible light. The excited CoPcS molecules inject electrons to the conduction band (CB) of TiO_2, the CB acts as a mediator and transfer electron acceptors on the TiO_2 surface, the valence band (VB) remains unaffected.

In the presence of hydrogen peroxide the activity of the photocatalyst may even be better (Figure 8). Almost 50% of 4-CP and 2,4-DCP are

Sample	SBET (m^2/g)	Pore radius (A°)	Pore volume (cm^3/g)
TiO_2	280	19	0.24
CoPcS-TNP (0.56)	133	35	0.26
CoPcS-TNP (1.12)	208	9	0.14
CoPcS-TNP (2.24)	141	38	0.29
CoPcS-TNP (3.36)	197	18	0.17

Table 4: Textural and structural parameters of the prepared samples.

Sample	4-CP elimination time (min)	2,4-DCP elimination time (min)
TiO_2	180	360
CoPcS-TNP (0.56)	150	330
CoPcS-TNP (1.12)	130	300
CoPcS-TNP (2.24)	90	180
CoPcS-TNP (3.36)	110	240

Table 5: Elimination time of chlorophenols in presence of the prepared samples.

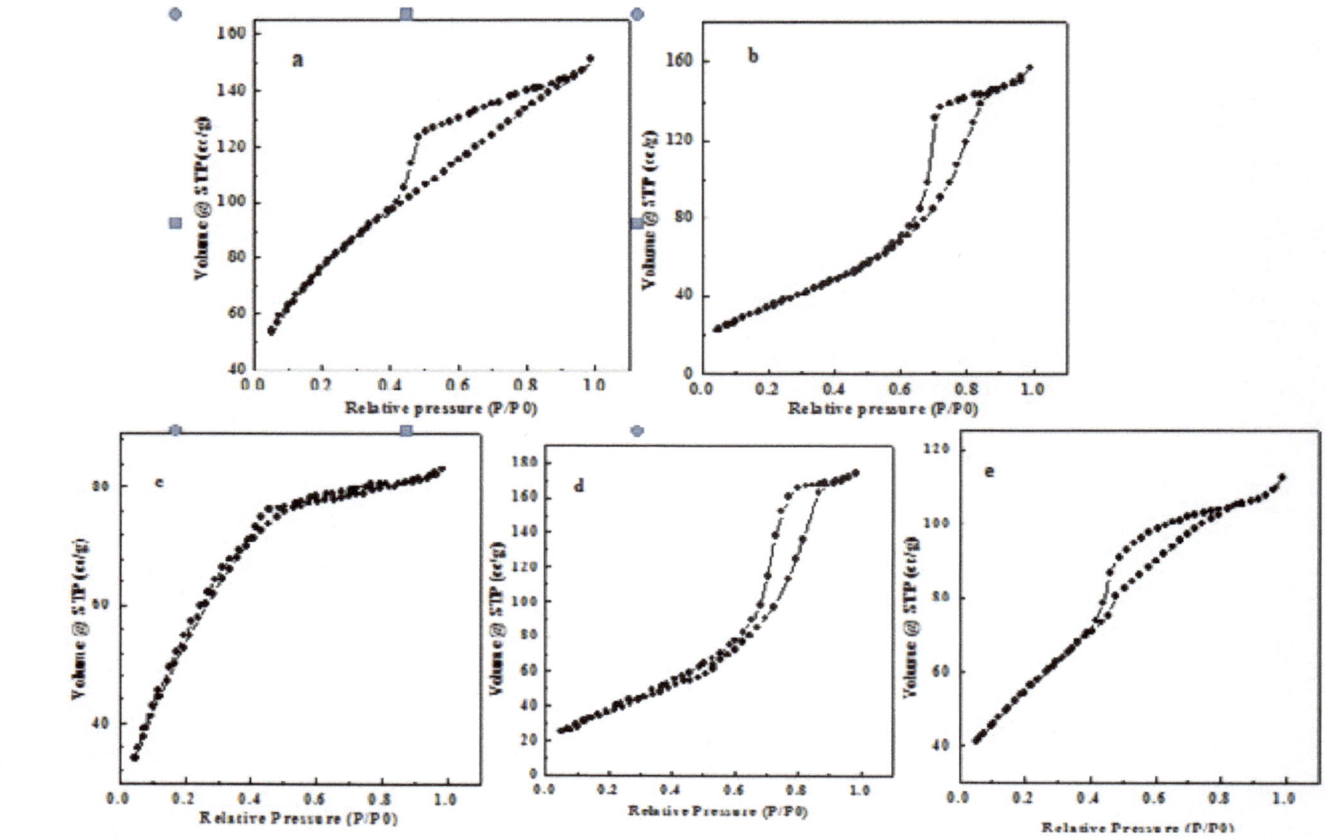

Figure 6: N_2 adsorption–desorption isotherms for (a) TiO_2, (b) CoPcS-TNP (0.56), (c) CoPcS-TNP (1.12), (d) CoPcS-TNP (2.24) and (e) CoPcS-TNP (3.36).

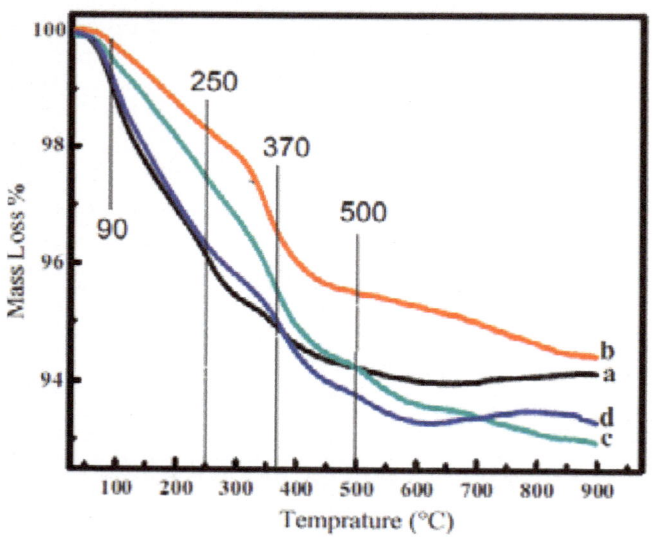

Figure 7: TG curves of (a) TiO_2, (b) CoPcS-TNP (0.56), (c) CoPcS-TNP (2.24) and (d) CoPcS-TNP (3.36).

degraded within 90 and 180 min, when there is no H_2O_2 added. More hydroxyl radicals are produced in the presence of H_2O_2 according to the below reaction:

$$H_2O_2+O_2^{\cdot-}\rightarrow OH^{\cdot}+OH^{-}+O_2 \quad (3)$$

This photocatalyst did not exhibit any photoactivity in the dark, neither in the absence nor in the presence of H_2O_2 (Figure 8), suggesting that it is necessary to photoexcite the sensitizer (CoPcS). The hydroxyl radicals are responsible for the degradation of the chlorophenols, yielding some intermediate compounds and mineralization products. The hydroxyl radical production was checked by Gao et al. proposed method [45,46], based on Nash's method [47]. It noticed that the production of hydroxyl radicals would produce a yellow color owing to the synthesis of diacetyldihydrolutidin [47]. The yellow color was observed when photocatalyst, H_2O_2 and the radiation were employed and the intensity of the color increased with increasing the time of irradiation. According to equation 3, the production of the hydroxyl radicals depends on the presence of superoxide radicals [46], therefore, the observation may suggest that the superoxide radicals are involved. During the photocatalytic degradation of chlorophenols, hydrogen peroxide could react with superoxide radicals to generate hydroxyl radicals.

GC/MS analysis: To obtain information about intermediates species formed during chlorophenols degradation process, GC/MS technique was employed. Figure 9 and Table 6 give major reaction intermediates containing carboxylic acid functional species. Formation of these intermediates can be explained as: After hydroxyl radicals formation, the hydroxyl radicals attack to chlorophenols, converting them to chlorocatechol and then to chlorobenzoquinone, then hydroxyl radicals break the aromatic ring of chlorobenzoquinone, transferring them into simple acids like oxalic acid and acetic acid as the final products [48-50]. Chaliha and et al. [49] and Bian et al. [50] identified

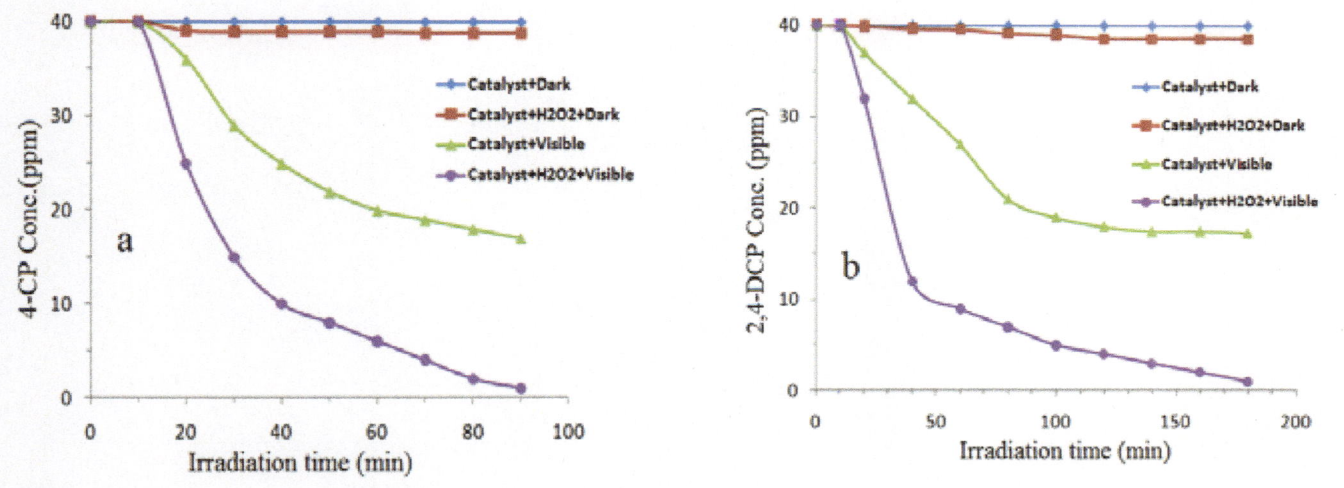

Figure 8: Concentration changes of a) 4-CP and b) 2, 4-DCP at various conditions.

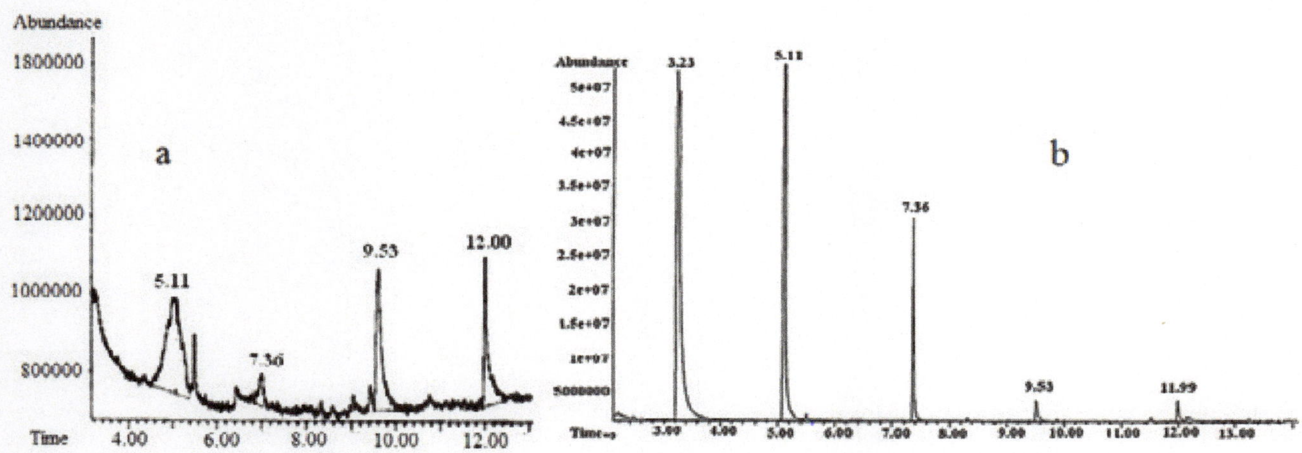

Figure 9: Chromatogram of the final products of the photocatalytic degradation of a) 4-CP and b) 2, 4-DCP by CoPcS-TNP (2.24).

Retention time (min)	Identified intermediate	Molecular formula
3.23	Methyl pyruvate	$H_3CCOCOOCH_3$
5.11	Methyl oxalate	$H_3COOCCOOCH_3$
7.36	Dimethyl malonate	$H_3COOCCH_2COOCH_3$

Table 6: The main intermediates of chlorophenols degradation determined by GC/MS.

the main ring-opened products resulting from 4-CP degradation by GC/MS and discussed the formation mechanism of them. In the proposed mechanism by Chaliha et al. [49], an electrophilic OH radical adds onto the 4-CP ring at an ortho- or para-position, leading to the formation of chlorocatechol and hydroquinone. On further oxidation, chlorocatechol is likely to be converted to aliphatic acids and hydroquinone to p-benzoquinone, which is easily oxidized to the corresponding dicarboxylic acid, 2,5-dioxo-3-hexenedioic acid. This acid produces maleic acid, which decomposes to oxalic acid, formic acid, acrylic acid, malonic acid and acetic acid. If there are no additional radical competitors in the solution, these low molecular weight organic acids can gradually be mineralized to carbon dioxide, the ultimate oxidation product. Also, a number of the intermediates identified in this work (Table 6) are in agreement with those reported in ref [51]. Benzoic acid formation with higher carbon compare chlorophenols can be illustrated according to the Rao et al. report (Figure 10) [52]. They proposed that radical I and a chloride radical ($Cl^•$) formed by homolytic cleavage of aryl-Cl bond or by interaction of 2-CP with electrons on the surface of TiO_2 [53]. In a second step, radical I may dimerize to yield [1,1'-biphenyl]-2,2'-diol or combine with formyl radical produced *in-situ* to give 4-hydroxybenzaldehyde and benzoic acid formed by oxidation of hydroxybenzaldehyde. Hydroxyl radicals can attack the aromatic moiety due to their electrophilic nature [54]. Progress of the photocatalytic degradation was detected by chloride ion measurement through $AgNO_3$ test method. The presence of Cl^- in the reaction media accounts up almost 100% for chlorophenols degradation if compared to the maximum achievable concentration.

Photocatalyst reuse: The stability of the anchored photocatalysts is very important for its application in environmental technology. Therefore, the effectiveness of photocatalyst reuse was examined for degradation of each chlorophenols during a four cycle experiment.

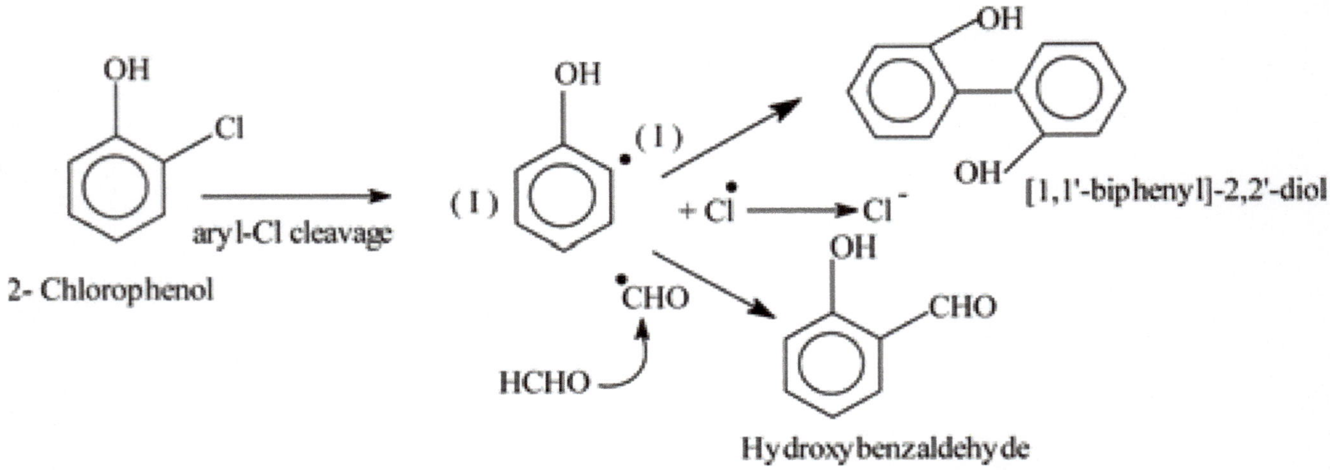

Figure 10: Rao proposed mechanism of formation of higher carbon intermediates [52].

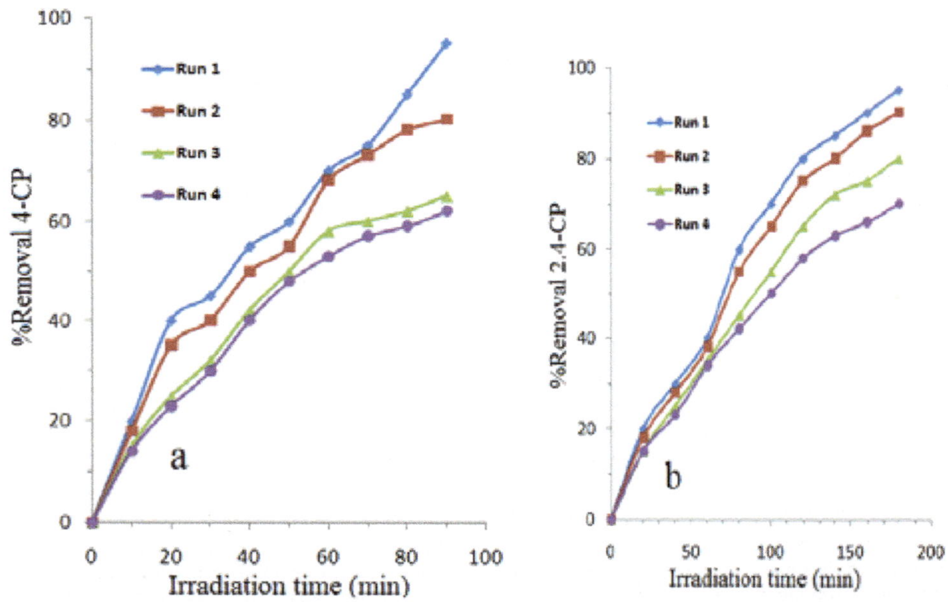

Figure 11: The recyclability of CoPcS-TNP (2.24) for a) 4-CP and b)2,4-DCP photocatalytic degradation for four successive cycles.

Each experiment was carried out under identical conditions (Figure 11). 50 mL of each chlorophenols with an initial concentration of 40 mg/L, 15 mg photocatalyst and 0.03 mol/L H_2O_2 and 90 and 180 min irradiation time under visible light were used. Recycling experiments showed a photocatalytic activity reduction from 95% to 60% after four catalytic cycles. The main reasons for reduction of photocatalytic activity are permanent adsorption of intermediate species or probable attack of OH radicals to the anchored phthalocyanine complex.

The effect of the initial concentration: The photocatalytic degradation of chlorophenols is a first-order reaction and and its kinetics may be expressed as $\ln(C_0/C)=k_{obs}t$. In this equation k_{obs} (min^{-1}) is the apparent rate constant, C_0 and C are the initial concentration and concentration at reaction time t of each chlorophenols, respectively. The k_{obs} are found from the slopes of the straight lines obtained by plotting $\ln(C_0/C)$ versus irradiation time. The reaction rates, rate constants and half-lifes ($t_{1/2}$) at various initial concentrations of each chlorophenols are given in Table 7. The results summarized in Table 7 show that the reaction rate of photocatalytic degradation of chlorophenols is faster at higher initial concentration. However, the rate constants decrease to some extent when the initial concentrations increase. The initial degradation rate of chlorophenols increased by increasing of chlorophenols concentration because the oxygen active species formed on the CoPcS surface have a limited lifetime and react with the chlorophenoles near the photocatalyst surface.

Conclusion

In this work, TiO_2 nanoparticles containing different amounts of CoPcS (CoPcS-TNP) were synthesized and characterized by different analysis such as: XRD, FTIR, TEM, TG, nitrogen physisorption and DRS. The nitrogen physisorption and FTIR analysis results indicate

4-CP				
Initial Conc. (µg/mL)	kobs (min^{-1})	Reaction rate (µg/mL. min)	t1/2 (min)	R^2
20	0.0539	1.1078	12.86	0.9985
30	0.0384	1.154	18.04	0.9992
40	0.0294	1.176	23.57	0.9995
60	0.0199	1.195	34.82	0.9989
80	0.0149	1.199	46.51	0.9991
2,4-DCP				
Initial Conc. (µg/mL)	kobs (min^{-1})	Reaction rate (µg/mL. min)	t1/2 (min)	R^2
20	0.0221	0.442	31.36	0.9986
30	0.0158	0.474	43.86	0.9975
40	0.0129	0.516	53.72	0.9965
60	0.0088	0.528	78.75	0.9991
80	0.0079	0.632	87.72	0.9978

Table 7: First-order reaction rate constants and correlation coefficients for different concentrations of chlorophenols.

that CoPcS molecules act as bridge ligand and attached on the surface of TiO_2 nanoparticles through SO_2-O-TiO_2 bonds. The CoPcS-TNP samples should contain an optimum amount of CoPcs anchored to the TiO_2 surfaces for better photocatalytic activity. The obtained results for degradation of chlorophenols showed that the incorporation of 2.24 wt% CoPcS provides the most photoactive sample. The kinetics of the reaction is first-order and the main intermediates of reaction identified by GC/MS.

Acknowledgements

The author wish to acknowledge the financial support of Iran National Science Foundation (Grant No: 92025598). Also, I am so thankful to University of Tehran for supporting of this research.

References

1. Peng JF, Liu JF, Hu XL, Jiang GB (2007) Direct determination of chlorophenols in environmental water samples by hollow fiber supported ionic liquid membrane extraction coupled with high performance liquid chromatography. J. Chromatogr. A 1139: 165-170.
2. Fukushima M, Tatsumi K (2001) Degradation pathways of pentachlorophenol byphoto-Fenton systems in the presence of iron (III), humic acid and hydrogenperoxide. Environ Sci Technol 35: 1771-1778.
3. Gonzalez LF, Sarria V, Sanchez OF (2010) Degradation of chlorophenols by sequential biological-advanced oxidative process using Trametes pubescens and TiO_2/UV. Bio Technol 101: 3493-3499.
4. Eriksson E, Baun A, Mikkelsen PS, Ledin A (2007) Risk assessment of xenobiotics in stormwater discharged to HarrestupAo, Denmark. Desalination 215: 187-197.
5. Oller I, Gernjak W, Maldonado MI, Perez ELA, Sanchez PJA, et al. (2006) Solar photocatalytic degradation of some hazardous water-soluble pesticides at pilot-plant scale. J. Hazard. Mater 138: 507-517.
6. Garcýa A, Amat AM, Arques A, Vicente R, Lopez MF, et al. (2007) Increased biodegradability of UltracidTMin aqueous solutions withsolar TiO_2 photocatalysis. Chemosphere 68: 293-300.
7. Ahmed S, Rasul MG, Martens WN, Brown R, Hashib MA (2010) Heterogeneous photocatalytic degradation of phenols in wastewater: A review oncurrent status and developments. Desalination 261: 3-18.
8. Goslich R, Dillert R, Bahnemann D (1997) Solar water treatment principles and reactors. Water Sci. Technol 35: 137-148.
9. Busca G, Berardinelli S, Resini C, Arrigi L (2008) Technologies for the removal ofphenol from fluid streams. A short review of recent developments J. Hazard. Mater 160: 265-288.
10. Liotta LF, Gruttadauria M, Di CG, Perrini G, Librando V (2009) Heterogeneous catalytic degradation of phenolic substrates: catalysts activity. J. Hazard. Mater 162: 588-606.
11. Augugliaroa V, Bellarditaa M, Loddoa V, Palmisanoa G, Palmisanoa L, et al. (2012) Overview on oxidation mechanisms of organic compounds by TiO_2 in heterogeneous photocatalysis. J. Photochem. Photobiol. C: Photochem. Rev 13: 224-245.
12. Gayaa UI, Abdullaha AH (2008) Heterogeneous photocatalytic degradation of organic contaminants overtitanium dioxide: A review of fundamentals, progress and problems. J. Photochem. Photobiol. C: Photochem. Rev 9: 1-12.
13. Agrios GA, Pichat P (2005) State of the art and perspectives on materials and applications of photocatalysis over TiO_2. J. Appl. Electrochem 35: 655-663.
14. Demeesterea K, Dewulf J, Van LH (2007) Heterogeneous photocatalysis as an advanced oxidation process for the abatement of chlorinated, monocyclicaromatic and sulfurous volatile organic compounds in air: State of the art. Crit. Rev. Environ. Sci. Technol 37: 489-538.
15. Pelaez M, Nolan NT, Pillai SC, Seery MK, Falaras P, et al. (2012) A review on the visible light active titanium dioxide photocatalysts for environmental applications. Appl. Catal. B: Environ. 125: 331-349.
16. Han F, Rao VS, Srinivasan M, Rajarathnam D, Naidu R (2009) Tailored titanium dioxide photocatalysts for the degradation of organic dyes in wastewater treatment: A review. Appl. Catal/ A: General 359: 25-40.
17. Choi W, Termin A, Hoffmann MR (1994) The role of metal ion dopants in quantumsizedTiO_2: correlation between photoreactivity and charge carrierrecombination dynamics. J. Phys. Chem 98: 13669.
18. Zhu J, Deng Z, Chen F, Zhang J, Chen H, et al. (2006) Hydrothermal doping method for preparation of Cr^{3+}-TiO_2 photocatalysts withconcentration gradient distribution of Cr^{3+}. Appl. Catal. B 62: 325-329.
19. Yuan S, Sheng Q, Zhang J, Chen F, Anpo M (2005) Synthesis of La^{3+} dopedmesoporoustitania with highly crystallized walls, Micropor. Mesopor. Mater 79: 93-99.
20. Asahi R, Morikawa T, Ohwaki T, Aoki K, Taga Y (2001) Visible-light photocatalysisin nitrogen-doped titanium oxides. Science 293: 269-271.
21. CongY, Zhang J, Chen F, Anpo M (2007) Synthesis and characterization of nitrogendopedTiO_2 nanophotocatalyst with high visible light activity. J. Phys. Chem. C 111: 6976-6982.
22. Bae E, Choi W, Park J, Shin HS, Kim SB et al. (2004) Effects of surface anchoringgroups (carboxylate vs. phosphonate) in ruthenium-complex-sensitized TiO_2on visible light reactivity in aqueous suspensions. J. Phys. Chem. B 108: 14093-14102.
23. Bae E, Choi W (2003) Highly enhanced photoreductive degradation of perchlorinatedcompounds on dye-sensitized metal/TiO_2 under visible light. Environ. Sci.Technol 37: 147-152.

24. Zhao J, Wu T, Wu K, Oikawa K, Hidaka H, et al. (1998) Photoassisteddegradation of dye pollutants. 3. Degradation of the cationic dye rhodamine Bin aqueous anionicsurfactant/TiO_2 dispersions under visible light irradiation: evidence for the need of substrate adsorption on TiO_2 particles. Environ. Sci.Technol. 32: 2394-2400.

25. Vargas E, Vargas R, Núñez O (2014) A TiO_2 surface modified with copper(II) phthalocyanine-tetrasulfonic acid tetrasodium salt as a catalyst during photoinduceddichlorvos mineralization by visible solar ligh. Appl. Catal. B: Environ. 156: 157 8-14.

26. Mele G, Del SR, Vasapollo G, Marci G, Garcia LE, et al. (2005) TRMC, XPS, and EPR characterizations of polycrystalline TiO_2 porphyrin impregnated powders and their catalytic activity for 4-nitrophenol photodegradation in aqueous suspension, J. Phys. Chem. B 109: 12347-12352.

27. Mele G, Del SR, Vasapollo G, Garcia LE, Palmisano L, et al. (2007) Photocatalytic degradation of 4-nitrophenol in aqueous suspension by using polycrystalline TiO_2impregnated with functionalized Cu(II)-porphyrin or Cu(II)-phthalocyanine Complexes. J. Catal. 117: 6581-6588.

28. Granados GO, Paez CAM, Martınez FO, Paez MEA (2005) Photocatalytic degradation of phenol on TiO_2 and TiO_2/Pt sensitized with metallophthalocyanines. Catal. Today 107: 589-594.

29. Li L, Xin B (2010) Photogenerated carrier transfer mechanism and photocatalysis properties of TiO2 sensitized by Zn(II) phthalocyanine. J. Central South University of Technol 17: 218-222.

30. Fa W, Zan L, Gong C, Zhong J, Deng K (2008) Solid-phase photocatalytic degradation of polystyrene with TiO_2 modified by iron (II) phthalocyanine. Appl. Catal. B: Environ 79: 216-223.

31. Ranjit KT, Willner I, Bossmann S, Braun A (1998) Iron(III) phthalocyanine-modified titanium dioxide: a novel photocatalyst for the enhanced photodegradation of organic pollutants. J. Phys. Chem. B 102: 9397-9403.

32. Wang Z, Mao W, Chen H, Zhang F, Fan X, et al. (2006) Copper(II) phthalocyaninetetrasulfonate sensitized nanocrystallinetitaniaphotocatalyst: synthesis in situ and photocatalysis under visible light. Catal. Commun. 7: 518-522.

33. Wang Z, Chen H, Tang P, Mao W, Zhang F, et al. (2006) Hydrothermal in situ preparation of the copper phthalocyaninetetrasulfonate modified titanium dioxide photocatalyst, Colloid. Surf. A: Physicochem. Eng. Aspects 289: 207-211.

34. Martin PC, Gouterman M, Pepich BV, Renzoni GE, Schindele DC (1991) Effects of ligands, solvent, and variable sulfonation on dimer formation of aluminum and zinc phthalocyaninesulfonates. Inorg. Chem. 30 3305-3309.

35. Iliev V, Alexiev V, Bilyarska L (1999) Effect of metal phthalocyanine complex aggregation on the catalytic and photocatalytic oxidation of sulfur containing compounds. J. Mol. Catal. A: Chem. 137: 15-22.

36. Kudo A, Miseki Y (2009) Heterogeneous photocatalyst materials for water splitting. Chem. Soc. Rev. 38: 253-278.

37. Seoudi R (2005) FTIR, TGA and DC electrical conductivity studies of phthalocyanine and its complexes. J. Mol. Struct 753: 119-126.

38. Karandikar P, Chandwadkar AJ, Agashe M, Ramgir NS, Sivasanker S (2006) Liquid phase oxidation of alkanes using Cu/Co-perchlorophthalocyanine immobilized MCM-41 under mild reaction conditions. Appl. Catal. A 297: 220-230.

39. Nabid MR, Asadi S, Shamsianpour M, Sedghi R, Osati S, et al. (2010) Oxidative polymerization of 3,4-ethylenedioxythiophene using transition-metal tetrasulfonatedphthalocyanine. React. Func. Polymers 70: 75-80.

40. Beyrhouty M, Sorokin AB, Daniele S, Hubert PLG (2005) Combination of two catalytic sites in a novel nanocrystalline TiO_2-iron tetrasulfophthalocyanine material provides better catalytic properties. NJ Chem 29: 1245-1248.

41. Sing KSW, Everett DH, Haul RAW, Moscou L, Pierotti RA (1985) Reporting physisorption data for gas/solid systems with special reference to the determination of surface area and porosity. Pure Appl. Chem. 57: 603-619.

42. Zhao Z, Fan J, Xie M, Wang Z (2009) Photo-catalytic reduction of carbon dioxide with in-situ synthesized CoPc/TiO_2 under visible light irradiation. J. Cleaner Product 17: 1025-1029.

43. Wei X, Du X, Chen D, Chen Z (2006) Thermal analysis study of 5,10,15,20-tetrakis (methoxyphenyl) porphyrins and their nickel complexes. Thermochim. Acta 440: 181-187.

44. Kluson P, Drobek M, Kalaji A, Zarubova S, Krysa J, et al. (2008) Singlet oxygen photogeneration efficiencies of a series of phthalocyanines in well-defined spectral regions. J. Photochem. Photobiol A: Chem. 199: 267-273.

45. Gao R, Stark J, Bahnemann DW, Rabani J (2002) Quantum yields of hydroxyl radicalsin illuminated TiO_2 nanocrystallite layers. J. Photochem. Photobiol. A: Chem 148: 387-391.

46. Machado AEH, Franc MD, Velani V, Magnino GA, Velani HMM, et al. (2008) Characterization and evaluation of theefficiency of TiO_2/zinc phthalocyanine nanocomposites as photocatalysts for wastewater treatment using solar irradiation. Int. J. Photoenergy 1-12.

47. Nash T (1953) The colorimetric estimation of formaldehyde by means of the Hantzschreaction. Biochem. J. 55: 416-421.

48. Chaliha S, Bhattacharyya KG (2009) Fe(III), Co(II) and Ni(II) impregnated MCM-41 for wet oxidative destruction of 2,4-dichlorophenol in water. Catal. Today 141: 225-233.

49. Chaliha S, Bhattacharyya KG, Paul P (2008) Catalytic destruction of 4□ chlorophenol in water. Clean, Soil, Air, Water 36: 488-497.

50. Bian W, Song X, Liu D, Zhang J, Chen X (2011) The intermediate products in the degradation of 4-chlorophenol by pulsed high voltage discharge in water. J Hazard Mater 192: 1330-1339.

51. Wu L, Li A, Gao G, Fei Z, Xu S, et al. (2007) Efficient photodegradation of 2,4-dichlorophenol in aqueous solution catalyzed by polydivinylbenzene supported zinc phthalocyanine. J. Mol. Catal. A 269: 183-189.

52. Rao NN, Dubey AK, Mohanty S, Khare P, Jain R, et al. (2003) Photocatalytic degradation of 2-chlorophenol: a study of kinetics, intermediates and biodegradability. J. Hazard. Mater. B 101: 301-314.

53. Konstantinou K, Sakkas VA, Albanis TA (2001) Photocatalytic degradation of the herbicides propanil and molinate over aqueous TiO_2 suspensions: identification of intermediates and the reaction pathway. Appl. Catal. B: Environ 34: 227-239.

54. Serpone N, Lawless D, Terzian R, Meisel D (1992) Electrochemistry in Colloids and Dispersions. VCH, New York, NY, USA, p: 399.

Modelling the Methodology for Crude Oil Bioremediation Decision Tree for an Integrated Environmental Management System

Ukpaka CP*

Department of Chemical/Petrochemical Engineering, Rivers State University of Science and Technology, Nkpolu Port Harcourt, Nigeria

Abstract

The application of Matrix Laboratory (MATLAB) computer programme language was used in the simulation approach to determine the best methodology for crude oil bioremediation using the concepts of decision tree mechanism with an integrated environmental management system to achieve high efficiency at end of the cleanup. Bioremediation is a process in which microorganisms metabolize contaminants either through oxidative or reductive processes. Under favorable conditions, microorganisms can oxidatively degrade organic contaminants completely into non-toxic by-products such as carbon dioxide and water or organic acids and methane. The area, volume, concentration of the soil, concentration of the liquid (crude oil) and the film mass transfer coefficients influence the effectiveness of bioremediation process. To obtain the parameters, $\mu_{max,}$ K_S, K_1 and y_x numerical solutions of these equations were performed using MATLAB. The method for the numerical modelling involves integration using fourth-order Runge Kutta numerical scheme to determine the trend and efficiency of the system.

Keywords: Modelling; Methodology; Crude oil; Bioremediation decision tree; Integrated environmental management system

Introduction

Nature has the ability to recycle and purify itself, but in recent years, the demand placed on the environment by huge amounts of anthropogenic pollution exceeds its capacity to recover, hence the need to save (remediate) the environment is necessary. Conventional techniques have been employed in environmental remediation [1-4]. Most recently, there is a need for bioremediation as this technique has proved to be more eco-friendly. Oil due to exploration and exploitation activities is a threat to the environment, particularly the mangrove ecosystem and agricultural land. World reserves of arable agricultural land are constantly diminishing and 25% of cultivated lands are affected by soil degradation due to man's activities. For example, a good percentage of oil spills that occurred on dry land between 1978 and 1979 in Nigeria affected farms in which crops such as rice, maize, yams, cassava, and plantain were cultivated. The primary concern of petroleum contamination of soil has been its effects on ground water, but its effect on the germination and growth of some plants have been reported [4-8].

Bioremediation technologies simply attempt to optimize microorganisms' natural capacity to degrade/recycle by supplying essential inorganic limiting reactants and minimizing abiotic stress. Recommendations have been advocated or the microbial seeding of oil spills, because bacteria and fungi are the only biological species which have the metabolic capacity of utilizing petroleum carbon for cell synthesis [1-3]. Bioremediation is the use of biological treatment systems to destroy or reduce the concentrations of hazardous wastes from a contaminated site. Such systems include the use of higher plants, which are efficient in removing or destroying the hazardous wastes [9-12]. These microbes and higher plants can also be variously modified through genetic engineering to become efficient and suitable for bioremediation. Bioremediation is an engineered technology that modifies environmental conditions (physical, chemical, biochemical, or microbiological) to encourage microorganisms to destroy or detoxify organic and inorganic contaminants in the environment [5]. The process can be applied above ground in land farms, tanks, biopiles, or other treatment systems (referred to as *ex situ*) or below ground in the soil or groundwater, referred to as in situ. In situ bioremediation of groundwater has become one of the most widely used technologies for contaminated site treatment because of its relatively low cost, adaptability to site-specific conditions, and efficacy when properly implemented [11].

Bioremediation techniques are versatile and can be utilized at various stages of treatment [7]. Applications include removal of contaminants from raw materials prior to processing, treatment of wastes before discharge, treatment of effluent streams, and decontamination of soils, sediments surface water and ground water. In environmental management, bioremediation techniques (including phytoremediation) have a variety of applications including the clean-up of ground water, soils, lagoons, sludge, and process water streams. One of the most important examples of large scale bioremediation is the shore-line clean-up efforts in Alaska, where in 1989; an environmental hazard resulted from a large oil-spill [10]. This led to clean-up operation covering 70 miles of shore-line. There are numerous other successful applications of bioremediation including clean-up of pollution due to chemical spills [2].

A decision tree can be used as a model for sequential decision problems under uncertainty [2]. A decision tree describes graphically the decisions to be made, the events that may occur, and the outcomes associated with combinations of decisions and events. Probabilities are assigned to the events, and values are determined for each outcome. A major goal of the analysis is to determine the best decisions. The bioremediation decision tree in this work was developed to aid interested parties (regulators, site owners and stakeholders) in evaluating sites as candidates for a particular type of bioremediation [6]. The background

***Corresponding author:** Ukpaka CP, Department of Chemical/Petrochemical Engineering, Rivers State University of Science and Technology, Nkpolu PMB 5080, Port Harcourt, Nigeria, E-mail: chukwuemeka24@yahoo.com information required in the decision process should be available from the site characterization data. Using the decision tree and the reference documents will assist regulators, site owners, technology vendors, the public and stakeholders in determining if a particular bioremediation type is applicable to a contaminated site [2-7]. This study is aimed at developing a bioremediation decision tree model for an integrated environmental management system.

Materials and Methods

In this research the materials and methods are based on the bioremediation decision tree concepts as well as the developed mathematical models in relation to functional parameters that influence bioremediation processes. These functional parameters are well classified and examined in terms of the integrated environmental management system for effective remediation of the polluted environment.

Model Development

The mathematical model was developed by considering the contamination of soil with polycyclic aromatic hydrocarbons (PAHs) which is a serious environmental issue because some polycyclic aromatic is toxic, carcinogenic and mutagenic.

The flux of substance from the particle interface to the bulk liquid phase can be described based on a mass balance over a certain thickness film.

$$N = \frac{D}{\delta}(C_S - C_L) \quad (1)$$

Where, N is the flux ($gm^{-2}h^{-1}$), δ is the film thickness (m), C_s is the saturation on the saturation in liquid phase (gm^{-s}), C_L is the dissolved concentration in liquid phase ($gm.^{-s}$). By definition the ratio of the diffusion coefficient to the film thickness can be expressed a

$$\frac{D}{\delta} = K_L \quad (2)$$

Where, K_L is the film mass transfer coefficient (mh^{-1})

The flux causes the change in substrate concentration with time that is observed in the bulk liquid phase giving

$$\frac{dC_t}{dt} = K_t \frac{A}{V}(C_s - C_t) \quad (3)$$

In which v is the liquid phase volume (m^{-3}) and A is a surface area of the particle (m^2).

Equation (1), (2) and (3) are combined and it will give

$$\frac{dC_t}{dt} = K_t \frac{A}{V}(C_s - C_t) \quad (4)$$

Since it is difficult to determine the surface area of particles, another form of mass transfer coefficient is often used:

$$\mu = \mu_{max} \frac{S}{Ks + S} \quad (5)$$

Where $(K_t a)_{SL}$ is the solid-liquid volumetric mass transfer coefficient (L^{-1}). The appropriate boundary conditions (C_L=0 at t=0) gives:

$$C_L = C_S\,[1 - \text{esp}\,\{-(K_L a)_{SL} \times t\}] \quad (6)$$

In this work, the metric mass transfer coefficient and saturation concentration. $(K_L a)_{SL}$ and C_S least square minimization in equation (6) to the experimental data using MATLAB.

Bioremediation Kinetics

Although the growth of microorganism is a complex phenomenon, it is a common practice to represent this growth by a relatively simple expression. One of the most widely employed expressions for the specific growth rate, μ, is the Monod equation which states that μ with the concentration of a single essential substrate. The Monod equation is

$$\mu = \mu_{max} \frac{S}{Ks + S} \quad (7)$$

Where, μ_{max} is the maximum specific growth rate (h^{-1}), K_S is known as the saturation constant ($mg\ L^{-1}$). Under such circumstances, the Haldane inhibition model has been proposed to express the specific growth rate.

$$\mu = \frac{\mu\,\text{maxs}}{Ks + S +^{s^2}/k_l}(K_1 >> K_s) \quad (8)$$

Microorganism is usually inhibited by these products whose effect can be added to the Monod equation as

$$\mu = \mu_{max}[\frac{s}{Ks + S}]\left(1 - \frac{p}{P_w}\right)y \quad (9)$$

Concentration increase and stoichiometric relationship between substrate utilization and biomass production. The bioremediation can be shown below

$$\frac{dX}{dt} = \mu \times X \quad (10)$$

$$\frac{dX}{dt} = -Y_{XS}\frac{ds}{dt}$$

To obtain the parameters, μ_{max}, K_S, K_1 and $y_{x/s}$ numerical solutions of these equations were performed using MATLAB. The method for the numerical modelling involves integration using fourth-order Runge Kutta numerical scheme.

Experimental

Sample collection

Crude oil obtained from an oil terminal in Niger Delta area of Nigeria was collected and transported to the department of Chemical/Petrochemical engineering laboratory in Rives State University of Science and Technology Port Harcourt for onward analysis. The soil used in this investigation was collected within the campus of Rivers State University of Science and Technology Nkpolu, Port Harcourt and then transported to the Department of Chemical/Petrochemical Engineering laboratory.

Experimental procedure

The crude oil and soil sample was measured using measuring cylinder with soil sample of 3.2 gm^2 mixed in a batch rector and 2.5 gm^{-2} of the crude oil, then both samples was mixed in a batch reactor A and 2.2 gm^{-2} of soil and 1.05 gm^{-2} crude oil in reactor B and finally 2.1 gm^{-2} of soil sample and 1.5 gm^{-2} of crude oil for reactor C. The volume and area of various sampled for this investigation is defined in terms of zone A, we have 7 m^3 in volume and 0.15 m^2 in area and for zone C we have 5 m^3 in volume and 0.12 m^2 in area. The experimental set up was examined to enable use determine the film mass transfer coefficient for each reactor (Figure 1). This was examined by considering the rate of diffusion of the crude oil in the solid medium subjected in the batch reactor. The process was monitored with variation in time. The samples were collected from the batch reactor and subjected into further examination to ascertain the penetration of the crude oil as well as the GC (Gas chromatography) was used to determine the concentration of the solid and liquid (soil and crude oil) and AS (Absorption spectrograph) used in examining the film thickness concentration. The result obtained from this investigation is

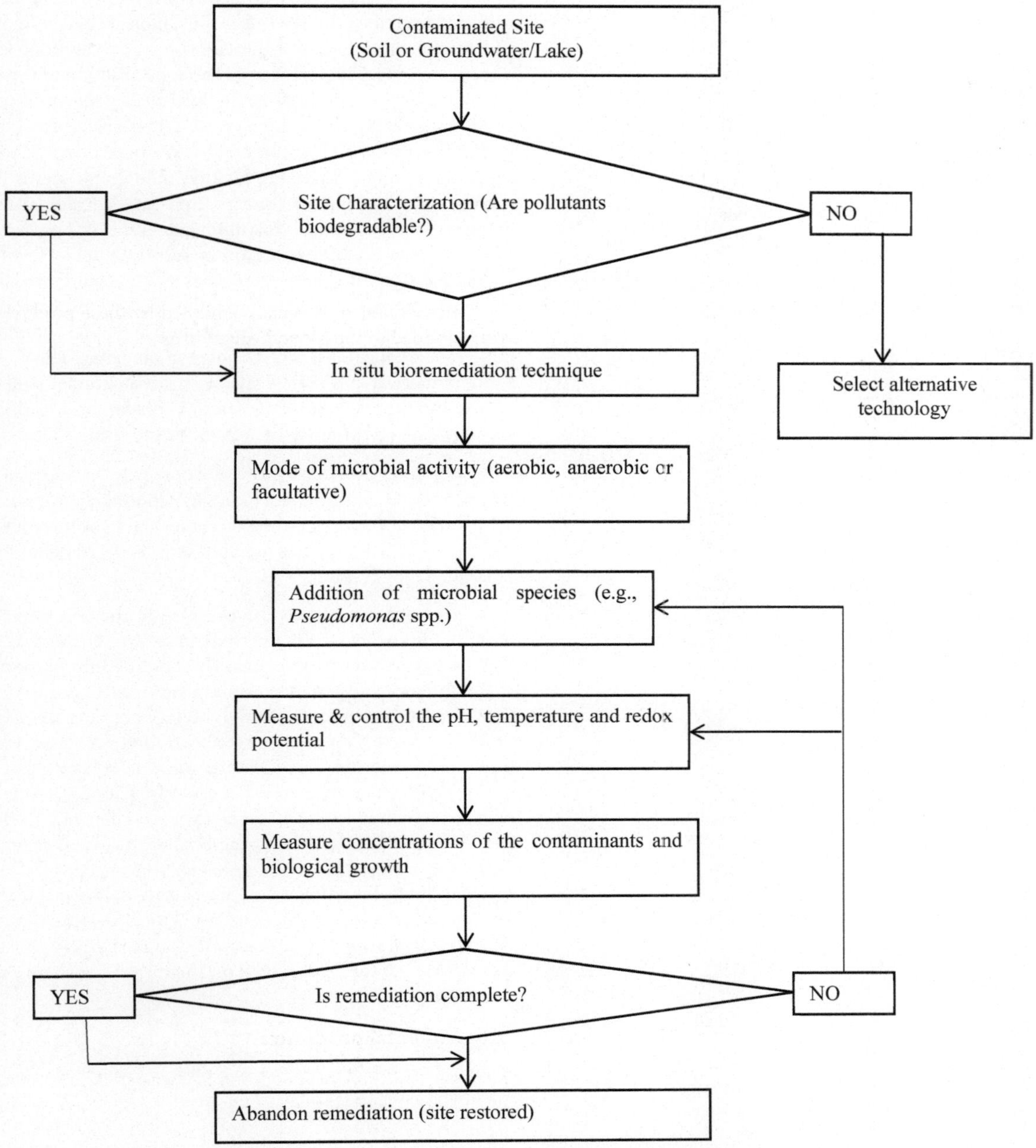

Figure 1: Decision Tree showing general approach to bioremediation.

simulated using matrix laboratory computer language programme. This application was useful to enable us determine the film mass transfer coefficient at interval as well as the degree on concentration of the pollutants upon the influence of diffusion in a batch reactor.

Results and Discussion

The results are presented in Tables 1-3, as defined in the computational procedures as well as in Figures as presented in the research work.

Figure 2 demonstrates the decrease in the chlorine concentration upon the influence of the microorganism in degradating the crude oil contaminated soil environment as a function of time. As the concentration of the microorganisms increases the substrate concentration decreases with respect time. In the case, the chlorine becomes the substrate and as shown in Figure 2 the chlorine concentration decreases with increase in time. The variation in the concentration of the substrate can be attributed to the variation in microbial concentration, integrated environmental management system put in place for high efficiency in the process as well as time effective monitoring of the process to check made other inhibiting factors. To obtain the parameters, μ_{max}, K_S, K_I and $y_{x/s}$

S. No	Parameters	Values	SI Unit
1	Area (A)	0.15	M^2
2	Volume (V)	7	M^3
3	Concentration of solid (C_S)	2.2	Gm^{-2}
4	Concentration of liquid (C_L)	1.05	Gm^{-2}
5	Film mass transfer coefficient (k_L)	2	Mh^{-1}

Table 1: Computational Parameters at various functional values.

S. No	Parameters	Values	SI Unit
1	Area (A)	0.14	M^2
2	Volume (V)	7	M^3
3	Concentration of solid (C_S)	3.2	Gm^{-2}
4	Concentration of liquid (C_L)	2.5	Gm^{-2}
5	Film mass transfer coefficient (k_L)	4	Mh^{-1}

Table 2: Computational Parameters at various functional values.

S. No	Parameters	Values	SI Unit
1	Area (A)	0.12	M^2
2	Volume (V)	5	M^3
3	Concentration of solid (C_S)	2.1	Gm^{-2}
4	Concentration of liquid (C_L)	1.5	Gm^{-2}
5	Film mass transfer coefficient (k_L)	2	Mh^{-1}

Table 3: Computational Parameters at various functional values.

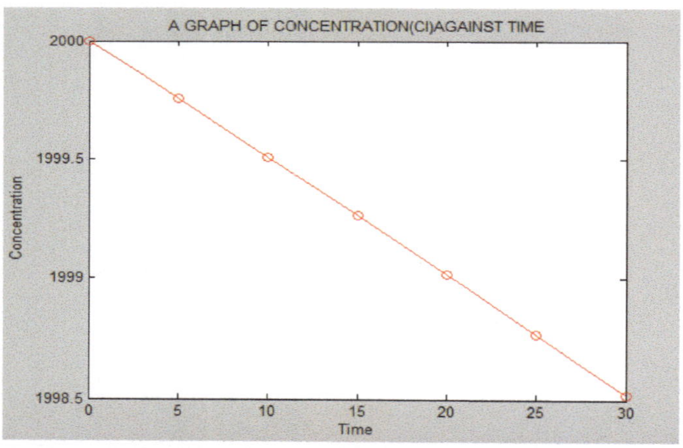

Figure 2: Matlab graphical solution showing decrease in concentration of pollutants in the contaminated site upon the influence of time for range of parameters.

numerical solutions of these equations were performed using MATLAB. The method for the numerical modelling involves integration using fourth-order RungeKutta numerical scheme. In this case, the functional parameters of the maximum specific rate of substrate degradation and the equilibrium constants are contributing factors to the process. The graph above shows a pilot of concentration against time putting into consideration the *in-situ* bioremediation of a soil contaminated site, it is observed that as the time increases the concentration of substrate decrease which shows a favourable sign for that methodology (*in-situ* bioremediation) chosen.

Figure 3 demonstrates the decrease in the chlorine concentration upon the influence of the microorganism in degradating the crude oil contaminated soil environment as a function of time. As the concentration of the microorganisms increases the substrate concentration decreases with respect time. In the case, the chlorine becomes the substrate and as shown in Figure 3 the chlorine concentration decreases with increase in time. The variation in the concentration of the substrate can be attributed to the variation in microbial concentration, integrated environmental management system put in place for high efficiency in the process as well as time effective monitoring of the process to check made other inhibiting factors. The graph above shows a pilot of concentration against time putting into consideration the *in-situ* bioremediation of a soil contaminated site, it is observed that as the time increases the concentration of substrate decrease which shows a favourable sign for that methodology (*in-situ* bioremediation) chosen. To obtain the parameters, μ_{max}, K_S, K_1. And $y_{x/s}$ numerical solutions of these equations were performed using MATLAB. The method for the numerical modelling involves integration using fourth-order Runge Kutta numerical scheme. In this case, the functional parameters of the maximum specific rate of substrate degradation and the equilibrium constants are contributing factors to the process.

The graph below shows a pilot of concentration against time putting into consideration the *in-situ* bioremediation of a soil contaminated site, it is observed that as the time increases the concentration of substrate decrease which shows a favourable sign for that methodology (*in-situ* bioremediation) chosen.

Figure 4 demonstrates the decrease in the chlorine concentration upon the influence of the microorganism in degradating the crude oil contaminated soil environment as a function of time. As the concentration of the microorganisms increases the substrate concentration decreases with respect time. In the case, the chlorine becomes the substrate and as shown in Figure 4 the chlorine concentration decreases with increase in time. The variation in the concentration of the substrate can be attributed to the variation in microbial concentration, integrated environmental management system put in place for high efficiency in the process as well as time effective monitoring of the process to check made other inhibiting factors. To obtain the parameters, μ_{max}, K_S, K_1. And $y_{x/s}$ numerical solutions of these equations were performed using MATLAB. The method for the numerical modelling involves integration using fourth-order Runge Kutta numerical scheme. In this case, the functional

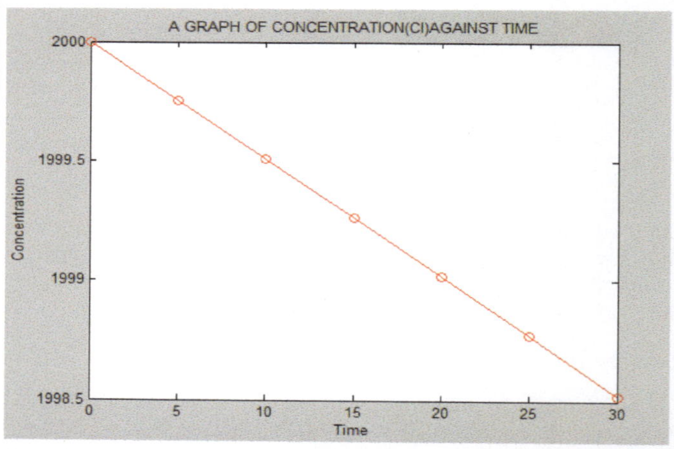

Figure 3: Matlab graphical solution showing decrease in concentration of pollutants in the contaminated site upon the influence of time for range of parameters.

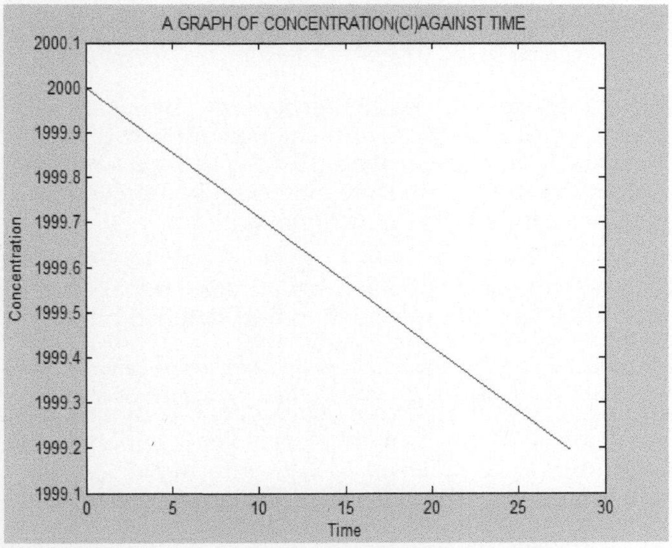

Figure 4: Matlab graphical solution showing decrease in concentration of pollutants in the contaminated site upon the influence of time for range of parameters.

parameters of the maximum specific rate of substrate degradation and the equilibrium constants are contributing factors to the process.

Conclusion

The following conclusion was drawn from the investigation such as:

1. Good integrated environmental management system will improve the bioremediation process.
2. Optimum remediation performance with low cost of project execution.
3. Low rate of inhibiting factors.
4. Less period required to accomplish the programme.
5. Reduction in constrain factors.

References

1. Cookson JT (1995) Bioremediation Engineering Design and Application. McGraw-Hill, New York, USA.
2. Environmental Protection Agency (2013) Introduction to in-situ bioremediation of groundwater. EPA 86.
3. Gupter PK (2006) Biotechnology and Genomics. Meerut, India.
4. Kogbara RB (2008) Ranking Agro-technical Methods and Environmental Parameters in the Biodegradation of Petroleum-Contaminated Soils in Nigeria. Electronic Journal of Environmental Biotechnology 11: 4.
5. Lee JY, Roh JR, Kum HS (1994). Metabolic Engineering of Pseudomonas Pulida for the Simultaneous Biodegradation of Benzene, Toluene and P-xylene Mixture. Biotech Bioeng 43: 1146-1152.
6. Sandro JB, Magali CC, Denize DCF (2005) Production off CO2, in Crude Oil Bioremediatin in Clay Soil. Brazilian Archives of Biology and Technology, p: 48.
7. US Environmental Protection Agency (1990) Interim Report, Oil Spill Bioremediation Project. US Environmental Protection Agency, Office of Research and Development, Washington, USA.
8. Amadi SA, Ukpaka CP (2015) Evaluation of the metals corrosion in soil environment of Niger Delta Area of Nigerian. Nigerian Journal of Oil and Gas Technology 1: 104-117.
9. Ikenyiri PN, Ukpaka CP (2016) Application of moringa dry stem for effective remediation of crude oil processing induced water environment using packed bed reactor. Nigerian Journal of Oil and Gas Technology 1: 58-73.
10. Igwe FU, Dokubo A, Ukpaka CP (2016) Overview of crude oil toxicity: a potential health and environmental hazard in the Niger Delta. Nigerian Journal of Oil and Gas Technology 2: 94-104.
11. Ukpaka CP, Farrow TS (2016). Simulation of Pneumatic Pressure Transducer Reliability on the Transfer Function of Bellow. Nigerian Journal of Oil and Gas Technology 2: 216-228.
12. Ukpaka C (2016). Evaluation of Olu-Obasanjo Groundwater Parameters: Application of Finite Element Concepts. Nigerian Journal of Oil and Gas Technology 2: 117-127.

Permissions

All chapters in this book were first published in JCEPT, by OMICS International; hereby published with permission under the Creative Commons Attribution License or equivalent. Every chapter published in this book has been scrutinized by our experts. Their significance has been extensively debated. The topics covered herein carry significant findings which will fuel the growth of the discipline. They may even be implemented as practical applications or may be referred to as a beginning point for another development.

The contributors of this book come from diverse backgrounds, making this book a truly international effort. This book will bring forth new frontiers with its revolutionizing research information and detailed analysis of the nascent developments around the world.

We would like to thank all the contributing authors for lending their expertise to make the book truly unique. They have played a crucial role in the development of this book. Without their invaluable contributions this book wouldn't have been possible. They have made vital efforts to compile up to date information on the varied aspects of this subject to make this book a valuable addition to the collection of many professionals and students.

This book was conceptualized with the vision of imparting up-to-date information and advanced data in this field. To ensure the same, a matchless editorial board was set up. Every individual on the board went through rigorous rounds of assessment to prove their worth. After which they invested a large part of their time researching and compiling the most relevant data for our readers.

The editorial board has been involved in producing this book since its inception. They have spent rigorous hours researching and exploring the diverse topics which have resulted in the successful publishing of this book. They have passed on their knowledge of decades through this book. To expedite this challenging task, the publisher supported the team at every step. A small team of assistant editors was also appointed to further simplify the editing procedure and attain best results for the readers.

Apart from the editorial board, the designing team has also invested a significant amount of their time in understanding the subject and creating the most relevant covers. They scrutinized every image to scout for the most suitable representation of the subject and create an appropriate cover for the book.

The publishing team has been an ardent support to the editorial, designing and production team. Their endless efforts to recruit the best for this project, has resulted in the accomplishment of this book. They are a veteran in the field of academics and their pool of knowledge is as vast as their experience in printing. Their expertise and guidance has proved useful at every step. Their uncompromising quality standards have made this book an exceptional effort. Their encouragement from time to time has been an inspiration for everyone.

The publisher and the editorial board hope that this book will prove to be a valuable piece of knowledge for researchers, students, practitioners and scholars across the globe.

List of Contributors

Inoue K, Ohto K and Kawakita H
Department of Applied Chemistry, Faculty of Science and Engineering, Saga University, Honjo-machi 1, Saga 840-8502, Japan

Kurata M, Atsumi K and Fukuda H
Research Laboratories, DENSO CORPORA-TION, Minamiyama 500-1, Komenoki, Nisshin, Aichi 470-0111, Japan

Khunathai K
Department of Applied Chemistry, Faculty of Science and Engineering, Saga University, Honjo-machi 1, Saga 840-8502, Japan
Research Laboratories, DENSO CORPORATION, Minamiyama 500-1, Komenoki, Nisshin, Aichi 470-0111, Japan

Alam S
Department of Chemical and Biological Engineering, College of Engineering, University of Saskatchewan, 57 Campus Drive, Saskatoon, SK, S7N 5A9, Canada

Hailu Z
Addis Ababa Institute of Technology, School of Chemical and Bio Engineering, Food Engineering, Addis Ababa University, Addis Ababa, Ethiopia

Sohan Shrestha
Nepal Academy of Science and Technology, Lalitpur, Bagmati Zone, Nepal

Kostiv IY
State Enterprise, Scientific-Research Gallurgi Institute, 5a, Fabrichna Str., 76000 Kalush, Ukraine

Basystyuk YI
Precarpathian National University named after V. Stephanyk 57, Shevchenko Str., 76025 Ivano-Frankivsk, Ukraine

Mustafa SA, El-Gayar DA and Sedahmed GH
Chemical Engineering Department, Faculty of Engineering, Alexandria University, Alexandria, Egypt

Ritesh Sojitra
ITM University, Gwalior, India

Zinabu Hailu and Derese Mekonnen
Department of Chemical Engineering, Adama Science and Technology University, Adama, Ethiopia

Khazangul Piriyeva B, Zulfiyya Ismailova R, Azer Gasimov A, Ulviyya Karimova Japharova N, Sevinj Hajizadeh M, Tunzala Aghayeva F and Rasim Japharov P
Department of Technology of Petrochemical Processes, Institute of Petrochemical Processes of Azerbaijan National Academy of Sciences, Baku, Azerbaijan

Fatin Amirah Ahmad Norddin
Department of Chemistry, Faculty of Science, Universiti Putra Malaysia, 43400 UPM Serdang, Selangor, Malaysia

Sharifah Nurfadhlin Afifah Syed Azhar and Siti Efliza Ashari
Department of Chemistry, Faculty of Science, Universiti Putra Malaysia, 43400 UPM Serdang, Selangor, Malaysia
Process Engineering Laboratory, Enzyme and Microbial Technology Research Center (EMtech), Faculty of Biotechnology and Biomolecular Sciences, Universiti Putra Malaysia, 43400 UPM Serdang, Selangor, Malaysia

El-Shamy AM
Physical Chemistry Department, Electrochemistry and Corrosion Lab, National Research Centre, El-Bohouth, Giza, Egypt

Hanaa A El-Boraey and El-Awdan HF
Faculty of Science, Menoufia University, Shebin El-Kom, Egypt

Alireza Mohammadi Ghahdarijani, Faramarz Hormozi and Ali Haghighi Asl
Faculty of Chemical, Gas and Petroleum Engineering, Semnan University, Semnan, Iran

Dmitry Vladimirovich Gradov, Arto Laari and Tuomas Koiranen
School of Engineering Science, Lappeenranta University of Technology, Lappeenranta, Finland

Gerardo González and Marko Vauhkonen
Department of Applied Physics, University of Eastern Finland, Kuopio, Finland

Kostiv IY
State Enterprise Scientific - Research Gallurgi Institute 5a, Fabrichna Str., 76000 Kalush, Ukraine

Ya I Basystiuk
Precarpathian National University named after V. Stephanyk 57, Shevchenko Str., 76025 Ivano Frankivsk, Ukraine

Vera M
Barcelona Research Center in Multiscale Science and Engineering, EEBE, Universitat Politècnica de Catalunya (UPC) - Barcelona TECH, c/Eduard Maristany, Barcelona, Spain
Aigües de Barcelona, Empresa Metropolitana de Gestió del Cicle Integral de l'Aigua, Gral. Barcelona, Spain

Martín-Alonso J and Mesa J
Aigües de Barcelona, Empresa Metropolitana de Gestió del Cicle Integral de l'Aigua, Gral. Barcelona, Spain

González S and Lefèvre B
CETAQUA, Water Technology Center, Ctra. D'Esplugues Cornellà de Llobregat, Spain

Gibert O and Cortina JL
Barcelona Research Center in Multiscale Science and Engineering, EEBE, Universitat Politècnica de Catalunya (UPC) - Barcelona TECH, c/Eduard Maristany, Barcelona, Spain
CETAQUA, Water Technology Center, Ctra. D'Esplugues Cornellà de Llobregat, Spain

Sadik Hameed
Department of Chemistry, College of Science, Wasit University, Kut, Wasit, Iraq

Sushil Kumar
Corporate R&D, BHEL, Hyderabad, Telangana, India

Gerd Kaupp
Institute of Applied Chemistry, Oldenburg University, Oldenburg, Germany

Nor Azizah Yacob
Department of Mathematics, Faculty of Computer and Mathematics Sciences, Universiti Teknologi MARA Pahang, Bandar Pusat Jengka, Pahang, Malaysia

Nor Fadhilah Dzulkifli
Department of Mathematics, Faculty of Computer and Mathematics Sciences, Universiti Teknologi MARA Pahang, Bandar Pusat Jengka, Pahang, Malaysia
Department of Mathematics and Institute for Mathematical Research, Universiti Putra Malaysia, UPM Serdang, Selangor, Malaysia

Norfifah Bachok, Norihan Md Arifin and Haliza Rosali
Department of Mathematics and Institute for Mathematical Research, Universiti Putra Malaysia, UPM Serdang, Selangor, Malaysia

Ioan Pop
Department of Mathematics, Babes-Bolyai University, Cluj-Napoca, Romania

Mohammad Hussein and Adam Donaldson
Process Engineering and Applied Science, Dalhousie University, Halifax, Nova Scotia, B3H4R2, Canada

Hossein Behniafar
Department of Chemistry, University of Damghan, Semnan Province, Damghan, Iran

Maryam Alipour Moghaddam
Department of Polymer Chemistry, Azad University of Damghan, Semnan Province, Damghan, Iran

Kanti Kumar Athankar, Kailas L Wasewar, Mahesh N Varma and Diwakar Z Shende
Advance Separation and Analytical Laboratory (ASAL), Department of Chemical Engineering, Visvesvaraya National Institute of Technology (VNIT), Nagpur (Maharashtra), India

Chidi Obi and Michelle Abonele Apemiye
Department of Pure and Industrial Chemistry, Faculty of Science, University of Port Harcourt, Rivers State, Nigeria

Deepshikha Singh and Shabina Khanam
Department of Chemical Engineering, Indian Institute of Technology Roorkee, Roorkee, India

Reznichenko AV, Rybalchenko VV, Badaev FZ, Ponomarev SG, Vasin AA and Smirnov AD
Federal State Educational Institution of Higher Education, Moscow Polytechnic University (Moscow Polytech), Moscow, Russia

Politaeva Natalia Anatolievna, Taranovskaya Elena Aleksandrovna, Slugin Vladimir Valentinovich and Prokhorov Vladislav Vladimirovich
Peter The Great St. Petersburg Polytechnic University, 194021, Polytechnic Street-29, Saint Petersburg, Russia

Mohammad Hussein and Adam Donaldson
Dalhousie University, Process Engineering and Applied Science, 1360 Barrington St. Halifax, Nova Scotia, B3H4R2, Canada

Mohamed N Nounou
Chemical Engineering Program, Texas A&M University at Qatar, Doha, Qatar

M Ziyan Sheriff
Chemical Engineering Program, Texas A&M University at Qatar, Doha, Qatar
Artie McFerrin Department of Chemical Engineering, Texas A&M University, College Station, TX 77843, USA

Ibrahim TK and Khalaf FH
Petroleum Engineering Department, Faculty of Engineering, Knowledge University, Kurdistan, Iraq

Abdulrahman RK
Chemical Engineering Department, Faculty of Engineering, Koya University Ribwar, Kurdistan, Iraq

Kamal IM
Chemical Engineering Department, Faculty of Engineering, Soran University, Kurdistan, Iraq

Azadeh Ebrahimain Pirbazari
Fouman Faculty of Engineering, College of Engineering, University of Tehran, Iran

Ukpaka CP
Department of Chemical/Petrochemical Engineering, Rivers State University of Science and Technology, Nkpolu Port Harcourt, Nigeria

Index